Neuroendocrine Molecular Biology

BIOCHEMICAL ENDOCRINOLOGY

Series Editor: Kenneth W. McKerns

Neuroendocrine Molecular Biology

Edited by

G. Fink *and*
A. J. Harmar

MRC Brain Metabolism Unit
Edinburgh, Scotland

and

Kenneth W. McKerns

The International Foundation for Biochemical Endocrinology
Blue Hill Falls, Maine

PLENUM PRESS • *NEW YORK AND LONDON*

Library of Congress Cataloging in Publication Data

International Symposium on Neuroendocrine Molecular Biology (1985: Edinburgh,
  Lothian)
  Neuroendocrine molecular biology.

  (Biochemical endocrinology)
  "Proceedings of an International Symposium on Neuroendocrine Molecular Biology,
held September 16–20, 1985, in Edinburgh, Scotland" – T.p. verso.
  Symposium was also the 13th annual meeting of the International Foundation for
Biochemical Endocrinology.
  Includes bibliographical references and index.
  1. Neuroendocrinology – Congresses. 2. Molecular biology – Congresses. 3. Neuro-
peptides – Congresses. 4. Pituitary hormones – Congresses. I. Fink, George. II.
Harmar, A. J. III. McKerns, Kenneth W. IV. International Foundation for Biochem-
ical Endocrinology. V. Title. VI. Series.
QP356.4.I587 1985   599′.0188   86-4964
ISBN 978-1-4684-5133-7   ISBN 978-1-4684-5131-3 (eBook)
DOI 10.1007/978-1-4684-5131-3

Proceedings of an International Symposium on Neuroendocrine Molecular Biology,
held September 16–20, 1985, in Edinburgh, Scotland

© 1986 Plenum Press, New York
Softcover reprint of the hardcover 1st edition 1986
A Division of Plenum Publishing Corporation
233 Spring Street, New York, N.Y. 10013

PREFACE

The 13th Annual Meeting of the Foundation was held in Edinburgh
during September 1985. The subject was neuroendocrine molecular
biology which brought together leading scientists in the fields of
molecular genetics, neuroendocrinology and developmental neurobiology. The conference was most stimulating and as the Proceedings
show, novel data presented was of the highest quality. The topics
presented were grouped under the headings, "Molecular Biology of the
Nervous System", "LHRH - New Perspectives", "Neuropeptides",
"Oxytocin and Vasopressin", "Transcriptional and Post-Translational
Regulation of Neuropeptide Synthesis", "Neuroendocrine Mechanisms at
the Cellular Level", "Receptors - Cellular and Molecular Biology" and
"Clinical Applications". The Chairpersons for the sessions were:
H. Niall, I. MacIntyre, G. Fink, J. Roberts, A.J. Harmar, B.A. Cross,
M. Jutisz, H. Kosterlitz, K. Brown-Grant, S.M. McCann and J. Morris.

G. Fink,
A.J. Harmar, and
K.W. McKerns

ACKNOWLEDGEMENTS

The Editors wish to express their special thanks to 'Ms. Jo Donnelly and other members of the Medical Research Council Brain Metabolism Unit without whose unstinting work the meeting in Edinburgh could not have been held.

We also wish to acknowledge with many thanks financial support obtained from: The Wellcome Trust; The Royal Society; The Medical Research Council; The Scottish Development Agency; The Society for Endocrinology; Bachem Inc., U.K. Branch; Amersham International; Cambridge Research Biochemicals; Du Pont (U.K.) Ltd., NEN Products Division; Scotlab Instrument Sales Ltd.; Peninsula Laboratories; RIA (U.K.) Ltd.; Cambridge University Press; Ciba-Giegy Pharmaceuticals; Merck Sharp and Dohme Ltd.; UCB Bioproducts; NBL Enzymes Ltd.; Sigma Chemical Company Ltd.

CONTENTS

MOLECULAR BIOLOGY OF THE NERVOUS SYSTEM

LHRH - NEW PERSPECTIVES

NEUROPEPTIDES

OXYTOCIN AND VASOPRESSIN

CLINICAL APPLICATIONS

STRUCTURE, FUNCTION AND EXPRESSION OF THE HUMAN CALCITONIN/α-CGRP GENE

Ian Marshall*, Susan Al-Kazwini*, Jenny J. Holman*,
Peter M. Broad[+], Mark R. Edbrooke[+] and Roger K. Craig[+]

*Department of Pharmacology and Therapeutics and [+]Cancer
Research Campaign Endocrine Tumour Molecular Biology Group
Courtauld Institute of Biochemistry, The Middlesex
Hospital Medical School, Mortimer Street, London W1P 7PN U.K.

INTRODUCTION

Calcitonin is a small peptide hormone (32 amino acids) synthesised
and secreted in mammals by the C-cells of the thyroid (Foster et al.,
1964). The physiological role of calcitonin in mammals appears to be the
protection of the skeleton in times of calcium stress such as growth,
pregnancy and lactation (Stevenson et al., 1979). In common with other
small peptide hormones, molecular cloning and nucleotide sequence analysis
of rat and human calcitonin mRNA (Amara et al., 1982; Craig et al., 1982;
Le Moullec et al., 1984) demonstrates that calcitonin mRNA encodes a precursor
polyprotein. Within this precursor, calcitonin is flanked by amino and
carboxy terminal peptides of as yet unknown function, from which calcitonin
is proteolytically cleaved and amidated prior to secretion (see Craig et
al., 1982). The human calcitonin gene is expressed at elevated levels
ectopically in lung carcinoma (Coombes et al., 1974), and at grossly elevated
levels in medullary thyroid carcinoma (Milhaud et al., 1974). Elegant
studies on the structure and expression of the rat calcitonin gene have
resulted in renewed interest in this gene. These demonstrate the generation
by RNA processing of alternative mRNA species from a single gene in an
apparently tissue-specific manner. The mRNAs encode polyproteins cleaved
by post-translational events to yield either calcitonin or the calcitonin
gene-related peptide (CGRP) (Amara et al., 1982). Subsequent immunocytochemical studies showed a wide distribution of rat CGRP-producing cells
within discrete regions of the central and peripheral nervous system (Rosenfeld
et al., 1983). This suggests a potential role as a neuromodulator or
neurotransmitter molecule, a supposition which has led to detailed analyses
in a number of laboratories of the structure, function and expression of
rat and human calcitonin/CGRP gene(s).

STRUCTURE AND EXPRESSION OF THE HUMAN CALCITONIN/α-CGRP GENE

Analysis of the structure of the human calcitonin gene by our own and
other laboratories (see Nelkin et al., 1984; Steenbergh et al., 1984,
Edbrooke et al., 1985; Jonas et al., 1985) demonstrates that the human
gene also encodes an alternative gene product, the human α-CGRP. The

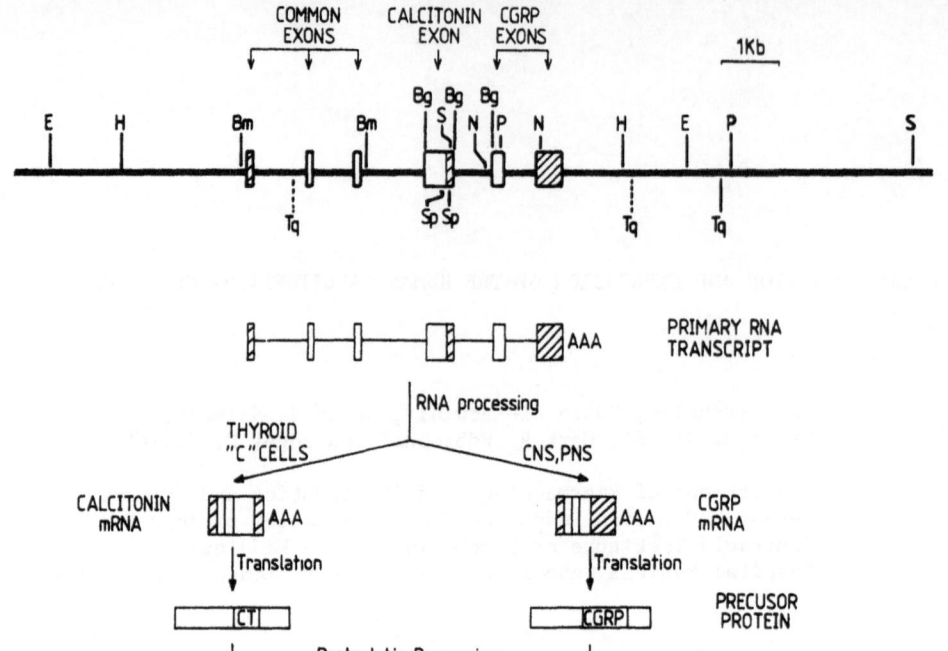

Fig 1 A: Structure of the human calcitonin/α-CGRP gene. Restriction enzyme
          sites are indicated by vertical lines; Bg, Bgl II; Bm, Bam HI;
          E, EcoRI: H, Hind III; N, NcoI; P, Pvu II; S, SstI; Sp, SphI;
          Tq, TaqI,.

     B: Tissue specific expression of the human calcitonin/α-CGRP gene.
          Alternative RNA processing pathways generate either calcitonin or
          CGRP mRNA from a common primary transcript. CNS and PNS: central
          and peripheral nervous systems. Hatched boxes represent noncoding
          exon sequences.

structure of the human gene is remarkably similar to that of the rat,
comprising six exons (see Fig 1). The first three exons encode sequence
expressed in both calcitonin and CGRP mRNA species; the fourth exon encodes
calcitonin, the carboxy terminal flanking peptide (PDN-21) and the 3'-
untranslated region of calcitonin mRNA; the fifth exon encodes CGRP and a
flanking tetrapeptide; and the sixth exon encodes the 3' untranslated
region of the CGRP mRNA. Human α-CGRP comprises 37 amino acids (see Fig
2) terminating in an amidated phenylalanine as judged by the presence of
an adjacent glycine residue (see Bradbury et al., 1982), and confirmed by the
isolation of the amidated peptide from medullary thyroid carcinoma (MCT: see
Morris et al., 1984). It is presumed that, as for calcitonin, the carboxy
terminal amide group is essential for biological activity (see below). Human
and rat α-CGRP differ at four amino acid positions (see Fig 2 and Table 1).
     The availability of calcitonin and α-CGRP gene probes, and of novel pre-
dicted peptide sequence, permits three lines of investigation. Firstly, gene
probes may be used to investigate the presence or otherwise of related
genes in the genome by Southern blotting, and to investigate the expression
of both gene products in medullary thyroid carcinoma, and ectopically in lung
carcinoma. Secondly, synthetic human α-CGRP may be used to raise antisera,
which in turn may be used to examine plasma CGRP levels in health and disease,
and the distribution of CGRP producing cells in various tissues. Thirdly,
the biological activity of the predicted peptide can be examined in vivo

```
                                          -6  -5  -4  -3 ┌-2  -1┐ 1   2   3   4   5   6   7   8
                                          Ile Ile Ala Gln│Lys Arg│Ala Cys Asp Thr Ala Thr Cys Val
CAG ATC TTC TCT TCT TTC TCC ATC CTG CAA ATC AGA ATC ATT GCC CAG│AAG AGA│GCC TGT GAC ACT GCC ACC TGT GTG
    │                                     │T   G       C  └      ┘         T       C A                C
    Bgl II          Splice Recognition Sequence ▼ Val Thr                 Ser     Asn
```

```
 9  10  11  12  13  14  15  16  17  18  19  20  21  22  23  24  25  26  27  28  29  30  31  32  33  34
Thr His Arg Leu Ala Gly Leu Leu Ser Arg Ser Gly Gly Val Val Lys Asn Asn Phe Val Pro Thr Asn Val Gly Ser
ACT CAT CGG CTG GCA GGC TTG CTG AGC AGA TCA GGG GGT GTG GTG AAG AAC AAC TTT GTG CCC ACC AAT GTG GGT TCC
C                                   G   G   A                   G                               C   T
                                                              Asp
```

```
35  36  37 ┌+1┐┌+2  +3  +4  +5┐┌+6  +7  +8  +9
Lys Ala Phe│Gly│Arg Arg Arg Arg│Asp Leu Gln Ala STOP
AAA GCC TTT│GGC│AGG CGC CGC AGG│GAC CTT CAA GCC TGA GCA GCT GAA CGA CTC AAG AAG GTC ACA ATA AAG CTG AAC
G           C   C   C          │         G   T      │
Glu                                              Pvu II
```

Fig 2:   Predicted amino acid sequence of human ⍺-calcitonin gene-related
          peptide (1-37) and flanking peptides. Boxes indicate amino acids
          involved in proteolytic processing of the precursor protein. The
          first 35 nucleotides are intron sequence. The positions at which
          the rat ⍺-CGRP mRNA nucleotide sequence and polyprotein amino acid
          sequence differ from the human sequences are indicated.

Table 1: Comparison of human and rat ⍺- and β-CGRP amino acid sequences.

```
           1                                                10
Human α   Ala Cys Asp Thr Ala Thr Cys Val Thr His Arg Leu Ala
Human β   Ala     Asn
Rat α     Ser     Asn
Rat β     Ser     Asn
```

```
                                    20
Human α   Gly Leu Leu Ser Arg Ser Gly Gly Val Val Lys Asn Asn
Human β                                   Met         Ser
Rat α                                     Val         Asp
Rat β                                     Val         Asp
```

```
               30
Human α   Phe Val Pro Thr Asn Val Gly Ser Lys Ala Phe-amide
Human β                               Lys
Rat α                                 Glu
Rat β                                 Lys
```

and in isolated perfused tissues once insight has been gained from localisation studies as to the potential physiological role of the molecule.

Analysis of human genomic DNA, using a CGRP specific hybridisation
probe, provides evidence for a second human CGRP gene (see Fig 3A, also
Edbrooke et al., 1985). This has recently been shown to be transcribed in
MCT tissue, and encodes a 37 amino acid human β-CGRP which differs from
human α-CGRP at three positions (Steenbergh et al., 1985; see Table 1).
An analogous rat β-CGRP has also been reported (see Rosenfeld et al., 1984).
We have also used calcitonin and α-CGRP specific gene probes to examine by
RNA blotting and S_1 mapping the differential expression of the human calcitonin gene in MCT and a number of lung carcinoma cell-lines (see Craig et al.,
1985; Edbrooke et al., 1985). Analysis of poly(A)-containing RNA from MCT
tissue using a calcitonin-specific probe shows (Fig 3B) an abundance of
mature calcitonin mRNA (1Kb) and four prominent higher molecular weight
precursors in the two preparations examined. Analysis using a CGRP-specific
probe showed lower and varying amounts of mature CGRP mRNA (1.1Kb) but
only three of the higher molecular weight precursor species. The 2.2Kb
precursor identified using a calcitonin-specific hybridisation probe did

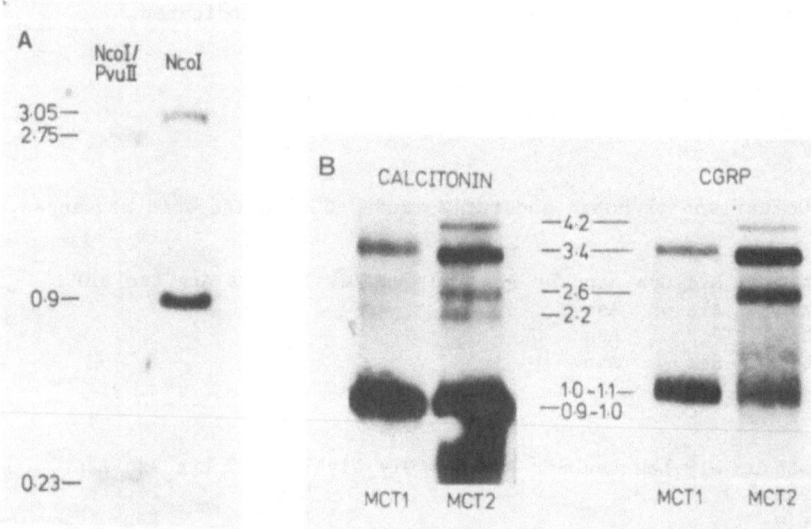

Fig 3 A: Genomic analysis of the human calcitonin/α-CGRP gene. Human placental
DNA was digested with NcoI/PvuII and NcoI, and analysed by Southern
blotting (see Edbrooke et al., 1985). Hybridisation with a CGRP-
specific cDNA probe identified the expected 900 bp and 230 bp
fragments from the calcitonin/α-CGRP gene. The additional band
in each digest indicates the presence of a second CGRP gene.

B: Northern blot analysis of poly(A)-containing RNA from two medullary
thyroid carcinomas (MCT). Parallel analysis of two separate
membranes with calcitonin or CGRP-specific cDNA hybridisation
probes demonstrates the presence of mature calcitonin mRNA
(0.9-1.0 kb) and mature CGRP mRNA (1.0 - 1.1 Kb) as well as a
series of larger precursor RNA species (see Edbrooke et al., 1985).

not contain CGRP sequences. These observations, a similar pattern of expression in lung carcinoma cell-lines producing high or low levels of
calcitonin, and the absence of a classic splice donor consensus sequence
at the 3' end of the calcitonin exon, have interesting implications
regarding the possible mechanisms involved in the differential expression
of two mRNA species from a single gene. Although we have discussed
these in detail elsewhere (see Edbrooke et al., 1985) it is apparent from
our own analyses, and analyses of the expression of the rat calcitonin/
α-CGRP gene (see Rosenfeld et al., 1984), that post-transcriptional
mechanisms determine the differential expression of calcitonin and
α-CGRP mRNA from a single gene. Thus preferential use of a splice acceptor
site gives rise to CGRP mRNA, whilst cleavage of a partially processed RNA
transcript followed by polyadenylation gives rise to calcitonin mRNA (see
Fig 4). The requirement for cleavage implicates the involvement of a
trans-acting gene product, and identifies a mechanism probably not unique
to the calcitonin gene by which differential exon usage gives rise to
different gene products dependent on the activation or otherwise of the
cleavage mechanism. Similar mechanisms may operate in the regulation of
immunoglobulin gene expression during B lymphocyte activation (Mather et
al., 1984), and during the maturation of the 3' termini of the sea urchin
histone mRNAs (see Birnstiel et al., 1985).

LOCALISATION OF THE CGRP PEPTIDE FAMILY

        Studies using antisera raised against the carboxy terminal fragment of
rat α-CGRP were first used to demonstrate the unexpected distribution of
CGRP producing cells. Immunohistochemical localisation studies (see Fig
5A, B) demonstrate an abundance of CGRP-containing nerve fibres in

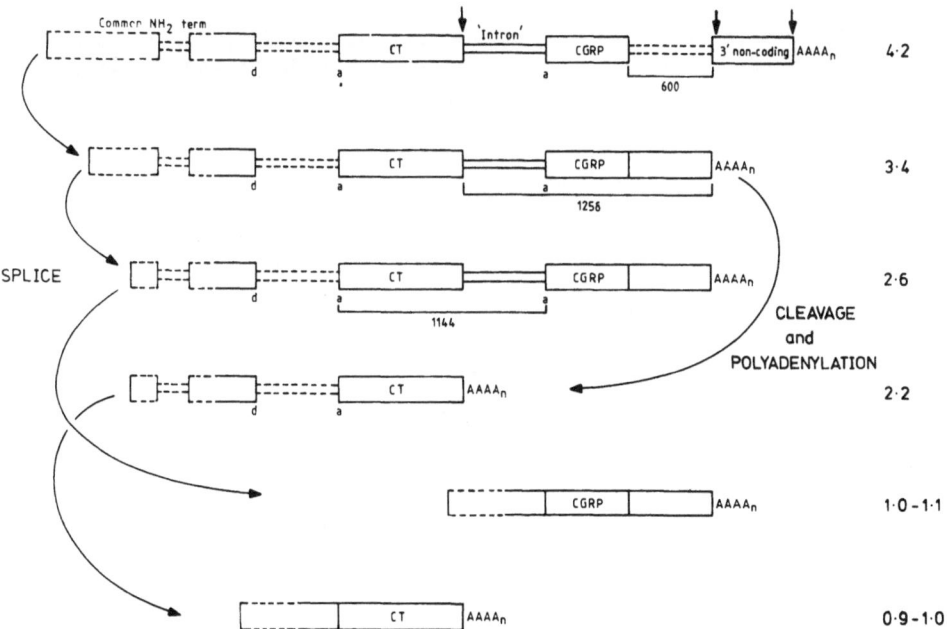

Fig 4: Differential processing pathways in the expression of the human
        calcitonin/α-CGRP gene. 'd' and 'a' denote donor and acceptor
        splice sites. Solid lines define regions of known nucleotide
        sequence, and potential poly(A)addition sites are indicated ($\downarrow$).

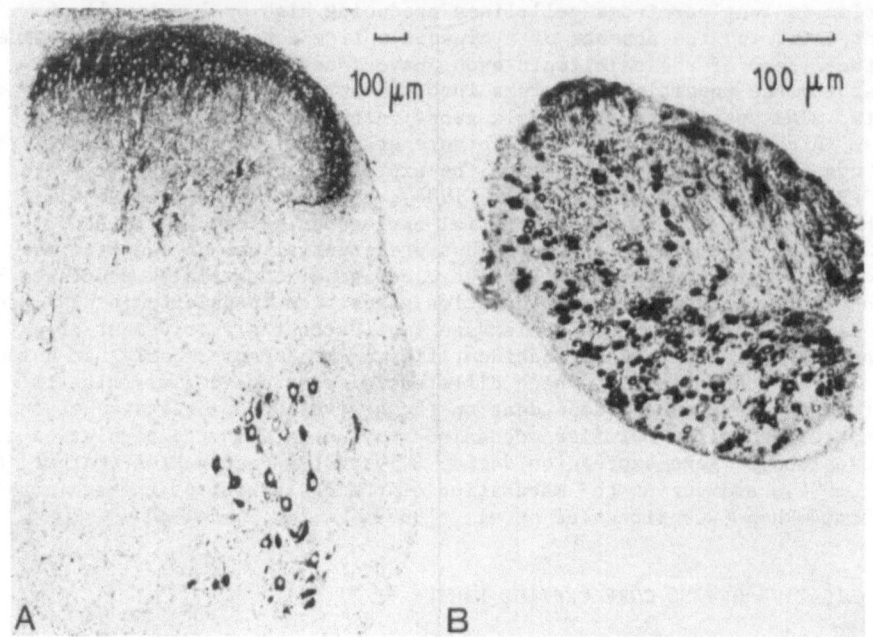

Fig 5 A: CGRP immunoreactivity in the lumbar spinal cord of the rat. Many
           immunoreactive fibres are present in the dorsal horn and
           positively stained motorneurones are visible in the ventral horn.
       B: CGRP immunoreactive cells in rat dorsal root of ganglia. Cryostat
           (20 M) sections, Bouin's fixation. Immunogold-silver intensi-
           fication method.
           (Figs 5A,B, from Immunocytochemistry 2nd Edition, Ed. J. Polak
           and S. Van Noorden, by kind permission of John Wright and Sons Ltd).

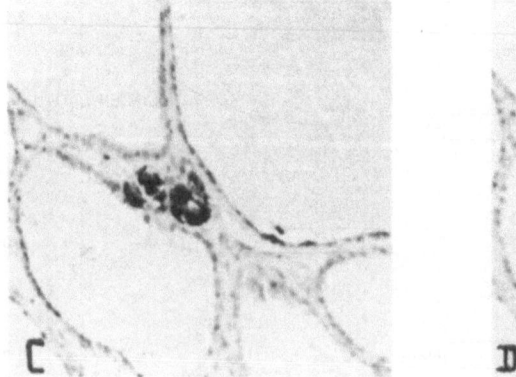
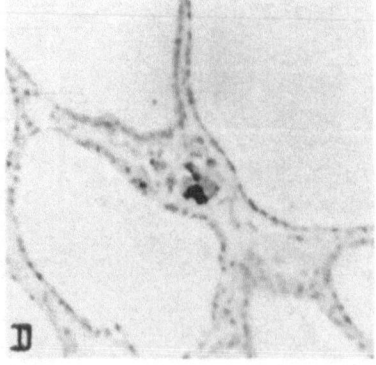

Fig 5 C and D: Immunoperoxidase staining of serial sections of a normal
                 thyroid, showing localisation of calcitonin immunoreactivity
                 (C) and CGRP immunoreactivity (D).

rat brain and central nervous system (Rosenfeld et al., 1983) and in nerve
fibres throughout the cardiovascular system (Mulderry et al., 1985). Our
own studies using antisera raised against human α-CGRP, demonstrate the
presence of elevated plasma CGRP levels in MCT (Craig et al., 1985; Edbrooke
et al., 1985) and lung carcinoma (Schifter, S. & Riley J. - unpublished),
and by immunohistochemical localisation, the presence of CGRP-staining
cells, in addition to calcitonin staining cells, in normal human thyroid
tissue and MCT tissue (Williams et al., 1986). Analysis of serial sections
of normal human thyroid tissue for the presence of CGRP and calcitonin
producing cells (Fig 5 C,D) demonstrates, not unexpectedly, that single
cells can produce both calcitonin and CGRP. The question however is which
CGRP - α and/or β? Available antibodies will not distinguish between
the two variant peptides; consequently it remains to be established by
using in situ hybridisation whether CGRP localised in the brain, CNS and
nerve fibres associated with the vasculature reflects the expression of
the calcitonin/α-CGRP gene, or of the related β -CGRP gene.

BIOLOGICAL ACTIVITY OF THE CGRP PEPTIDE FAMILY

        Immunohistochemical localisation studies suggest a role for the CGRP
peptide family as neurotransmitter or neuromodulator molecules. Their
distribution in nerve fibres associated with the vasculature points to a
major role in cardiovascular regulation, a conclusion substantiated by Fisher
and co-workers(1983), who demonstrated that synthetic rat α-CGRP had a
profound effect on blood pressure in vivo. We have used synthetic human
and rat α-CGRP, and human β-CGRP to investigate their role in cardiovascular
regulation both in vivo and in vitro in a number of species. In addition we
have investigated the potential role of CGRP as a neuromodulator using the
twitch response in the isolated vas deferens as a model system.

Cardiovascular Effects of CGRP in vivo

        Human α-CGRP, human β -CGRP and rat α-CGRP have been studied in a number
of cardiovascular systems (Table 2). When given intravenously, human α-CGRP
and rat α-CGRP evoke marked falls in blood pressure of rapid onset
(maximal within $1\frac{1}{2}$ min) in anaesthetised rats or dogs. For example, in the
rat, 1 nmol kg^{-1} of human α-CGRP lowered mean arterial pressure from 125 ± 8
mm Hg (mean ± s.e.mean) to 68 ± 9 mm Hg (Fig. 6;see also p.11). Falls in
blood pressure were accompanied by small but significant increases in
heart rate in the rat but not in the dog. In the latter species 3 nmol kg
i.v. of human α-CGRP and rat α-CGRP evoked falls in mean arterial pressure
from 128 ± 5 to 74 ± 7 mm Hg and from 150 ± 13 to 85 ± 3 mm Hg respectively,
and a significant reduction in blood pressure was maintained for at least
40-60 mins (Craig et al., 1986).

Cardiovascular effects of CGRP in vitro

        A number of in vitro models have been employed to separate direct
effects on the heart from those on the vasculature. CGRP produced
vasodilatation in perfused mesenteric vasculature in the presence of the
vasoconstrictor agent noradrenaline (10^{-5}M) (Fig. 7) and in the perfused
coronary vasculature where the vasconstrictor was arginine vasopressin
(10^{-6}M) (Fig. 8). These effects were observed in tissues from rat, rabbit
and man(Table 2).
        The onset of the vasodilator effect of CGRP and the speed of recovery
appeared quicker in the rat and rabbit perfused mesentery than in the human
preparation. For example, the maximum effect of 3 nmoles of human α-CGRP
occurred within 1 min in the rat mesenteric vasculature with recovery largely
completed in 10 min. By contrast the maximum effect of a single 3nmoles dose
of human α-CGRP in human perfused mesenteric vasculature took as much as

Table 2. Potency relative to human α-CGRP (= 1) of human β-CGRP and rat α-CGRP in preparations from a variety of species.

Preparation	Response	human α-CGRP	human β-CGRP	rat α-CGRP
(a) In vivo				
Anaesthetised rat	Hypotension	1	-	1
Anaesthetised dog	Hypotension	1	-	1
(b) In vitro				
Perfused mesenteric vasculature				
Rat	Vasodilatation	1	10	10
Rabbit	Vasodilatation	1	-	5
Man	Vasodilatation	1	-	-
Mesenteric artery rings				
Rat	Relaxation	1	3	3
Man	Relaxation	1	-	-
Perfused coronary vasculature				
Rat	Vasodilatation	1	3	1
Rabbit	Vasodilatation	1	-	3
Perfused isolated heart				
Rat	Positive chronotropism	1	10	3
Rabbit	Positive chronotropism	*	-	*
Isolated right atria				
Rat	Positive chronotropism	1	-	1
Rat	Positive inotropism	1	-	1
Guinea-pig	Positive chronotropism	1	-	10
Guinea-pig	Positive inotropism	1	-	1
Vas deferens				
Mouse	Inhibition of nerve stimulation	1	-	1
Mouse	Inhibition of acetyl-choline contraction	1	-	1
Rabbit	Inhibition of nerve stimulation	*	-	*
Rabbit	Inhibition of acetyl-choline contraction	*	-	*

- not tested * CGRP ineffective at doses/concentrations eliciting effects in the same preparation from other species.

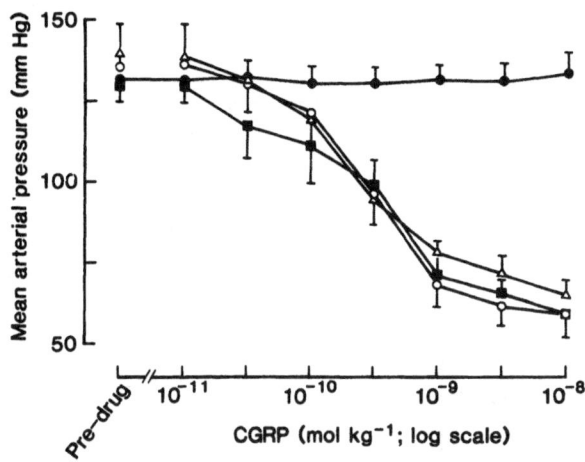

Fig. 6 The effect on mean arterial pressure of the pentobarbitone
anaesthetised rat of the cumulative administration of saline
(•), human α-CGRP (○), rat α-CGRP (▲) or human α-CGRP after
pretreatment with mepyramine 12.4 μmol.kg^{-1} s.c. and cimetidine
59.5 μmol.kg^{-1} i.v. infusion over 30 min (■). Points represent
mean ± s.e.m from 4-6 animals

10 min to develop, and was then maintained for an hour or even longer.
However, while these results suggest a longer duration of action for
CGRP in the human mesentery compared with other species, the possibility
that this may be a consequence of drug treatments, or the disease state of
the individual from whom the preparations were obtained, cannot be excluded.
In ring preparations of superior mesenteric artery of the rat and man,
CGRP relaxed tone produced by the α-adrenoceptor agonist noradrenaline
(10^{-7}M). Acetylcholine and sodium nitroprusside were more potent and
more effective at relaxing the arterial rings than any of the three CGRP
compounds (Fig 9). This contrasts with results from the rat perfused
mesenteric vasculature where the peptides were usually at least as potent
and with similar maxima to sodium nitroprusside.

While there are differences in relative potency of the three CGRP compounds, they all exerted qualitatively similar effects in these vascular
preparations. However, this finding contrasts with their effects on the
rate of beating of isolated perfused hearts. Human and rat α-CGRP increased
the rate of beating of the rat heart (e.g. increases of 75 ± 23 and 40 ± 22
b.min^{-1} by rat and human α-CGRP, 10^{-9} moles respectively), but did not
alter the heart rate of rabbit hearts (Al-Kazwini et al., 1985). The lack
of a positive chronotropic effect of the peptides occurred in the same
rabbit preparations in which CGRP evoked dose-dependent falls in coronary
perfusion pressure. The force of contraction of isolated rat and rabbit
hearts was unaltered by CGRP suggesting that the peptides have relatively
little effect on the ventricles. In right atrial preparations from the
rat and the guinea-pig, human and rat α-CGRP increased both the rate and
the force of contraction (Table 2). For example, in the rat isolated right
atrium, human and rat α-CGRP (100 nM) increased the rate of beating (from
around 190 b.min^{-1}) by 62 ± 7 and 66 ± 6 b. min^{-1} respectively.

Site of action of CGRP

The isolated vas deferens preparation is a sympathetically innervated
smooth muscle tissue which contracts in response to the release of neurotransmitter elicited by field stimulation or in response to a variety of

9

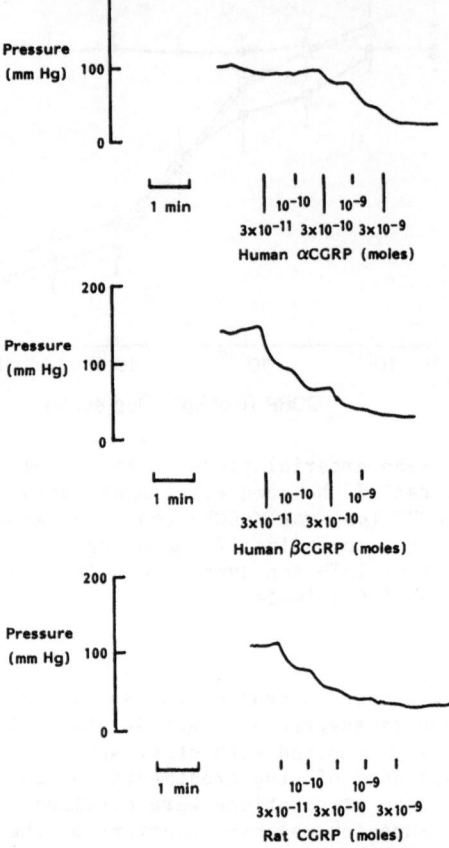

Fig. 7. An example of the falls in perfusion pressure evoked by the cumulative administration of human α-CGRP, human β-CGRP and rat α-CGRP in rat isolated mesenteric vasculature perfused at constant flow (5.8ml.min^{-1}) with Krebs solution containing noradrenaline (10^{-5}M) as a vasoconstrictor agent.

exogenous agonists including acetylcholine. Human and rat α-CGRP both inhibited contractions of the mouse vas deferens evoked either by field stimulation or by acetylcholine (Table 2). These findings suggest that the effects of the peptides are mediated post-junctionally. This mechanism may have a wider application than the vas deferens and potentially provides a way in which CGRP could modulate the actions of neurotransmitters and hormones elsewhere in the body.

In contrast to these findings in the mouse vas deferens, neither human nor rat α-CGRP inhibited contractions of the rabbit vas deferens to either field stimulation or acetylcholine. One explanation of this species difference is that there may be no CGRP 'receptors' present, or alternatively that they may differ in structure from those in the mouse vas deferens. A similar species difference was noted above for the lack of effect of the peptides on the heart rate of rabbit but not rat hearts (Table 2). Taken together these results suggest that the rabbit differs from a number of other species in some of its responses to CGRP. However, in some rabbit tissues, for instance in the perfused mesenteric and coronary vasculature, CGRP evokes the same response as in other species.

Evidence for CGRP receptors

In several preparations the possibility that CGRP might act indirectly

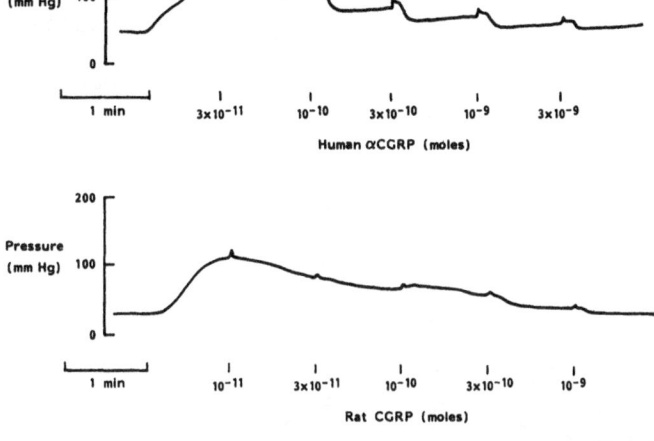

Fig. 8. An example of the falls in perfusion pressure evoked by the cumulative administration of human α-CGRP and rat α-CGRP in rabbit isolated hearts perfused via the coronary arteries at constant flow (12.0ml.min^{-1}) with Krebs solution. The initial rise in pressure follows the addition of arginine vasopressin (10^{-6}M) to the Krebs solution perfusing the tissue.

via another endogenous mediator has been studied. CGRP is a slightly basic molecule and it is known that basic peptides can release histamine (Goth, 1973) which, in the anaesthetised rat could result in a fall in blood pressure and an increase in heart rate. However, the combination of a histamine H_1-receptor antagonist mepyramine and H_2-receptor antagonist cimetidine failed to alter the hypotensive effect of human α-CGRP(see Fig. 6). It has been suggested that the positive chronotropic and inotropic effects of human and rat α-CGRP in rat isolated atria can be antagonised by propranolol, 7.7 μM (Tippins et al., 1984). However, the effect was non-competitive and might reflect the local anaesthetic effect of this concentration of propranolol in atria (Morales-Aguilera & Vaughan Williams, 1965). A lower concentration of propranolol (300 nM) has been found to abolish increases in the rate and force of contraction of rat atria evoked by isoprenaline without altering responses to CGRP. A similar lack of effect of the β-adrenoceptor antagonist propranolol was found against CGRP responses in guinea-pig atria and the mouse vas deferens. Thus, so far, there is no convincing evidence for an indirect action of CGRP.

Radioligand binding sites for CGRP in the cardiovascular system have yet to be described. However, specific binding sites for CGRP have been identified in the human central nervous system, pituitary and spinal cord (Tschopp et al., 1985). Indirect evidence in support of CGRP receptors has been provided in some of the present work. Use of selective antagonists has eliminated the possibility that CGRP acts in a number of preparations at either histamine receptors or β-adrenoceptors. In addition, if CGRP was an agonist at α_1-adrenoceptors it would elicit vasoconstriction instead of being a vasodilator. The lack of effect of the α_2-adrenoceptor antagonist idazoxan against CGRP inhibition of field stimulated contractions of the mouse vas deferens eliminates the involvement of α_2-adrenoreceptors, whilst the inability of CGRP to inhibit acetylcholine contractions in the rabbit vas deferens shows that CGRP has little affinity for muscarinic cholinoceptors. Thus at present there is no evidence for an action of these peptides through known receptors. Indirect evidence for CGRP receptors in

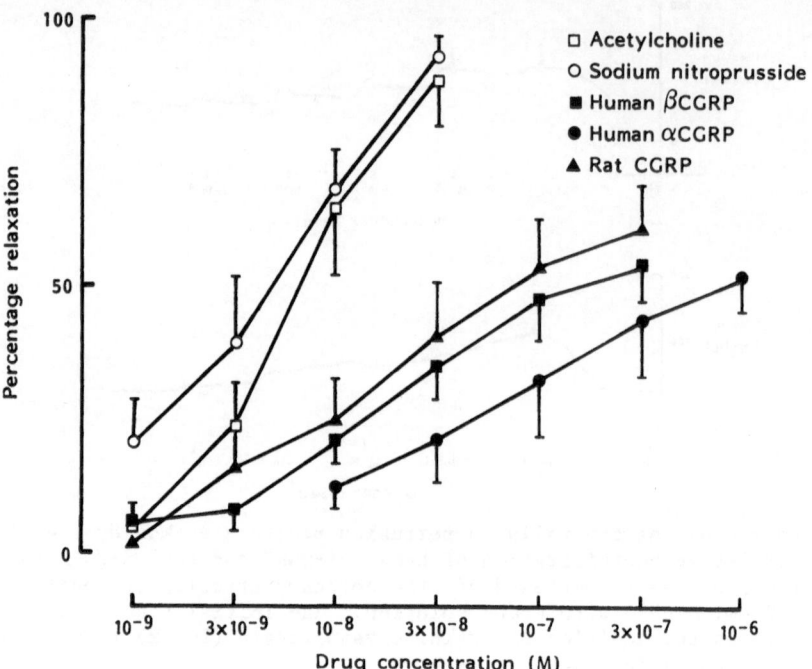

Fig. 9. Relaxation produced by the cumulative addition of acetylcholine, sodium nitroprusside, rat α-CGRP, and human α- and β-CGRP in rings of rat mesenteric artery contracted with noradrenaline (10^{-7}M). The response to acetylcholine shows that the arterial rings had not had their endothelial cells removed. Points represent the mean$^\pm$ s.e.m. from at least 4 experiments.

the cardiovascular system comes from the finding that the peptides are vasodilators whether the tone of the vasculature is increased by noradrenaline (mesentery) or by arginine vasopressin (coronary).

A further argument supporting the concept of CGRP receptors is the lack of effect of the peptides in some, but not all, rabbit tissues. In the perfused isolated heart of the rabbit, CGRP evoked vasodilation of the coronary vessels with no change in heart rate in contrast to the positive chronotropic effect in the rat isolated heart. The results in the rabbit suggest that 'receptive areas' were present for CGRP in the vasculature of the heart, but in not the atria. This differential distribution in response to CGRP in different species and within a single organ is in line with a different distribution of CGRP receptors. While these findings do not prove the existence of CGRP receptors, they are consistent with the hypothesis that CGRP acts through its own receptors in the cardiovascular system.

The presence of a particular receptor may often be inferred from structure-activity relationships, with small structural changes giving rise to large changes in biological activity. The relative potency of the 3 CGRP molecules examined so far reveals fairly modest alterations in activity. In this connection the activity of salmon calcitonin, which has some structural homology with α-CGRP, is of interest. It has been found that salmon calcitonin competes for high affinity binding sites of ^{125}I-rat α-CGRP in rat spinal cord (Goltzman & Mitchell, 1985). In the rat perfused mesenteric vasculature preparation, both salmon calcitonin and human calcitonin had less than a hundredth of the potency of human α-CGRP as a vasodilator (Fig. 10), reinforcing the view that the effects of CGRP are distinct from those of calcitonin.

Mechanism of action of CGRP in the vasculature

The work of Furchgott has led to the separation of vasodilators into two main groups (Furchgott et al., 1984). One group acts on the endothelium of the vasculature to initiate relaxation of the smooth muscle via the release of an as yet unidentified endothelial dependent relaxing factor. An example of this type of vasodilator is acetylcholine. The second group, typified by sodium nitroprusside, act independently of the endothelium. In rings of rat mesenteric artery contracted by noradenaline (10^{-7}M), nitroprusside ($10^{-9}-3\times10^{-8}$M) evoked concentration-dependent relaxation whether the endothelium was present or absent. In contrast, acetylcholine ($10^{-9}-10^{-7}$M) relaxed endothelial-containing tissues by over 90% at 10^{-7}M but had no effect when the endothelium had been removed. Human α-CGRP, like acetylcholine, evoked a concentration-dependent relaxation ($10^{-9}-10^{-7}$M) of up to 50% of noradrenaline-induced tone with the endothelium present but was totally ineffective over the same concentration range when the endothelium was absent. Similar endothelial-dependent relaxation was found for rat α-CGRP, in agreement with the results of Brain et al.(1985). It is possible that circulating CGRP could act at endothelial sites to regulate blood pressure. Such a mechanism would have pharmacological significance and could be of physiological significance.

The characteristics of vasodilatation for a number of vasodilators vary with the spasmogen. For example calcium antagonist drugs are more potent at inhibiting contractions of arterial muscle evoked by potassium depolarisation than against noradrenaline induced contractions while the reverse is true for sodium nitroprusside and atrial natriuretic factor (Cauvin et al., 1983; Winquist et al., 1984). The potassium depolarisation evokes contraction of vascular smooth muscle mainly via the influx of calcium ions whereas that of noradrenaline is more dependent on the release of intracellular calcium (Cauvin et al., 1983). In the rat perfused mesenteric vasculature, human α- and β-CGRP and rat α-CGRP were more than 30 times less potent as vasodilators (and with a lower maximum effect) when the pressure was raised by K^+ 60 mM than when pressure was raised by noradrenaline 10^{-5}M. The potency of sodium nitroprusside was also reduced in the high K^+ experiments. These results demonstrate that CGRP does not act in the vasculature in a similar fashion to the calcium antagonists.

The idea that CGRP might have a physiological role in regulating vascular tone is supported by the finding of CGRP immunoreactivity in the plasma of human subjects (Morris et al., 1984; Craig et al.,1985; Edbrooke et al., 1985., 1985; Girgis et al., 1985). The circulating peptide could then act via the endothelium to elicit vasodilatation. There is evidence for CGRP immunoreactivity in the cardiovascular system of the rat where the highest concentrations in the heart are in the right atrium and in the vasculature within the superior mesenteric artery, renal artery, abdominal artery and in some veins (Mulderry et al., 1985). The peptide was in nerve fibres, mainly in the adventitia. The finding that the relaxant effect of CGRP did not occur in the absence of endothelium could suggest either that peptide released from nerve terminals in the advertitia does not act locally or that it may be involved, at least partly, in producing other biological effects in these tissues.

Calcitonin gene-related peptides have been found to be very potent vasodilators in a number of preparations from a variety of species. The increase in heart rate is not seen in all species. These peptides undoubtedly have many other biological effects apart from those in the cardiovascular system. From studying the action of synthetic CGRP there is evidence for a role in inflammation (Brain et al.,1985), an inhibition of gastric acid release mediated both peripherally and centrally (Hughes et al.,1984; Tache et al., 1984) and an antinociceptive effect (Bates et al.,1984). Finally, the widespread distribution of CGRP and CGRP receptors in the central nervous system (Rosenfeld et al.,1983;Tschopp et al.,1985) suggests a role for the peptide in the regulation of a wide variety of functions.

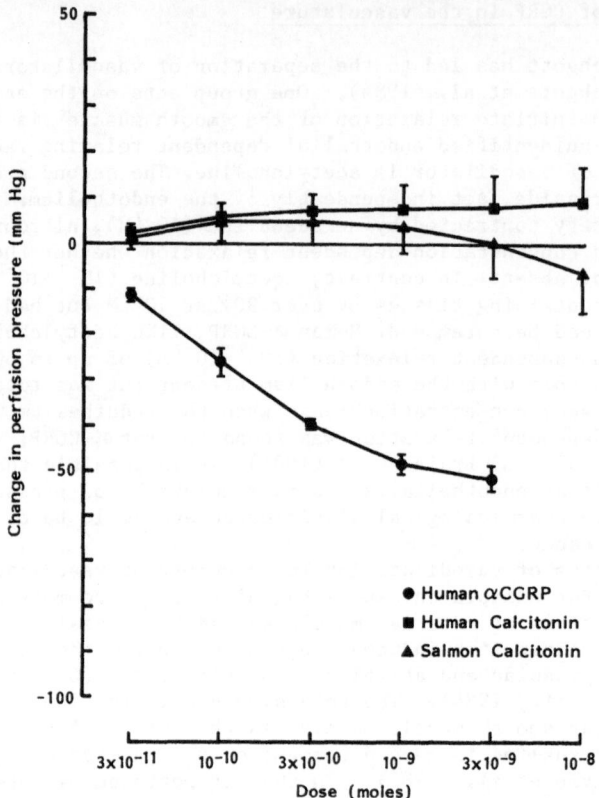

Fig. 10. Changes in perfusion pressure evoked by the cumulative administration of human α-CGRP, or human and salmon calcitonin in rat isolated mesenteric vasculature perfused at constant flow (5.8ml.min^{-1}) with Krebs solution containing noradrenaline (10^{-5}M) as a vasoconstrictor agent. Points represent the mean $\pm$ s.e.m. from at least 4 experiments.

SUMMARY

Studies on the structure and expression of the rat and human calcitonin gene provide a remarkable example by which a combination of molecular, immunocytochemical and pharmacological techniques have led to the identification of a novel gene product, calcitonin gene-related peptide, which has an unexpected tissue distribution and biological activity. Future studies at the molecular level will provide insight into post-transcriptional mechanisms by which a single gene can give rise to discrete mRNAs in a tissue-specific manner. The identification of the sites and mechanisms of action of the CGRP peptide family within the vasculature will also add significantly to our understanding of molecular mechanisms involved in the modulation of cardiovascular function in man. In addition, the role and mechanism of action of the CGRP peptide family in the central and peripheral nervous systems remains to be elucidated.

ACKNOWLEDGEMENTS

We thank the Cancer Reseach Campaign, Wellcome Trust, Medical Research Council and Celltech Ltd. for supporting different aspects of our work. We are also grateful to Professor Julia Polak for permission to reproduce immunohistochemical localisation studies of the spinal cord, and we thank Professor Kees Lips for communicating the predicted amino acid sequence of human β-CGRP prior to publication.

REFERENCES

Al-Kazwini, S.J., Craig, R.K., Holman, J.J. and Marshall, I.(1985). Human and rat calcitonin gene-related peptides are vasodilators in coronary and mesenteric vasculature. Br. J. Pharmac., Proc. Suppl., in press.

Amara, S.G.,Jonas, V., Rosenfeld, M.G., Ong, E.S., and Evans, R.M. (1982). Alternative RNA processing in calcitonin gene expression: mRNAs encoding different polypeptide products. Nature, 298:240-244.

Bates, R.F.L., Buckley, G.A. and McArdle, C.A. (1984). Comparison of the antinociceptive effects of centrally administered calcitonin and calcitonin gene-related peptide. Br. J. Pharmac.,82:295.

Birnstiel, M.L., Busslinger, M. and Strub, K. (1985). Transcription termination and 3' processing: the end is in site. Cell, 41:349-359.

Brain, S.D., Williams, T.J., Tippins, J.R., Morris, H.R. and MacIntyre, I. (1985). Calcitonin gene-related peptide is a potent vasodilator. Nature, 313:54-56.

Cauvin, C., Loutzenhiser, R. and van Breemen, C.(1983). Mechanisms of calcium antagonist induced vasodilation. Annu. Rev. Pharmacol., 23:373-396.

Coombes,R.C., Greenberg, P.B., Hillyard, C. and MacIntyre, I.(1974). Plasma-immunoreactive-calcitonin in patients with non-thyroid tumours. Lancet, 1:1080-1083.

Craig, R.K., Hall, L., Edbrooke, M.R., Allison, J. and MacIntyre, I. (1982). Partial nucleotide sequence of human calcitonin precursor mRNA identifies flanking cryptic peptides. Nature, 295:345-347.

Craig, R.K., Edbrooke, M.R., Riley, J.H., McVey, J.H. and Parker, D. (1985). Differential expression of the human calcitonin/CGRP gene in medullary thyroid carcinoma and lung carcinoma cell-lines. Recent Results in Cancer Research, 99: 71-78.

Craig, R.K., Marshall, I., Paciorek, P. & Shepperson, N.B. (1986). Cardiovascular effects of human and rat calcitonin gene-related peptides in anaesthetised dog. Br. J. Pharmac., Proc. Suppl.

Edbrooke, M.R., Parker, D., McVey, J.H., Riley, J.H., Sorenson, G.D., Pettengill, O.S. and Craig, R.K. (1985). Expression of the human calcitonin/CGRP gene in lung and thyroid carcinoma. EMBO J., 4:715-724.

Fisher, L.A., Kikkawa, D.O., Rivier, J.E., Amara, S.G., Evans, R.M., Rosenfeld, M.G., Vale, W.W. and Brown, M.R. (1983). Stimulation of noradrenergic sympathetic outflow by calcitonin gene-related peptide. Nature, 305: 534-536.

Foster, G.V., Baghdiantz, A., Kunar, M.A., Slack, E., Soliman, H.A. and MacIntyre, I. (1964). Thyroid origin of calcitonin. Nature, 202:1303-1305.

Furchgott, R.F., Cherry, P.D., Zawadzki, J.V. & Jothianandan, D. (1984). Endothelial cells as mediators of vasodilation of arteries. J. Cardiovascular Pharm., 6:S336-S343.

Girgis, S.I., MacDonall, D.W., Stevenson, J.C., Bevis, P.J.R., Lynch, C., Wimalawansa, J.S., Self, C.H., Morris, H.R. & MacIntyre I. (1985). Calctonin gene-related peptide: potent vasodilator and major product of calcitonin gene. Lancet ii:14-16.

Goltzman, D. & Mitchell, J. (1985). Interaction of calcitonin and calcitonin gene-related peptide at receptor sites in target tissues. Science, 227:1343-1345

Goth, A. (1973). Histamine released by drugs and chemicals. In Histamine and antihistamines ed. Schachter, M. Vol. I Int. Encyclopedia of Pharmacology and Therapeutics, section 74, Pergamon Press, London,p25.

Hughes, J.J., Levine, A.S., Morely, J.E., Gosnell, B.A. & Silivis, S.E. (1984). Intraventricular calcitonin gene-related peptide inhibits gastric acid secretion. Peptides 5:665.

Jonas, V., Lin, C.R., Kawashima, E., Semon, D., Swanson, L.W., Mermod, J.-J., Evans, R.M. and Rosenfeld, M.G. (1985). Alternative RNA processing events in human calcitonin/CGRP gene expression. Proc. Natl. Acad. Sci. U.S.A. 82: 1994-1998.

Le Moullec, J.M., Jullienne, A., Chenais, J., Lasmoles, F., Guliana, J.M., Milhaud, G. and Moukhtar, M.S. (1984). The complete sequence of human preprocalcitonin. FEBS Lett. 167: 93-97.

Mather, E.L., Nelson, K.J., Haimovich, J. and Perry, R.P. (1984). Mode of regulation of immunoglobulin μ and δ chain expression varies during B-lymphocyte maturation. Cell, 36:329-338.

Milhaud, G., Calmette, C., Taboulet, J., Julienne, A. and Moukhtar, M.S. (1974). Hypersecretion of calcitonin in neoplastic conditions. Lancet, i 462-463.

Morales-Aguilera, A. & Vaughan-Williams, E.M. (1965). The effects on cardiac muscle of β-receptor antagonists in relation to their activity as local anaesthetics. Br. J. Pharmac., 24:332.

Morris, H.R., Panico, M., Etienne, T., Tippins, J., Girgis, S.I. and MacIntyre, I. (1984). Isolation and characterisation of human calcitonin gene-related peptide. Nature, 308:746-748.

Mulderry, P.K., Ghatel, M.A., Rodrigo, J., Allen, J.M., Rosenfeld, M.G., Polak, J.M. and Bloom, S.R. (1985). Calcitonin gene-related peptide in cardiovascular tissues of the rat. Neuroscience, 14: 947-954.

Nelkin, B.D., Rosenfeld, K.I., de Bustros, A., Leong, S.S., Roos, B.A. and Baylin, S.B. (1984). Structure and expression of a gene encoding human calcitonin and calcitonin gene-related peptide. Biochem. Biophys. Res. Comun., 123:648-655.

Rosenfeld, M.G., Mermod, J-J., Amara, S.G., Swanson, L.W., Sawchenko, P.E., Rivier, J., Vale, W.W. and Evans, R.M. (1983). Production of a novel neuropeptide encoded by the calcitonin gene via tissue-specific RNA processing. Nature, 304: 129-135.

Rosenfeld, M.G., Amara, S.G. and Evans, R.M. (1984). Alternative RNA processing: determining neuronal phenotype. Science, 225: 1315-1320.

Steenbergh, P.H., Höppener, J.W.M., Zandberg, J., Van de Ven, W.J.M., Jansz, H.S. and Lips, C.J., (1984). Calcitonin gene-related peptide coding sequence is conserved in the human genome and is expressed in medullary thyroid carcinoma. J. Clin. Endocrinol. Metab., 59:358-360.

Stevenson, J.C., Hillyard, L.J., MacIntyre, I., Cooper, M. and Whitehead, M.I. (1979). A physiological role for calcitonin: protection of the maternal skeleton. Lancet, ii 769-770.

Tache, Y., Pappas, T., Lauffenburger, M., Goto, Y., Walsh, J.H. & Debas, H. (1984). Calcitonin gene-related peptide: potent peripheral inhibitor of gastric acid secretion in rats and dogs. Gastroenterology, 87:344.

Tippins, J.R., Morris, H.R., Panico, M., Etienne, T., Bevis, P., Girgis, S., MacIntyre, I., Azria, M. & Attinger, M. (1984). The myotropic and plasma calcium modulating effects of calcitonin gene-related peptide. Neuropeptides, 4: 425.

Tschopp, F.A., Henke, H., Petermann, J.B., Tobler, P.H., Janzer, R., Hokfelt, T., Lundberg, J.M., Cuello, C. & Fischer, J.A. (1985). Calcitonin gene-related peptide and its binding sites in the human central nervous system and pituitary. Proc. Natl. Acad. Sci. U.S.A., 82:248.

Williams, F.D., Ponder, B.A.J. and Craig, R.K. (1986). CGRP in human medullary carcinoma and C-cell hyperplasia - an immunohistochemical study. Clin. Endocrinol. submitted.

Winquist, R.J., Faison, E.P. & Nutt, R.F. (1984). Vasodilator profile of synthetic atrial natriuretic factor. Eur. J. Pharmac., 102:169-173.

NEURONAL SPECIFIC GENE EXPRESSION IN THE MURINE HIPPOCAMPUS

M.C. Wilson, P.L. Branks and G.A. Higgins

Department of Molecular Biology
Research Institute of Scripps Clinic
10666 North Torrey Pines Road
La Jolla, CA 92037 USA

INTRODUCTION

The mammalian brain has been traditionally examined through combinations of anatomical, electrophysiological and pharmacological techniques. Together, these studies have correlated the phenotype of neurologic function, be it neuronal networks or choice of neuroactive transmitters, with the underlying structure of the brain. Another approach, which has been more recently explored, is the identification of specific gene products which may distinguish brain structures or neuronal groups defined by such classical studies. Implicit in this approach is the assumption that the differential expression of sets of genes help determine function in different neuronal cell populations.

The results of several laboratories demonstrate that the repertoire of expressed mRNAs in brain exceeds that of other tissues. Hybridization of single copy DNA to an excess of brain polyA$^+$ polysomal mRNA indicates that as much as 7-8% of the rodent genome is expressed as mRNA in brain, while only 4% of the genome is expressed as polyA$^+$ polysomal mRNA in liver and other tissues ((Bantle and Hahn, 1976; Grouse et al., 1978; Chikaraishi, 1979). In terms of mRNAs of average 1500 nucleotide length these estimates can be converted to approximately 8.3×10^4 and 4.3×10^4 different mRNAs in brain and liver, respectively. This two-fold difference in complexity is maintained when total polyA$^+$ nuclear and cytoplasmic RNA is compared (Bantle and Hahn, 1976; Chikaraishi et al.; Kaplan et al., 1978), suggesting that the major regulatory event generating tissue specific mRNAs is prior to polyadenylation of nuclear transcripts, probably at the level of transcription. The vast majority of these brain-specific mRNAs are present in low copy number within the general population of whole brain mRNA. Considering the diverse function and morphological organization of the mammalian brain, it is likely that abundance of a particular RNA species is in fact significantly higher in restricted neuronal subpopulations.

The hippocampus as a model of neural-specific gene expression

The hippocampal formation provides a valuable model to explore the significance of regional and neuronal-specific gene expression in the development and function of the mammalian brain. A major attribute of hippocampus is its relatively simple laminar organization, composed primarily of only a few neuronal subtypes that populate distinct morphological regions of the structure. These include the granule neurons of the dentate gyrus and large pyramidal neurons of the hippocampus proper, or Ammon's horn, as well as more peripheral interneurons. Within the pyramidal layer of the hippocampus proper, the neurons, which can again be distinguished in their intrinsic morphology, constitute the major fields CA1, CA2, CA3 and CA4 that are organized in a linear manner. This organization of neurons is reflected, in part, in the electrophysiology of the hippocampus. For example, a major intrinsic pathway of innervation within the hippocampal formation, initiated in the entorhinal area, proceeds in a well defined manner through the dentate gyrus and is progressively relayed by the pyramidal neurons of the three major fields of the hippocampus proper: CA4, CA3 and CA1 (reviewed in Swanson et al., 1982)). Moreover, the utilization of neurotransmitters both intrinsic and extrinsic in origin also appears to demarcate neuronal subgroups within the hippocampus. Although the major excitatory neurons of the hippocampus, including granule cells of the dentate gyrus and pyramidal cells of the hippocampus proper, appear to primarily utilize the amino acids glutamine and aspartic acid (Nadler et al.; Storm-Mathisen, 1977; Fonnum and Walaas, 1978), localization is evident for other neuroactive compounds. For example, inhibitory GABAergic interneurons, fusiform and dentate pyramidal basket cells, are associated with the granule layer of the dentate gyrus and present in the area pyramidal of the hippocampus (Fonnum and Walaas, 1978)). Recently McGinty et al, (McGinty et al., 1982) have presented evidence for two enkephalin pathways in the hippocampus: dynorphin containing mossy fiber efferents to the CA_3/CA_4 pyramidal cells from the dentate gyrus and a distinct, possibly Met^5, enkephalin innervation from the entorhinal cortex. Similarly the non-random distribution of different hippocampal interneurons expressing acetylcholinesterase (AChE), cholecystokinin (CCK), vasoactive intestinal peptide (VIP) and somatostatin (SRIF) has been reported (Walaas, 1983). These observations strongly support the premise that neurons of the hippocampus can be distinguished by biochemical as well as morphological criteria.

To identify cloned cDNA probes that distinguish either cell type or regional specific gene expression in brain, we have developed two strategies. The initial approach was to generate a cDNA library from total murine brain and establish screening criteria that would select clones most likely to represent relatively abundant, regionally expressed mRNAs. The specific expression of these mRNAs within the hippocampal formation and in other regions of brain was then established by in situ hybridization. In a second approach, cDNA sequences were enriched for hippocampal-specific mRNA prior to cloning, by subtractive hybridization of sequences shared with cerebellum or cerebral cortex. Our results employing these strategies indicate that cDNA clones to mRNAs differentially expressed by neurons of the hippocampal formation have been identified.

Identification of cDNA clones to brain-specific mRNAs

A collection of cDNA clones taken from a recombinant DNA library of adult murine brain poly(A) containing cytoplasmic RNA was screened for cloned sequences that specifically hybridize to brain mRNA with respect to either non-neuronal tissue, such as liver or kidney or to mRNA expressed by established cell lines of neural origin. Although neuronal and glioma cell lines exhibit differentiated properties and contain much of the complexity of brain mRNA sequence not found in non-neural tissue (Kaplan, 1978; Grouse et al., 1979; Beckman et al., 1981), we considered that those mRNA sequences expressed in brain but not in cultured neural cell lines might be enriched for those that encode proteins whose function are important for the organization of brain morphology or function. As the initial screen was hybridization of labeled brain and liver polyadenylated RNA to colony blots of cloned cDNA, the kinetic constraints of this hybridization required that the positive clones were of moderate to high abundant mRNAs, greater than 0.1% of the brain mRNA population (Gergen et al., 1979; Dworkin and Dawid, 1980). Of the 2000 cDNA clones analysed by these criteria, fifteen were identified that detected mRNAs which appeared to be differentially regulated or brain-specific.

Hybridization to blots of size fractionated RNA demonstrates the specificity of expression of the mRNAs detected by cDNA clones pMuBr2, 3, 8 and 85 (Figure 1). The cDNA sequences pMuBr2 and 8 each hybridize to single mRNA transcript 3.0 and 2.2 kb in length, respectively, expressed in brain and not in liver or kidney or in the neural cell lines examined. In contrast, cDNA sequence pMuBr85 hybridizes to at least four major polyadenylated transcripts 7.5, 6.8, 6.0 and 2.6 kb in

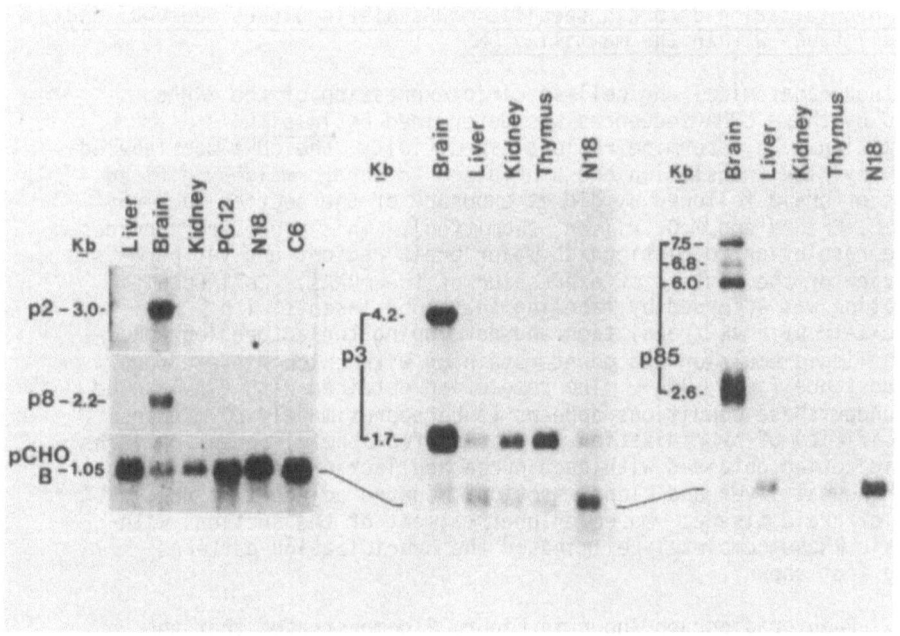

Fig. 1. Brain specific expression of pMuBr2, 3, 8 and 85. PolyA$^+$ enriched RNA of each sample was fractionated on a formaldehyde agarose gel blotted and hybridized with ^{32}P-labeled cDNA. Brain, liver and kidney polyA$^+$ RNA was prepared from total cellular RNA; the polyA$^+$ RNA of neural cultured cells (N18, PC12 and C6) was obtained from cytoplasmic RNA of NP-40 detergent lysed cells. Molecular weight determination was provided by parallel fractionation of denatured Hind III and Hind I cleaved pBR322 DNA (not shown). Both pMuBr3 and 85 hybridizations were performed with gel purified cDNA inserts, confirming that the single cDNA sequence was present in multiple polyA$^+$ transcripts.

length that appear specific to murine brain. Hybridization of this cDNA sequence to restriction endonuclease cleaved genomic DNA resolves a single Eco RI fragment and two Bam HI fragments (data not shown) suggestive that these multiple transcripts are products of a single genetic locus and result from alternative splicing, selection of polyadenylation sites or a combination of both. The two mRNA species hybridizing to pMuBr3, 4.2 and 1.7 kb in length, though more abundant in brain are present in lesser amounts in other tissues. The abundance of these two transcripts, however, appear to be independently regulated. Densometric measurement of the autoradiogram depicted in Figure 1 demonstrates that in brain the 4.2 kb transcript is of nearly equal abundance with the 1.7 kb transcript while in other tissues or in the N18 neuroblastoma cell line, the 4.2 kb transcript is more than ten fold reduced with respect to the 1.7 kb species. Normalizing the hybridization of the pMuBr3 probe to that obtained with a control probe pCHOB that detects a housekeeping mRNA transcript equally abundant in all tissues (Harpold et al., 1979; Levy et al., 1982) demonstrates further that the abundance of the 1.7 kb MuBr3 species is 20 fold reduced in liver, and neural cell lines but only five fold reduced in thymus. The highly regulated expression of these transcripts, abundant in brain but reduced in other tissues is consistent with a class of brain mRNAs, termed class II, by Milner and Sutcliffe (1984).

In situ hybridization of brain specific mRNAs distinguishes neuronal and glial cell types within the mammalian CNS

The neuroanatomical and cell-specific expression of the mRNAs detected by these cDNA sequences was determined by in situ hybridization. To determine regional specificity, the cDNA was labeled with ^{32}P by nick translation and hybridized to paraformaldehyde fixed sections of brain followed by direct exposure of the section to X-ray film (P.L. Branks and M.C. Wilson, submitted). This procedure provides adequate resolution to distinguish major brain regions and yet permits a global view of the pattern of expression of the mRNAs. Cell specific localization was afforded by labeling the probe inserts with ^{35}S-thio-αATP by nick translation and developing the autoradiographic image on liquid emulsion and counterstaining with thionin stain for Nissl substance (i.e., RNA). The resolution obtained with ^{35}S-labeled probes under these conditions appears to be approximately 50 microns. The specificity of hybridization is evident from the different patterns of hybridization obtained with each probe and lack of discernible signal over non-neural liver and kidney sections mounted adjacent to the section of brain tissue. Moreover pretreatment of the sections with pancreatic RNase completely eliminated the hybridization patterns observed (not shown).

The ^{32}P-autoradiograph shown in Figure 2 demonstrates that the mRNAs detected by the cDNA clones correspond to different cell populations within the brain. For example the hybridization of pMuBr2 corresponds to oligodendrocytes closely associated with myelinated fiber tracts throughout the CNS. These include the white matter of the cerebellum, corpus callosum, fornix, anterior commissure and optic chiasm. Recently we have established that the nucleotide sequence of pMuBr2 is homologous to the terminal 530 nucleotides of the 3' untranslated region of the mRNA encoding rat myelin proteolipid protein (Milner et al., 1985). In contrast pMuBr3, 8 and 85 appear to hybridize exclusively to RNA of neuronal cell bodies and not to oligodendrocytes or axonal processes of the fiber tracts as seen with pMuBr2. This is most clearly seen in the cerebellum and regions surrounding the hippocampus, where all three putative neuronal-specific cDNAs hybridize to discrete laminar arrays of neurons, but not to adjacent regions such

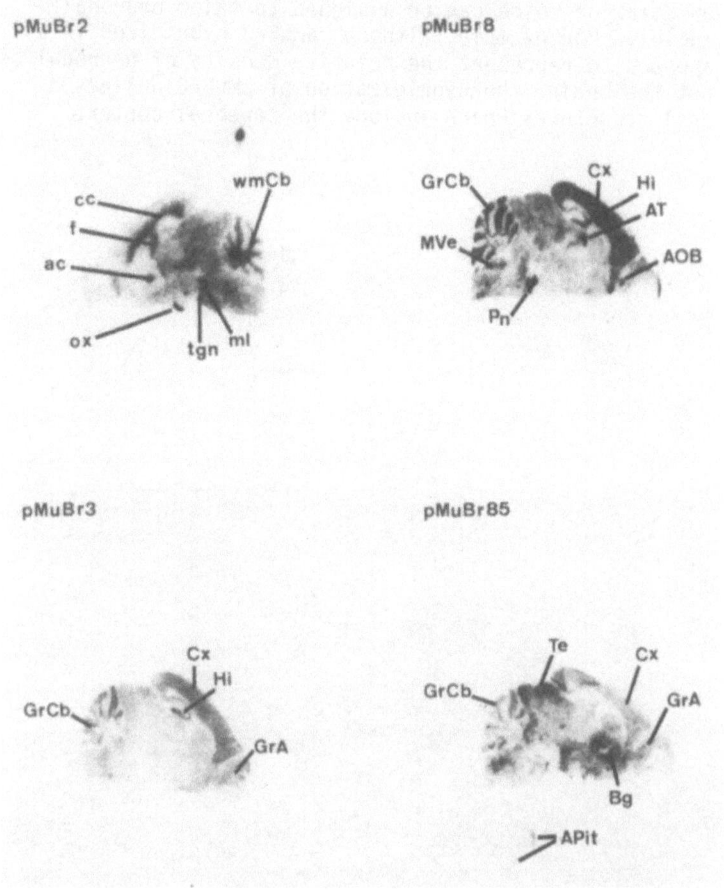

Fig. 2. Distinct patterns of mRNA expression in brain
distinguished by in situ hybridization with
^{32}P-labeled cDNA probes.
Mid-sagittal sections 4-6 microns in thickness and
0.5-1.5mm from the midline were cut from frozen
brains of young adult male mice. Similar sections of
kidney and liver, in which mouse pituitary gland was
embedded, were placed adjacent to the brain sections
as a control for non-specific hybridization. The
sections were fixed in paraformaldehyde and
hybridized to ^{32}P-labeled probes and exposed to
X-ray film for 3 days. The abbreviations used: cc,
corpus callosum; f, fornix; ac, anterior commissure;
ox, optic chiasm; tgn, trigeminal nerve; ml, medial
lemniscus; wmCb, white matter of the cerebellum;
GrCb, granular layer of cerebellum; MVe, medial
vestibular nucleus; Pn, pontine nucleus; AOB,
accessory olfactory bulb; At anterior thalamus; Hi,
hippocampus; Cx, cerebral cortex; GrA, granular
layer of AOB; Te, tectum, Bg, basal ganglia; APit,
anterior lobe of pituitary gland.

as the white matter of the cerebellum or corpus callosum and fornix of the hippocampal formation. Elsewhere these probes distinguish different patterns of hybridization which can be assigned to major neuronal structures of nuclei. For example, although pMuBr3 hybridized in a pattern which appears to represent the relative density of neuronal cell bodies throughout the brain, the hybridization of pMuBr8 defines a subset of neuronal structures which include the cerebral cortex,

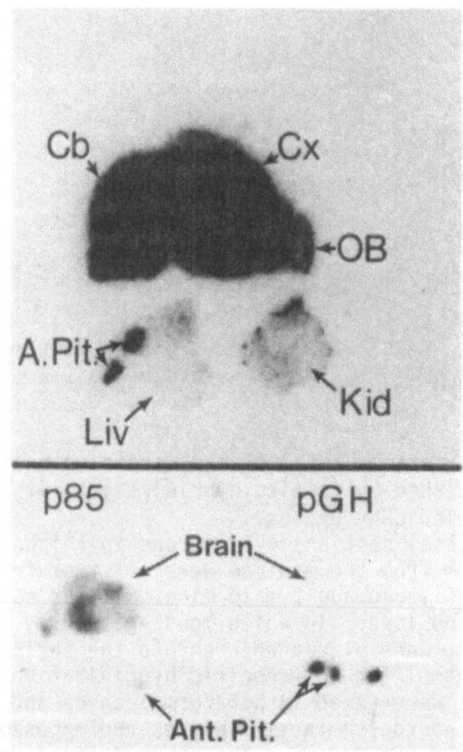

Fig. 3. Hybridization of pMuBr85 and pGH to sagittal sections of murine brain and cross section of pituitary. ^{32}P-labeled cDNA sequences of pMuBr85, pGH (rat growth hormone, Harpold et al., 1978) were hybridized and the sections exposed to X-ray film. The upper panel is an overexposed autoradiogram to define the pMuBr85 hybridization to the anterior region of the pituitary. The lower panel depicts the strength and specificity of hybridization to the anterior pituitary and brain of pMuBr85 and pGH under identical hybridization conditions and autoradiographic exposure.

hippocampus, anterior thalmus and pontine nuclei, as well as the accessory olfactory bulb and medial vestibular nucleus. In contrast, pMuBr85 hybridizes most strongly to mRNA of the tectum and basal ganglia and less to the hippocampus and neocortex.

Of the four probes examined pMuBr85 hybridized to RNA of sections of pituitary. Shown in Figure 3 is a lengthened exposure of a ^{32}P-autoradiograph of pMuBr85 hybridization of brain pituitary. Also shown is a comparison of hybridization obtained with pMuBr85 and a cDNA probe to rat growth hormone (Harpold et al., 1978). From these autoradiographs it is apparent that both pMuBr85, as well as the probe to growth hormone mRNA, hybridizes to mRNA expressed by cells within the anterior lobe of the pituitary. In an attempt to define further the expression of pMuBr85 hybridizing transcripts we hybridized this sequence to RNA blots of size fractionated RNA isolated from bovine and rat pituitary and the rat somatomammotropic cell line GH$_4$. In contrast to the prominent hybridization to murine transcripts, the cDNA probe pMuBr85 hybridizes poorly to rat brain and pituitary RNA, detecting only the 7.5 kb species (M.C. Wilson, unpublished results). No hybridization is seen to either polyadenylated bovine pituitary RNA or rat GH$_4$ RNA. This suggests that the cDNA pMuBr85 may reflect a poorly conserved region, perhaps of 3' untranslated sequence, which is not present in abundant transcripts of lower molecular weight in rat brain and pituitary, except as part of the larger 7.5 kb transcript. As the 7.5 kb transcript is not detected in GH$_4$ cells we suggest that these mRNA species may be expressed by gonadotrophin cells of the anterior pituitary.

To determine precisely the cellular distribution of these mRNAs and therefore distinguish whether they are expressed in functionally distinct neuronal subsets, ^{35}S-labeled probes were used to provide sufficient resolution at the microscopic level. To begin detailed analysis of the cellular specificity of these probes, we focused our attention on the expression of these mRNAs in the hippocampal formation (see Figure 4). In sagittal sections, the hippocampus is bordered by myelinated nerve fiber tracts, the corpus callosum and fornix where a strong autoradiographic signal can be seen over these tracts with the oligodendrocyte-specific pMuBr2 cDNA probe (4e). Scattered clusters of grains are also apparent within the pyramidal cell layer of the hippocampus indicating the presence of an occasional oligodendroglial cell cluster among these densely packed neuronal cell bodies. In contrast cDNAs pMuBr3, 8 and 85 hybridize to distinct neuronal subgroups which can be defined by morphological criteria and their interconnectivity. pMuBr8 hybridizes predominantly to mRNA in the CA3 and CA4 fields of the hippocampus (Figure 4b). Greatly reduced hybridization and hence expression of the homologous mRNA is seen to either the pyramidal neurons of the CA1 region or the granule cells of the dentate gyrus. Hybridization of pMuBr8 is found to the RNA of neuronal cell bodies of the subiculum and neocortex. Examination of the CA3 region under higher magnification shows grains evenly dispersed over the principle pyramidal neurons (Figure 4f) suggesting that within the CA3 region the mRNA of pMuBr8 is uniformly expressed by these cells. Occasionally regions essentially free of grains are associated with distinctive darkly staining cells which may be interspersed glial cells.

Hybridization of cDNA probes pMuBr3 and 85 to the mRNA of pyramidal neurons appears uniform throughout the length of hippocampus proper and does not discriminate between CA3 and CA1 neurons (Figure 4c and d). These two probes, however, differ in hybridization to the granule cells of the dentate gyrus. pMuBr3, in keeping with its more generalized pattern of hybridization, hybridized uniformly to both the RNA of

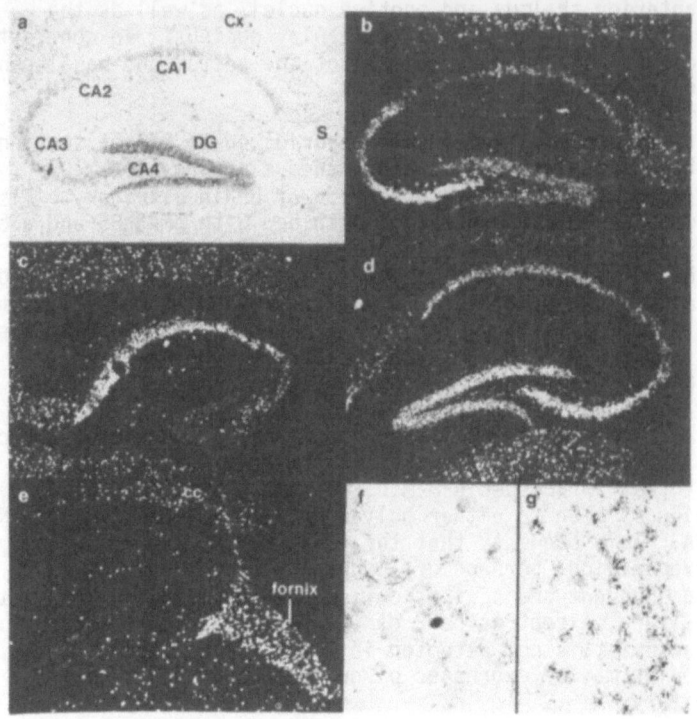

Fig.4. In situ hybridization of cDNA probes to sagittal view of the murine hippocampal formation. Photomicrographs of the hippocampal formation after hybridization with ^{35}S-labeled pMuBr8 (panels a and b), pMuBr85 (panel c), pMuBr3 (panel d) and pMuBr2 (panel e) taken at a magnification of 63X. Panel a shows the distribution of cell bodies stained with thionin Nissl stain and indicates the major components of the hippocampal formation: CA4, CA3, CA2, CA1, S (subiculum), DG (dentate gyrus) and Cx (cerebral cortex). Panels b-e are shown under dark field illumination to distinguish the silver grains exposed by the hybridized cDNA probes. Panels f and g are bright field photomicrographs taken at high magnification (630X) to show directly the distribution of silver grains resulting from hybridization of pMuBr8 (f) in the CA3 region and pMuBr85 (g) in the CA1 region of the hippocampus proper.

dentate gyrus and hippocampus. pMuBr85 hybridizes to mRNA highly abundant in the pyramidal neurons of the hippocampus but is virtually undetectable in the granule cells of the dentate gyrus. The distribution of large neurons within the hippocampus expressing pMuBr85 appears to be uniform when viewed at higher magnification (Figure 4g). As seen with pMuBr8 the more densely stained cells are free of grains indicating that the expression of the sequence is limited to the pyramidal neurons of the hippocampus. Both pMuBr3 and 85 do hybridize to RNA of neuronal cell bodies of the subiculum and neocortex.

In order to document further the specific expression of these mRNAs, their expression was also examined in the major neuronal and glial cell types of the cerebellum (see Figure 5). As expected, pMuBr2 hybridizes extensively and exclusively to cells within the medullary layer consistent with the expression by the prevalent oligodendrocytes throughout this layer, while the three cDNA clones pMuBr3, 8 and 85 hybridize differentially to neurons of the cerebellum. pMuBr3 hybridizes to mRNA of cells bordering the external surface of the granular layer. On higher magnification this hybridization is seen to be specific to Purkinje cells. In contrast, pMuBr8 hybridizes uniformly over the granule cell layer (Figure 5c) and does not hybridize to mRNA of the adjacent Purkinje cells. Hybridization of pMuBr85 appears to detect the homologous mRNA in both granule and Purkinje neurons. As evidenced by both dark (Figure 5b) and bright field microscopy this sequence is expressed in greater abundance in Purkinje cells. In the molecular layer, grains are also clustered over large interneurons. The

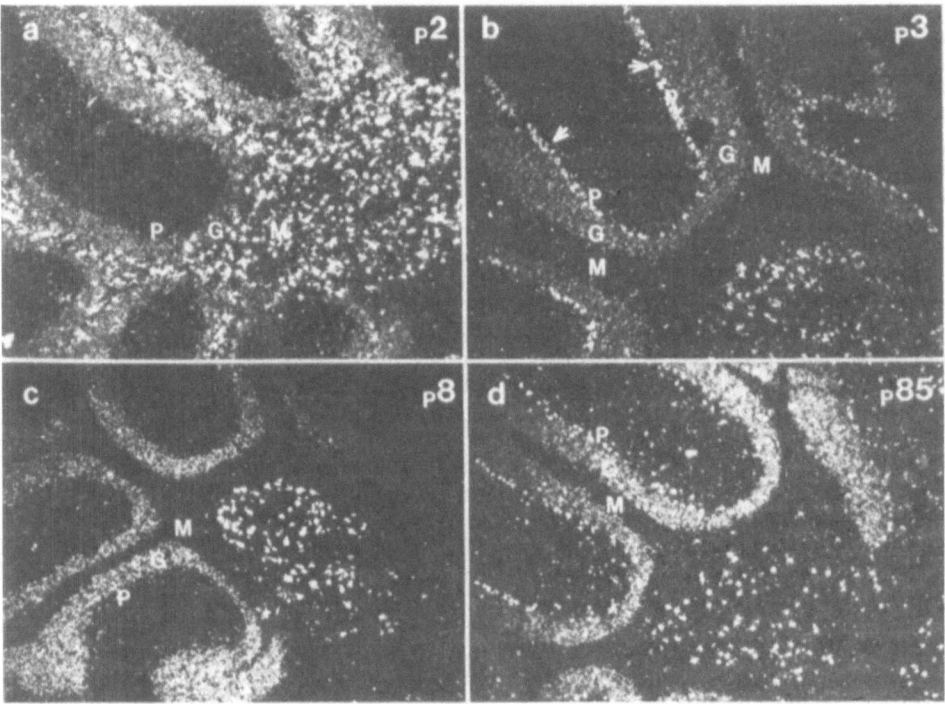

Fig. 5. In situ hybridization of ^{35}S-labeled cDNA probes to sagittal view of murine cerebellum.
Dark-field illumination photomicrographs of mid-sagittal brain sections hybridized with cDNA probes labeled by nick-translation with ^{35}S-dCTP demonstrate the cell-specific expression in the cerebellum of mRNAs homologous to pMuBr2 (panel a), pMuBr3 (panel b), pMuBr8 (panel c) and pMuBr85 (panel d). The major cell layers are indicated: P, Purkinje cell layer; G, granule cell layer and M, medullary layer or white matter of the cerebellum. Intense hybridization to the Purkinje cells by pMuBr3 is indicated by arrows in panel b. Exposure was to NTB2 liquid emulsion for 14 days. The photomicrographs were taken under dark field illumination at a magnification of 63X.

close proximity of these interneurons to the Purkinje cells suggest that these neurons expressing MuBr85 mRNA are basket cells. In the deep nuclei, within the central region of the cerebellum, each of the neuron-specific cDNA clones hybridize to scattered cells in a pattern clearly unlike the glial specific probe pMuBr2. At higher magnification, both pMuBr8 and pMuBr85 can be seen to hybridize selectively to mRNA of neurons of the deep cerebellar nuclei. It is likely that these probe sequences are therefore differentiating between neurons in the deep nuclei of the cerebellum.

As demonstrated by in situ hybridization of cloned cDNA probes, the expression of mRNAs MuBr3, 8 and 85 is independently regulated between neurons of the CNS. However, correlation between expression of these mRNAs and cellular phenotype is not apparent. For example, transcripts of both pMuBr3 and 85 are expressed by hippocampal neurons that primarily utilize the excitatory amino acid transmitters glutamic or aspartic acid (Nadler et al.; Storm-Mathisen, 1977; Fonnum and Wallaas, 1978) while in the cerebellum these RNAs are abundant in inhibitory Purkinje and molecular layer interneurons that express GABA (McLaughlin et al., 1974; McGeer et al., 1975). The correlation between expression of MuBr8 mRNA and neuronal function is also probably not due to the choice of neurotransmitter by these cells. Recent evidence has suggested that in the hippocampus, dynorphin is specifically localized in the mossy fiber/dentate pathway innervating the CA3 neurons that preferentially express MuBr8 mRNA (McGinty et al., 1982). However, evidence for opioid transmitters in the cerebellum, and in particular in the cerebellar granule cells, of the adult mouse, is lacking. The high abundance of this mRNA in the anterior thalamus and pons, moreover, does not fit into a simple pattern of known anatomical or functional correspondence; although the anterior thalamus is a major recipient of hippocampal formation output, the neurons of the pons are not (Swanson et al., 1982). Similarly, we observe that MuBr85 mRNA sequences, expressed in the brain and particularly in the pyramidal but not granule neurons of the hippocampus, are also synthesized in the anterior pituitary.

Gee et al., (1983) have demonstrated that the mRNA encoding the hormone precursor POMC is expressed in neurons of the rat hypothalamus, although clearly not to the extensive regions of the brain found for the expression of mRNA sequence MuBr85. As described above the cDNA sequence pMuBr85 hybridizes to multiple RNA species but appears to be encoded by a single genetic loci. Thus, the large 7.5-6.0 kb forms may represent particularly abundant nuclear precursors to the 2.6 kb mRNA, alternatively these species may be mature mRNAs differentially processed from a common primary transcript. In this regard it is noteworthy that two distinct mRNAs, transcribed from a single transcription unit, encoding calcitonin and a calcitonin related peptide (CGRP), are differentially expressed in neuronal and endocrine tissues (Amara et al., 1982). Whether or not the polyadenylated RNAs containing pMuBr85 sequence encode different proteins in the brain and anterior pituitary, the expression of pMuBr85 represents another example of genes specifically coexpressed by neurons of the CNS and by the endocrine system. Efforts are now in progress in our laboratory to isolate and sequence full length cDNAs of these particularly interesting mRNAs, to resolve their encoded proteins, and to determine their role in brain and neuroendocrine function.

Isolation and characterization of cDNA clones to hippocampal mRNAs

Although the cDNA clones described above hybridize to mRNAs expressed by neurons of the hippocampus, they also recognize mRNAs in a

26

variety of other brain regions. Similarly, other recently described neuronal mRNAs, characterized on the basis of brain-specificity, have also been found to be expressed in many different brain nuclei (Milner and Sutcliffe, 1985). In order to identify gene products which are expressed more specifically by neuronal subpopulations within the hippocampal formation, and thus will be useful for understanding developmental and organization features specific to this structure, we have constructed hippocampal cDNA libraries depleted of sequences shared with mRNA populations of other brain regions, such as cerebellum and cerebral cortex. With subtractive hybridization, mRNA sequences present on the order of one or more copies per cell, on average, are removed as cDNA/mRNA hybrids. We realize that many mRNA sequences will be present at a lower frequency, due to the complexity of cell types in brain and therefore, remain in the non-hybridized cDNA fraction and cloned as recombinants. Subsequent colony hybridization, however, detects only those sequences present in the subtracted but heterogenous probe at a frequency of 0.05% or more. Thus the final positive screen identifies relatively abundant sequences present in the hippocampal mRNA population and greatly reduced in cerebellum or cerebral cortex.

As described by M. Davis and colleagues (Davis et al., 1984), cDNA was prepared after oligo dT priming of polyA$^+$ enriched RNA with AMV reverse transcriptase in the presence of 100μg/ml Actinomycin D to reduce synthesis of double stranded cDNA. In these studies cDNA was synthesized from both polyA$^+$ RNA of rabbit hippocampus. Removal of relatively abundant sequences common to other brain regions was accomplished by hybridization to a twenty fold excess of mRNA of either cerebellum or cerebral cortex. Following liquid hybridization, remaining single-stranded cDNA was isolated by hydroxyapatite chromatography. Significant differences in the relative amount of unhybridized cDNA were found whether challenged with excess mRNA of cerebellum or cerebral cortex. Hybridization of hippocampal cDNA with cerebral cortex RNA depleted more mass of cDNA than did cerebellum RNA (Table 1). These data are consistent with the observation that cerebral cortex polyA$^+$ RNA has nearly twice the complexity of cerebellar mRNA.

The hippocampal enriched cDNA was rendered double-stranded and cloned into pBR322 to produce 2 cDNA libraries: Hippocampus-Cerebellum (H-Cb) and Hippocampus-cerebral cortex (H-Cx). These libraries were screened by colony hybridization with a twice-subtracted hippocampal-specific cDNA probe and/or plus-minus screening with whole hippocampal and cerebellar RNA. Approximately 5% of these clones were identified as positives in colony hybridization, and were assayed for regional distribution in the central nervous system by RNA blotting.

TABLE 1

Subtracted Libraries	Amt. cDNA remaining single-stranded after hydroxyapatite chromotography	No. of clones
Hippocampus-cerebellum		
(H-CB)	22%	312
Hippocampus-cerebral cortex		
(H-Cx)	18%	234

Hippocampal cDNA inserts were labeled with ^{32}P by nick translation and hybridized to RNA blots in which polyA+ RNA from hippocampus amygdala, cerebellum, and cerebral cortex and other regions, plus whole brain and liver, had been fractionated under denaturing conditions on adjacent lanes of an agarose gel and transferred to nitrocellulose. The specificity of these cDNAs was confirmed by hybridization to hippocampal RNA but not to transcripts present in the subtracted region, either cerebellum or cerebral cortex. We chose also to examine expression in amygdala because it is highly interconnected to, and shares functional and biochemical features with the hippocampal formation.

An example of hybridization of a H-Cb cDNA to polyA+ RNA of a variety of different brain regions is shown in Figure 6a. This cDNA, a 263 nucleotide insert from pH-CbE3, hybridizes to a 4.3 kb mRNA expressed at greater abundance in the hippocampus and amygdala, at reduced levels in the cerebral cortex, and virtually absent from the

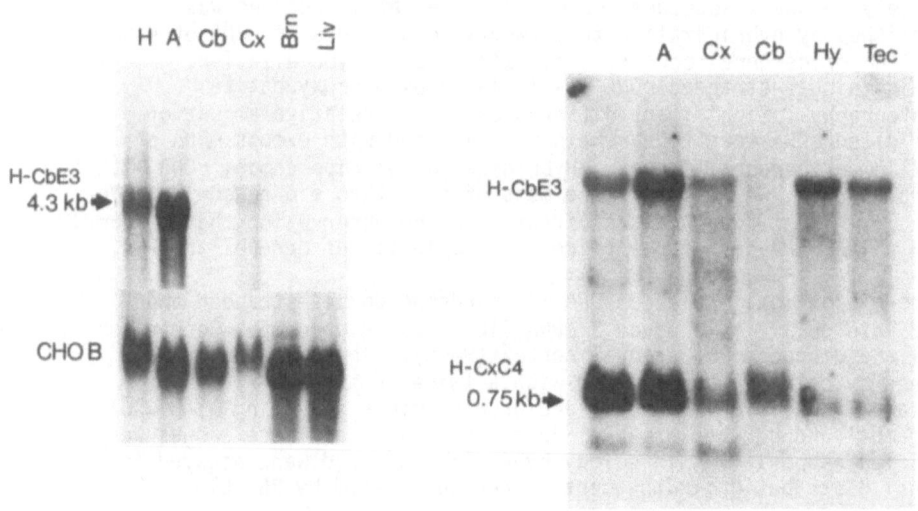

Fig. 6. Example of regionally distributed mRNA transcripts identified by hippocampal cDNAs.
(a) RNA blot showing hybridization of the 263 bp insert of pH-CbE3 to a 4.3 kb mRNA present at high abundance in rabbit hippocampus (H) and amygdala (A), lower abundance in cerebral cortex (Cx), and absent from rabbit cerebellum (Cb), whole mouse brain (Brn) and liver (Liv). The relative abundance of the 4.3 kb transcript in these different brain regions was determined by comparison to hybridization of a "house-keeping" cDNA, CHO B. Approximately 5 ug of polyA+ RNA was loaded in each lane, fractionated on a 1.25% denaturing agarose gel, blotted onto nitrocellulose, hybridized with a ^{32}P-labeled cDNA insert.
(b) RNA blot showing the distribution of H-CxC4 and H-CbE3 mRNAs in several rabbit brain regions, including hypothalamus (Hy) and tectum (Tec). Same conditions as in (a).

subtracted region, cerebellum. Sequencing of the cDNA reveals a putative open reading frame with no homology to proteins in the Protein Identification Resource (PIR). Thus it is possible that this cDNA recognizes a novel mRNA species which is present in hippocampus and other brain regions, but not cerebellum. Although not truly hippocampal-specific, its presence in other structures such as amygdala and hypothalamus (see Figure 6b) suggests that it may encode a protein related to neurotransmitter systems not present, or at greatly reduced levels in the cerebellum.

Similarly, pH-CxC4 hybridizes to a small, 750 nucleotide mRNA, which is found at higher levels in hippocampus and amygdala than in cerebral cortex (Figure 6b). However, this mRNA is expressed in varying amounts in all of the regions which were examined by RNA blotting. Thus, although hippocampus and amygdala appear to contain the greatest abundance of H-CxC4 mRNA, reduced amounts are also found in cerebellum, hypothalamus, tectum, and cerebral cortex. Thus, while not representing purely "region-specific" mRNAs, the subtractive hybridization method does appear to enrich for cDNA depleted of the appropriate sequences.

CONCLUDING REMARKS

A major goal in the application of recombinant DNA technology to a complex system such as the mammalian brain has been to distinguish regional differences of gene expression. The ability to perform subtractive hybridization to remove cDNA sequences complementary to mRNA expressed by cells of two different phenotypes can be employed to enrich the cDNA population and hence a recombinant cDNA library for sequences of interest. The use of recombinant cDNA to provide probes for neuronal specific gene express moreover provides advantages not easily attained by the more conventional approach of generating monoclonal antibodies to identify brain-specific gene products. First, a cDNA library will be equally representative of the input mRNA population and thus does not discriminate against weakly immunogenic proteins. This may be especially important in addressing the origin and regulation of rapidly processed proteins such as neuroendocrine intercellular messengers. Second, the structure of the protein precursor can be rapidly determined by direct DNA sequence analysis. Third, this information allows the generation of antibodies with defined sequence specificity for immunohistochemistry to localize the protein produce. Fourth, in situ hybridization with the cloned cDNA sequence itself allows rapid identification of cells expressing the mRNA which may or may not be the same as the protein. Finally, the cloned DNA also provides a hybridization probe for mRNA blots, genomic DNA blots and the isolation of the gene and homologous sequences from a recombinant genomic DNA library.

With the cDNA clones presented here and others now being generated in our laboratory and those of others, it is reasonable that the identification of novel and intrinsically interesting proteins will be identified that are germane to the function of the mammalian brain. How these proteins function to provide the complex interactions of brain function will require continued interaction between the disciplines of molecular genetics and classical neuroanatomy and neurophysiology.

ACKNOWLEDGEMENTS

This work was supported in part by a grant from the National Institutes of Health. G.A. Higgins is supported by an NIH NRSA award No. 5 F32 NS07528-02. We wish to thank Peggy Graber for assistance in preparing the manuscript.

REFERENCES

Amara, S. G., Jonas, V., Rosenfeld, M. G., Ong, E. S. and Evans, R. M., 1982 Alternative RNA processing in calcitonin gene expression generates mRNAs encoding different polypeptide products, Nature 298: 240-244.

Bantle, J. A. and Hahn, W. E., 1976 Complexity and characterization of polyadenylated RNA in the mouse brain, Cell 8:193-150.

Beckman, S. L., Chikaraishi, D. M., Deeb, S. S. and Sueoka, N., 1981 Sequence complexity of nuclear and cytoplasmic ribonucleic acids from clonal neurotumor cell lines and brain sections of the rat, Biochemistry 20:2684-2692.

Chikaraishi, D. M., 1979 Complexity of cytoplasmic polyadenylated and nonpolyadenylated rat brain ribonucleic acids, Biochemistry 18:3249-3256.

Chikaraishi, D. M., Deeb, S. S. and Sueoka, N. Sequence complexity of nuclear RNAs in adult rat tissues, Cell 13:111-120.

Davis, M. M., Cohen, D. I., Nielsen, E. A., Steinmetz, M., Paul, W. E. and Hood, L., 1984 Cell-type specific cDNA probes and the murine I region: the localization and orientation of A^d, Proc. Natl. Acad. Sci. USA 81: 2194-2198.

Dworkin, M. B. and Dawid, I. B., 1980 Use of a cloned library for the study of abundant poly(A)$^+$ RNA during Xenopus laevis development, Developmental Biol. 76:449-464.

Fonnum, F. and Walaas, I., 1978 The effect of intrahippocampal kainic acid injections and surgical lesions on neurotransmitters in hippocampus and septum, J. Neurochem. 31:1173-1181.

Gee, C. E., Chen, C.-L. G., Roberts, J. L., Thompson, R. and Watson, S. J., 1983 Identification of proopiomelanocortin neurones in rat hypothalamus by in situ cDNA-mRNA hybridization, Nature 306: 374-376.

Gergen, J. P., Stern, R. H. and Wensink, P. C., 1979 Filter replicas and permanent collections of recombinant DNA plasmids, Nucleic Acids Res. 7:2115-2136.

Grouse, L. D., Schrier, B. K., Bennet, E. L., Rosenzweig M. R. and Nelson, P. G., 1978 Sequence diversity studies of rat brain RNA: effects of environmental complexity on rat brain RNA diversity, J. Neurochem. 30:191-203.

Grouse, L. D., Letendre, C. H. and Schrier, B. K., 1979 Sequence complexity and frequency distribution of poly(A)-containing messenger RNA sequences from the glioma cell line C6, J. Neurochemistry 33:583-585.

Harpold, M. M., Evans, R. M., Salditt-Georgieff, M. and Darnell, J. E., 1979 Production of mRNA in Chinese hamster cells: relationship of the rate of synthesis to the cytoplasmic concentration of nine specific mRNA sequences, Cell 17:1025-1035.

Harpold, M. M., Dobner, P. R., Evans, R. M. and Bancroft, F. C., 1978 Construction and identification by positive hybridization-translation of a bacterial plasmid containing a rat growth hormone structural gene sequence, Nucl. Acids Res. 5: 2039-2053.

Kaplan, B. B., Schacter, B. S., Osterburg, H. H., de Vellis, J. S. and Finch, C. E., 1978 Sequence complexity of polyadenylated RNA obtained from rat brain regions and cultured rat cells of neural origin, Biochemistry 17:5516-5524.

Levy, D. E., Lerner, R. A. and Wilson, M. C., 1982 A genetic locus regulates the expression of tissue specific mRNAs from multiple transcription units, Proc. Natl. Acad. Sci. 79:5823-5827.

McGeer, P. L., Hattor, T. and McGeer, E. G., 1975 Chemical and autoradiographic analysis of -aminobutyric acid transport in Purkinje cells of the cerebellum, Exp. Neurol. 47: 26-41.

McGinty, J. F., Henricksen, S. J., Goldstein, A., Terenius, L. and Bloom, F. E., 1982 Dynorphin is contained within hippocampal mossy fibers: immunochemical alterations after kainic acid administration and colchicine-induced toxicity, Proc. Natl. Acad. Sci. USA 79:6747-6751.

McLaughlin, B. J., Wood, J. G., Saito, K. K., Barber, R., Vaughn, J. E. Roberts, E. and Wu, J. Y., 1974 The fine structural localization of glutamate decarboxylase in synaptic terminals of rodent cerebellum, Brain Res. 76:377-391.

Milner, R. J. and Sutcliffe, J. G., 1984 Gene expression in rat brain, Nuc. Acid Res. 11:5497-5520.

Milner, R. J., Lai, C., Nave, K.-A., Lenoir, D., Ogata, J., Bloom, F. E. and Sutcliffe, J. G., 1985 Nucleotide sequences of two mRNAs for rat brain myelin proteolipid protein, Cell in press.

Nadler, J. V., Vaca, K. W., White, W. F., Lynch, G. S. and Cotman, C. W. Aspartate and glutamate as possible transmitters of excitatory hippocampal afferents, Nature 260:538-540.

Storm-Mathisen, J., 1977 Glutamic acid and excitory nerve endings: reduction of glutamic acid uptake after axotomy, Brain Res., 120:379-386.

Swanson, L. W., Taylor, T. J. and Thompson, R. F., 1982 Hippocampal long-term potentiation: mechanisms and implications for memory, Neurosciences Res. Program Bulletin 20: 624-634.

Walaas, I., 1983 The hippocampus, in: Chemical Neuronanatomy, ed., P. C. Emsen, Raven Press, New York, pp. 337-358.

HYBRIDIZATION HISTOCHEMISTRY - LOCATING GENE EXPRESSION

J.P. Coghlan, P. Aldred, A. Butkus, R.J. Crawford,
I.A. Darby, J. Haralambidis, J.D. Penschow, P.J. Roche,
C. Troiani and G.W. Tregear

Howard Florey Institute of Experimental Physiology and
Medicine, University of Melbourne, Parkville 3052, Australia

A technique we have called hybridization histochemistry has been developed for the location of specific mRNA populations in specially prepared sections of tissue (Hudson et al 1980; Coghlan et al 1981; Coghlan et al 1984; Hudson et al 1981; Jacobs et al 1983; Coghlan et al 1984). Later the same approach has been used by others to identify specific neurones in the hypothalamus (Gee et al 1983), to study the origin and fate of identified neurones in aphysia (McAllister et al 1983), location of specific genes in Drosophila embryos (McGinnis et al 1984), and enkephalin in the adrenal (Block et al 1984).

The idea of using the property of base pairing between mRNA/DNA and DNA for tissue localization has been around for some time (Coghlan et al 1984). Indeed in situ tissue hybridization or hybridization histochemistry is simply a logical extension of standard procedures used in molecular biology. HYBRIDIZATION HISTOCHEMISTRY is thus a procedure in which cells or tissue sections are treated so as to immobilize and protect the naturally occurring DNA or mRNA for in situ hybridization while at the same time retaining sufficient cellular morphology for accurate histological location.

We chose to use this particular nomenclature to focus on the similarity of the technique to immunohistochemistry, but at the same time distinguish the procedure from the many in situ hybridization procedures which are the corner-stone of molecular biology.

In our studies we have used ^{32}P labelled cDNA or synthetic oligonucleotide probes. Whilst ^{32}P labelling does not provide the resolution of ^{3}H, ^{35}S or ^{125}I labelling when steps are taken to minimize the problem of ^{32}P adequate resolution for our needs down to a single cell can be obtained.

^{32}P-labelling of probes
CDNA
_____

- nick translation
- random primers
- riboprobe SP6

Synthetic DNA

- 5' with polynucleotide kinase
  DNA polymerase

DNA polymerase is used also to incorporate nucleotide labelled with ^{3}H, ^{35}S and ^{125}I.

The specific activities of probes/unit mass made by the various methods finish up more or less of the same order in our hands. Whereas with a single labelled nucleotide substitution the nick translation and random primer methods should give one nucleotide in four labelled and thus very high specific activity, the actual efficiency of labelling is far from ideal. This inefficient labelling is probably due to variable chain lengths and difficulties arising from competing strands and other problems. The theoretical advantage of the riboprobe system in our experience has been offset by the high concentrations of reagents (lower initial specific activity) required for good efficiency with RNA polymerase. Although the synthetic oligonucleotides when end labelled have only one label per mole, the increased mass available for hybridization histochemistry with this approach confers considerable advantages, and we would consider them superior. The specific activities are of the order $1-6 \times 10^8$ cpm/µg.

An outline of the procedure is set out in table 1.

Table I   HYBRIDIZATION HISTOCHEMISTRY

DAY 1

1.   Freeze tissue in hexane/dry ice at -70°C.
2.   Cut frozen sections.
3.   Fix glutaraldehyde + ethylene glycol at 4°C.
4.   Soak in hybridization buffer containing salt, DNA formamide denharts.
5.   Rinse-ethanol
6.   Apply ^{32}P-labelled probe in hybridization buffer.
7.   Incubate at 30-50°C (Depending on probe length homology etc.)

DAY 2

8.   Wash off probe in salt solutions at 30-50°C.
9.   Rinse-ethanol
10. Count sections with geiger counter.
11. Apply fast X-ray film. Expose 4-24 hours.
12. Develop film, evaluate result, estimate exposure time for emulsion.

DAY 3

13. Apply emulsion and/or high resolution X-ray film. Expose 1-21 days.

One considerable advantage of the use of ^{32}P-labelled probes is the fact that the hybridized sections can be previewed after exposure to fast X-ray film. With experience these crude autoradiographs allow the prediction of the appropriate exposure time when using emulsion which reduces the wastage in time and effort commonly associated with this type of autoradiographic end point. In addition, failed experiments where labelling has not occurred or background is too high can be aborted early. The fast X-ray film is also useful when screening large numbers of sections for the location of gene expression, for example, in serial sections of the brain.

It is difficult to describe the technique and results without extensive illustrative material. However, a variety of cDNA probes were used to illustrate the versatility of the technique. Sensitivity down to single cells can be demonstrated for most probes, and illustrations of this degree of resolution were shown using cDNA probes for human prolactin and human calcitonin.

The major thrust of the orgal presentation was to illustrate the use of synthetic oligonucleotides in the procedure. The advantages of these are listed in table II,and some disadvantages in table III.

Table II   Synthetic DNA Probes

Advantages over Cloned Probes

1. Probes can be made in a few days.
2. They are much easier to prepare and label in the ordinary laboratory.
3. Overall better labelling of tissue results because unlimited amounts of probe available.
4. Consistent specific activity of probes is more easily obtainable.
5. Pure synthetic probes give lower backgrounds.
6. Discriminating sequences for similar genes may be made.
7. Known sequences may be altered for different species using prepferred codons.
8. In some cases probes may be synthesized from amino acid sequences when DNA sequence is unknown.
9. Shorter probes may be more accessible to cellular mRNA.
10. Absence of poly(T) tails eliminates one source of background.
11. Bacterial sequences in cloned probes labelling gut or infected tissues no longer a problem.

Table III   Synthetic DNA Probes

Disadvantages over Cloned Probes

1. Fewer labels available at the moment.
2. Stability of probes yet to be fully explored.
3. Establishment of optimal conditions for short probes.
4. Cross species hybridizations more likely to be successful with longer probes.
5. Typographical proof-reading or sequencing errors in published sequences lead to "wild goose chases".

Synthetic oligonucleotide probes have allowed the widest application
of the technique. Synthetic probes can be constituted which allow differentiation to be made between expression of genes even within gene
families, eg. between oxytocin and arginine vasopressin, or between
members of the kallikrein family.

Figures 1, 2 and 3 illustrate the results obtained using synthetic
oligonucleotide probes for arginine vasopressin (AVP) and corticotrophin
releasing factor (CRF) on sections of rat and sheep hypothalamus and
cortex. Also included are results using a cDNA probe for somatostatin.
Although black and white reproduction somewhat disadvantages the presentation of the findings the figures do provide an indication of the degree
of resolution that can be obtained with the technique.

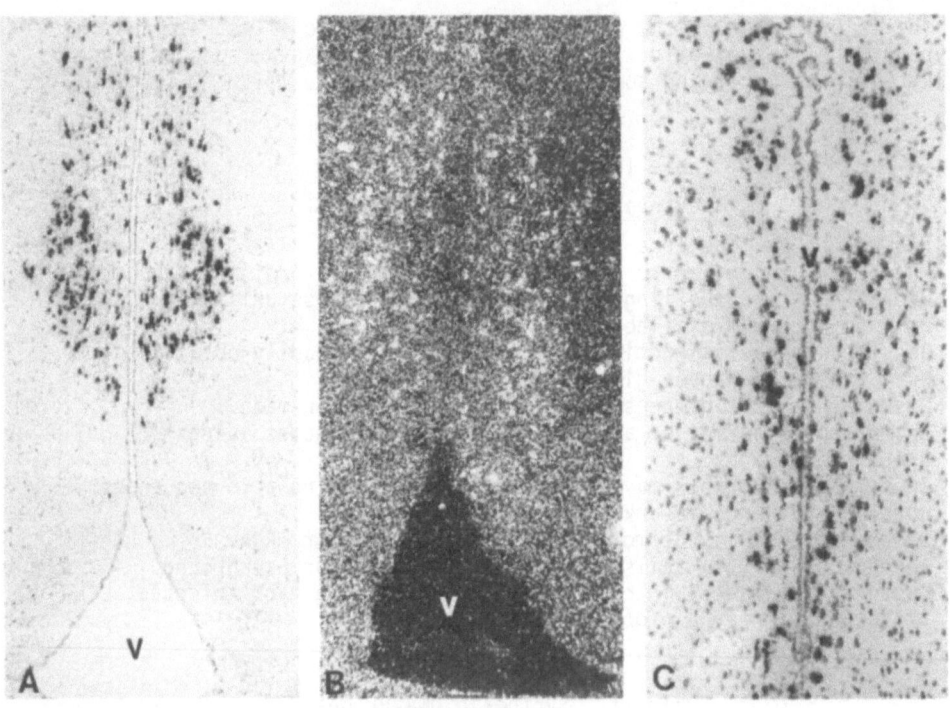

Fig. 1 Liquid emulsion autoradiographs of frozen 10μm coronal sections
of sheep (A,B) and rat (C) hypothalamus after hybridization with ^{32}P-
labelled DNA probes. 3rd ventricle (V) bisects field.
a) mRNA for arginine vasopressin-neurophysin II (AVP-Np II) in magnocellular neurones of the paraventricular nucleus located by a specific
30 mer synthetic oligodeoxyribonucleotide corresponding to amino acids
125-134 of bovine pre-pro AVP-Np II. Mag. X 20
(Uhl et al 1985)
b) Dark field photomicrograph showing neurones of the paraventricular
nucleus containing mRNA for corticotrophin releasing factor (CRF) located
by a 25 mer synthetic oligodeoxyribonucleotide corresponding to nucleotides 430-454 of the ovine CRF precursor. Mag. X 20
c) mRNA for somatostatin in neurones of the periventricular hypothalamus
area located by a 550 b.p. mouse somatostatin cDNA probe.
Mag. X 200. Probe made available by R.J. Crawford and C. Trioani.

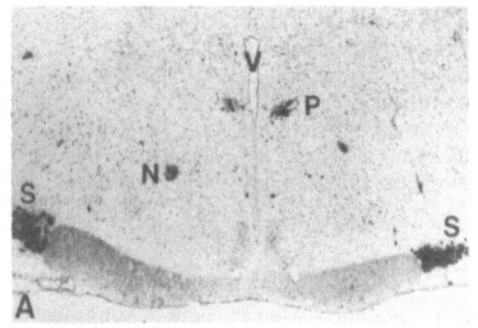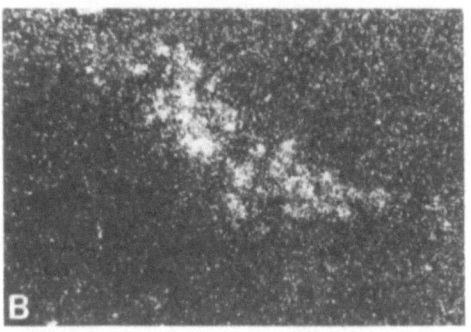

Fig. 2 Liquid emulsion autoradiographs of frozen 20 μm coronal sections through the hypothalamus of rat (A) and sheep (B) brain after hybridization with 32P-labelled synthetic oligodeoxyribonucleotide probes.
a) mRNA for AVP-Np II mRNA in the supra-optic (S) and paraventricular (P) nuclei, nucleus circularis (N) and scattered neurones lateral to the hypothalamus located by a 30 mer probe corresponding to amino acids 123-132 inclusive of rat pre-pro AVP-Np II. Mag. X 20
b) Dark-field photomicrograph showing magno-cellular neurones of the supra optic nucleus containing mRNA for CRF located by the probe described in Fig. 1 (B). Mag. X 80

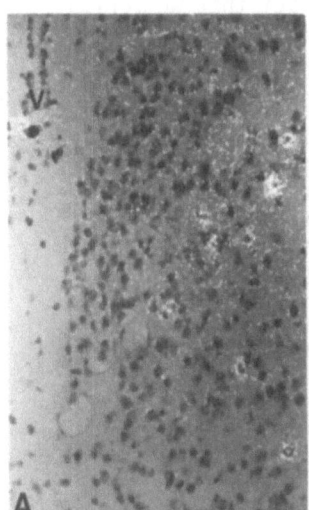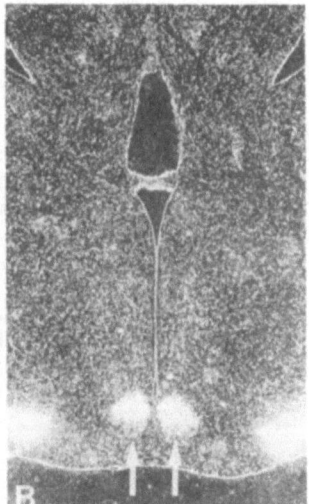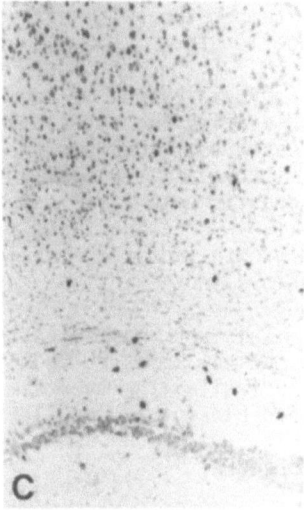

Fig. 3 Liquid emulsion autoradiographs of 10 μm frozen sections through rat hypothalamus (A,B) and cortex (C) after hybridization with 32P-labelled DNA probes.
a) Photomicrograph by polarized epi-illumination of the supra-chiasmatic nucleus showing neurones containing mRNA for somatostatin located by the probe described in Fig. 1 (C). 3rd ventricle (V). Mag. X 280
b) Dark-field photomicrograph showing mRNA for AVP-Np II in supra-optic and supra-chiasmatic (arrowed) nuclei located by the probe described in Fig. 2 (A). 3rd ventricle bisects field. Mag. X 30
c) Neurones containing somatostatin mRNA in the cortex and hippocampus located by the cDNA probe described in Fig. 1 (C) Mag. X 80

Synthetic oligo nucleotide probes can also be used to provide a semi-quantitative measure of mRNA levels in different sections of tissue by using accurately measured amounts of unlabelled probe to dilute the labelled binding to extinction. As yet we have not used this type of assessment extensively. Obviously a single section is so unrepresentative of a large neuronal pool that precise statements about mRNA are not very meaningful without some estimate of whether or not more neurones expressing the gene have been recruited.

Synthetic DNA can also be used when the cDNA is not available, but the sequence published. We have obtained preliminary data using a synthetic 36 mer probe scanning the angiotensin I sequence of rat renin substrate that has given some positive binding in the arcuate nucleus of the normal rat.

Our own papers in this field (Hudson et al, 1980; Coghlan et al 1981; Coghlan et al 1984; Hudson et al 1981; Jacobs et al 1983; Coghlan et al 1984; Coghlan et al 1985; Coghlan et al 1984) provide information not possible to include in this brief survey. Coghlan et al 1985 reviews recent advances in hybridization histochemistry and includes better illustrative material. Varndell et al 1984 have used recently, a novel approach to the production and use of non-radioactively labelled probes. Uhl et al 1985 have published recently their experiences with synthetic AVP probes. They found a two-fold increase in an AVP gene expression during dehydration not dissimilar to our own findings, but we were not prepared to make a definite statement. They also report an increase in tissue hybridization with probe length from 8-75 mer, a finding not shared with us. Perhaps longer hybridization times and the fact that we varied the temperature with each probe made our procedure less variable with probe length.

Hybridization histochemistry has already proved a valuable technique for locating the sites of gene expression. Clearly it is a valuable research tool complementary to existing techniques such as immunohisto-chemistry, and especially so in the neuroendocrine context. Its vast diagnostic potential is yet to be tapped in infectious disease, especially viral, in creating a new taxonomy of malignant tumours based on oncogene and growth factor expression and possibly even where parasitic vectors are involved.

This work was supported by grants-in-aid from the National Health and medical Research Council of Australia, the National Heart Foundation of Australia; the myer Family Trustees; the Ian Potter Foundation; The Howard Florey Biomedical Foundation (USA).

REFERENCES

1. Block, B., Milner, J.R., Baird, A., Grubber, W., Raymond, C., Bohlen, P., Le Guellec, D. and Bloom, F.E. 1984. Detection of mRNA coding for preproenkephalin-A in bovine adrenal by in situ hybridization. Regulatory Peptides, in press.
2. Coghlan, J.P., Butkus, A., Hudson, P.J., Niall, H.D., Penschow, J.D., Shine, J., Ryan, G. and Walsh, J. 1981. Hybridization Histochemistry: Use of Recombinant DNA as a "homing probe" for tissue localization of specific mRNA populations. Proc. Aust. Endocrine Soc., 24: 58.
3. Coghlan, J.P., Penschow, J.D., Hudson, P.J. and Niall, H.D., 1984. Hybridization Histochemistry: Use of Recombinant DNA for tissue localization of specific mRNA populations. Clin. and Exp. Hypertension, A2, 63-78.
4. Coghlan, J.P., Aldred, P., Haralambidis, J. Niall, H.D., Penschow,

J.D. and Tregear, G.W. Hybridization Histochemistry. 1985. Analytical Biochem. 249.

5.   Coghlan, J.P., Penschow, J.D., Tregear, G.W. and Niall, H.D. 1984. Hybridization Histochemistry: Use of complementary DNA for tissue localization of specific mRNA populations. In Receptors, Membranes and Transport Mechanisms in Medicine (ed.) Mendelsohn, F. 1-11, Excerpta Medica, Amsterdam.

6.   Furutani, Y., Morimoto, Y., Shibahara, S., Noda, M., Takahashi, H., Hirose, T., Asai, M., Inayama, S., Hayahida, H., Miyata, T. and Numa, S., 1983, Nature, 301, 537-540.

7.   Gee, C.E., Chen, C.L.C., Roberts R.L., Tho,pson and Watson, S.J. 1983. Identification of pro-opiomelanocortin neurones in rat hypoth-amalmus. Nature (Lond.) 306: 374-376.

8.   Hudson, P.J., Penschow, J.D., Shine, J., Ryan, G., Niall, H.D. and Coghlan. J.P. 1981. Hybridization Histochemistry: Use of Recombinant DNA as a homing probe for tissue localization of specific mRNA populations. Endocrinology, 108: 353-356.

9.   McAllister, L.B., Schiller, R.H., Kandel, E.R. and Richard, A. 1983. In situ hybridization to study the origin and fate of identified neurones. Science 222 : 800-808.

10. McGinnis, W., Levine, C.H.S., Hafen, E., Kurowa and Gehring, W.J. 1984. A conserved DNA sequence in homoeotic genes of the Drosophila Antennapedia and bothorax complex. Nature 308: 428-433.

11. Uhl, G.R., Zingy, H.A. and Habener, J.F. 1985. Vasopressin mRNA in situ, Proc. Nat. Acad. Sci. 82: 5555-5559.

12. Varndell, I.M., Polak, J.M., Sikri, K.L., Minth, C.D., Bloom S.R. and Dinson, J.E. 1984. Visualization of mRNA directing peptide synthesis by in situ hybridization using a novel single stranded cDNA probe. Histochemistry, 91: 597-601.

THE IDENTIFICATION AND FUNCTION OF DIFFERENTIATION ANTIGENS ON PRIMARY

SENSORY NEURONS

Jane Dodd

Dept. of Physiology and Cellular Biophysics
Columbia University, College of Physicians and Surgeons
New York, NY 10032, USA

The perception of different qualities of cutaneous stimuli depends, initially, on the activation of specific classes of primary sensory fibres that have their cell bodies located in the dorsal root ganglia (DRG) (Iggo, 1973; Perl, 1983). Anatomical and physiological studies have shown that more than a dozen classes of cutaneous afferents can be identified by their peripheral receptive properties, axonal conduction velocity and central terminal organization (Brown, 1981, 1982; Perl, 1983).

Reconstruction of the central projections of afferent fibres after intra-axonal injection of HRP has revealed that functionally distinct classes of sensory neurons exhibit stereotyped and restricted patterns of axon collateral arborization in the dorsal horn (Brown, 1981). Thus, the majority of the central terminals of nociceptive cutaneous afferents are confined to the superficial dorsal horn (laminae I and II) (Rethelyi, 1977; Light and Perl, 1979; Ribiero Da Silva and Coimbra, 1982), while the terminals of low-threshold mechanoreceptive afferents are found in deeper regions (laminae III-IV) (Brown et al., 1977; Semba et al., 1983; Ralston et al., 1984). The synapses formed between the central terminals of sensory afferents and neurons in the dorsal horn of the spinal cord represent the first site at which incoming sensory information is processed and integrated. An examination of the molecular mechanisms that regulate the differentiation of DRG neurons and the formation of selective sensory synapses in the dorsal horn may, therefore, lead to an understanding of the initial events in central processing of sensory information. The identification of developmentally regulated surface molecules expressed by defined classes of DRG neurons would provide specific markers with which to examine the molecular events associated with the differentiation of neuroblasts and the acquisition of final sensory neuron phenotypes. Cell surface molecules present on sensory neurons and their cellular targets may also contribute to the specificity of cellular interactions during the development of sensory projections in the spinal cord.

Studies on several cell types have demonstrated that cell surface antigenic variation during development and differentiation can occur through modifications in the length and branching patterns of oligosaccharide side chains associated with glycoproteins or glycolipids (Feizi, 1981; Hakomori, 1981). Oligosaccharide structures associated with blood group determinants are expressed as developmentally regulated cell-surface molecules on pre- and peri-implantation mouse embryos (Solter and Knowles, 1979). We have therefore used monoclonal antibodies that recognize defined carbohydrate epitopes associated with blood group antigens and embryonic stage-specific antigens to determine whether these structures are expressed by functionally distinct subpopulations of sensory neurons.

The localization of two classes of oligosaccharide structures, the globoseries and the lactoseries carbohydrates, has been studies in detail (Dodd et al., 1984; Jessell and Dodd, 1984; Dodd and Jessell, 1985). The globoseries carbohydrates are characterized by the backbone oligosaccharide structure $GalNAc\beta1-3Gal\alpha1-4Gal\alpha1-R$, whereas many lactoseries carbohydrates are derived from $Gal\beta1-4GlcNAc$ (Type 2).

In the rat DRG, globo- and lacto- series epitopes are expressed by distinct subsets of neurons. Two antibodies that recognize globoseries carbohydrate epitopes constituting the stage-specific embryonic antigens (SSEA's) -3 and -4 (Shevinsky et al., 1982; Kannagi et al., 1983) label cytoplasmic and surface determinants on a subset of DRG neurons in neonatal and adult rat (Dodd et al., 1984). The SSEA epitopes are present in the cytoplasm of approximately 10% of DRG neurons. These neurons are large and do not contain any of the cytoplasmic markers that may be related to transmitter function in small neurons, such as substance P (SP), somato-statin (SRIF) and fluoride-resistant acid phosphatase (FRAP). The central terminals of SSEA[+] neurons are located within laminae III and IV of the dorsal horn of the spinal cord and correlate with the distribution of low threshold afferent fibres (Dodd et al., 1984; Brown, 1981). Lamina I also receives a sparse projection, suggesting that some high threshold mechanoreceptors express the antigen, whereas lamina II, the region in which unmyelinated afferents terminate, is SSEA[-].

We have identified a second family of antigenic determinants that appear to be restricted to populations of snall- and intermediate-diameter DRG neurons (Jessell and Dodd, 1984; Dodd and Jessell, 1985). Approximately 50% of DRG neurons express the backbone lactoseries structure ($Gal\beta1-4GlcNAc$) recognized by the MAb A5. Immunoreactive labelling is associated with vesicular and granular structures in the cytoplasm and with the neuronal plasma membrane. A5[+] neurons exhibit varying levels of intensity of labelling, suggesting that the total population may represent more than one subpopulation (Dodd and Jessell, 1985). This idea is supported by the distribution of the central terminals of A5[+] neurons. A dense plexus of A5[+] afferent fibres and terminals is present within laminae I, IIA and IIBd. Furthermore, the segregation of globoseries and lactoseries carbohydrate antigens to different populations of DRG neurons is confirmed by the location of the central terminals of these neurons in the spinal cord.

MAbs directed against complex epitopes derived from type 2 lactoseries carbohydrates identify structures that are restricted to functional subsets within the small- and intermediate-diameter DRG neuronal population (Jessell and Dodd, 1984; Dodd and Jessell, 1985). Approximately 25% of DRG neurons are labelled with four separate MAbs, typified by the antibody LD2 that recognizes epitopes in the cytoplasm of neurons in sections of DRG and on the surface of DRG neurons grown in dissociated culture (Dodd and Jessell, 1985). These neurons project exclusively to lamina IIA and IIBd.

Evidence for a differential distribution of complex lactoseries oligosaccharides on subsets of DRG neurons is obtained in studies with a further MAb, LA4 (Dodd and Jessell, 1985). The LA4 epitope is expressed in the cytoplasm and on the surface of 45-50% of DRG neurons, the central terminals of which are located predominantly within lamina IIBd, with a few fibres and terminals also apparent in lamina IIBv.

The selective expression of carbohydrate epitopes by DRG neurons that project to distinct lamina domains in the dorsal horn indicates that complex globo- and lacto-series carbohydrates identify functional subsets of DRG neurons. A role as differentiation antigens is further supported by the finding that the expression of other cytochemical markers, such as FRAP or the peptides SP and SRIF, that may be involved in neurotransmission by functional subsets of sensory neurons, appears to be correlated with carbohydrate phenotype. Essentially all SP[+], SRIF[+] and FRAP[+] DRG neurons expressed lactoseries glycoconjugates (Jessell and Dodd, 1984; Dodd and Jessell, 1985). A5 labels all the SRIF-containing DRG neurons, approximately 50% of the SP[+] population and a subset of FRAP[+] neurons, providing further support for the notion that the type 2 lactoseries backbone structure is expressed by several functional subsets of sensory neurons. However, all SRIF[+] but none of the SP[+] DRG neurons express the more complex antigens recognized by the MAb LD2 whereas, while SRIF-containing DRG neurons are weakly LA4[+], over 75% of LA4[+] neurons are FRAP[+] and these constitute more than 90% of the FRAP[+] population (Dodd and Jessell, 1985).

Some DRG neurons express the type 1 lactoseries structures (backbone structure Galβ1-3GalNAc) recognized by the MAbs FC 10.2 and AH7-229A (Lewis a) (Dodd and Jessell, 1985). FC 10.2-immunoreactivity is present in scattered fibres and terminals in laminae I, II and X (surrounding the central canal). Many of the FC 10.2[+] fibres in the dorsal horn and in lamina X express SP-immunoreactivity. This is in contrast to the distribution of A5[+]/SP[+] fibres that are found only in superficial dorsal horn. The Lewis a structure is present in the cytoplasm of about 20% of small-diameter DRG neurons and the relationship of the Lewis a determinant with SP-, SRIF- and FRAP- containing neurons is strikingly different from that of the complex type 2 lactoseries structures. There is no overlap between Lewis a- and SRIF- containing DRG neurons whereas some Lewis a[+] DRG neurons contain SP and the rest contain FRAP (Dodd and Jessell, 1985).

These studies have begun to provide a molecular description of subsets of DRG neurons that convey different qualities of sensory information from the periphery to the CNS. The expression of globo- and lacto-series structures appears to be related to the embryologic origin of sensory neurons. Globoseries carbohydrate structures may be associated preferentially with the large light cell population in DRG and with placodally-derived cranial sensory neurons whereas the distribution of lactoseries antigens is consistent with a preferential association with small dark DRG neurons and neural crest-derived cranial sensory neurons (Dodd and Jessell, 1985). Since the glycosyl transferases that are required for the synthesis of carbohydrate antigens are expressed at very early stages of development, sensory neuron-specific carbohydrate antigens may represent useful markers for studying early lineage and differentiation of DRG neurons.

Although the function of globo- and lacto-series carbohydrate antigens on functional subsets of sensory neurons is not clear, the expression of these structures on early embryonic cells (Gooi et al., 1981; Kapadia et al., 1981; Knowles et al., 1982; Kannagi et al., 1983a,b; Fenderson et al., 1984), raises the possibility that molecules bearing these determinants might play similar roles in the organization of sensory projections in the developing spinal cord. Since most of the globo- and lacto-series structures we have examined are restricted to sensory neurons, interactions with spinal cord target cells may be mediated by the binding of sensory neuron carbohydrates to complementary cell-surface molecules. Carbohydrate binding proteins have been isolated from several vertebrate species (Barondes, 1984) and represent one class of molecules with the potential for interaction with carbohydrates on DRG neurons. The availability of the MAbs described above permits experimental approaches to test the possibility that interactions between carbohydrate differentiation antigens and complementary carbohydrate-binding proteins contribute to the laminar segregation of sensory terminals in the dorsal horn of the spinal cord.

REFERENCES

Barondes SH (1984) Soluble lectins: A new class of extracellular proteins. Science 223: 1259-1264.

Bird JM and Kimber SJ (1984) Oligosaccharides containing fucose linked $\alpha(1,3)$ and $\alpha(1,4)$ to N-acetylglucosamine cause decompaction of mouse morulae. Dev. Biol. 104: 449-460.

Brown AG (1981) Organization in the Spinal Cord: The Anatomy and Physiology of Identified Neurones. Berlin, Heidelberg, New York: Springer Verlag.

Brown AG (1982) The dorsal horn of the spinal cord. QJ Exp Physiol. 67: 193-212.

Brown AG, Rose PK and Snow PJ (1977) The morphology of hair follicle afferent fibre collaterals in the spinal cord of the cat. J. Physiol. Lond. 272: 779-797.

Dodd J, Solter D and Jessell TM (1984) Monoclonal antibodies against carbohydrate differentiation antigens identify subsets of primary sensory neurons. Nature 311: 469-472.

Dodd J and Jessell TM (1985) Lactoseries carbohydrates specify subsets of dorsal root ganglia neurons projecting to the superficial dorsal horn of the spinal cord. J Neurosci In Press.

Feizi T (1981) Carbohydrate differentiation antigens. Trends Biochem. Sci. 11: 333-335.

Fenderson BA, Zehavi U and Hakomori SI (1984) A multivalent lacto-N-fucopentaose III-lysyllysine conjugate decompacts preimplantation mouse embryos, while the free oligosaccharide is ineffective. J Exp Med 160: 1591-1596.

Gooi HC, Feizi T, Kapadia A, Knowles BB, Solter D and Evans MJ (1981) Stage specific embryonic antigen involves 1,3 fucosylated type 2 blood group chains. Nature 292: 156-158.

Hakomori SI (1981) Glycosphingolipids in cellular interaction, differentiation and oncogenesis. Ann Rev Biochem 50: 733-764.

Iggo A (Ed) (1973) In Handbook of Sensory Physiology Vol 2 Springer-Verlag.

Jessell TM and Dodd J (1984) Structure and expression of differentiation antigens on functional subsets of primary sensory neurons. Philos Trans R Soc Lond (Biol) 208: 271-281.

Kannagi R, Levery SB, Ishigami F, Hakomori SI, Shevinsky LH, Knowles BB and Solter D (1983a) New globoseries glycolipids in human teratocarcinoma reactive with the monoclonal antibody directed to a developmentally regulated antigen: stage specific antigen 3. J Biol Chem 258: 8934-8942.

Kannagi R, Cochran N, Ishigami F, Hakomori SI, Andrews PW, Knowles BB and Solter D (1983b). Stage-specific embryonic antigens (SSEA-3 and SSEA-4) are epitopes of a unique globoseries ganglioside isolated from human teratocarcinoma cells. EMBO J 2: 2355-2361.

Kapadia A, Feizi T and Evans MJ (1981) Cahnges in the expression and polarization of blood group I and i antigens in post-implantation embryos and teratocarcinomas of mouse associated with cell differentiation. Exp Cell Res 131: 185-195.

Knowles BB, Rappaport J and Solter D (1982) Murine embryonic antigen SSEA-1 is expressed on human cells and structurally related human blood group antigen I is expressed on mouse embryos. Dev Biol 93: 54-58.

Light AR and Perl ER (1979) Spinal termination of functionally identified primary afferent fibres with slowly conducting myelinated fibres. J Comp Neurol 186: 133-150.

Mollicone R, Davies R, Evans B, Dalix AM and Oriol R (1985) Cellular expression and genetic control of ABH antigens in primary sensory neurons of marmoset baboon and man. J. Neuroimmunol. In Press.

Perl ER (1983) Characterization of nociceptors and their activation of
neurons in the superficial dorsal horn: First steps for the sensation of
pain. Adv Pain Res Ther 6: 23-51.
Ralston JH, Light AR, Ralston DD and Perl ER (1984) Morphology and synaptic
relationships of physiologically identified low-threshold dorsal root axons
stained with intra-axonal horseradish peroxidase in the cat and monkey.
J Neurophysiol 51: 777-792.
Rethelyi M (1977) Preterminal and terminal axon arborizations in the
substantia gelatinosa of cats' spinal cord. J Comp Neurol 172: 511-528.
Ribiero Da Silva A and Coimbra A (1982) Two types of synaptic glomeruli
and their distribution in laminae I-III of the rat spinal cord. J Comp
Neurol 209: 176-189.
Semba K, Masarachia P, Melamed S, Jacquin M, Harris S, Yang G and Egger MD
(1983) An electron microscopic study of primary afferent terminals from
slowly adapting type 1 receptors in the cat. J Comp Neurol 221: 466-481.
Shevinsky LH, Knowles BB, Damjanov I and Solter D (1982) Monoclonal
antibody to murine embryos defines a stage-specific embryonic antigen
expressed on mouse embryos and human teratocarcinoma cells. Cell 30: 697-705.
Solter D and Knowles BB (1979) Developmental stage specific antigens during
mouse embryogenesis. Curr Top Dev Biol 13: 139-165.
Stern P (1984) Differentiation antigens of teratomas and embryos. Brit
Medical Bulletin 40: 218-223.

THE PHYSIOLOGICAL ACTIONS OF LHRH – EVIDENCE FROM THE HYPOGONADAL (hpg)

MOUSE

H. M. Charlton

Dept of Human Anatomy
University of Oxford
South Parks Road, Oxford, UK

GENERAL INTRODUCTION

The ultimate actions of a hormone can vary at different times during development and also at different sites in the body at any particular point in time. Provided that there is provision to prevent an overproduction of hormone from one site causing inappropriate physiological function at other sites then there is no reason to deny that the same molecule may indeed have widely disparate final actions at different sites in the body. The varied geography of the sites of somatostatin production and the different end physiological functions of this tetradecapeptide are a case in point.

The production of a particular hormone in different tissues may depend upon the transcription of a single gene with the possibility of post transcriptional and post translational modifications yielding different molecular end products. If there is more than one copy of a gene then it may be that different flanking sequences etc may dictate that although one particular copy of this gene is switched off in one tissue, that tissue may utilise another copy for the purpose of final messenger RNA production and hormone synthesis.

The proof that one group of cells is producing a hormone obviously cannot depend upon the evidence of radioimmunoassay alone because this method of assay indicates similar molecular conformation. Direct methods of chemical analysis are essential for the final confirmation that what is made in tissue A is indeed the same molecule as in tissue B. The presence of an identical mRNA for a putative peptide hormone in two tissues would also provide strong evidence that they were producing the same molecule.

The analysis of hormone action can be undertaken in vivo, and in vitro. In most cases such experiments indicate what processes a hormone molecule is capable of influencing at the concentration used and/or in its mode of presentation. In particular, in vitro experiments can occasionally be grossly non-physiological in terms of local hormone concentration. The binding of a hormone to a particular tissue may again only really indicate a receptor recognising similar molecular conformation rather than an absolute chemical similarity with the "normal" ligand.

Classically, studies on the actions of hormones have been investigated in vivo by observing the results of the surgical removal of endocrine secretory cells, by observing the physiological perturbations when there may be an overproduction of hormone, for example in the case of an endocrine tumour and by observing the effects of hormone replacement either by injections or tissue transplantation. Surgical removal of endocrine tissue will, in nearly every case, remove more than one type of secreting cell and a more recent approach has been to use immunological suppression of circulating hormones either by passive or active immunisation. These techniques have proved extremely useful but may not result in abolishing any local paracrine effects of a hormone.

A natural failure in the production of a hormone occurs occasionally in the form of a genetic mutation, but the usefulness of such models in possibly identifying the direct and indirect actions of a missing peptide will depend upon an initial analysis of the exact genetic lesion and on a knowledge of the number of gene copies and any possible transcriptional and translational modifications there may be in the processing of the gene product.

The hypogonadal (hpg) mouse has been shown to be deficient in the production of the decapeptide LHRH in neural tissue both by radioimmunoassay and by immunocytochemistry (Cattanach et al 1977; Lyon et al 1981). This has been backed up by an analysis of hpg hypothalamic extracts using HPLC methods and also by a failure of such extracts to elicit LH release from pituitary tissue (G Fink, personal communication). More conclusively cell free translation of hpg hypothalamic mRNAs do not result in the formation of an LHRH like peptide (Curtis et al 1983) and finally it has been demonstrated that the hpg mouse lacks two exons in the hpg gene (P Seeburg, personal communication). As a consequence of the mutatation, the pituitary gland is severely depleted in its stores of LH and FSH and the gonads fail to develop post-natally.

At the present moment the evidence available suggests that there is only one copy of the LHRH gene and so it would appear that hpg mice will be incapable of making the LHRH decapeptide in any tissue and that therefore this mutant represents an ideal model in which to ask questions about the possible varied direct and indirect physiological roles of the hormone.

POSSIBLE DIRECT AND INDIRECT PHYSIOLOGICAL ROLES FOR LHRH

In the general introduction the point has been made that a hormone may have different functions at different times during embryonic and adult development. At the present moment the following list of a variety of functions that may be dependent upon LHRH can be drawn up:

The differentiation of pituitary cell types
The differentiation of the gonadal tract
Local paracrine effects in the placenta
Sexual differentiation of the brain
Pituitary stimulation and pubertal development
Local paracrine effects in the gonads
Adult behavioural effects
Interplay with pheromones
Other extra pituitary actions

This list is not meant to be exhaustive and is further complicated by the fact that at least two tissues are known to transcribe and translate the LHRH gene, the CNS and placenta, (Lee et al 1981; Tan & Rousseau 1982) and that LHRH has been claimed to be found in milk (Smith-White & Ojeda 1983; Sarda & Nair 1981). Inside the CNS the olfactory system appears to have a separate set of LHRH secretory neurones with the bulk of secretory activity confined to the medial pre-optic area - albeit in a diffuse pattern (Witkin et al 1982; Witkin & Silverman 1983).

Accepting the caveat that the argument is confined to mice and accepting the evidence that there is only one LHRH gene and that this is grossly deficient in hpg mice, then a study of the untreated hpg male and female allows certain conclusions to be made with regard to some of the physiological functions outlined above that have been attributed to the molecule.

The differentiation of different pituitary cell types

It has been claimed that LHRH may be essential for the differentiation of both gonadotrophs and prolactotrophs (Begeot et al 1983; Aubert et al 1983). However both of these cell types can be readily identified in electron micrographs of hpg pituitary sections (McDowell et al 1982) and LH, FSH and prolactin have been shown to be present in hpg pituitary extracts (Charlton et al 1983a). The decapeptide may well enhance the division of gonadotrophs, but it does not seem that its presence is a prerequisite for their initial cytogenesis.

The differentiation of the gonadal tract

We know that this depends upon the production of hormones by the foetal testis and that the hypothalamic-pituitary system does not seem to be involved in this early testicular stimulation (Josso, 1981). The fact that hpg male mice have fully developed internal and external genitalia, but that there is a failure only in post-natal gonadal growth is strong additional evidence that whatever initiates testicular secretion in utero at the time of tract differentiation does not depend absolutely upon LHRH.

What then causes the foetal testis to secrete androgens? A simple explanation has been to involve placental gonadotrophins but recent genetic analysis has shown that rodents may not possess chorionic gonadotrophin genes, and there is, as yet, no evidence that the LH gene is activated in placental tissue (Tepper & Roberts, 1984; Carr & Chin, 1985). There can be no doubt that the human placenta is capable of manufacturing LHRH and it may be that this LHRH could stimulate the foetal pituitary to produce LH and therefore stimulate the testis. However in the hpg mouse there is no LHRH gene and therefore we have the situation where the foetal testis must have produced androgens at the correct time without perhaps any command from the placenta or pituitary. The rabbit testis has been shown to secrete androgens in culture at the correct time to influence sexual differentiation (Wilson 1981) and the above argument could suggest that this may also be true for the mouse.

Local paracrine effects in the placenta

The exact function of LHRH in placental tissue is not known, and its presence has yet to be demonstrated in rodent placentae. Should the rodent placenta produce LHRH then whatever its local or distant functions may be, they are not absolutely essential for foetal development because hpg foetuses are carried to term and cannot readily be identified from their normal littermates until weaning.

Sexual differentiation of the brain

It is now well established that sexual differentiation of the rat and mouse brain is accomplished during the first few post natal days of life and that normal male differentiation depends upon testicular secretions at this time (Gorski 1979). It is just possible that foetal hormones are still circulating in the early post-natal period, but a more readily acceptable explanation would be that the hypothalamic pituitary axis is now activated and that this early testicular activity depends ultimately upon LHRH secretion. What is the evidence from the hypogonadal mouse?

Untreated adult male hpg mice show no sexual behaviour even when given testosterone implants which stimulate full spermatogenesis and seminal vesicle growth. However hpg males given androgen injections on day 1 after birth and then given testosterone pellets when adult demonstrate full male behaviour and in fact are capable of siring several litters (Ward 1980; Lyon et al 1981). These experiments would argue that hypothalamic LHRH is essential for the early stimulation of pituitary and testicular activity resulting in male neural sexual differentiation. These arguments presuppose that LHRH can only reach the foetal pituitary from a hypothalamic source; however several groups claim that LHRH may be concentrated in or secreted into milk and therefore reach the pituitary via absorption from the intestinal tract (Smith-White & Ojeda 1983). If this were to be the case then all members of a litter may be expected to absorb this "LHRH" and later testosterone treatment of hpg males would be expected to result in male sexual behaviour. This does not happen.

Pituitary stimulation and pubertal development

At present our interpretation of the development of the pituitary gonadal system up to puberty and in normal adulthood is that hypothalamic output of LHRH occurs in a pulsatile manner and that pulse frequency may be important in controlling gonadotrophic hormone output (Carmel et al 1976; Yen et al 1972).

It has been shown that compared with the effect of single daily injections of LHRH, multiple injections result in a more normal function of the pituitary-gonadal-accessory organ system in hpg mice of either sex (Charlton et al 1983b). It has also been shown that the hpg pituitary contains receptors for LHRH but that their numbers are greatly reduced and that receptor numbers are rapidly increased by LHRH injections (Young et al 1983).

Controversy still reigns as to whether there is more than one releasing factor for both LH and FSH (Mizunuma et al 1983; Powers & Johnson 1984; Mills et al 1983). The fact that both LH and FSH are depleted in the hpg mouse and that pituitary synthesis and release of both hormones are elicited by injections of synthetic LHRH may seem to favour a single releasing factor hypothesis. However we can never rule out that the injections have in some way allowed normal development of an FSH.RH system which depends upon previous LHRH stimulation either directly or indirectly for its genesis or activation.

Intragonadal control mechanisms

In vivo and in vitro studies have suggested that LHRH or an LHRH-like molecule is secreted locally in the testis and ovaries of certain species where it may act as part of a local paracrine control mechanism (Hsueh & Jones 1983; Sharpe 1984). As there is, as yet, no evidence of such a system even in normal mice the hpg mutant would appear to have little to add to this argument. Oxytocin is a further addition to the number of hormones supposedly secreted by the testis and the chapter by Pickering et al in this symposium describes a lack of oxytocin in the hpg testis but with the stimulation of oxytocin synthesis by LH and testosterone. LHRH would seem to be an essential early switch mechanism in the cascade of events leading to testicular oxytocin production.

Adult Sexual Behaviour

Several groups have argued that LHRH may be involved locally in the brain as a neuropeptide directly affecting sexual behaviour in both sexes (Dorsa et al 1981; Sakuma & Pfaff 1983; Riskind & Moss 1983a, b; Dudley et al 1983). In mice this may not be an absolute requirement because hpg females exhibit full mating behaviour with oestrogen and progesterone treatment alone. However LHRH treatment also increases mating behaviour to normal levels in females primed only with oestrogen. The evidence therefore suggests that LHRH may have a role, but that it is not an absolute requirement that the hormone is present (Ward & Charlton 1981). Hypogonadal male mice, as already mentioned, exhibit full male mating behaviour with testosterone treatment alone as adults (provided a neonatal priming injection of testosterone was given) and so in the male any effect of LHRH may be part of an overall control system, but without being an absolute requirement.

Interplay with pheromones

There is evidence that olfactory cues in behaviour may be partly controlled by LHRH produced locally in the olfactory system (Dluzen et al 1981; Dluzen & Ramirez 1983) The fact that hpg mice mate normally with appropriate steroid treatment again argues that these olfactory interactions are part of a wider fail/safe mechanism ensuing adequate mating behaviour in both sexes, but that any LHRH part of the system is not an absolute requirement.

Other extra pituitary actions

Comparative endocrinological studies have argued a role for GnRH as a neurotransmitter in the sympathetic nervous system in some lower vertebrates (Jan & Jan 1983; Akasu et al 1983). As yet there is no evidence for such an action in mammals and indeed it is difficult to assess what the phenotypic results of a lack of LHRH in sympathetic ganglia would be. The usefulness of the hpg mutant in this regard is, therefore, impossible to assess.

EXPERIMENTAL USE OF THE HPG MOUSE: hormone injections

LHRH receptor values are only 30-50% of normal in pituitaries of hpg mice, but within 3 days of pulsatile injections receptor numbers are normalized (Young et al 1983). Both single and multiple daily injections result in a massive stimulation of pituitary FSH content with a marginal increase in pituitary LH content in the pulsatile injected group (Charlton et al 1983b). Both modes of injection result in an increased number of gonadotrophs in the pituitary gland, and these cells are also hypetrophied and contain more secretory granules than saline controls. A pathological side effect of both injection regimes results in about 40% of gonadotrophs containing large lipid droplets (McDowell et al 1982b; Megson et al 1983).

At first sight the single and pulsatile injections do not seem to differ greatly in their effect upon pituitary structure and hormone content in the hpg male. The effects on the testes are, however, different. Both single and multiple injections cause an increase in testis weight and all stages of spermatogenesis are found in the semiferous epithelium, but only in the multiple daily injection group was seminal vesicle growth stimulated. Whilst single injections seemingly produced enough androgen for local testicular actions, only with the episodic treatment was enough androgen produced for export to stimulate the accessory sexual tissue (Charlton et al 1983b). The hpg mutant may also be used to investigate sites of steroid feedback and in mutant males given a silastic implant of testosterone which rapidly inhibits LH and FSH secretion in normal mice and then given GnRH injections, pituitary FSH was elevated to normal adult male levels. There was no difference between the androgen group and control empty capsule animals. This would suggest that a major site of androgen negative feedback may be at the hypothalamic level or higher (Charlton et al 1983b).

In female mutants single daily injections resulted in a massive increase in pituitary FSH to levels five times those found in normal females, with little or no increase in LH content. Multiple daily injections, on the other hand, elevated both FSH and LH content into the normal female range (Charlton et al 1983b). In the latter experiments the episodic mode of injection resulted in vaginal opening, uterine growth and the stimulation of ovarian follicular development, whereas single daily injections had no effect upon the ovaries (Halpin 1984). When ovariectomised hpg females were given multiple injections of LHRH, then pituitary FSH content was elevated to the levels found in normal males. However, implanting an estradiol 17B capsule under the skin prevented this massive FSH synthesis. These experiments provide evidence that a major site of negative feedback of estradiol in the mouse is at the level of the pituitary gland (Charlton et al 1983b).

EXPERIMENTAL USE OF THE HPG MOUSE: neural grafting

A second method of supplying LHRH to the hpg mouse has been to graft foetal medial preoptic area (MPOA) from normal animals into the third ventricle of the mutants. The grafts survive well and result in dramatic changes in pituitary and gonadal function in both sexes (Krieger et al 1982; Young et al 1985). Within 26-40 days of grafting, pituitary LH content and LHRH receptors were in the normal range with pituitary and plasma FSH reaching 50% of normal values (Young et al 1985). At the ultrastructural level the number of gonadotrophs had increased, the cells were larger and contained more secretory granules, and this time there was no evidence of lipid droplet inclusions in the cytoplasm (Morris & Charlton 1983). Testicular and seminal vesicle weight increased more than 10-fold by 40 days after implantation and testicular LH receptors were 70% of normal, compared with the 8% of normal value found in untreated hpg males (Young et al 1985). Although mean testis weight never reached full adult values, even in long term grafted animals, full spermatogenesis had been stimulated (Krieger et al 1982; Charlton 1985).

In hpg female mice bearing normal foetal hypothalamic grafts pituitary LH and FSH content was normal within 40 days and serum FSH was in the normal range. Vaginal opening occurred as early as 14 days post-operation and the females exhibited periods of prolonged oestrus smears with no evidence of cyclicity (Gibson et al 1984a; Young et al 1985). LH receptors in the ovaries increased into the normal female range but there has, so far, been no evidence of ovulation and corpora lutea in the ovaries (Charlton 1985). The neural grafts have therefore resulted in almost normal pituitary and ovarian function except for a lack of cyclicity and ovulation. This may reflect a lack of appropriate neural control of the MPOA tissue removed from its normal anatomical site.

Although none of the females spontaneously ovulated, it is not perhaps surprising with the evidence of vaginal cornification that many of them mated with normal males. The surprising observation was that a significant proportion in fact became pregnant (Gibson et al 1984b; Charlton et al 1985). The most obvious explanation of this phenomenon is that the graft had somehow responded to the mating stimulus by releasing a surge of LHRH which itself elicited a surge of LH and therefore reflex ovulation. These females took their pregnancies to term and reared their litters normally.

Thus full fertility was elicited in female hpg mice in some cases within 60 days of neural transplantation.

The injection experiments indicated that episodic release of GnRH may be the natural mode of pituitary stimulation. At the present moment we have no evidence in what manner this grafted tissue is releasing the hormone. It may be that males with more stimulated testes either possess more grafted LHRH neurones, or that the release parameters are more propitious in terms of LH secretion, or indeed both of these together. With up to 80 grafted females studied the failure of spontaneous ovulation probably argues that the grafts are not secreting a surge of LHRH at regular intervals and that therefore whatever inputs influence cyclic release of the decapeptide are not present in the graft. A more detailed analysis of the neuroanatomical problems may come from experiments using purified cell suspensions grafted directly into the brain parenchyma.

Neurobiological aspects of POA grafting

A fascinating additional bonus of this endocrine research has been the observation that over 90% of LHRH fibres leaving the ventricular grafts innervate the median eminence no matter where the cell bodies are situated. More specifically the growing fibres seem to follow a normal course through the lateral sulcus of the median eminence. The hpg mutant may therefore also offer a model for an investigation of neurotrophic factors involved in axon guidance (Charlton et al 1985).

ACKNOWLEDGEMENTS

The ideas put forward in the first part of this paper are entirely those of the author and no blame should be placed upon research colleagues for their expression. The research reviewed in the second half has been the result of widespread collaboration with the following colleagues in the UK and USA:

Clayton, R., Davies, T., Detta, A., Gibson, M., Ferin, M., Fink. G., Halpin, D., Harrison, C., Iddon, C., Jones, A., Krieger, D., Kokoris, G., Levy, G., Lewis, C., McDowell, I., Megson, A., Morris, J., Parry, D., Perlow, M., Pledge, P., Rosie, R., Sheward, J., Silverman, A., Speight, A., Ward, B., Young, L.

Financial assistance has been forthcoming from the MRC, Wellcome Trust, the EPA Cephalosporin Trust and the NIH (Grant No. 20338).

REFERENCES

Akasu, T., Kojima, M., Koketsu, K. (1983) LHRH modulates nicotinic ACh-receptor sensitivity in amphibian cholinergic transmission (frog). Brain Res., 279: 347-51.
Aubert, M.L., Begeot, M., Winiger, B.P., Morel, G., Sigoneako, P.C. & Dubois, P.M. (1985) Ontogeny of hypothalamic LHRH and pituitary GnRH receptors in fetal and neonatal rats. Endocrinology, 116: 1565-76.
Begeot, M., Hemming, F.J., Martinat, N., Dubois, M.P. & Dubois P.M. (1983) GnRh stimulates immunoreactive lactotrope differentiation. Endocrinology, 112: 2224-6.
Carmel, P.W., Arai, S., & Ferin, M. (1976) Pituitary stalk portal blood-collection in rhesus monkeys: evidence for pulsatile release of gonadotrophin-releasing hormone (GnRH). Endocrinology, 99: 243-248.

Carr, F.E. & Chin, W.W. (1985) Absence of detectable chorionic gonadotrophic submit messenger ribonucleic acids in the rat placenta throughout gestation. Endocrinology, 116: 1151-7.

Cattanach, B.M., Iddon, C.A., Charlton, H.M., Chiappa, S.A., & Fink, G. (1977) Gonadotrophin-releasing hormone deficiency in a mutant mouse with hypogonadism. Nature, 269: 338-340.

Charlton, H.M. (1985) The use of neural transplants to study neuroendocrine mechanisms. Frontiers in Neuroendocrinology, Vol. 9. Ed. W.F. Gangong & L. Martin. Raven Press (In press).

Charlton, H.M., A. Speight, D.M.G. Halpin, A. Bramwell, W.J. Steward & G. Fink (1983b) Prolactin measurements in normal and hypogonadal (hpg) mice: developmental and experimental studies. Endocrinology, 113: 545-548.

Charlton, H.M., D.M.G. Halpin, C. Iddon, R. Rosie, G. Levy, I.F.W. McDowell, A. Megson, J.F. Morris, A. Bramwell, A. Speight, B.J. Ward, G. Davey-Smith & G. Fink (1983b) The effects of daily administration of single and multiple injections of gonadotrophin-releasing hormone on pituitary and gonadal function in the hypogonadal (hpg) mouse. Endocrinology, 113: 535-544.

Charlton, H.M., Parry, D.M. & Jones, A. (1985) Thy 1.1 and GnRH fibre output and physiological function in female hpg mice given pre-optic area grafts from foetal AKR mice. In: Neural grafting in the mammalian CNS, edited by A. Bjorklund, U. Stenevi, Elsevier. (637-641).

Curtis, A., Lyons, V., & Fink, G. (1983) The human hypothalamic LHRH precursor is the same size as that in rat and mouse hypothalamus. Biochem. Biophys. Res. Comm., 117: 872-877.

Dluzen, D.E., Ramirez, V.D., Carter, C.S. & Getz, L.L. (1981) Male vole urine changes LH-RH and norepinephrine in female olfactory bulb. Science, 212: 573-5.

Dluzen, D.E. & Rameriz, V.D., (1983) Localised and discrete changes in neuropeptides (LHRH & TRH) and neurotransmitters (NE & DA) concentrations within the olfactory bulbs of male mice as a function of social interaction. Hormones of Behaviour, 17: 139-145.

Dorsa, D.M., Smith, E.R. & Davidson, J.M. (1981) Endocrine and behavioural effects of continuous exposure of male rats to a potent LHRH agonist: evidence for CNS actions of LHRH. Endocrinology, 109: 729-35.

Dudley, C.A., Vale, W., Rivier, J. & Moss, R.L. (1983) Facilitation of sexual receptivity in the female rat by a fragment of the LHRH decapeptide AC LHRH 5-10. Neuroendocrinology, 36: 486-8.

Gibson, M.J., Charlton, H.M., Perlow, M.J., Zimmerman, E.A., Davies, T.F & Krieger, D.T. (1984a) Preoptic area brain grafts in hypogonadal (hpg) female mice abolish effects of congenital hypothalamic gonadotrophin-releasing hormone (GnRH) deficiency. Endocrinology, 114: 1938-1940.

Gibson, M.J., Krieger, D.T., Charlton, H.M., Zimmerman, E.A., Silverman, A.J. & Perlow, M.J. (1984b) Mating and pregnancy can occur in genetically hypogonadal mice with preoptic area brain grafts. Science, 225, 949-951.

Gorski, R.A. (1979) The neuroendocrinology of reproduction: an overview. Biol. Reprod., 20: 111-127.

Halpin, D.M.G. (1984) Aspects of ovarian function in the mouse. D.Phil Thesis Oxford.

Hsueh, A.J.W. & Jones, P.B.C. (1983) GnRH extrapituitary actions and oaracrine control mechanisms. Ann. Rev. Physiol., 45: 83-94.

Jan, Y.N. & Jan, L.Y. (1983) A LHRH-like peptidergic neurotransmitter capable of 'action at a distance' in autonomic ganglia. Trends Neurosci., 6: 320-4.

Josso, N. (1981) Differentiation of the genital tract: stimulators and inhibitors. In Mechanisms of Sex Differentiation in Mammals, Ch. 5: 165-204. Eds. C.R. Austin & R.G. Edwards, Academic Press, London.

Krieger, D.T., Perlow, M.J., Gibson, M.J., Davies, T.F., Zimmerman, E.A, Ferin, M., & Charlton, H.M. (1982) Brain grafts reverse hypogonadism of gonadotrophin releasing hormone deficiency. Nature, 298: 468-471.

Lee, J.N., Seppala, M. & Chard, T. (1981) Characterization of placental LH-releasing factor-like material. Acta Endocr., Copenh. 96: 394-7.

Lyon, M.F., Cattanach, B.M. & Charlton, H.M. (1981) Genes affecting sex differentiation in mammals. In: Mechanisms of Sex Differentiation in Animals and Man: Ch. 9, pp. 367-374. Eds. C.R. Austin & R.G. Edwards, Academic Press, London.

McDowell, I.F.W., Morris, J.F. & Charlton, H.M. (1982a) Characterization of the pituitary gonadotroph cells of hypogonadal (hpg) male mice: comparison with normal mice. J. Endocrinol., 95: 321-330.

McDowell, I.F.W., Morris, J.F., Charlton, H.M. & Fink, G. (1982b) Effects of luteinizing hormone releasing hormone on the gonadotrophs of hypogonadal (hpg) mice. J. Endocrinol., 95: 311-340.

Megson, A., Lewis, C.E., Morris, J.F., Charlton, H.M. & Fink, G. (1983) Heterogeneous responses to 'pulsatile' GnRH administration among the gonadotrophs of intact and castrate hpg mice. J. Anat., 137: 819-820, Abstr.

Mills, T.M., Copland, J.A., Coy, D.H. & Schally, A.V. (1983) Is the postovulatory release of FSH in the rabbit mediated by LH-RH? Endocrinology, 113: 1020-4.

Mizunuma, H., Sampson, W.K., Lumpkin, M.D. & McCann, S.M. (1983) Evidence for an FSH-releasing factor in the posterior portion of the rat median eminence. Life Sci., 33: 2003-10.

Morris, J.F. & Charlton, H.M. (1983) Responses of gonadotrophs in female hypogonadal (hpg) mice to intracerebroventricular implantation of preoptic area tissue from normal mice. J. Anat., 137: 806-807, Abstr.

Powers, C.A. & Johnson, D.C. (1984) Correlation of LH, FSH and GnRH release during various brain-pituitary coincubations. Horm. Res., 19: 117-26.

Riskind, P., & Moss, R.L. (1983) Midbrain LHRH infusions enhance lordotic behavior in ovariectomized estrogen-primed rats independently of a hypothalamic responsiveness to LHRH. Brain Res. Bull., 11: 481-5.

Riskind, P. & Moss, L.R. (1983b) Effects of lesions of putative LHRH-containing pathways and mid-brain nuclei on lordotic behavior and LH release in ovariectomized rats. Brain Res. Bull., 11: 493-500.

Sakuma, Y. & Pfaff, D.W. (1983) Modulation of the lordosis reflex of female rats by LHRH its antiserum and analogues in the mesencephalic grey. Neuroendocrinology, 36: 218-24.

Sarda, A.K. & Nair, R.M.G. (1981) Elevated levels of LRH in human milk. J. Clin. Endocr. Metab., 52: 826-8.

Sharpe, R.M. (1984) Intratesticular factors controlling testicular function. Biol. Reprod., 30: 29-49.

Smith-White, S. & Ojeda, S.R. (1983) Maternal modulation of infantile ovarian development via milk' LHRH. Fed. Proceeding., 42: 978.

Tan, L. & Rousseau, P. (1982) The chemical identity of the immunoreactive LHRH peptide biosynthesised in the human placenta. Biochem. Biophys. Res. Comm., 109: 1061-71.

Tepper, M.A. & Roberts, J.L. (1984) Evidence for only one B luteinising hormone and no B chorionic gonadotrophin gene in the rat. Endocrinology, 115: 385-391.

Ward, B.J. (1980) Some aspects of the reproductive physiology and behaviour of the hypogonadal (hpg) mouse. M.Sc. Thesis, Oxford.

Ward, B.J. & Charlton, H.M. (1981) Female sexual behaviour in the GnRH deficient hypogonadal (hpg) mouse. Physiol. & Behavior., 27: 1107-1109.

Wilson et al. (1981) The role of gonadal steriods in sexual differentiation. Rec. Prog. Horm. Res., 37: 1-39.

Witkin, J.W. & Silverman, A.J. (1983) LH-RH in rat olfactory systems. J. Comp. Neurol., 218: 426-32.

Witkin, J.W., Paden, C.M. & Silverman, A.J. (1982) The LHRH systems in the rat brain. Neuroendocrinology, 35: 429-38.

Yen, S.S.C., Tsai, C.C., Naftolin, F., Vanden Berg, G. & Ajabor, L. (1972) Pulsatile patterns of gonadotrophin release in subjects with and without ovarian function. J. Clin. Endocrinol. Metab. 34: 671-675.

Young, L.S., Detta, A., Clayton, R.N., Jones, A. & Charlton, H.M. (1985) Pituitary and gonadal function in hypogondadotrophic hypogonadal (hpg) mice bearing hypothalamic implants. J. Reprod. Fert., In Press.

Young, L.S., Speight, A., Charlton, H.M. & Clayton, R.N. (1983) Pituitary gonadotrophin-releasing hormone receptor regulation in the hypogonadotropic hypogonadal (hpg) mouse. Endocrinology, 113, 55-61.

Young, L.S., A. Detta, R.N. Clayton, A. Jones & H.M. Charlton (1985) Pituitary and gonadal function in hypogonadotrophic hypogonadal (hpg) mice bearing hypothalamic implants. J. Reprod. Fert., 74: 247-255.

THE BIOSYNTHETIC PRECURSOR OF GONADOTROPIN-RELEASING

HORMONE: A MULTIFUNCTIONAL PROHORMONE

Karoly Nikolics and Peter H. Seeburg

Genentech, Inc.
460 Point San Bruno Boulevard
South San Francisco, CA 94080

INTRODUCTION

Mammalian reproduction is regulated by the pituitary hormones luteinizing hormone (LH), follicle stimulating hormone (FSH), and prolactin. The secretion of these proteins from the anterior pituitary is controlled by humoral factors of the central nervous system (CNS) and by feedback signals in the form of hormonal factors from the gonads. LH stimulates ovulation and corpus luteum formation in females and androgen secretion in males. FSH stimulates the growth and maturation of ovarian follicles in females and spermatogenesis in males. Prolactin stimulates milk production and in certain species has direct gonadal effects.

Since the discovery of gonadotropin-releasing hormone (GnRH, also called LH-releasing hormone, LHRH), this decapeptide has been regarded as the key molecule in the CNS control of mammalian reproduction (Matsuo et al., 1971; Schally et al., 1971; Burgus et al., 1972). GnRH is produced in hypothalamic neurosecretory cells and released in a pulsatile pattern into the hypothalamo-hypophyseal portal circulation (Pohl and Knobil, 1982; Krey and Silverman, 1983). Through this unique mechanism, GnRH triggers the release of LH and FSH from the anterior pituitary which in turn control the function of the gonads. Even though the discovery of GnRH represented a major step towards our understanding of the CNS control of gonadotropin secretion, certain physiological conditions involving gonadotropins could not be explained by the stimulatory effect of a single hypothalamic hormone (McCann et al., 1984). LH and FSH secretion do not always coincide during reproductive cycles and a complex regulation of gonadotropin and prolactin secretion has been found where levels of these hormones are inversely related in many reproductive states (Fink et al., 1983; Leong et al., 1983).

The biosynthesis of GnRH has been under investigation for more than a decade, with little progress in terms of precursor structure (Millar et al., 1981). However, using antibodies against GnRH, the site of production in the CNS and elsewhere has been mapped. The highest density of GnRH-producing neuronal cell bodies which are of hypophysiotropic significance has been found in the septo-preoptic region of the mammalian brain, yet other areas also synthesize the decapeptide (Krey and Silverman, 1983; Shivers et al., 1983). Extrahypothalamic tissues producing and/or containing GnRH are the pineal gland, placenta, gonads, and lactating mammary gland (Piekut and Knigge, 1981; Khodr and Siler-Khodr, 1980; Ying et al., 1982; Hedger et al., 1985; Bhasin et al., 1983; Baram et al. 1977).

Reports on precursor forms of the decapeptide have been contradictory with size estimates ranging from a few thousand to 60,000 daltons (Millar et al., 1981; Gautron et al., 1981; Curtis and Fink, 1983; Charli et al., 1984).

BIOSYNTHESIS OF GnRH

Using a recombinant DNA approach, we isolated cDNA coding for GnRH from human placental material (Seeburg and Adelman, 1984). Correct translation of the corresponding mRNA could be predicted to result in the synthesis of a precursor of approximately 10,000 Daltons having a simple structure of three elements, as shown in Fig. 1. A signal peptide precedes the GnRH decapeptide along with a typical enzymatic precursor processing and C-terminal amidation site, which is followed by a sequence of 56 amino acid residues. After loss of the signal peptide and upon enzymatic processing, this precursor can be expected to give rise to two peptides: GnRH and the 56 amino acid C-terminal fragment, which we called GAP for GnRH-associated peptide. GAP is not homologous with other known protein sequences.

As this cDNA was derived from placental tissue we had to consider the possibility that a different precursor for GnRH might be synthesized in the hypothalamus. This possibility, however, was unlikely since DNA analysis suggested the existence of a single human gene per haploid genome, which has since been located on the short arm of chromosome 8 in collaboration with Theresa Yang-Feng and Uta Francke (Yang-Feng et al., 1985). The existence of the same GnRH-GAP precursor in CNS was established by two independent approaches. In collaboration with Heidi Phillips and Dale Branton, immunocytochemical mapping of rat brain revealed colocalization of GnRH and GAP-related antigen as discussed below (Phillips et al., 1985). Independently, hypothalamic cDNA coding for the same precursor has been characterized from human and rat

```
-23                         -1
MKPIQKLLAGLILLTSCVEGCSS-


1          10
QHWSYGLRPG-GKR-


14                                              69
DAENLIDSFQEIVKEVGQLAETQRFECTTHQPRSPLRDLKGALESLIEEETGQKKI
```

Fig. 1. Structure of the human GnRH precursor in the one-letter code
        as deduced from the cDNA sequence. The top line represents
        the signal region with negative numbers relative to GnRH. The
        middle line contains the primary structure of GnRH and the
        processing site, and the sequence of GAP is shown in the lower
        line.

material supporting the notion of the existence of a single GnRH gene in these species (Adelman et al., 1985). Fig. 2 shows the comparison of peptide sequences deduced from human and rat cDNAs coding for the GAP part of the GnRH precursor. The two sequences are highly homologous, with GnRH being identical in both species. Even though there are approximately 15% nonconservative differences between human and rat GAP sequences, their three-dimensional structure as predicted by different methods (Chou and Fasman, 1974; Garnier et al., 1978; Finer-Moore and Stroud, 1984)) is greatly conserved. The molecule appears to be constructed of two strongly helical domains (N- and C-terminal 20-22 amino acids) and a middle domain which is more flexible or not highly ordered (R. Stroud, personal communication).

```
1        10        20        30        40        50    56

DAENLIDSFQEIVKEVGQLAETQRFECTTHQRPSPLRDLKGALESLIEEETGQKKI


--E-L-DSFQE--KE--Q-AE-Q-FECT-H-PRSPLRDL-GALE-LIEEE-GQKK-
```

Fig. 2. Comparison of the primary structures of human (top) and rat (middle) GAP. The two 56 amino acid sequences are shown in the one-letter code together with a consensus sequence (third line). Dashes in this consensus sequence signal differences in the corresponding positions between human and rat GAP.

ANTIBODIES TO GAP-RELATED ANTIGENS

We have raised polyclonal antibodies in rabbits against synthetic peptides representing several regions of GAP, as summarized in Fig. 3. Antiserum 39A was generated against GAP-27-40 resembling the human sequence. Antisera 24A, 56A and KN-1 were raised against peptides corresponding to the rat GAP sequence. Antigen for antiserum 24A was rat GAP-1-11, for 56A rat GAP-20-43, and for KN-1 rat GAP-39-53, respectively. These peptides correspond approximately to the three principal structural domains of GAP as discussed above. The synthetic peptide antigens were coupled to bovine serum albumin or soybean trypsin inhibitor via an N-terminal cysteine. The same Cys-extended peptides were also used for the derivatization with p-hydroxyphenyl-maleimide which served for radioiodination (Burnier, J., in preparation).

All these antisera have been found suitable for the development of radioimmunoassays, for immunocytochemistry, and for the preparation of immuno-affinity chromatography supports. Antibody 39A was used for Western blots and immunoblots in the course of the preparation and purification of GAP from a bacterially expressed fusion protein (Nikolics et al., 1985). Radioimmunoassays based on all of the above antibodies are able to detect GAP in concentrations of approximately 100 pg/ml in physiological samples, as shown in Fig. 4, and have been used to measure GAP in hypophyseal stalk blood and rat median eminence (Clarke et al., in preparation). Radioimmunoassays were carried out by described procedures, either by double-antibody methods or by separating bound and free ligand using charcoal.

```
1          10          20          30          40          50    56
NTEHLVDSFQEMGKEEDQMAEPQNFECTVHWPRSPLRDLRGALERLIEEEAGQKKM
```

```
     _____            _____
        AS 24A                         AS 56A

                                             _____
                                                  AS KN-1
```

Fig. 3. Regions of rat GAP represented by synthetic peptides used for
         generating antisera in rabbits. The peptides were conjugated
         to soybean trypsin inhibitor and injected i.m. into rabbits.
         Antiserum 39A (not shown) was raised against GAP-27-40 of the
         human sequence.

IMMUNOCYTOCHEMICAL LOCALIZATION OF GAP-RELATED ANTIGENS IN RAT BRAIN

     Initially, antiserum 39A was used to investigate whether placental
GAP-related antigens existed in the CNS (Phillips et al., 1985). As
shown in the sequence comparison between rat and human (Fig. 2), the
region between GAP positions 27-40 is highly conserved and contains
only one major change. Immunocytochemical staining of sections of rat
brain demonstrated that GAP-related antigens were localized in exactly
the same areas known from earlier studies to contain GnRH-related
antigen (Shivers et al., 1983; Krey and Silverman, 1983). Dual
immunofluorescence staining with antiserum against GnRH and GAP
antiserum 39A demonstrated coexistence of the two peptides in neuronal
cell bodies and axons in the septo-preoptic region and median eminence
(Phillips et al., 1985). These studies strongly suggested that the
precursor form of GnRH in the CNS is identical to the placental form,
and were confirmed independently by molecular cloning of hypothalamic
cDNAs for proGnRH as described earlier.

     Immunocytochemical studies at the ultrastructural level showed that
GAP immunoreactivity was localized in large dense-cored vesicles in
terminal and preterminal varicosities of the median eminence (Phillips
et al., 1985). Earlier observations on GnRH processing had shown that
the mature form of the decapeptide (i.e., pGlu and Gly-NH$_2$ termini)
appears along axonal transport and that secretory granules in the median
eminence also contained the decapeptide form of GnRH (King and Anthony,
1983). Similar immunocytochemical staining patterns to those obtained

Fig. 4. Radioimmunoassays with antisera 24A (a), 56A (b) and 39A (c).
         The antisera were used in final dilutions of 12,000, 18,000
         and 10,000, respectively. The Cys (p-hydroxyphenyl-maleimide)
         derivatives of rat GAP-1-11, rat GAP-20-43 and human
         GAP-27-40, respectively, were radioiodinated in the presence
         of chloramine T and purified by gel filtration. Assay
         procedures were identical to those described (Phillips et al.,
         1985). Displacement curves represent the following peptides:
         human GAP (●), rat GAP-1-11 (△), rat GAP-20-43 (0) and rat
         GAP-39-53 (□).

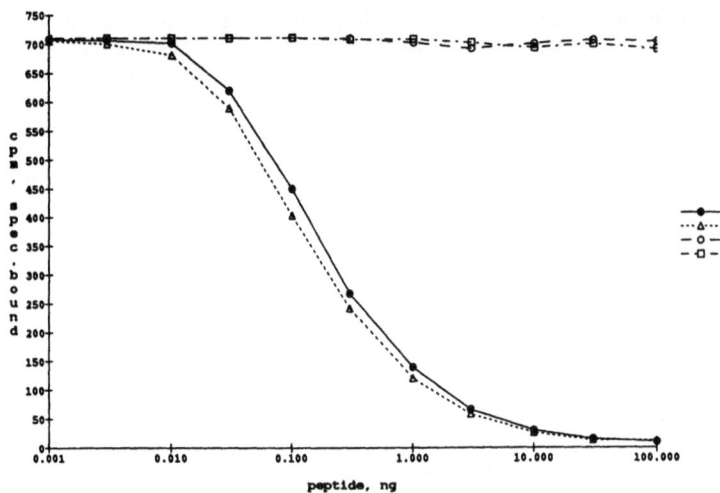

RIA with AS 24A

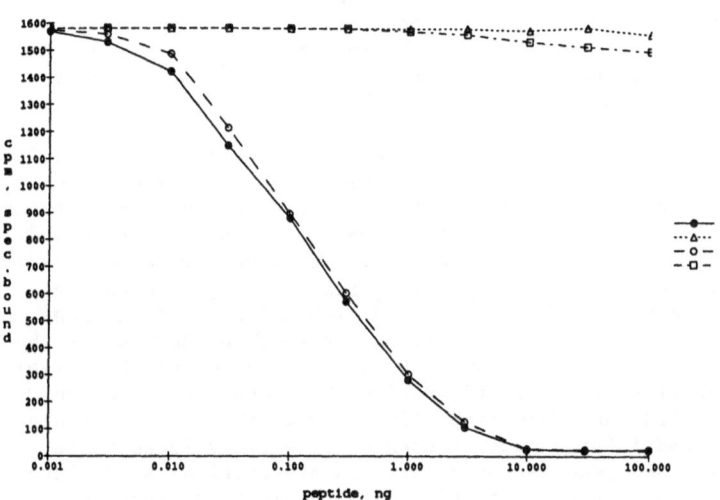

RIA with AS 56A

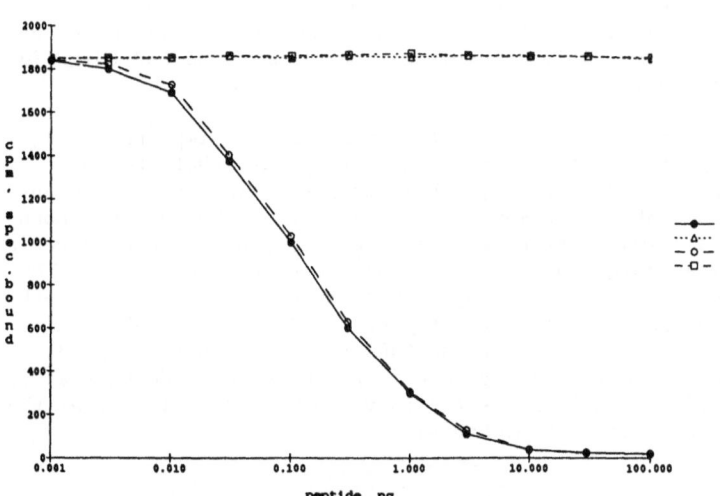

RIA with AS 39A

with antiserum 39A have been obtained with antisera 24A, 56A and KN-1 specific for the N-terminal, middle and C-terminal regions of GAP, respectively, supporting the view that the processed form of rat GAP is the 56 amino acid peptide, although possible cleavage of the C-terminal Lys-Lys-Ile (Met) tripeptide cannot be excluded.

Using radioimmunoassays with antisera 24A and 56A, we were able to detect GAP in sheep hypophyseal portal blood samples, confirming that the peptide is secreted in the median eminence and reaches the anterior pituitary (Clarke et al., in preparation).

GAP IS A HYPOPHYSIOTROPIC HORMONE

The demonstration of mRNA encoding the GnRH-GAP precursor in the hypothalamus, the presence of GAP-related antigen in hypothalamic cell bodies, axons and secretory vesicles within nerve terminals as well as in portal plasma strongly suggested the possibility that GAP might exert an effect on the anterior pituitary.

To investigate this, we prepared the 56-amino acid peptide via a fusion protein by bacterial expression (Nikolics et al., 1985). The fusion protein produced by E. coli was cleaved by cyanogen bromide to generate free GAP. The peptide was obtained in homogeneous form after a series of purification steps and its structure was confirmed by composition and sequence analysis.

In rat anterior pituitary cell cultures GAP was found to stimulate the secretion of gonadotropins and to inhibit the basal secretion of prolactin. More detailed analyses of these effects revealed that GAP showed preference in stimulating FSH secretion vs. LH secretion and that it had very high potency in inhibiting prolactin release (Nikolics et al., 1985). Following these in vitro results we monitored the levels of prolactin in rabbits immunized with GAP-related antigens, i.e. peptides resembling the three primary structural domains of the 56-amino acid molecule. As shown in Fig. 5, these animals had highly elevated levels of serum prolactin during the immunization period. Prolactin concentrations were proportional to antibody titers against GAP. In hypogonadal (hpg) mice (Cattanach et al., 1977) which carry a lesion in their GnRH gene (Mason et al., in preparation) serum levels of prolactin are higher than normal. As a result of a two-week treatment of sc. injections every 3 hours with GAP, serum prolactin levels decreased to normal range. We also observed a significant increase of serum gonadotropin levels which was most dramatic in the group treated with an equimolar combination of GnRH and GAP (Nikolics et al., in preparation).

These findings clearly suggest that GAP plays an important role in the central regulation of reproduction. It appears that the GnRH prohormone synthesized in the hypothalamus gives rise to two hypophysio-tropic hormones controlling mammalian reproductive physiology. Both factors exert a stimulatory effect on pituitary gonadotropin secretion and one inhibits prolactin secretion. Although the secretion of LH, FSH and prolactin is influenced by a great number of factors of both hypothalamic and peripheral origin (Leong et al., 1983; Fink et al., 1983), our findings suggest that the GnRH prohormone and its products play a very important regulatory role in reproduction. Further studies utilizing specific animal models and in vitro methods will reveal the dynamics of this regulatory mechanism.

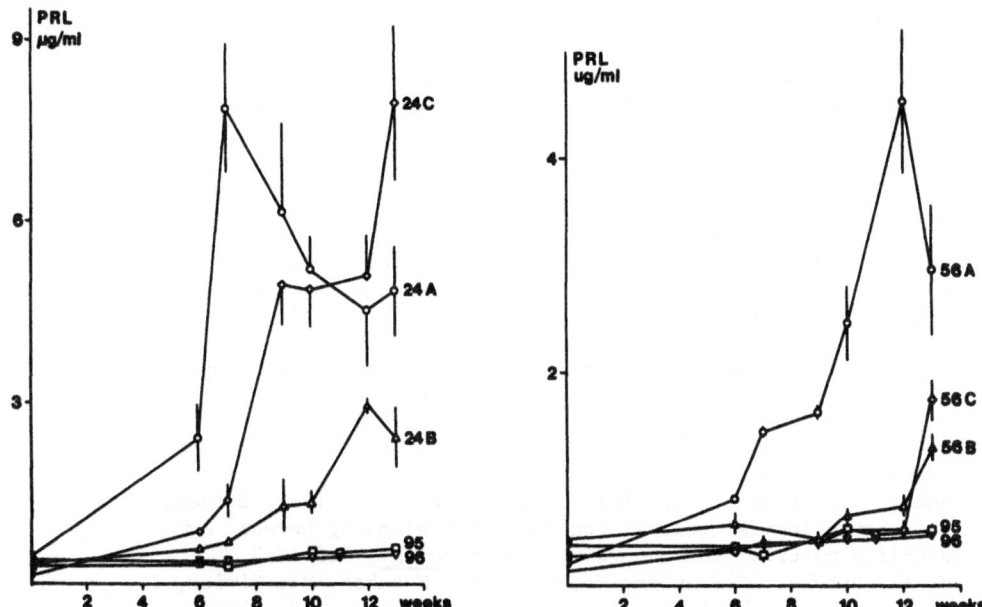

Fig. 5. Effect of immunization on serum prolactin levels in rabbits
with GAP-1-11 (24A,B,C), GAP-20-43 (56A,B,C) and human insulin
receptor-1-12 (95,96). Rabbits were injected intramuscularly
with the soybean trypsin inhibitor conjugates of the peptides
at 0, 3, 5, 7, 10 and 13 weeks. Values are expressed in
microgram prolactin/ml serum ± s.e.m.

CONCLUSION

Like other hypophysiotropic hormones, GnRH is present in extrahypothalamic tissues. Using antibodies against GAP which also recognize
proGnRH, we detected the presence of GAP-related antigen in testis and
ovary, placenta, mammary gland and extrahypothalamic brain areas
(Goldsmith et al., in preparation). The same tissues have been known
to contain GnRH as well. The finding that GnRH and GAP have identical
tissue/cell distribution provides further support for the view that
mammals have a single gene and gene product related to the GnRH
decapeptide.

GnRH appears to have extrahypophysiotropic functions, including
behavioral effects (Moss and McCann, 1973; Pfaff, 1973) and local
gonadal effects (Sharpe et al., 1982; Birnbaumer et al., 1985),
placental effects (Khodr and Siler-Khodr, 1980). GnRH in maternal milk
may have a role in infantile reproductive development (Smith and Ojeda,
1984). The extrapituitary effects of GnRH are not as well understood
as the pituitary function of the hormone and even further possible
roles of the decapeptide are being investigated. The fact that GAP
shows identical tissue and cellular distribution with GnRH suggests
that GAP may have extrapituitary functions also involved in the
regulation of reproductive functions. The availability of GAP and
antisera against the 56-amino acid peptide opens new ways to investigate
such possibilities.

ACKNOWLEDGMENTS

We thank our colleagues Anthony Mason, John Burnier, Eva Szonyi and J. Ramachandran for their collaboration and stimulating discussions. We thank Jeanne Arch for preparation of the manuscript.

REFERENCES

Adelman, J.P., Mason, A.J., Hayflick, J. and Seeburg, P.H., 1985, Isolation of gene and hypothalamic cDNA for common precursor of gonadotropin releasing hormone and prolactin inhibiting factor in human and rat, Proc. Natl. Acad. Sci. USA, in press.

Baram, T., Koch, Y., Hazum, E. and Fridkin, M., 1977, Gonadotropin-releasing hormone in milk, Science, 198:300.

Bhasin, S., Heber, D., Peterson, M. and Swerdloff, 1983, Partial isolation and characterization of testicular GnRH-like factors, Endocrinology, 112:1144.

Birnbaumer, L., Shahabi, N., Rivier, J. and Vale, W., 1985, Evidence for a physiological role of gonadotropin-releasing hormone or GnRH-like material in the ovary, Endocrinology, 116:1367.

Burgus, R., Butcher, M., Amoss, M., Ling, N., Monahan, M., Rivier, J., Fellows, R., Blackwell, R., Vale, W. and Guillemin, R., 1972, Primary structure of the ovine hypothalamic luteinizing hormone-releasing factor (LRF), Proc. Natl. Acad. Sci. USA, 69:278.

Cattanach, B.M., Iddon, C.A., Charlton, H.M., Chiappa, S.A. and Fink, G., 1977, Gonadotropin releasing hormone deficiency in a mutant mouse with hypogonadism, Nature, 269:338.

Charli, J.L., Cohen, S., Diaz de Leon, L., Millar, R.P., Arimura, A. and Joseph-Bravo, P., 1984, Molecular weight of a putative LHRH precursor synthesized in cell free system. 7th Int. Cong. Endocrinol. Abstract No. 411.

Chou, P.Y. and Fasman, G.D., 1974, Prediction of protein conformation, Biochemistry 13:222.

Curtis, A. and Fink, G., 1983, A high molecular weight precursor of LH-RH from rat hypothalamus, Endocrinology, 112:390.

Finer-Moore, J. and Stroud, R., 1984, Amphipathic analysis and possible formation of the ion channel in an acetylcholine receptor, Proc. Natl. Acad. Sci. USA, 81:155.

Fink, G., Stanley, H.F. and Watts, A.G., 1983, Central nervous system control of sex and gonadotropin release: Peptide and nonpeptide transmitter interactions, in: "Brain Peptides", Krieger, D.T., Brownstein, M.J. and Martin, J.B., eds., Wiley, New York.

Garnier, J., Osguthorpe, D.J., and Robson, B., 1978, Analysis of the accuracy and implications of simple methods for predicting the secondary structure of globular proteins, J. Mol. Biol., 120:97.

Gautron, J.P., Pattou, E. and Kordon, C., 1981, Occurrence of higher molecular forms of LHRH in fractionated extracts from rat hypothalamus, cortex and placenta, Mol. Cell. Endocrinol., 24:1.

Hedger, M.P., Robertson, D.M., Browne, C.A. and de Kretser, D.M., 1985, The isolation and measurement of luteinizing hormone-releasing hormone from the rat testis, Mol. Cell Endocrinol., 42:163.

Khodr, G.S. and Siler-Khodr, T.M., 1980, Placental luteinizing hormone-releasing factor and its synthesis, Science, 207:315.

King, J.C. and Anthony, E.L.P., 1983, Biosynthesis of LHRH: Inferences from immunocytochemical studies, Peptides, 4:963.

Krey, L.C. and Silverman, A.J., 1983, Luteinizing hormone-releasing hormone, in: "Brain Peptides", Krieger, D.T., Brownstein, M.J. and Martin, J.B., eds., Wiley, New York.

Leong, D.A., Frawley, L.S. and Neill, J.D., 1983, Neuroendocrine control of prolactin secretion, Ann. Rev. Physiol., 45:109.

Matsuo, H., Baba, Y., Nair, R.M.G., Arimura, A. and Schally, A.V., 1971, Structure of the porcine LH and FSH releasing hormone, I. Proposed amino acid sequence, Biochem. Biophys. Res. Commun., 43:1334.

McCann, S.M., Lumpkin, M.D., Mizunuma, H., Khorram, O., and Samson, W.K., 1984, Recent studies on the role of brain peptides in control of anterior pituitary hormone secretion, Peptides, 5:3.

Millar, R.P., Wegener, I. and Schally, A.V., 1981, Putative prohormonal luteinizing hormone-releasing hormone, in: Neuropeptides: Biochemical and Physiological Studies, Churchill Livingstone, Edinburgh.

Moss, R.L. and McCann, S.M., 1973, Induction of mating behavior in rats by luteinizing hormone-releasing factor, Science, 181:177.

Nikolics, K., Mason, A.J., Szonyi, E., Ramachandran, J. and Seeburg, P.H., 1985, A prolactin-inhibiting factor within the precursor for human gonadotropin-releasing hormone, Nature, 316:517.

Pfaff, D.W., 1973, Luteinizing hormone-releasing factor potentiates lordosis behavior in hypophysectomized ovariectomized female rats, Science, 182:1148.

Phillips, H.S., Nikolics, K., Branton, D. and Seeburg, P.H., 1985, Immunocytochemical localization in rat brain of a prolactin release-inhibiting sequence of gonadotropin-releasing hormone, Nature, 316:542.

Piekut, D.T. and Knigge, K.M., 1981, Immunocytochemical analysis of the rat pineal gland using antisera generated against luteinizing hormone-releasing hormone, J. Histochem. Cytochem., 29:616.

Pohl, C.R. and Knobil, E., 1982, The role of the central nervous system in the control of ovarian function in higher primates, Ann. Rev. Physiol., 411:583.

Schally, A.V., Arimura, A., Kastin, A.J., Matsuo, H., Baba, Y., Redding, T.W., Nair, R.M.G. and Debeljuk, L., 1971, Gonadotropin-releasing hormone: One polypeptide regulates secretion of luteinizing and follicle-stimulating hormones, Science, 173:1036.

Seeburg, P.H. and Adelman, J.P., 1984, Characterization of cDNA for precursor of human luteinizing hormone releasing hormone, Nature, 311:666.

Sharpe, R.M., Fraser, H.M., Cooper, I. and Rommerts, F.F.G., 1982, The secretion, measurement and function of a testicular LHRH-like factor, Ann. N.Y. Acad. Sci., 383:272.

Shivers, B.D., Harland, R.E. and Pfaff, D.W., 1983, Reproduction: The central nervous system role of luteinizing hormone releasing hormone, in: "Brain Peptides", Krieger, D.T., Brownstein, M.J. and Martin, J.B., eds., Wiley, New York.

Smith, S.S. and Ojeda, S.R., 1984, Material modulation of infantile ovarian development and available ovarian luteinizing hormone-releasing hormone receptors via milk LHRH, Endocrinology, 115:1973.

Yang-Feng, T.L., Seeburg, P.H. and Francke, U., 1985, The human luteinizing hormone releasing hormone gene is located on the short arm of chromosome 8. Somat. Cell Genet., in press.

Ying, S.Y., Ling, N., Bohlen, P. and Guillemin, R., 1982, Gonadocrinins: Peptides in ovarian follicular fluid stimulating the secretion of pituitary gonadotropins, Endocrinology, 108:1206.

GONADOTROPIN-RELEASING HORMONE: DIFFERENTIATION OF STRUCTURE AND

FUNCTION DURING EVOLUTION

Nancy M. Sherwood

Biology Department
University of Victoria
V:ctoria, B.C., Canada

INTRODUCTION

Gonadotropin-releasing hormone (GnRH) has been important in the short
history of peptide neurobiology. The sequence of GnRH isolated from
mammalian brains was established in 1971 and 1972 (Matsuo et al., 1971;
Burgus et al., 1972). An identical peptide was shown to exist in human
placenta (Tan and Rousseau, 1982). However, GnRH was not the first peptide
to be sequenced. For example, the structures of oxytocin, vasopressin and
thyroid-releasing hormone in vertebrates and eledoisin in octopus were known.
But the central role of GnRH in reproduction led to extensive studies of its
function, conformation and location. In the 14 years since the structure of
GnRH was elucidated, over 2,000 analogues have been synthesized and tested
(Struthers et al., 1985). Also it is now clear that GnRH is a member of a
family of homologous peptides, has multiple functions, is located both
within and out of the central nervous system and has well-studied neurotransmitter actions.

GnRH offers several advantages for the study of neuropeptides. For
biochemical characterization, GnRH is located in the vertebrate brain where
relatively large amounts of material can be collected compared with inver-
tebrate neural tissue. For physiological studies, GnRH can be tested in
reasonably simple systems; the frog sympathetic ganglion, the isolated
fish retina and the cultured or fragmented anterior pituitary offer such
systems. These vertebrate preparations have similar advantages to the less
complex invertebrates. For anatomical studies, many antisera have been made
against native mammalian GnRH and its analogues. The antisera make it possible to identify GnRH in specific brain areas and other tissues by immunocytochemistry and radioimmunoassay (RIA). One of the antisera, which recognizes the 3-dimensional structure of GnRH, was used in place of bioassay
for purification of salmon GnRH (Sherwood et al., 1983). For pharmacological and conformational studies, many GnRH analogues are available. The
role of the ten amino acids in mammalian GnRH has been studied to determine
their importance in conformation, receptor binding and gonadotropin (GtH)
release.

Only a modest beginning, however, has been made concerning the role of:
(1) phylogenetically older GnRH molecules and their amino acid substitutions,
(2) multiple forms of GnRH in one species, (3) differentiation of GnRH
function in vertebrates, (4) changes in GnRH receptors during evolution and

(5) possible structural links of the molecule in species evolving before vertebrates. These topics are the subject of this review. Other reviews have considered the classical role of mammalian GnRH in releasing LH and FSH from the mammalian pituitary (Schally et al., 1973; Schally, 1978; Fink et al., 1982).

CHANGES IN GnRH PRIMARY STRUCTURE DURING EVOLUTION

The internal segment of the GnRH molecule has been the site of amino acid substitutions during evolution. This is illustrated by the four GnRH structures that have been determined from material isolated from mammals, birds and fish (Figure 1); each is a decapeptide with the same termini. Mammalian GnRH was originally isolated from porcine (Matsuo et al., 1971) and ovine (Burgus et al., 1972) hypothalami. A peptide with the same primary structure was purified from human placenta (Tan and Rousseau, 1982). To date only a single form of GnRH has been isolated from mammals.

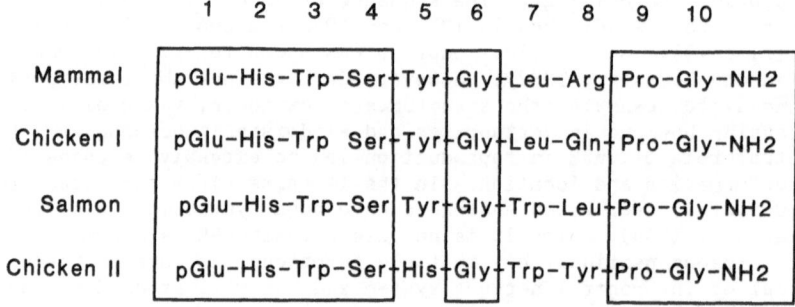

Fig. 1. Comparison of amino acid sequences for 4 identified gonadotropin-releasing hormone (GnRH) family members (top). Boxes enclose regions of identical sequence for the decapeptides. Shown below is the minimum number of nucleotide base changes required for the amino acid substitutions among mammalian (M), chicken I (C-I), chicken II (C-II) and salmon (S) GnRH.

One amino substitution in position 8 separates mammalian from chicken I GnRH. The latter was isolated from chicken hypothalami (King and Millar, 1982a,b; Miyamoto et al., 1982, 1983). It is necessary to postulate only one nucleotide base change to account for the Arg to Gln[8] codon change. In chicken hypothalami, a second form of GnRH was also found (Miyamoto et al., 1984). This peptide contains 3 amino acid substitutions compared with chicken I or mammalian GnRH (Figure 1). The minimum base changes for the conversions between chicken II and other forms of GnRH are 4 with chicken I, 4 with mammalian and 3 with salmon GnRH. Salmon GnRH, isolated from whole brains, requires minimum base changes of 2, 2 and 3 for the interchanges to chicken I, mammalian and chicken II GnRH, respectively. Hence chicken II GnRH has the greatest separation compared with the other three GnRH peptides. This opens the possibility that chicken II is more ancient than either chicken I or mammalian GnRH. The common Trp^7 in salmon and chicken II GnRH suggests this change preceded the change to Leu^7 in chicken I and mammalian GnRH. However, chicken II alone contains His^5 whereas the others have Tyr^5. There is no indication as to whether the His^5 substitution preceded or followed Tyr^5.

The most variable position is 8. Each molecule has a different amino acid: mammals have Arg^8, chicken I has Gln^8, chicken II has Tyr^8 and salmon has Leu^8. In the oxytocin-vasopressin family of peptides, the 8 position is also highly variable. Neutral substitutions (Leu^8, Ile^8, Val^8 or Gln^8) occur in the oxytocin-like peptides, whereas basic substitutions (Arg^8 or Lys^8) occur in vasoactive-antidiuretic peptides such as vasopressin. This functional differentiation based on the hydrophobicity of the eighth amino acid does not appear to apply to the GnRH family. Rather, the speciation of GnRH is the evolution of four homologous peptides which possess at least one common function, release of GtH, in vertebrates. Whether the nature of the eighth amino acid in GnRH is the basis for the differentiation of other GnRH functions is not clear; it is important in receptor binding.

During evolution the length and termini of the GnRH family members have been conserved. The primary structure of lamprey GnRH shows this principle extends even to one of the most primitive vertebrates (Sherwood et al., submitted). Likewise the functional center of His^2-Trp^3 or possibly $pGlu^1$-His^2-Trp^3 has remained stable. The proof that these amino acids are essential for release of gonadotropins is that analogues with certain substitutions in positions 1-3 become competitive antagonists which bind the GnRH receptors, but do not release gonadotropins in mammals. The changes in GnRH during evolution to date are in positions 5, 7 and 8, important for receptor binding. The evidence is that salmon and chicken I GnRH have less than 5% binding to mammalian receptors compared to mammalian GnRH (Sherwood et al., 1983; Hasegawa et al., 1984).

PHYLOGENY OF PARTIALLY CHARACTERIZED FORMS OF GnRH

It is now clear that more forms of GnRH exist than shown in Figure 1. The primary structure is not determined, but amino acid composition shows the same amino acids are present in rat and frog GnRH compared with mammalian GnRH (Figures 2 and 3; Böhlen et al., 1981; Rivier et al., 1981). The amino acid composition of a second form of GnRH in lamprey, distinct from the other characterized forms of GnRH, is also known (Sherwood et al., submitted).

High performance liquid chromatography and immunological cross-reactivity have been used to partially identify other forms of GnRH (Figure 2). In birds the ostrich has two forms similar to chicken I and II GnRH (Millar, 1985). In reptiles two forms of GnRH are also present in the slider turtle; to date the two forms have not been separated by chromatographic or immunological means from chicken I and II (Sherwood and Whittier, unpublished). In the same study, snake brains contained only one form of GnRH which was similar to chicken II. A similar story exists for alligator in which Millar found chicken I-and II-like GnRH molecules (personal communication; Millar and King, 1984). In evolutionary terms, the reptiles and birds separated recently; this may explain the similarity to date of the GnRH forms. Lizard is anomalous in that salmon GnRH is reported to be present in the brain (King and Millar, 1984). However, structural analysis of the isolated forms of GnRH from reptiles is needed to be certain of their identity.

Amphibians, both newts and frogs, contain mammalian GnRH in their brains as the dominant form. Very small amounts of two other forms of GnRH are chromatographically and immunologically similar to the two forms of GnRH found in salmon and several other teleosts. In frogs a small amount of salmon-like GnRH is present in the brains of tadpoles and adults (Branton et al, 1982), but mammalian-like GnRH appears in the brain and steadily increases during metamorphosis to reach high levels in adults (Branton et al., 1982; King and Millar, 1981; Whalen and Crim, 1985). In newts, however, mammalian- and salmon-like GnRH are both detected in the larval and neotenic forms of Ambystoma gracile and in the adults of Taricha granulosa. The difference in GnRH in the premetamorphic state of newts and frogs may reflect the less specialized nature of newts. It can be speculated that the mammalian- and salmon-like forms of GnRH were also present in the common phylogenetic ancestor, the labyrinthodont amphibian (Sherwood et al., in press). The mammalian form has not been detected in classes below amphibians.

In teleosts the two recognizable forms of GnRH are similar to the two forms in salmon by chromatographic and immunological methods. In fish the primary structure of salmon I GnRH is known (Sherwood et al., 1983), but another less hydrophobic form, salmon II GnRH, can be separated on HPLC. Figure 2 shows the HPLC elution of this second form for trout, a member of the salmonid family. A similar pattern of two peaks which elute in the same position as immunoreactive (ir) GnRH from salmon whole brain extracts has been detected for several other teleosts: goldfish, herring, milkfish, mullet, siganids and sea bass (Sherwood et al., 1984; Sherwood and Harvey, in press; Sherwood, in press-a). Hake (King and Millar, 1984) and cod (Wu and Jackson, 1985) are also reported to contain salmon I GnRH. Again the primary sequence is necessary to determine the relationship and function of salmon I and II.

In a primitive bony fish, sturgeon, a unique form of GnRH exists based on both its HPLC elution pattern and immunological profile. It is likely to be a GnRH family member as its cross-reactivity with antiserum R-42 shows it is a decapeptide and probably has similar termini to other GnRH molecules (Sherwood, Carolsfeld and Doroshov, unpublished).

In cartilaginous fish, separation of ratfish and dogfish during evolution occurred approximately 400 million years ago. Brain extracts of these fish contain a form of GnRH which elutes in a similar position to salmon II (Figure 2; Sherwood and Carolsfeld, unpublished). Additionally, dogfish has a small amount of ir-GnRH which elutes in a distinct position compared with the 4 synthetic forms of GnRH.

The most primitive vertebrates alive today are lamprey and hagfish. Lamprey has a distinct GnRH molecule, but nonetheless a recognizable GnRH family member (Sherwood and Sower, 1985; Sherwood et al., submitted). A second form of lamprey GnRH in extracts of whole brains represents approximately 10% of the total ir-GnRH. Hagfish, the only other living member of the class Agnatha, does not have a clearly recognizable form of ir-GnRH. In hagfish, GnRH was not detected in several studies by a large number of antisera including R-42 which detects GnRH in all other vertebrates tested

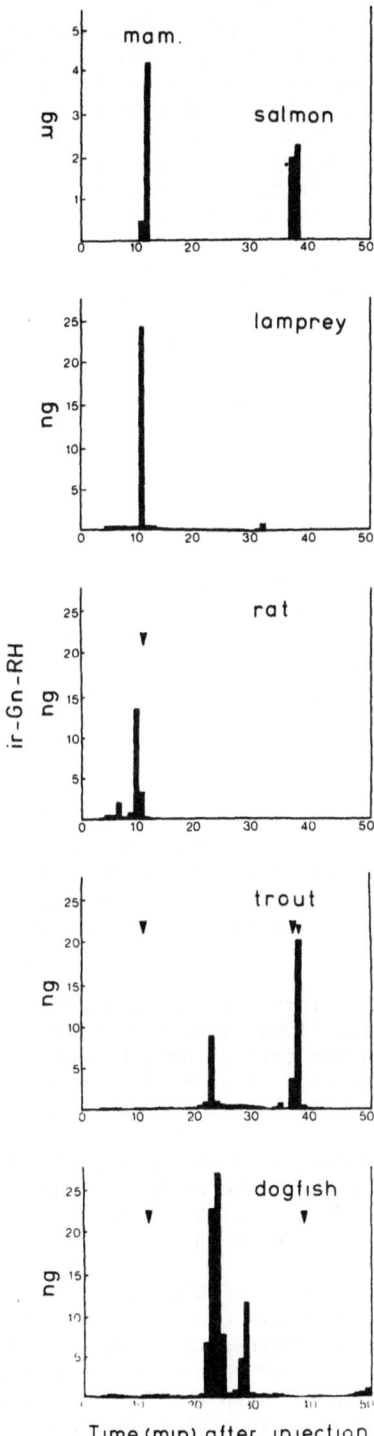

Fig. 2. Reverse phase HPLC of synthetic mammalian (mam.) and synthetic
salmon GnRH (top). The elution pattern of lamprey, rat, trout
and dogfish brain extracts containing immunoreactive gonadotropin
is shown below. 'The arrows in the lower figures mark the elution
of the standards run shortly after the brain extracts. (From
Sherwood and Sower, 1985, with permission of Neuropeptides. The
dogfish data is from Sherwood and Carolsfeld, unpublished.

71

(Crim et al., 1979a; Nozaki and Kobayashi, 1979; Sherwood and Sower, 1985); two other studies did report the presence of a small amount of ir-GnRH in hagfish brains (Jackson, 1980; King and Millar, 1980). Hagfish may never have evolved GnRH, may have lost GnRH by brain degeneration (Gorbman, 1980) or may have a form fundamentally different from other vertebrate GnRH molecules. Even lamprey, in which GnRH is clearly established, remains a mystery in terms of the pathway between GnRH axon terminals and the pituitary. The agnathans have neither a hypothalamohypophysial portal system nor axon terminals which end in the pituitary as in most teleosts. Lamprey GnRH fibers terminate on a layer of connective tissue above the pars distalis or in the neurohypophysis (Crim et al., 1979a,b; Nozaki and Kobayashi, 1979; Nozaki et al., 1984). Lamprey GnRH may reach the anterior pituitary by diffusion or by an unknown pathway (Gorbman, 1980).

A protochordate, the tunicate, contains a form of GnRH which has been detected by immunological means only (Georges and Dubois, 1980); GnRH has not been detected in invertebrates.

MULTIPLE FORMS OF GnRH IN ONE SPECIES

More than one member of the GnRH family has been reported to be present in a single species within the classes Aves, Reptilia, Amphibia, Osteichthyes, Chondrichthyes and Agnatha (Figures 2 and 3). Multiple forms of GnRH have not yet been demonstrated in ratfish, snake or mammals. However, only a few species, often at one reproductive stage, have been screened for different forms of GnRH. Also it is not known if the multiple forms exist within one individual or sex because brain extracts are usually prepared from pooled tissue.

One theory for the origin of multiple forms of GnRH in a species is based on gene duplication. One copy of the gene may produce a product which serves the original function; the redundant copy may change and then produce a peptide product with an altered structure and function. There is less likelihood that the organism will be harmed if the original peptide is still produced.

The existence of multiple forms has not yet been confirmed by genetic analysis. Seeburg and Adelman (1984) isolated cloned genomic and cDNA sequences encoding the precursor form of GnRH by using human placenta. The DNA sequence coded in order: a 23 amino acid signal peptide, the 10 amino acid GnRH peptide, a Gly-Lys-Arg sequence for enzymatic cleavage and amidation of GnRH, a 53 amino acid associated peptide and at the end a Lys-Lys-Ile for the final cleavage. Only one copy of the GnRH gene was found as predicted for mammals, but the authors do not exclude the idea that other copies may yet be found. Similar studies are needed for a species containing multiple forms of GnRH.

The functions of multiple forms of GnRH in one species will only become clear when the synthetic forms are available for bioassays. Chicken I and II have both been synthesized. Chicken II is considerably more potent in releasing both LH and FSH in cultured chicken or rat pituitary cells compared with chicken I GnRH (Millar and King, 1984 ; Miyamoto et al., 1984). Whether differentiation of function has occurred for chicken I and II remains to be tested.

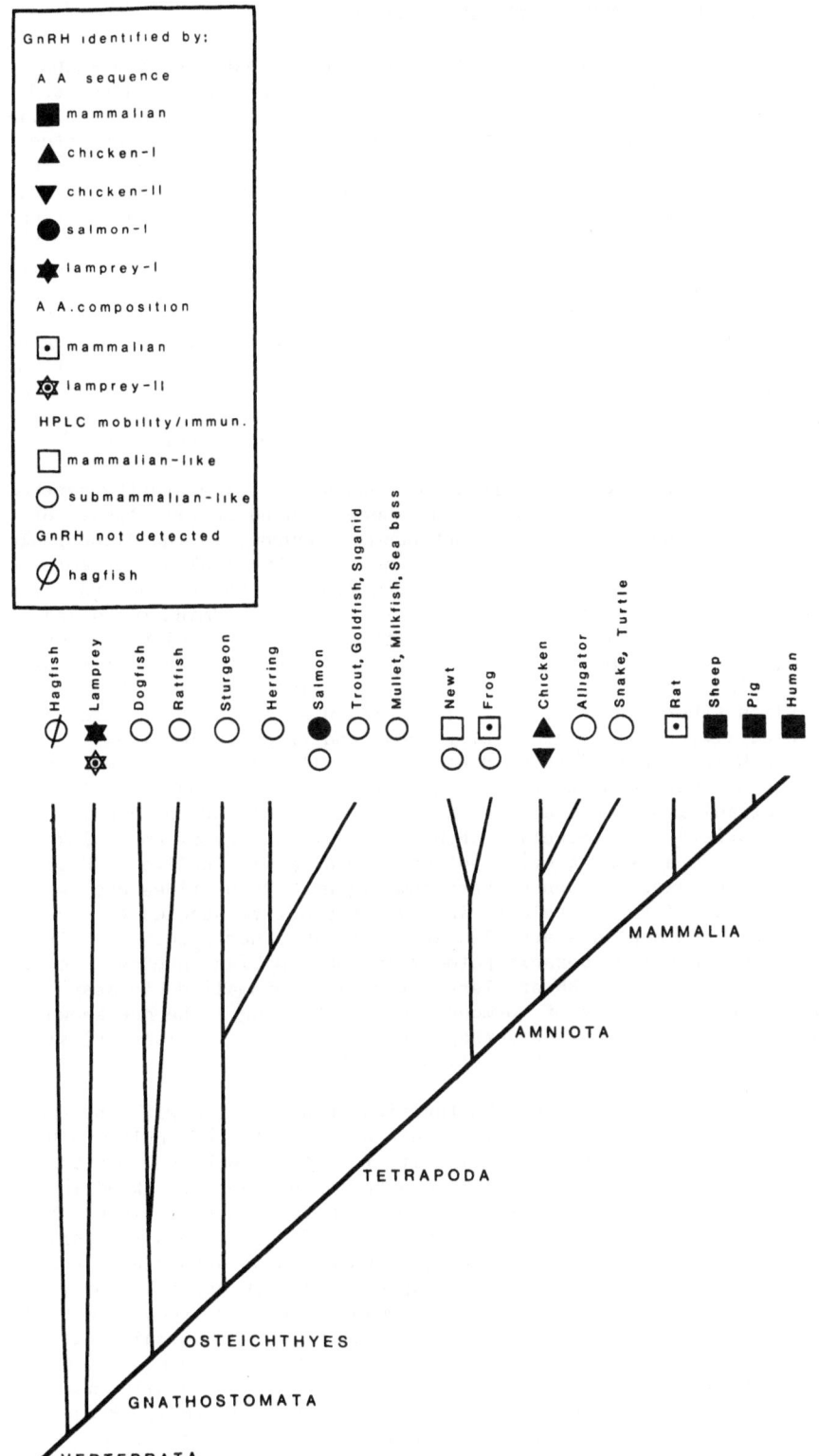

Fig. 3. Changes in the structure of GnRH during evolution. (Adapted from Sherwood, in press-b, and reproduced with permission of the Editor of Hormones and Reproduction in Fishes, Amphibians and Reptiles.)

Gonadotropin-releasing hormone was named on the basis of its known function. In a review article in 1973, the only GnRH function discussed was the stimulation of synthesis and release of LH and FSH from the pituitary (Schally et al., 1973). It was known that mammalian GnRH was effective in both mammals and birds. In 1978 in a subsequent review, Schally noted that mammalian GnRH also released GtH in amphibians and fish, was present in extrahypothalamic areas suggesting a neuromodulator role, and modified sexual behavior in rats as shown by Moss and McCann (1973). Also, in 1978 the antifertility effects of pharmacological doses of GnRH had been observed. In 1985 it is clear that GnRH, like most neuropeptides, occurs in parts of the body in addition to the brain to act as a possible neurotransmitter, neuromodulator or local hormone. Iversen (1985) notes that "the biological significance of using the same chemical for apparently quite different purposes remains obscure". However, most of the diverse functions of GnRH are directly or indirectly involved with reproduction raising the possibility that some coordination among the various GnRH functions might exist.

The location of the GnRH-staining neurons which stimulate the synthesis and release of GtH from the pituitary has been studied in all classes of vertebrates. The GnRH cells are a continuum of neurons in the diencephalon and telencephalon with axon terminals in the basal hypothalamus or anterior pituitary depending on the species. For functional studies, the synthetic form of each of the identified GnRH molecules has been administered and shown to be effective for GtH release in its respective species: mammalian GnRH in a variety of mammals (see Schally, 1978), chicken I and II in birds (Millar and King, 1983; Hasegawa et al., 1984; Hattori et al., 1985), salmon GnRH in salmon (Sherwood et al., 1983; Donaldson et al., 1984), goldfish (MacKenzie et al., 1984; Peter et al., 1985) and trout (Crim, 1984) and lamprey I GnRH in lamprey (Sherwood et al., submitted). In lamprey neither a homologous nor heterologous RIA exists for lamprey GtH and hence an increase in estradiol was used as an indirect measure of GnRH potency. In addition, a variety of superactive agonists exist for mammalian, chicken and salmon GnRH; many show greater potency compared with native GnRH in their respective species. Hence these homologous GnRH peptides are isofunctional in regard to GtH release in their respective species (Figure 4). To some extent GnRH peptides are also effective in other species. Synthetic mammalian GnRH shows considerable potency in submammalian species. However, chicken I, II and salmon GnRH are less potent in a mammalian bioassay suggesting some specificity of hormone-receptor binding. The one known exception of a GnRH role in gonadotropin release is the hagfish. Little or no GnRH has been detected in the hagfish brain.

A more recently recognized GnRH function is associated with the presence of the peptide in the terminal nerve near the olfactory bulbs. The importance of the system is the possibility that these GnRH neurons translate olfactory signals into reproductive behavior or events. Chondrichthyans to mammals are known to have a terminal nerve (Larsell, 1950), but the presence of GnRH in discrete cell bodies and fibers within the terminal nerve is a recent observation. The projection of these GnRH fibers rostrally into the olfactory bulbs varies with species (Witkin and Silverman, 1983). Terminal nerve axons project caudally to medial forebrain areas (Demski and Northcutt, 1983) and in some species to the retina. GnRH staining within the terminal nerve has been reported for mammals (macaque, new world monkey, prosimian, rabbit, guinea pig, rat and hamster; see Krey and Silverman, 1983), in avians (chicken and pheasant; see Krey and Silverman, 1983), in amphibians (anurans; Stell, personal communication), in a variety of teleosts (Münz et al., 1981, 1982; Stell et al., 1984) and in chondrichthyans (dogfish; Stell, 1984). GnRH was not detected in lamprey in the telencephalon near the olfactory region (Nozaki et al, 1984). A functional relationship between olfaction and GnRH was shown for voles (Dluzen et al., 1981). In the female vole, estrus and ovulation are initiated by brief contact with the male or male urine. A significant increase in GnRH occurs in the posterior olfactory bulb and in LH in the serum after one drop of male urine is put on the upper lip of the female. In goldfish stimulation of the terminal nerve axons in males results in milt release (Demski and Northcutt, 1983); transection of the area prevents the reproductive behavior normally triggered in males by ripe females (Stacey and Kyle, 1983). It is largely speculative, but plausible that the terminal nerve system detects pheromones, then modulates sexual behavior by GnRH release at synapses in the forebrain.

A novel function for GnRH is that of neurotransmitter or perhaps neuromodulator in the retina. This function is closely tied to the terminal nerve; some of the axon terminals from the terminal nerve project caudally from GnRH cell bodies in the paraolfactory region, via the optic nerve, to the retina. The retinal projections from the terminal nerve have been well documented in teleosts by both retrograde labelling and immunocytochemistry with GnRH antisera (Münz et al., 1982; Stell et al., 1984), but are "minor or absent in dogfish and have not been observed yet in mammals" (Stell, personal communication). Likewise, GnRH-like material was identified by RIA in the retinae of carp, goldfish, trout and bullfrogs, but not turtles, chickens, rats, guinea pigs or monkeys (Figure 4; Eiden et al., 1982). In amphibians two forms of GnRH, eluting with mammalian or carp GnRH, are present in the retina of bullfrog (Eiden et al., 1982). In goldfish the spiking activity in retinal ganglion cells was modified by application of salmon GnRH, but rarely by mammalian GnRH or its analogs. Stell has suggested "Because its neurons may be primary olfactory receptors, the nervus terminalis may therefore constitute a direct pathway by which specific olfactory stimuli can alter the visual responsiveness of specific retinal outflow neurons" (Stell et al., 1984).

There is considerable evidence that GnRH acts as a neurotransmitter in the lumbar sympathetic ganglia of the bullfrog. Jan, Jan and Kuffler (1979) detected a GnRH-like substance in these ganglia by RIA; stimulation of the preganglionic sympathetic fibers released ir-GnRH into the medium. A late, slow excitatory postsynaptic potential (epsp) developed in the postganglionic neurons after stimulation of preganglionic fibers or application of synthetic GnRH, but not after application of a GnRH antagonist. C-type postganglionic cells appear to be stimulated across a synapse by the GnRH-like peptide whereas B-type postganglionic cells are affected by diffusion of GnRH from some distance (Jan and Jan, 1982). In this case GnRH is not a conventional neurotransmitter in the ganglion in that it diffuses a distance much greater than the synapse to affect the B-type cells, acts slowly and remains active for a long time. Eiden and Eskay (1980) showed the form of ganglionic GnRH was immunologically and chromatographically distinct from mammalian GnRH. Sympathetic ganglionic GnRH is more hydrophobic, eluting late from a C-18 column as does salmon I GnRH (Eiden et al., 1982). A similar form of GnRH was detected in the adrenal gland, retina, tadpole brain, and adult brain of amphibians. Other supporting evidence that sympathetic-adrenal GnRH is similar to salmon GnRH is that synthetic salmon GnRH and its analogs are more potent than mammalian forms in duplicating the neurotransmitter actions of ganglionic GnRH (Jan et al., 1983; Jones et al., 1984). However, salmon is only equipotent to mammalian GnRH in producing a maximal increase in pulse and mean arterial pressure (Wilson, 1985). The physiological significance of GnRH in the sympathetic ganglia in frogs may be that GnRH mobilizes catecholamines mainly from the adrenal and raises blood pressure (Wilson et al., 1984). The authors suggest that endogenous GnRH-like peptides may "coordinate the pituitary, nervous and cardiovascular mechanisms which prepare toads for seasonal reproductive activity". That GnRH is present in the sympathetic

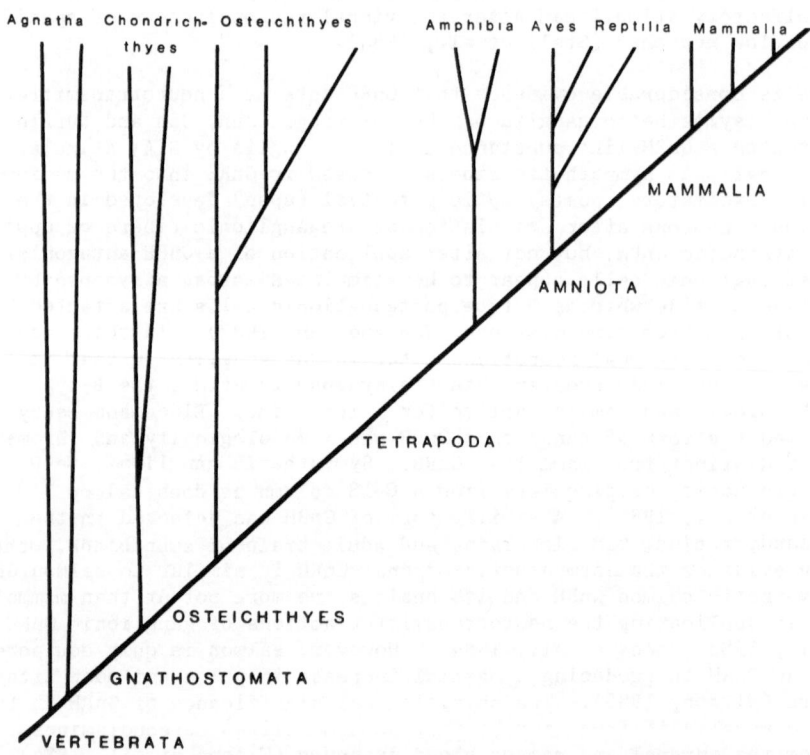

Releases and enhances synthesis of gonadotropins in pituitary

May translate olfactory signals into reproductive events (Terminal nerve)

May act as neurotransmitter in retina

Acts as neurotransmitter in sympathetic ganglia

Releases catecholamines, increases blood pressure

Enhances sexual receptivity

May act as local hormone in gonads

May act as local hormone in placenta

May activate pituitary in neonates via milk

Agnatha Chondrich- Osteichthyes   Amphibia Aves Reptilia Mammalia
          thyes

MAMMALIA

AMNIOTA

TETRAPODA

OSTEICHTHYES

GNATHOSTOMATA

VERTEBRATA

Fig. 4. Differentiation of GnRH functions during evolution. Horizontal
lines represent classes of animals in which the function has
been established or proposed for representatives within the class.

ganglia only in amphibians (Figure 4) is supported by the fact that ir-GnRH could not be detected in adrenals of rat, guinea pig or monkey (Eiden et al., 1982).

The enhancement of sexual receptivity by GnRH was first reported for rats (Moss and McCann, 1973), latter for doves, lizards and frogs (Figure 4; see Shivers et al., 1983). Administration of GnRH to several brain areas, especially central gray midbrain, facilitates the lordotic response. Also, in the midbrain ir-GnRH fibers are present, iontophoretic application of GnRH inhibits single neurons especially in steroid-treated females and infusion of GnRH antisera inhibits lordotic behavior (see Chan et al., 1985).

The presence of GnRH in extrahypothalamic areas is more widespread than the terminal nerve and central gray midbrain. The location of GnRH varies with species, but is found in cortex, amygdala, midbrain, habenula, subfornical organ and posterior pituitary (Krey and Silverman, 1983; King and Anthony, 1984; King et al., 1984). The GnRH extrahypothalamic fiber pathways have been reviewed in detail (Eiden and Brownstein, 1981). The extrahypothalamic GnRH fibers are speculated to (1) be involved in sexual behavior possibly by integration of limbic and motor control, (2) modulate GnRH release by activating reciprocal innervation with GnRH neurons and (3) release GnRH into cerebral spinal fluid (King and Anthony, 1984). GnRH neurons do not concentrate estradiol (Shivers et al., 1983), but the close proximity of GnRH and steroid concentrating neurons in some brain areas may lead to interaction of the two cell types in control of reproductive behavior. The scarcity of GnRH receptors in the brain further clouds the role of extrahypothalamic GnRH.

Novel functions of GnRH are striking in mammals in that GnRH-like peptides have been reported in the placenta and milk (Figure 4). This is an example of a peptide conserved in evolution, but expressed and apparently functional in two newly evolved tissues in mammals. It is also remarkable that these tissues and functions are part of the reproductive system; many peptides act in seemingly unrelated functions. The presence of GnRH receptors in ovary and testes and of GnRH-like material in the same organs suggests the possibility of another nonclassical, but reproduction-related function. The GnRH-gonadal connection may not be exclusively in mammals, but has been best studied in this class.

The evidence that GnRH is a local hormone in the placenta rests on the fact that (1) the level of GnRH varies with the stage of pregnancy, (2) low affinity placental receptors for GnRH exist (Belisle et al., 1984) and (3) synthetic GnRH incubated with placental tissues releases human chorionic gonadotropin and steroids (Khodr and Siler-Khodr, 1978; Belisle et al., 1984).

GnRH detected in milk of rat, cow and man (Baram et al., 1977) was shown indirectly to have the same structure as hypothalamic and placental GnRH. Smith-White and Ojeda (1984) found that one function of milk GnRH may be to inhibit ovarian development in the suckling young; the effects in males were not studied, but may be more complicated as testosterone secretion in young male rats is important for differentiation of the "male" brain. The origin of milk GnRH is not clear as normal breast tissue does not contain GnRH by RIA. However, mammillary gland tumors (ductal carcinoma) do contain ir-GnRH (Seppälä and Wahlström, 1980). GnRH and its agonists bind to low affinity receptors in certain breast tumor cells, then repress growth of these cells in culture (Miller et al., 1985).

GnRH-like factors may be present in the gonads, but have not been fully characterized to date. Indirect evidence suggests the gonads may locally produce a GnRH-like substance: (1) GnRH-like factor can be measured by receptor- and immuno-assay (Sharpe and Fraser, 1980; Sharpe et al.,1981),

(2) both ovary and testes contain GnRH receptors (Clayton et al., 1979, 1980), (3) circulating levels of GnRH are very low or not detectable and (4) pharmacological doses of GnRH and its agonists act directly on the gonads to inhibit reproduction (Hsueh and Erickson, 1979a,b). From ovary a GnRH-like gonadocrinin was isolated (Ying et al., 1981), but could not be repeated (Guillemin, 1982). Studies with testicular extracts suggest the GnRH-like factors may be extended at the N-terminus and have molecular weights of 6,000 and 68,000. One of these factors may be a glycoprotein (Swerdloff et al., 1984). The factor may also be more lipophilic compared with mammalian GnRH (Arimura and Turkelson, 1984). The testicular extracts inhibited LH-stimulated testosterone production (Swerdloff et al., 1984) suggesting that the GnRH-like factor functions as a modulator of Leydig cell response to gonadotropins.

Functional diversification of a peptide or homologous peptides such as the GnRH family may occur in two ways. First, a peptide with identical structure may perform several different functions. An example is mammalian GnRH which is active in GtH release, sexual behavior and possibly in mediation of olfactory stimuli and in local control of the placenta. Alternatively, diversification may occur by the presence of multiple homologous peptides within an individual. To date the presence of multiple forms of GnRH has been demonstrated, but differentiation of function has not been shown, only differences in potency for chicken I and II GnRH.

PHYLOGENY OF GnRH RECEPTORS

The 4 identified forms of GnRH show considerable differences in their ability to bind to pituitary receptors in different species. Generally mammalian GnRH has higher binding and potency in submammalian species than the reverse. This synthetic mammalian GnRH shows potency in releasing gonadotropins in vitro and in vivo in birds (see Hattori et al., 1985) and in fish (see Sherwood, in press-b). However, chicken I and salmon GnRH have less than 5% potency and binding in mammalian cells compared with mammalian GnRH. Chicken II has 32% and 41% potency in releasing LH and FSH compared with mammalian GnRH (Miyamoto et al., 1984). These data suggest that the receptor in addition to the peptide has evolved. The ability of GnRH to use different target tissues such as the pituitary, placenta, ovary, testes and breast tumors shows many cells have the potential to express receptors. The receptors are not necessarily identical as shown by high and low affinity receptors. It is possible that a family of related receptors exists for GnRH.

STRUCTURAL LINKS OF GnRH FAMILY TO OLDER MOLECULES

GnRH-like material has been reported for protochordates (Georges and Dubois, 1980), but not for invertebrates. However, Hunt and Dayhoff (1979) noted that mammalian GnRH has 6 amino acids which may be homologous to those in yeast α_1 mating factor (Trp-His-Trp-Leu-Gln-Leu-Lys-Pro-Gly-Gln-Pro-Met-Tyr). Yeast mating factor may be both structurally and functionally homologous to GnRH as Loumaye and associates (1982) found that mating factor binds specifically to rat pituitary cells and releases LH. This opens the possibility that connecting links with phylogenetically older molecules may be found.

REFERENCES

Arimura, A., and Turkelson, C. M., 1984, LHRH-like substance in the rat testis, Ann. N. Y. Acad. Sci., 438: 390.

Baram, T., Koch, Y., Hazum, E., and Fridkin, M., 1977, Gonadotropin-releasing hormone in milk, Science, 198: 300.

Belisle, S., Guevin, J.-F., Bellabarba, D., and Lehoux, J.-G., 1984, Luteinizing hormone-releasing hormone binds to enriched human placental membranes and stimulates in vitro the synthesis of bioactive human chorionic gonadotropin, J. Clin. Endocrinol. Metab., 59: 119.

Böhlen, P., Castillo, F., Yin, S. Y., Brazeau, P., Baird, A., and Guillemin, R., 1981, A general approach to the microisolation of peptides, in "Peptides: Synthesis-Structure-Function," Proc. 7th American Symposium, D. H. Rich and E. Gross, eds., Pierce Chemical Co., Rockford, Illinois p. 777.

Branton, W. D., Jan, L. Y., and Jan, Y. N., 1982, Non-mammalian luteinizing hormone-releasing factor (LRF) in tadpole and frog brain, Soc. Neuroscience Abstracts, p.14.

Burgus, R., Butcher, M., Amoss, M., Ling, N., Monahan, M., Rivier, J., Fellows, R., Blackwell, R., Vale, W., and Guillemin, R., 1972, Primary structure of the ovine hypothalamic luteinizing hormone-releasing factor (LRF), Proc. Natl. Acad. Sci. USA, 69: 278.

Chan, A., Dudley, C. A., and Moss, R. L., 1985, Hormonal modulation of the responsiveness of midbrain central gray neurons to LH-RH, Neuroendocrinology, 41: 163.

Clayton, R. N., Harwood, J. P., and Catt, K. J., 1979, Gonadotropin-releasing hormone analogue binds to luteal cells and inhibits progesterone production, Nature, 282: 90.

Clayton, R. N., Katikineni, M., Chan, V., Dufau, M. L., and Catt, K. J., 1980, Direct inhibition of testicular function by gonadotropin-releasing hormone: Mediation by specific gonadotropin-releasing hormone receptors in interstitial cells, Proc. Natl. Acad. Sci. USA, 77: 4459.

Crim, J. W., Urano, A., and Gorbman, A., 1979a, Immunocytochemical studies of luteinizing hormone-releasing hormone in brains of agnathan fishes I. Comparisons of adult Pacific lamprey (Entosphenus tridentata) and the Pacific hagfish (Eptatretus stouti), Gen. Comp. Endocrinol., 37: 294.

Crim, J. W., Urano, A., and Gorbman, A., 1979b, Immunocytochemical studies of luteinizing hormone-releasing hormone in brains of agnathan fishes II. Patterns of immunoreactivity in larval and maturing Western brook lamprey (Lampetra richardsoni), Gen. Comp. Endocrinol., 38: 290.

Crim, L. W., 1984, A variety of synthetic LHRH peptides stimulate gonadotropin hormone secretion in rainbow trout and the landlocked Atlantic salmon, J. Steroid Biochem., 20: B1390 (Abstract).

Demski, L. S., and Northcutt, R. G., 1983, The terminal nerve: A new chemosensory system in vertebrates?, Science, 220: 435.

Dluzen, D. E., Ramirez, V. D., Carter, C. S., and Getz, L. L., 1981, Male vole urine changes luteinizing hormone-releasing hormone and norepinephrine in female olfactory bulb, Science, 212: 573.

Donaldson, E. M., Van Der Kraak, G., Hunter, G. A., Dye, H. M., Rivier, J., and Vale, W., 1984, Teleost GnRH and analogues: Effect on plasma GtH concentration and ovulation in coho salmon (Oncorhynchus kisutch), Gen. Comp. Endocrinol., 53: 458 (Abstract).

Eiden, L. E. and Brownstein, M. J., 1981, Extrahypothalamic distributions and functions of hypothalamic peptide hormones, Federation Proc.,40: 2553

Eiden, L. E., and Eskay, R. L., 1980, Characterization of LRF-like immunoreactivity in the frog sympathetic ganglia: Non-identity with LRF decapeptide, Neuropeptides, 1: 29.

Eiden, L. E., Loumaye, E., Sherwood, N., and Eskay, R. L., 1982, Two chemically and immunologically distinct forms of luteinizing hormone-releasing hormone are differentially expressed in frog neural tissues, Peptides, 3: 323.

Fink, G., Aiyer, M., Chiappa, S., Henderson, S., Jamieson, M., Levy-Perez, V., Pickering, A., Sarkar, D., Sherwood, N., Speight, A., and Watts, A., 1982, Gonadotropin-releasing hormone.Release into hypophyseal portal blood and mechanism of action, in "Hormonally Active Brain Peptides," K. W. McKerns and V. Pantić, eds., Plenum Press, New York, p. 397.

Georges, D., and Dubois, M. P., 1980, Mise en evidence par les techniques d'immunofluorescence d'un antigene de type LH-RH dans le systeme nerveux de Ciona intestinalis (Tunicier ascidiace), C. R. Acad. Sci. Paris, Serie D, 290: 29.

Gorbman, A., 1980, Evolution of the brain-pituitary relationship: Evidence from the Agnatha, Can. J. Fish. Aquat. Sci., 37: 1680.

Guillemin, R., 1982, Gonadal peptides involved in reproduction, in The role of peptides and proteins in control of reproduction, Workshop sponsored by National Institutes of Health, Bethesda, Maryland, 15-16 Feb., 1982.

Hasegawa, Y., Miyamoto, K., Igarashi, M., Chino, N., and Sakakibara, S., 1984, Biological properties of chicken luteinizing hormone-releasing hormone: Gonadotropin release from rat and chicken cultured anterior pituitary cells and radioligand analysis, Endocrinology 114: 1441.

Hattori, A., Ishii, S., Wada, M., Miyamoto, K., Hasegawa, Y., Igarashi, M., and Sakakibara, S., 1985, Effects of chicken (Gln^8)- and mammalian (Arg^8)-luteinizing hormone-releasing hormones on the release of gonadotrophins in vitro and in vivo from the adenohypophysis of Japanese quail, Gen. Comp. Endocrinol., 59: 155.

Hsueh, A. J. W., and Erickson, G. F., 1979a, Extrapituitary action of gonadotropin-releasing hormone: Direct inhibition of ovarian steroidogenesis, Science, 204: 854.

Hsueh, A. J. W. and Erickson, G. F., 1979b, Extra-pituitary inhibition of testicular function by luteinising hormone releasing hormone, Nature, 281: 66.

Hunt, L. T., and Dayhoff, M. O., 1979, Structural and functional similarities among hormones and active peptides from distantly related eukaryotes, in "Peptides: Structure and Biological Function," Proc. Sixth American Peptide Symposium, E. Gross and J. Meienhofer, eds., Pierce Chemical Co., Rockford, Illinois, p. 757.

Iversen, L., 1985, Chemicals to think by, New Scientist, No. 1458: 11.

Jackson, I. M. D., 1980, Distribution and evolutionary significance of the hypophysiotropic hormones of the hypothalamus, Front. Horm. Res. 6: 35.

Jan, Y. N., Bowers, C. W., Branton, D., Evans, L., and Jan, L. Y., 1983, Peptides in neuronal function: Studies using frog autonomic ganglia, Cold Spring Harbor Symp. Quant. Biol. XLVIII: 363.

Jan, L. Y., and Jan, Y. N., 1982, Peptidergic transmission in sympathetic ganglia of the frog, J. Physiol., 327: 219.

Jan, Y. N., Jan, L. Y. and Kuffler, S. W., 1979, A peptide as a possible transmitter in sympathetic ganglia of the frog, Proc. Natl. Acad. Sci. USA, 76: 1501.

Jones, S. W., Adams, P. R., Brownstein, M. J., and Rivier, J. E., 1984, Teleost luteinizing hormone-releasing hormone: Action on bullfrog sympathetic ganglia is consistent with role as neurotransmitter. J. Neuroscience, 4: 420.

Khodr, G. S., and Siler-Khodr, T. M., 1978, The effect of luteinizing hormone-releasing factor on human chorionic gonadotropin secretion, Fertil. Steril., 30: 301.

King, J. A., and Millar, R. P., 1980, Comparative aspects of luteinizing hormone-releasing hormone structure and function in vertebrate phylogeny, Endocrinology, 106: 707.

King, J. A., and Millar, R. P., 1981, TRH, GH-RIH, and LH-RH in metamorphosing Xenopus laevis, Gen. Comp. Endocrinol., 44: 20.

King, J. A., and Millar, R. P., 1982a, Structure of chicken hypothalamic luteinizing hormone-releasing hormone. I. Structural determination on partially purified material, J. Biol. Chem., 257: 10722.

King, J. A., and Millar, R. P., 1982b, Structure of chicken hypothalamic luteinizing hormone-releasing hormone. II. Isolation and characterization, J. Biol. Chem., 257: 10729.

King, J. A., and Millar, R. P., 1984, Isolation and structural characterization of chicken hypothalamic luteinizing hormone releasing hormone, J. Exp. Zool., 232: 419.

King, J. C., and Anthony, E. L. P., 1984, LHRH neurons and their projections in humans and other mammals: Species comparisons, Peptides, 5: Suppl. 1: 195.

King, J. C., Anthony, E. L. P., Gustafson, A. W., and Damassa, D. A., 1984, Luteinizing hormone-releasing hormone (LH-RH) cells and their projections in the forebrain of the bat Myotis lucifugus lucifugus, Brain Res., 298: 289.

Krey, L. C., and Silverman, A. J., 1983, Luteinizing hormone releasing hormone (LHRH), in "Brain Peptides," D. T. Krieger, M. J. Brownstein, and J. B. Martin, eds., John Wiley & Sons, New York, p. 687.

Larsell, O., 1950, The nervus terminalis, Ann. Otol. Rhinol. Laryngol., 59: 414.

Loumaye, E., Thorner, J., and Catt, K. J., 1982, Yeast mating pheromone activates mammalian gonadotrophs: Evolutionary conservation of a reproductive hormone?, Science, 218: 1323.

MacKenzie, D. S., Gould, D. R., Peter, R. E., Rivier, J., and Vale, W. W., 1984, Response of superfused goldfish pituitary fragments to mammalian and salmon gonadotropin-releasing hormones, Life Sci., 35: 2019.

Matsuo, H., Baba, Y., Nair, R. M. G., Arimura, A., and Schally, A. V., 1971, Structure of the porcine LH- and FSH-releasing hormone. I. The proposed amino acid sequence, Biochem. Biophys. Res. Communic., 43: 1334.

Millar, R. P., and King, J. A., 1983, Synthesis, luteinizing hormone-releasing activity, and receptor binding of chicken hypothalamic luteinizing hormone-releasing hormone, Endocrinology, 113: 1364.

Millar, R. P., and King, J. A., 1984, Structure-activity relations of LHRH in birds, J. Exp. Zool., 232: 425.

Millar. R. P., King, J. A., and Milton, R. deL, 1985, Structural and functional evolution of GnRH and its receptors, 10th International Symposium on Comparative Endocrinology, Copper Mountain, Colorado, 21-26 July, 1985 (Abstract).

Miller, W. R., Scott, W. N., Morris, R., Fraser, H. M., and Sharpe, R. M., 1985, Growth of human breast cancer cells inhibited by a luteinizing hormone-releasing hormone agonist, Nature, 313: 231.

Miyamoto, K., Hasegawa, Y., Minegishi, T., Nomura, M., Takahashi, Y., Igarashi, M., Kangawa, K., and Matsuo, H., 1982, Isolation and characterization of chicken hypothalamic luteinizing hormone-releasing hormone, Biochem. Biophys. Res. Communic., 107: 820.

Miyamoto, K., Hasegawa, Y., Igarashi, M., Chino, N., Sakakibara, S., Kangawa, K., and Matsuo, H., 1983, Evidence that chicken hypothalamic luteinizing hormone-releasing hormone is (Gln^8)-LH-RH, Life Sci., 32: 1341.

Miyamoto, K., Hasegawa, Y., Nomura, M, Igarashi, M., Kangawa, K., and Matsuo, H., 1984, Identification of the second gonadotropin-releasing hormone in chicken hypothalamus: Evidence that gonadotropin secretion is probably controlled by two distinct gonadotropin-releasing hormones in avian species, Proc. Natl. Acad. Sci. USA, 81: 3874.

Moss, R. L., and McCann, S. M., 1973, Induction of mating behavior in rats by luteinizing hormone-releasing factor, Science, 181: 177.

Münz, H., Stumpf, W. E. and Jennes, L., 1981, LHRH systems in the brain of platyfish, Brain Res., 221, 1.

Münz, H., Claas, B., Stumpf, W. E., and Jennes, L., 1982, Centrifugal innervation of the retina by luteinizing hormone releasing hormone (LHRH)-immunoreactive telencephalic neurons in teleostean fishes, Cell Tissue Res., 222: 313.

Nozaki, M., and Kobayashi, H., 1979, Distribution of LHRH-like substance in vertebrate brain as revealed by immunohistochemistry, Arch. Histol. Jap., 42: 201.

Nozaki, M., Tsukahara, T., and Kobayashi, H., 1984, Neuronal systems producing LHRH in vertebrates, in: "Endocrine Correlates of Reproduction," K. Ochiai, Y. Arai, T. Shioda, and M. Takahashi, eds., Japan Sci. Soc. Press, Tokyo/Springer-Verlag, Berlin, p. 3.

Peter, R. E., Nahorniak, C. S., Sokolowska, M., Chang, J. P., Rivier, J. E., Vale, W. W., King, J. A., and Millar, R. P., 1985, Structure-activity relationships of mammalian, chicken, and salmon gonadotropin releasing hormone in vivo in goldfish, Gen. Comp. Endocrinol., 58: 231.

Rivier, J., Rivier, C., Branton, D., Millar, R., Spiess, J., Vale, W., 1981, HPLC purification of ovine CRF, rat extra hypothalamic brain somatostatin and frog brain GnRH, in: "Peptides: Synthesis-Structure-Function," Proc. 7th American Peptide Symposium, D. H. Rich, and E. Gross, eds., Pierce Chemical Co., Rockford, Illinois, p. 771.

Schally, A. V., 1978, Aspects of hypothalamic regulation of the pituitary gland, Science, 202: 18.

Schally, A. V., Arimura, A., and Kastin, A. J., 1973, Hypothalamic regulatory hormones, Science, 179: 341.

Seeburg, P. H., and Adelman, J. P., 1984, Characterization of cDNA for precursor of human luteinizing hormone releasing hormone, Nature, 311: 666.

Seppälä, M., and Wahlström, T., 1980, Identification of luteinizing hormone-releasing factor and alpha subunit of glycoprotein hormones in ductal carcinoma of the mammary gland, Int. J. Cancer, 26: 267.

Sharpe, R. M., and Fraser, H. M., 1980, Leydig cell receptors for luteinizing hormone releasing hormone and its agonists and their modulation by administration or deprivation of the releasing hormone, Biochem. Biophys. Res. Communic., 95: 256.

Sharpe, R. M., Fraser, H. M., Cooper, I., and Rommerts, F. F. G., 1981, Sertoli-Leydig cell communication via an LHRH-like factor, Nature, 290: 785.

Sherwood, N. M., in press-a, Evolution of a neuropeptide family: Gonadotropin-releasing hormone, Amer. Zool.

Sherwood, N., in press-b, Gonadotropin-releasing hormones in fishes, in: "Hormones and Reproduction in Fishes, Amphibians and Reptiles," D. O. Norris, and R. E. Jones, eds., Plenum Press, New York.

Sherwood, N. M., and Harvey, B., in press, Topical absorption of gonadotropin-releasing hormone (GnRH) in goldfish, Gen. Comp. Endocrinol.

Sherwood, N. M., and Sower, S. A., 1985, A new family member for gonadotropin-releasing hormone, Neuropeptides, 6: 205.

Sherwood, N. M., Eiden, L., Brownstein, M., Spiess, J., Rivier, J., and Vale, W., 1983, Characterization of a teleost gonadotropin-releasing hormone, Proc. Natl. Acad. Sci. USA, 80: 2794.

Sherwood, N. M., Harvey, B., Brownstein, M. J., and Eiden, L. E., 1984, Gonadotropin-releasing hormone (Gn-RH) in striped mullet (Mugil cephalus), milkfish (Chanos chanos), and rainbow trout (Salmo gairdneri): Comparison with salmon Gn-RH, Gen. Comp. Endocrinol., 55: 174.

Sherwood, N. M., Sower, S. A., Marshak, D. R., Fraser, B. A., and Brownstein, M. J., submitted, Primary structure of gonadotropin-releasing hormone from lamprey brain.

Sherwood, N. M., Zoeller, R. T., and Moore, F. L., in press, Multiple forms of gonadotropin-releasing hormone in amphibian brains, Gen. Comp. Endocrinol.

Shivers, B. D., Harlan, R. E., and Pfaff, D. W., 1983, Reproduction: The central nervous system role of luteinizing hormone releasing hormone, in: "Brain Peptides," D. T. Krieger, M. J. Brownstein, and J. B. Martin, eds., John Wiley and Sons, New York, p. 389.

Smith-White, S. S., and Ojeda, S. R., 1984, Maternal modulation of infantile ovarian development and available ovarian luteinizing hormone-releasing hormone (LHRH) receptors via milk LHRH, Endocrinology, 115: 1973.

Stacey, N. E., and Kyle, A. L., 1983, Effects of olfactory tract lesions on sexual and feeding behavior in goldfish, Physiol. and Behav., 30: 621.

Stell, W. K., 1984, Luteinizing hormone-releasing hormone (LHRH)- and pancreatic polypeptide (PP)-immunoreactive neurons in the terminal nerve of the spiny dogfish, Squalus acanthias, Anat. Rec., 208: 173A.

Stell, W. K., Walker, S. E., Chohan, K. S., and Ball, A. K., 1984, The goldfish nervus terminalis: A luteinizing hormone-releasing hormone and molluscan cardioexcitatory peptide immunoreactive olfactoretinal pathway, Proc. Natl. Acad.Sci. USA, 81: 940.

Struthers, R. A., Rivier, J., and Hagler, A. T., 1985, Molecular dynamics and minimum energy conformations of GnRH and analogs. A methodology for computer-aided drug design, Ann. N. Y. Acad. Sci., 439: 81.

Swerdloff, R. S., Bhasis, S., and Sokol, R. Z., 1984, GnRH-like factors in the rat testis and human seminal plasma, Ann. N. Y. Acad. Sci., 438: 382.

Tan, L., and Rousseau, P., 1982, The chemical identity of the immunoreactive LHRH-like peptide biosynthesized in the human placenta. Biochem. Biophys. Res. Communic., 109: 1061.

Whalen, R., and Crim, J. W., 1985, Immunocytochemistry of luteinizing hormone-releasing hormone during spontaneous and thyroxine-induced metamorphosis of bullfrogs, J. Exp. Zool., 234: 131.

Wilson, J. X., 1985, Conjugated catecholamines and pressor responses to angiotensin, luteinizing hormone-releasing hormone and prazosin in conscious toads, Br. J. Pharmac., 85: 647.

Wilson, J. X., Van Vliet, B. N. and West, N. H., 1984, Gonadotropin-releasing hormone increases plasma catecholamines and blood pressure in toads, Neuroendocrinology, 39: 437.

Witkin, J. W., and Silverman, A.-J., 1983, Luteinizing hormone-releasing hormone (LHRH) in rat olfactory systems, J. Comp. Neurol., 218: 426.

Wu, P., and Jackson, I. M. D., 1985, LHRH in cod-fish brain: Evidence of identity with salmon LHRH, 67th Annual Meeting of the Endocrine Society, 1985, Abstract.

Ying, S.-Y., Ling, N., Böhlen, P., and Guillemin, R., 1981, Gonadocrinins: Peptides in ovarian follicular fluid stimulating the secretion of pituitary gonadotropins, Endocrinology, 108: 1206.

THE BIOSYNTHESIS OF LHRH

Ann Curtis, Roberta Rosie, Valerie Lyons and George Fink

MRC Brain Metabolism Unit
Royal Edinburgh Hospital
Morningside Park, Edinburgh, EH10 5HF

Many small peptides, possessing hormonal or neurotransmitter properties are synthesised as part of large precursor forms which often incorporate the sequences of more than one biologically active molecule. For example, β-preprotachykinin incorporates the sequences of both substance P and substance K (Nawa et al., 1983), and similarly oxytocin and arginine-vasopressin are synthesised within the same precursor molecule as their respective associated proteins, neurophysins I and II (Land et al., 1983). Two common features of precursor molecules which have thus far been identified are the presence of a short hydrophobic signal sequence at the N-terminus and the presence of paired basic amino acids at the boundaries of the active peptide sequences which form cleavage sites for further processing of the large molecules.

Studies on the biosynthesis of the decapeptide LHRH were initiated by attempting to identify a possible high molecular weight precursor. Evidence that LHRH is synthesised as part of a larger molecule which is subsequently processed into intermediate and final products had previously come from several studies using hypothalamic extracts. Millar, Wegener & Schally (1981) demonstrated the presence of a 5,000 M_r LHRH immunoreactive species in ovine hypothalamic extracts. Treatment with carboxy- and amino-peptidases indicated that there were both N and C terminal extensions to the decapeptide and these larger forms were processed by trypsin-like enzymes during passage along the axons. Gautron, Pattou & Kordon (1981) detected LHRH immunoreactive species with molecular weights ranging from 1,800 to 26,000 daltons in axoplasmic and cytoplasmic supernatants of rat hypothalamus, cerebral cortex and placenta. Native LHRH was almost exclusively recovered from synaptosomal fractions which were devoid of high molecular weight forms.

Our studies (Curtis & Fink, 1983; Curtis, Lyons & Fink, 1983) involving in vitro translation of mRNA have demonstrated that LHRH is initially synthesised by way of a precursor which is twenty times the size of the active peptide having a molecular weight of 28,000. The size of this precursor is comparable to the 26,000 M_r polypeptide allowing for the cleavage of a putative signal sequence. Using similar techniques, Ivell & Richter (personal communication) detected large immunoreactive forms of LHRH in the molecular weight range 21,000 to 28,000.

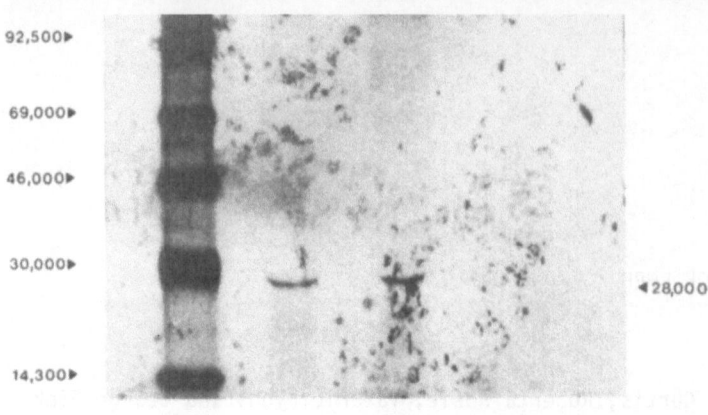

Figure 1 Gel electrophoresis of the LHRH precursor following mRNA
translation from the rat hypothalamus and immunopreciptitation.
Lane 1 shows a series of ^{14}C-labelled molecular weight markers.
Lane 4 shows the position of a single polypeptide with a molecular
weight of 28,000 which was immunoprecipitated from a translation mix
supplemented with ^{3}H-leucine by a specific LHRH antiserum.   This
band was not evident when a translation mix supplemented with
[35]S-methionine is treated in an identical manner (lane 3), but is
present if both ^{3}H-leucine and ^{35}S-methionine are present in the
translation reaction (lane 2).   Lanes 5 and 6 show that no
immunoprecipitation bands appeared in translation mixes using
[35]S-methionine and ^{3}H-leucine respectively, to which no exogenous
mRNA was added but which were treated with the LHRH antiserum in an
identical manner to that used in lanes 2 to 4 and 7.   Lane 7 shows a
competition experiment where a ^{3}H-leucine immunoprecipitation was
carried out in the presence of 10µg of unlabelled LHRH.   This amount
was sufficient excess to compete 100% for the antiserum.

Total RNA was extracted from the hypothalami of 60 adult rats, 60
adult mice and two post-mortem human brains as previously described
(Curtis & Fink, 1983; Curtis, Lyons & Fink, 1983).   The messenger
fractions were translated in vitro using an amino acid depleted rabbit
reticulocyte system and a mixture of tritiated amino acids.   Newly
synthesised proteins were observed by electrophoresing under denaturing
conditions on polyacrylamide gels (O'Farrell, 1975) followed by
autoradiography (Bonner & Laskey, 1974; Laskey & Mills, 1975).
Proteins spanning a broad molecular weight range were synthesised,
indicating that high molecular weight mRNA species do not become
degraded during the extraction procedure, and in the case of human
material, large messengers do remain intact during the post-mortem
interval.

Aliquots of the translation mixes containing a minimum of 10^6 acid
precipitate counts were treated with a specific LHRH antiserum (HC-6) in
an attempt to immunoprecipitate any newly synthesised molecules
incorporating an immunoreactive LHRH amino acid sequence.   The pellet
was electrophoresed on polyacrylamide gels and autoradiographed as
described previously.

Figure 1 shows that the LHRH antiserum specifically recognised a
protein of 28,000 M_r among the translation products of the rat
hypothalamic mRNA when ^{3}H-leucine was incorporated as the radioactive

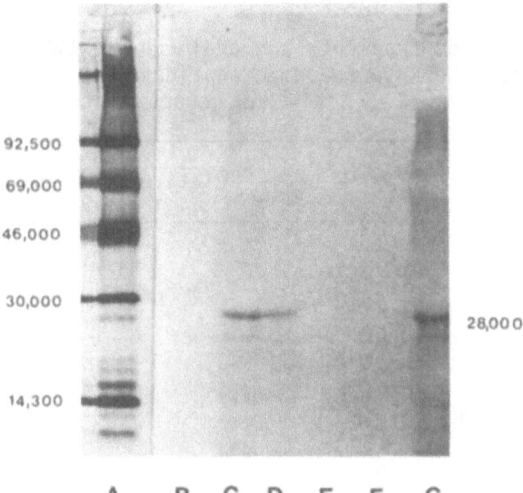

Figure 2 Polyacrylamide gel electrophoresis and autoradiography
following immunoprecipitation of rat and human hypothalamic mRNA
translation products with an anti-LHRH serum.   Lanes B, C and D show
the rat hypothalamic translation products immunoprecipitated in the
presence of excess cold LHRH (lane B), using either the second
antibody (lane C) or the Staphylococcal protein A
immuno-precipitation technique (lane D).   A 28,000 molecular weight
polypeptide was immunoprecipitated by both methods.   Lane E shows
that no polypeptides were immunoprecipitated with a translation mix
without exogenous mRNA was treated in an identical manner.   Lane G
shows that the main immunoprecipitation product in an human
translation mix was a 28,000 MW polypeptide; no immunoprecipitation
was seen in the presence of excess cold LHRH (lane F).   Lane A shows
a series of ^{14}C-labelled standard molecular weight markers.

amino acid but not if ^{35}S-methionine was used as the label indicating
that methionine is absent, or present at very low amounts in the LHRH
precursor.   A protein of identical molecular weight was
immunoprecipitated from the translation products of the human
hypothalamic mRNA (fig. 2) and similarly from that of the normal mouse
(data not shown).   This large precursor could not be detected among the
translation products of the rat, mouse or human mRNA when an excess of
cold LHRH was added before immunoprecipitation, showing that the
decapeptide is able to compete for the antiserum and block its binding
with the precursor completely.   However, we could not identify any high
molecular weight LHRH immunoreactive species among the translation
products of the LHRH deficient hypogonadal (hpg) mouse hypothalamic mRNA
when it was treated in an identical way.

More recently, Seeburg & Adelman (1984) have employed recombinant DNA
techniques to show that LHRH is synthesised by way of a 10,000 M_r
precursor in the human placenta.   Placental cDNA was cloned into a
bacteriophage λgt10 library.   An oligonucleotide probe detected two
clones which incorporated a sequence coding for the LHRH decapeptide.
The precursor was shown to possess a typical signal sequence and the
actual decapeptide sequence was followed by a lys-arg cleavage site.
Also incorporated within the LHRH precursor was a 56 amino acid peptide
which has been shown to possess potent prolactin inhibiting properties
(Nikolics et al., 1985).

TABLE 1

Cross reaction of fragments and analogues of LHRH
and other hypothalamic peptides
with the HC-6 anti-LHRH serum.

Peptide		Relative cross-reactivity (%)
LHRH		100%
LHRH fragments:	1-9	< 0.05
	1-8	< 0.05
	1-6	< 0.05
	1-2 & 9-10	< 0.05
	3-10	2.8
	3-9	< 0.05
	7-10	3.3
$[Gly-OH^{10}]$-LHRH		< 0.002
$[Trp^7,Leu^8,Gly-OH^{10}]$-LHRH		0.07
$[D-Phe^2,Pro^3,D-Phe^6]$-LHRH		23
$[D-pGly^1,D-Phe^2,D-Trp^{3,6}]$-LHRH		3.5
$[D-Lys^6]$-LHRH		63
$[D-Ala^6]$-LHRH		100
Des-Gly$^{10}[Pro^9]$-LHRH ethylamide		0.15
Des-Gly$^{10}[D-Trp^6,Pro^9]$-LHRH ethylamide		0.39
Des-Gly$^{10}[D-leu^6,Pro^9]$-LHRH ethylamide		0.11
TRH		< 0.001
CRF		< 0.001
GHRH		< 0.001

The studies described above illustrate the controversy which
surrounds the nature of the LHRH precursor and hence its biosynthesis.
The identification of a 28,000 high molecular weight LHRH immunoreactive
polypeptide within the hypothalami of three different mammalian species
and its complete absence from the hpg mouse shows that LHRH is
synthesised by way of a precursor in a similar manner to other
hypothalamic peptides.   The placental protein, consisting of 92 amino
acids, having a molecular weight of 10,000 as shown by the recombinant DNA
studies is clearly different from the hypothalamic precursor detected by
immunoprecipitation.

The HC-6 antiserum used in the immunoprecipitation studies is very
specific for LHRH.   It did not cross react with many LHRH fragments and
analogues and various other hypothalamic peptides as shown in table 1.
The only cross reactivity observed was with analogues involving amino
acid changes at the '6' position, that is, within the centre of the

```
     1      2    3    4    5    6    7    8    9    10
   p-glu – his – trp – ser – tyr – gly – leu – arg – pro – gly – NH₂
```

$$3'-T_C^T \quad GT_G^A \quad ACC \quad AG_G^A \quad AT_G^A \quad CC-5'$$

Figure 3 Sequences of the 16-base synthetic oligonucleotides which were used to screen a rat hypothalamic cDNA library constructed in pBR322.   The amino acid sequence of LHRH is shown directly above the deduced possible nucleotide sequence.

active peptide.   It is, therefore, unlikely that the 28,000 Mr protein is non-specifically binding to the antiserum.

The possibility that the different forms arise due to the expression of different genes in different tissues must be considered.   The normal pulsatile release of LHRH from hypothalamic neurones influences the release of LH and FSH from the anterior pituitary.   However, other functions are also implied since it is found in the gonads, other parts of the CNS and the mammary gland as well as the placenta.   The expression of different genes in these tissues is unlikely since Southern analysis suggests the presence of a single gene within the human genome (Seeburg & Adelman, 1984).   Different precursor forms can arise from a single gene by way of differential processing of the primary RNA transcript as occurs in the case of calcitonin and CGRP (see Craig, this volume).   Different sets of axons are spliced in a tissue specific manner from the same gene transcript to yield different messengers, though the final mRNA molecules and hence the precursor proteins, would be expected to include some common domains and, therefore, may cross-react with certain antisera.

1.   20µg total RNA
2.   4µg mRNA
3.

Figure 4 Northern hybridisation of the LHRH-like cDNA insert to 20µg rat hypothalamic total RNA (lane 1) and 4µg mRNA (lanes 2 and 3). RNA was electrophoresed on 0.75% agarose gels and blotted directly onto nitrocellulose paper.   This was probed with the LHRH-like cDNA insert labelled to a specific activity of 10^8 cpm/µg by nick translation with $\alpha-^{32}P$-dCTP, and exposed to film for two weeks.

We have cloned total rat hypothalamic cDNA in the bacterial plasmid pBR322 using a procedure which gives very high numbers of transformants (Gubler & Hoffman, 1983). A single positive clone was obtained after screening 82,000 recombinants with a mixture of short synthetic oligonucleotides whose sequence is shown in fig. 3. This positive cDNA insert was used in Northern Type analysis as a hybridisation probe to investigate hypothalamic mRNA. Figure 4 shows that this rat hypothalamic cDNA sequence hybridises to an RNA in the messenger fraction of approximately 700 nucleotides, but also hybridises to much larger forms in the total RNA. The placental mRNA of the human is much larger than this (at least 1,500 nucleotides in length) but has an unusually long 5'-end, so it is possible that transcription takes place from different promotors in different tissues.

Having presented some of the evidence for the existence of more than one high molecular weight form of LHRH detected using immunological methods, it is now necessary to await the results of DNA sequencing studies currently under investigation, which will determine unequivically, the similarities and differences between the human placental LHRH precursor and that of the rat hypothalamus.

REFERENCES

Bonner, W.M. & Laskey, R.A. (1974). A film detection method for titium labelled proteins and nucleic acids in polyacrylamide gels. Eur. J. Biochem. 46: 83–88.

Curtis, A. & Fink, G. (1983). A high molecular weight precursor of luteinizing hormone releasing hormone from rat hypothalamus. Endocrinology 112: 390–392.

Curtis, A., Lyons, V. & Fink, G. (1983). The human hypothalamic LHRH precursor is the same size as that in rat and mouse hypothalamus. Biochem. Biophys. Res. Commun., 117: 872–877.

Gautron, J.P., Patton, E. & Kordon, C. (1981). Occurrence of higher molecular forms of LHRH fractionated extracts from rat hypothalamus, cortex and placenta. Mol. Cell. Endocrinol. 24: 1–15.

Gubler, U. & Hoffman, B.J. (1983). A simple and very efficient method for generating cDNA libraries. Gene 25: 263–269.

Land, H., Grez, M., Ruppert, S., Sichmale, H., Rehbein, M., Richter, D. & Schutz, G. (1983). Deduced amino acid sequence from the bovine oxytocin-neurophysin I precursor cDNA. Nature 302: 342–344.

Laskey, R.A. & Mills, A.D. (1975). Quantitative film detection of ^{3}H and ^{14}C in polyacrylamide gels by fluorography. Eur. J. Biochem. 56: 335–341.

Miller, R.P., Wegener, I. & Schally, A.V. (1981). Putative prohormonal luteinizing hormone-releasing hormone. In: Neuropeptides, Biochemical and Physiological Studies. Ed: R.P. Miller, Pub: Churchill Livingstone.

Nawa, H., Hirose, T., Takashima, H, Inayamas, S. & Nakanishi, S. (1983). Nucleotide sequences of cloned cDNAs for two types of bovine brain substance P precursor. Nature, 306: 32–36.

Nikolics, K., Mason, A.J., Szonyi, E., Ramachandran, J. & Seeburg, P.H. (1985). A prolactin inhibiting factor within the precursor for human gonadotropin-releasing hormone. Nature, 316: 511–517.

O'Farrell, P.H. (1975). High resolution 1–dimensional electrophoresis of proteins. J. Biol., Chem., 250: 4007–4021.

Seeburg, P.H. & Adelman, J.P. (1984). Characterisation of cDNA for precursor of human luteinizing hormone releasing hormone. Nature, 311: 666–668.

EVIDENCE FOR SEX DIFFERENCES IN GnRH RECEPTORS AND MECHANISM OF ACTION

R. Mitchell, S.-A. Ogier, M. Johnson, A. Cleland,
J. Bennie, G. Fink

MRC Brain Metabolism Unit
University Department of Pharmacology
1 George Square, Edinburgh EH8 9JZ

INTRODUCTION

Over recent years, a number of hypotheses have been proposed for the biochemical mechanism of action of Gonadotropin-Releasing Hormone (GnRH). Although adenosine 3',5'-monophosphate (cAMP) has been implicated as a second messenger in GnRH action (Borgeat et al, 1972), this has been refuted (Conn et al., 1979) and Cronin et al (1983) have shown that changes in Luteinizing Hormone (LH) secretion are not temporally coupled to changes in cAMP levels. An apparently direct effect of cAMP derivatives on GnRH receptors (Smith et al, 1982) may have contributed to some of these results.

In contrast, there is evidence that Ca^{2+} is necessary for the post-receptor process coupling LH secretion to occupancy of GnRH receptors. While most groups have described a strict dependence of GnRH-stimulated LH release on extracellular Ca^{2+} concentration (Marian and Conn, 1979; Pickering and Fink, 1979; Naor et al, 1980; Bourne and Baldwin, 1980; Borges et al, 1983), it has been reported that this is less marked than the Ca^{2+}-dependence of depolarisation-induced release (Hopkins and Walker, 1978). Radioactive ion flux experiments have indicated that GnRH can mobilise Ca^{2+} from extracellular-, but probably also intracellular sources (Williams, 1976; Hopkins and Walker, 1978; Conn et al, 1981). A number of studies have reported marked inhibition of GnRH-stimulated LH secretion by verapinoid calcium antagonists, although in many instances this appears to be incomplete, even at very high antagonist concentrations (Hopkins and Walker, 1978; Marian and Conn, 1979; Pickering and Fink, 1979; Naor et al, 1980), again implying a component of intracellular Ca^{2+} mobilisation. An electrophysiological study on (trypsin-dispersed) gonadotrophes of the ovine pars tuberalis, (Mason and Waring, 1985) has reported membrane voltage fluctuations in response to GnRH which were suggested to reflect the activation of a Ca^{2+} channel or a Ca^{2+}-dependent channel.

Recent evidence suggests that inositol 1,4,5-trisphosphate (formed in the metabolic turnover of inositol phospholipids) can act as a second messenger signalling for Ca^{2+} mobilisation (Streb et al, 1984; Berridge and Irvine, 1984). GnRH has been reported to stimulate phospholipid metabolism in gonadotrophes, with increased turnover of phosphatidyl inositol (Snyder and Bleasdale, 1982; Naor et al, 1984). This response is likely to be multifunctional, in that diacylglycerol (an activator of protein kinase C) is also formed (Nishizuka, 1984). Both diacylglycerols and phorbol esters (which also activate protein kinase C) can stimulate LH secretion (Smith and Vale, 1980; Catt et al, 1984; Conn et al, 1985). Arachidonic acid is also liberated in the GnRH-induced turnover of inositol phospholipids and whilst the possible involvement of prostaglandins in GnRH-induced LH secretion appears to have been eliminated, a number of novel epoxygenated metabolites have been recently implicated (Snyder et al, 1983). A synergistic interaction between protein kinase C and elevated intracellular Ca^{2+} concentration as LH secretagogues has recently been suggested (Harris et al, 1985).

Despite all this evidence implicating Ca^{2+} as a second messenger in GnRH action, it is important to consider that mobilisation of Ca^{2+} is likely to be a ubiquitous requirement in secretory response processes (Moriarty, 1978). Great caution is required therefore, in attributing to this the role of the primary mechanism transducing the signal of GnRH receptor occupancy by an agonist.

Extensive electrophysiological analysis of GnRH action at a neuronal site has been carried out using the bullfrog sympathetic ganglion. Although the GnRH-like peptide found there endogenously may be teleost GnRH rather than the mammalian sequence (Eiden et al, 1982), it is clear that exogenous GnRH can mimic the synaptically-mediated response of a late slow depolarisation of the principal cells of the ganglion (Jan et al, 1980). Adams and Brown (1980) have reported that the observed depolarisation is associated with inhibition of the potassium M-current, which appears to show quite distinct electrical. kinetic and pharmacological properties (Adams et al, 1982a,b). A minority of GnRH-sensitive cells, either additionally or separately, show conductance changes that cannot be accounted for by M-current blockade (Katayama and Nishi, 1982).

Clearly therefore, there may be a number of membrane ion channels involved in GnRH action, especially potassium M-currents, which have not so far been investigated in gonadotrophes. It is not clear whether any one ionic or chemical signal will prove to be the primary co-ordinating event ensuing from GnRH receptor activation, but it now seems likely that a number of components contribute to the overall response observed.

Our approach was to attempt an assessment, by pharmacological means, of the role of membrane K^+- and Ca^{2+}-channels in the secretory response of gonadotrophes to GnRH. It is feasible that the receptors could interact relatively directly with particular ion channels within some membrane molecular complex, so we also sought any allosteric effects of appropriate agents on ligand binding to receptor sites. In relation to our other studies on the GnRH priming response, we were interested to compare the mechanism of the basic LH secretory response in male and female gonadotrophes with the premise that subtle but distinct differences may be concerned with conferring the property of "primability".

TECHNIQUES

Rapid superfusion of anterior pituitary tissue in vitro

Adult Wistar rats were used for all studies; either males, or females pooled in equal proportions from metoestrus and dioestrus days of the 4 day cycles.

Animals were stunned, decapitated, and the pituitaries were removed. The neurointermediate lobe was discarded and the anterior lobe was mechanically chopped at 500µm intervals in two directions at 90°. Tissue was then incubated in Krebs Bicarbonate medium (composition in mM: NaCl, 127; KCl, 3.83; $CaCl_2$, 1.8; KH_2PO_4, 1.18; $MgCl_2$, 1.18; $NaHCO_3$, 20; with $2gl^{-1}$ glucose, 30mg l^{-1} bacitracin and 0.1% BSA, pH 7.4) under $95\%O_2/5\%$ CO_2 at 37°C. After one hour of static incubation, tissue equivalent to one pituitary gland transferred into a superfusion chamber, where it was supported on a Millipore AP20 prefilter. Tissue was then superfused with oxygenated medium at 0.5ml/min, at 37°C, and 2 minute fractions were collected. After 60 min of superfusion, inlet lines were changed to medium containing drugs or ions as appropriate. Subsequent challenges with GnRH were carried out after some 45 min in the presence of the relevant substances. Luteinizing Hormone was estimated by double antibody radioimmunoassay.

Binding of [^{125}I] buserelin to GnRH receptors

[D-Ser(tBu)^6] des Gly^{10}-GnRH ethylamide (buserelin) was iodinated using the Chloramine-T procedure and the product separated from reactants according to Sandow and Konig (1979) with modifications.

Anterior pituitary glands were homogenised in 100 volumes of cold Tris HCl buffer (25mM), pH 7.6. Membrane fragments were washed once by centrifugation (48,000g, 10 min) and resuspension in assay buffer, (which additionally contained 0.1% BSA). Assays were carried out in triplicate or duplicate in a total volume of 500µl, comprising: ~0.5-1mg tissue equivalent, [^{125}I] buserelin to a final concentration of ~25pM (~30,000-50,000 cpm), and drugs or ions as appropriate, with 300nM GnRH for determination of non-specific binding. After 90min at 4°C, bovine γ-globulin (to 0.025%) and polyethylene glycol 8000 (to 15%) were added, tubes vortexed and left for 15 min at 4°C before centrifugation and aspiration of the supernatant.

RESULTS AND DISCUSSION

GnRH-stimulated LH secretion

The superfusion protocol allowed a consistently steady baseline of LH secretion to be reached, with basal levels of 2-4ng/ml from male and female tissue. GnRH (1-1000µM) caused a prompt stimulation of LH secretion, in a concentration-dependent manner, from both male and female tissue. Near-maximal responses of around 80-100% increase were obtained at a concentration of 1000nM and a standard stimulus of 300nM was used in all experiments here (Table 1).

Membrane Ca^{2+} channels and GnRH action

Verapinoid Ca^{2+} channel antagonists, but not 1,4 dihydropyridines or diltiazem, have been previously reported to inhibit GnRH-evoked LH secretion in female weanling rats (Conn et al., 1983). Ro5-4864 (4'-chlorodiazepam) belongs to a new class of putative Ca^{2+} channel antagonist (Mestre et al., 1985). Surprisingly, both (±) verapamil and

Table 1. GnRH–stimulated LH secretion.

Time after addition of 300nM GnRH (min)	LH secretion [a] (% of mean control baseline)	
	male	female
-4	99 ± 2	100 ± 1
-2	101 ± 2	98 ± 1
0	100 ± 2	99 ± 1
2	162 ± 8	170 ± 8
4	161 ± 9	174 ± 8
6	162 ± 8	178 ± 9
8	156 ± 8	177 ± 15
10	158 ± 8	180 ± 12
12	148 ± 6	188 ± 15
14	151 ± 6	187 ± 16

[a] Mean ± S.E.M., n = 5 - 10 in all experiments.

Table 2. Effects of Ca^{2+} channel antagonists on GnRH–stimulated secretion

drug (concentration)	% of control [a] response to 300nM GnRH	
	male	female
GnRH + (±) verapamil (100µM)	14 ± 7* **	51 ± 7*
GnRH + Ro5–4864 (5µM)	41 ± 10* **	112 ± 15

[a] values for 10 min after GnRH addition
* Significant inhibition of response
** significant difference between extent of inhibition in male/female.
($P < 0.05$, Mann–Whitney U test).

Ro5-4864 were significantly more effective at blocking the GnRH response in male rather than female gonadotrophes (Table 2), suggesting that there may be a greater involvement of membrane Ca^{2+} channels in the male response. Barium ions are reported to enhance the unitary conductance through membrane Ca^{2+} channels* (Hagiwara and Byerly, 1981) and so might be expected to facilitate any response involving them. A profound secretagogue action of Ba^{2+} is widely reported (Douglas et al., 1983) which presumably also involves such a mechanism. Male gonadotrophes are anomalous in that they fail to show any secretagogue response to Ba^{2+} (Mitchell and Anderson, 1985), although such a response is seen with female weanling gonadotrophes (Conn et al., 1980). In the present experiments, Ba^{2+} released LH from adult female but not male tissue (Table 3); a result apparently at variance with the Ca^{2+} channel antagonist data.

Membrane K^+ channels and GnRH action

One explanation of these apparently disparate results might be that an action of Ba^{2+}, not directly related to Ca^{2+} channels, is involved. One such action is its powerful and selective blockade of potassium M-currents (Adams et al., 1982b), equivalent to the proposed mechanism of GnRH action in bullfrog sympathetic ganglion. Another selective blocker of M-currents, uridine 5'-triphosphate (Adams et al., 1982b) was also able to produce a small increase in LH secretion from female but not male tissue. It may be therefore, that closure of potassium M-currents (as mimicked by Ba^{2+}) could account for a part of the secretory response to GnRH in female but not male gonadotrophes. This would be in accordance with a significant proportion of the female gonadotrophe response being resistant to Ca^{2+} channel antagonists.

Table 3. Effects of some K^+ channel antagonists
on basal LH secretion

drug concentration	LH secretion (peak % of mean control baseline)	
	male	female
Ba^{2+} (3mM)	106 ± 4	164 ± 7* **
4-AP (1mM)	441 ± 25*	170 ± 7* **

  * significant increase
 ** significant difference between extent of increase in
    male/female.
(P < 0.05, Mann-Whitney U test)

* A recent report (Nilius et al., 1985) describes a novel class of Ba^{2+}-insensitive Ca^{2+} channel.

When GnRH stimuli were applied to gonadotrophes that had been previously exposed to Ba^{2+} or UTP, again a marked difference was seen between responses of male and female tissue. The amplitude of stimulus induced release was profoundly enhanced in female tissue but patently unaffected in male (Table 4). If Ba^{2+}, UTP and GnRH are all acting on female but not male gonadotrophes to inhibit potassium M-currents, then it implies that the huge potentiation seen here represents a synergistic interaction of the modes by which GnRH and the other agents influence M-currents. Alternatively, it is possible that M-currents serve only as an inhibitory modulator rather than mediator of GnRH action in female gonadotrophes. Detailed analysis of the interaction will be required to resolve this.

Other K^+ currents, such as the transient outward current (I_A), the delayed rectified (I_K) and Ca^{2+}-activated currents (including the classical I_C), (Thompson, 1977) could potentially be involved. Preliminary experiments however, indicated no major effects of tetraethylammonium or quinine on either basal or GnRH-stimulated LH secretion, suggesting that currents of the latter two categories may not be significantly involved. Nevertheless, the selective inhibitor of A-currents, 4-aminopyridine (4-AP) (Thompson, 1977), caused both prominent changes in baseline LH secretion and a modification of the GnRH response. At a concentration of 1mM, 4-AP caused a vast, but transient increase in LH secretion from male gonadotrophes, with a much more minor effect on female tissue (Table 3). This perhaps suggests that more K^+ A-current channels are normally present in an activated state in the male gonadotrophe. After the baseline had recovered (within 30 min), the response to GnRH of male but not female gonadotrophes was significantly attenuated in the initial rapidly

Table 4. Effects of some K^+ channel antagonists on GnRH-stimulated LH secretion

drug (concentration)	% of control response to 300nM GnRH	
	male	female
GnRH + Ba^{2+}(3mM)	[a] 89 ± 9	345 ± 51*
GnRH + UTP(100μM)	[a] 82 ± 9	226 ± 25*
GnRH + 4-AP (1mM)	[b] 44 ± 4*	126 ± 15

[a] Values for 10 min. after GnRH addition

[b] Values for 2 min after GnRH addition

* significantly different from control response (P < 0.05, Mann-Whitney U test.

peaking phase of the response (Table 4). This is consistent with closure of an ongoing A-current representing a component of the initial response to GnRH of male but not female gonadotrophes. Although experiments on bullfrog sympathetic ganglion failed to implicate A-currents in GnRH action (Adams et al., 1982a), the A-current there is unusual in being insensitive to 4-AP and other (as yet unidentified) currents have been considered to contribute to the response (Jan et al., 1980).

Effects of ions on GnRH receptors

Degradation-resistant analogues of GnRH bind almost entirely to a single high affinity site in anterior pituitary membranes (Clayton et al., 1979). A number of cations can substantially influence the binding of [125I]-labelled ligands (Marian and Conn, 1980; Hazum, 1981). In none of these experiments however, was a comparison made of membranes from male and female anterior pituitaries. We addressed this problem and found that while some cations (Na^+, Ca^{2+}) had identical effects on [125I]buserelin binding, there appeared to be a sex-related difference in the effect of K^+ ions (Table 5). Under these conditions, any allosteric influences from other membrane components closely associated with GnRH receptors may be detectable. The selective inhibitory effect of K^+ on binding to the male GnRH receptor parallels the selective inhibition by a potassium A-current blocker of part of the male secretory response to GnRH. It may be therefore, that in the male, the effect on [125I] buserelin binding represents an allosteric influence of an element of the ion channel mediating A-currents on GnRH receptors, thereby reflecting the close functional interaction of these sites.

Table 5. Effects of some cations on [125I] buserelin binding to pituitary GnRH receptors.

ion (concentration)	% of control specific binding	
	male	female
Ca^{2+} (3mM)	a 72 ± 6*	74 ± 5*
Na^+ (50mM)	94 ± 3	94 ± 4
K^+ (50mM)	56 ± 6*	85 ± 5

* significant inhibition of specific binding
(P < 0.05, Mann Whitney U-test)

(3mM Ca^{2+} also inhibited non-specific binding to a similar extent)

Summary

Whilst it is clear that Ca^{2+} is involved in GnRH action, it is now apparent that membrane K^+ channels are also of major importance. There is evidence for a role of potassium A- and M- currents both in contributing to the basal membrane state of gonadotrophes and in either mediating (or modulating) GnRH responses. At each of these levels, there appears to be a quantitative or qualitative difference between male and female gonadotrophes. While Ca^{2+} channel opening and potassium A-current blockade appear to account for a large part of the male response to GnRH, potassium M-current blockade and to a lesser extent Ca^{2+} channel opening appear important in the female. Experiments on properties of the receptors as labelled by [125I] buserelin suggested an allosteric interaction between potassium A-channels and GnRH receptors in the male but not female, and no evidence of such close molecular interactions with Ca^{2+} channels. We believe this sex difference in the ionic/biochemical mechanism of action of what is empirically a single receptor, to be unprecedented.

ACKNOWLEDGEMENTS

We are grateful to Hoechst AG for the gift of unlabelled buserelin and to Drs. G.D. Niswender, L.E. Reichert Jr. and the Pituitary Hormone Distribution Agency of the NIADDK, Baltimore, U.S.A., for the gift of radioimmunoassay materials.

REFERENCES

Adams, P.R. and Brown, D.A., 1980, Luteinizing Hormone-Releasing Factor and muscarinic agonists act on the same voltage-sensitive K^+ current in bullfrog sympathetic neurones, Br. J. Pharmac., 68: 353.
Adams, P.R., Brown, D.A. and Constanti, A., 1982a, M-currents and other potassium currents in bullfrog sympathetic neurones. J. Physiol. (Lond.), 330: 537.
Adams, P.R., Brown, D.A. and Constanti, A., 1982b, Pharmacological inhibition of the M-current, J.Physiol. (Lond.), 332: 223.
Berridge, M.J. and Irvine, R.F., 1984, Inositol trisphosphate, a novel second messenger in cellular signal transduction, Nature, 312: 315.
Borgeat, P., Chavancy, G., Dupont, A., Labrie, F., Arimura, A. and Schally, A.V., 1972, Stimulation of adenosine 3'5'-cyclic monophosphate accumulation in anterior pituitary gland by synthetic luteinizing hormone-releasing hormone, Proc. Natl. Acad. Sci. USA, 69: 2677.
Borges, J.L.C., Scott, D., Kaiser, D.L., Evans, W.S. and Thorner, M.O., 1983, Ca^{2+}-dependence of Gonadotropin-Releasing Hormone - stimulated Luteinizing Hormone secretion: in vitro studies using continuously perifused dispersed rat anterior pituitary cells, Endocrinology, 113: 557.
Bourne, G.A. and Baldwin, D.M., 1980, Extracellular Ca^{2+}-independent and -dependent components of the biphasic release of LH in response to Luteinizing Hormone-Releasing Hormone in vitro, Endocrinology, 107: 780.
Catt, K.J., Loumaye, E., Wynn, P., Suarez-Quian, C., Kiesel, L., Iwashita, M., Hirota, K., Morgan, R. and Chang, J., 1984, Receptor-mediated activation mechanisms in the hypothalamic control of pituitary-gonadal function, in: "Endocrinology", F.Labrie, L. Proulx, eds., Elsevier.

Clayton, R.N., Shakespear, R.A., Duncan, J.A., Marshall, J.C., 1979, Radioiodinated non-degradable Gonadotropin-Releasing Hormone analogs: new probes for the investigation of pituitary Gonadotropin-Releasing Hormone receptors, Endocrinology, 105: 1369.

Conn, P.M., Ganong, B.R., Ebeling, J., Staley, D., Neidel, J.E. and Bell, R.M., Diacylglycerols release LH: structure-activity relations reveal a role for protein kinase C, Biochem. Biophys. Res. Comm., 126: 532.

Conn, P.M., Marian, J., McMillian, M. and Rogers, D., 1980, Evidence for calcium mediation of Gonadotropin-Releasing Hormone action in the pituitary, Cell Calcium, 1: 7.

Conn, P.M., Marian, J., McMillian, M., Stern, J., Rogers, D., Hamby, M., Penna, A. and Grant, E., 1981, Gonadotropin-Releasing Hormone action in the pituitary: a three-step mechanism, Endocrine Rev., 2: 174.

Conn, P.M., Morrell, D.V., Dufau, M.L. and Catt, K.J., 1979, Gonadotropin-Releasing Hormone action in cultured pituicytes: independence of Luteinizing Hormone release and adenosine 3'5'-monophosphate production, Endocrinology, 104: 448.

Conn, P.M., Rogers, D.C. and Seay, S.C., 1983, Structure-function relationship of calcium ion channel antagonists at the pituitary gonadotrope, Endocrinology, 113: 1592

Cronin, M.J., Evans, W.S., Hewlett, E.L. Rogol, A.D. amd Thorner, M.O., 1983, Luteinizing Hormone secretion is enhanced by pertussis toxin, cholera toxin and forskolin, Neuroendocrinology, 37: 161.

Douglas, W.W., Taraskevich, P.S. and Tomiko, S.A., 1983, Secretagogue effect of barium on output of Melanocyte-Stimulating Hormone from pars intermedia of the mouse pituitary, J. Physiol., 338: 243.

Eiden, L.E., Loumaye, E., Sherwood, N. and Eskay, R.L., 1982, Two chemically and immunologically distinct forms of Luteinizing Hormone-Releasing Hormone are differentially expressed in frog neural tissues, Peptides, 3: 323.

Hagiwara, S. and Byerly, L., 1981, Calcium channel, Annu. Rev. Neurosci. 4: 69.

Harris, C.E., Staley, D. and Conn, P.M., 1985, Diacylglycerols and protein kinase C: potential amplifying mechanism for Ca^{2+}-mediated Gonadotropin-Releasing Hormone-stimulated Luteinizing Hormone release, Mol. Pharmacol., 27: 532.

Hazum, E., 1981, Some characteristics of GnRH receptors in rat pituitary membranes: differences between an agonist and an antagonist, Mol. Cell. Endocrinol., 23: 275.

Hopkins, C.R. and Walker, A.M., 1978, Calcium as a second messenger in the stimulation of Luteinizing Hormone secretion, Mol. Cell. Endocrinol., 12: 189.

Jan, Y.N., Jan, L.J. and Kuffler, S.W., 1980, Further evidence for peptidergic transmission in sympathetic ganglia, Proc. Natl. Acad. Sci. USA, 77: 5008

Katayama, Y. and Nishi, S., 1982, Voltage-clamp analysis of peptidergic slow depolarisations in bullfrog sympathetic ganglion cells, J.Physiol. (Lond.), 333: 305

Marian, J. and Conn, P.M., 1979, Gonadotropin-Releasing Hormone stimulation of pituitary cells requires calcium, Mol. Pharmacol., 16: 196.

Marian, J. and Conn, P.M., 1980, The calcium requirement in GnRH-stimulated LH release is not mediated through a specific action on receptor binding, Life Sci., 27: 87.

Mason, W.T. and Waring, D.W., 1985, Electrophysiological recordings from gonadotrophes, Neuroendocrinology, 41: 258.

Mestre, M., Carriot, T., Uzanm A., Renault, C., Dubroeucq, M.C., Gueremy, C., Doble, A. and LeFur, G., 1984, Electrophysiological and pharmacological evidence that peripheral type benzodiazepine receptors are coupled to calcium channels in the heart, Life Sci., 36: 391.

Mitchell, R. and Anderson, R.A., 1985, Selective secretagogue action of Ba^{2+} ions at lactotrophes but not gonadotrophes, Biochem. Soc. Trans., 13: 1186.

Moriarty, C.M., 1978, Role of calcium in the regulation of adenohypophysial hormone release, Life Sci., 23: 185.

Naor, Z., Leifer, A.M. and Catt, K.J., 1980, Calcium-dependent actions of Gonadotropin-Releasing Hormone on pituitary guanosine 3'5'-monophosphate production and gonadotropin release, Endocrinology, 107: 1438.

Naor, Z., Molcha, J., Zilberstein, M. and Sakut, H., 1984, Phospholipid turnover in Gonadotropin-Releasing Hormone target cells: comparative studies, in: "Hormonal Control of the Hypothalamo-Pituitary-Gonadal Axis", McKernes, K.W., Naor, Z., eds., Plenum, New York.

Nilius, B., Hess, P., Lansman, J.B. and Tsien, R.W., 1985, A novel type of cardiac calcium channel in ventricular cells, Nature, 316: 443.

Nishizuka, Y., 1984, The role of protein kinase C in cell surface signal transduction and tumour promotion, Nature, 308: 693

Pickering, A.J-M.C. and Fink, G., 1979, Priming effect of Luteinizing Hormone-Releasing Factor in vitro: role of protein synthesis, contractile elements, Ca^{2+} and cyclic AMP, J. Endocr., 81: 223.

Sandow, J. and Konig, W., 1979, Studies with fragments of a highly active analogue of Luteinising Hormone Releasing Hormone, J.Endocrinol., 81: 175.

Smith, M.A., Perrin, M.H. and Vale, W.W., 1982, Interaction of adenosine 3'5'-monophosphate derivatives with the Gonadotropin-Releasing Hormone receptor on pituitary and ovary, Endocrinology, 111: 1951.

Smith, M.A. and Vale, W.W., 1980, Superfusion of rat anterior pituitary cells attached to Cytodex beads: validation of a technique, Endocrinology, 107: 1425.

Snyder, G.D. and Bleasdale, J.E., 1982, Effect of LHRH on incorporation of $[^{32}P]$-orthophosphate into phosphatidyl inositol by dispersed anterior pituitary cells, Mol. Cell. Endocrinol., 28: 55.

Snyder, G.D., Capdevila, J., Chacos, N., Manna, S. and Falck, J.R., 1983, Action of Luteinizing Hormone-Releasing Hormone: Involvement of novel arachidonic acid metabolites, Proc. Natl. Acad. Sci. USA, 80: 3504.

Streb, H., Irvine, R.F., Berridge, M.J. and Schulz, I., 1984, Release of Ca^{2+} from a non-mitochondrial intracellular store in pancreatic acinar cells by inositol-1,4,5-triphosphate, Nature, 306: 67.

Thompson, S.H., 1977, Three pharmacologically distinct potassium channels in molluscan neurones, J. Physiol., 265: 465.

Williams, J.A., 1976, Stimulation of $^{45}Ca^{2+}$ efflux from rat pituitary by LHRH and other pituitary stimulators, J. Physiol., 260: 105.

THE ROLE OF BRAIN PEPTIDES IN THE CONTROL OF ANTERIOR PITUITARY HORMONE SECRETION

S.M. McCann, W.K. Samson, M.C. Aguila, J. Bedran de Castro, N. Ono, M.D. Lumpkin and O. Khorram

University of Texas Health Science Center of Dallas
Southwestern Medical School
Dallas, Texas 75235

Introduction

A host of peptides act within the hypothalamus and also directly on the pituitary to control the secretion of anterior pituitary hormones. The classical releasing and inhibiting hormones act directly on the pituitary to inhibit or stimulate the release of anterior pituitary hormones. They appear also to act within the brain to modulate their own release and that of other releasing factors. With the discovery of many additional brain peptides, most of which were found initially in the gastrointestinal tract, many of these have also been examined and it is apparent that there are important hypothalamic actions of a number of these peptides to alter pituitary hormone secretion (1). In this paper, we will review some of our recent work in this area beginning with an examination of possible additional peptidic releasing or inhibiting factors which may directly alter pituitary hormone secretion, continuing with an examination of the intrahypothalamic action of the various peptides to alter their own release and that of pituitary hormones.

Evidence for an FSH-Releasing Factor

Since LH-releasing hormone (LHRH) will also release FSH, albeit to a lesser extent, the concept has developed that there is only one gonadotropin-releasing hormone (GNRH) which controls the release of FSH and LH and is the decapeptide LHRH (2). Overwhelming evidence now supports the concept that there must be a separate FSH-releasing factor which is yet to be isolated. The evidence for this concept is the following. Electrochemical (3,4) or prostaglandin E2 (5,6) stimulation of the dorsal anterior hypothalamic area and regions extending ventrally and caudally from this area, particularly the caudal median eminence, evoked selective FSH secretion, whereas stimulations in the preoptic area evoked selective LH-release (3). Lesions in the dorsal anterior hypothalamic area produced a partial blockade of FSH secretion in the castrate and in the steroid-primed animal (7) and eliminated pulsatile FSH secretion in the castrate while LH pulsations proceeded apace (8). Conversely lesions in the suprachiasmatic-preoptic area blocked the steroid-induced LH but not FSH release (9). These observations are consistent with the concept that the LH-controlling area is in the preoptic area whereas the FSH-controlling area is, at least in part, localized to the dorsal anterior hypothalamic area.

Antisera against LHRH have recently been reported to inhibit selectively LH but not FSH release acutely (10). Intragastric administration of alcohol in the castrate rat produced a complete blockade of LH pulsations and the basal level of plasma LH fell, whereas pulsations of FSH and the basal level of the hormone were maintained (11).

Extracts of the posterior median eminence contained more FSH-releasing activity than could be accounted for by the content of LHRH (12), and an even more pronounced exaggeration of FSH-releasing activity was found in extracts of the organum vasculosum lamina terminalis (13). Recently we have completed experiments employing gel filtration on Sephadex G25 of sheep hypothalamic extracts and observed elution of FSH-releasing activity measured by an in vivo assay in ovariectomized, steroid-blocked rats in the same tubes in which we observed it 20 years ago by bioassay of FSH (14). These tubes contained no detectable radioimmunoassayable LHRH and had no bioassayable LHRH activity. Following these tubes, LHRH was eluted in the region previously found before as measured by radioimmunoassay of the peptide and also by bioassay of LH-releasing activity. This LH-releasing zone off the column contains some FSH-releasing activity, probably attributable to the LHRH present. With the thought that the LHRH recently isolated from chick and fish hypothalami might possibly represent the elusive FSH-releasing factor, we have assayed these peptides but have not found any selective FSH-releasing activity (14). The FSH-releasing factor remains to be isolated and synthesized; however, the weight of evidence is almost overwhelmingly in favor of its existence.

In addition, several peptides other than LHRH have been found capable of stimulating gonadotropin release by dispersed pituitary cells perifused in vitro. Examples of this are provided by gastric inhibitory polypeptide (15) and neuropeptide Y (16). Whether these effects are of physiologic significance remains to be determined.

Evidence for a peptidic prolactin-inhibiting factor

It has now become apparent that many factors can act directly on the lactotroph to alter prolactin secretion (17). A number of inhibitors have been discovered which include dopamine, gamma aminobutyric acid and acetylcholine. The evidence is conclusive that dopamine exercises a physiologic role in this regard. Early evidence suggested the presence of a peptidic prolactin-inhibiting factor in hypothalamic extracts (18). We have recently examined the fractions off of the Sephadex column just described for prolactin release-inhibiting activity as measured by the inhibition of prolactin release from dispersed anterior pituitary cells of adult male or estrogen-primed, ovariectomized rats. Using this system, prolactin-inhibiting activity was detected in the same tubes in which we had found it many years ago assaying for the activity by in vivo bioassay. These fractions also contained LHRH and somatostatin; however, these peptides had no prolactin- inhibiting activity in the quantities present. Neither dopamine nor GABA was detected in the active fractions by radioenzymatic and fluorophotoenzymatic assay, respectively. In addition, receptor blockers for dopamine or GABA did not interfere with the PIF activity (19). In other experiments we have found a pronase-labile PIF activity in crude rat hypothalamic extracts (20). Taken together these findings indicate that this PIF cannot be either dopamine or GABA and is presumably a peptide. Recently, Adelman et al.(21), have detected PIF activity of a 58 amino acid fragment of the preproLHRH molecule. This fragment is presumably larger than the molecule which has been shown to have PIF activity in the present experiments. It is possible that PIF may represent a fragment of this 58 amino acid peptide. The evidence is clearly mounting that a peptidic PIF in fact exists and probably plays a physiologically significant role in the control of prolactin secretion.

In addition to multiple PIFs, there are multiple peptides with prolactin-releasing activity (17). These include thyrotropin-releasing factor, oxytocin, vasoactive intestinal polypeptide, peptide histadine isoleucine, neurotensin, substance P, and angiotensin II. In the case of oxytocin the PRF activity is most readily demonstrated in perifused columns of pituitary cells (Lumpkin, Samson and McCann, In preparation) and it appears that oxytocin has a physiologically significant role in the suckling and proestrus-induced prolactin release on the basis of passive neutralization studies with antibodies directed against oxytocin (Samson, in preparation).

Intrahypothalamic Actions of Peptides to Alter Gonadotropin Secretion

We have now evaluated the effects of a number of peptides on the secretion of gonadotropins (22). Space does not permit a complete elaboration of all of these effects. It is apparent that very minute doses of a number of these peptides can inhibit gonadotropin secretion in the ovariectomized rat. Larger doses are required in some cases. The important point is the physiologic significance of these effects which must be examined by the use of suitable antagonists to the peptides and/or antisera directed against the peptide.

Cholecystokinin (CCK) is an extremely potent inhibitor of LH release in the castrate female rat (23). The introduction of only 4 ng into the third ventricle of conscious ovariectomized rats was sufficient to inhibit LH but not FSH release in these animals. We have recently tested the CCK antagonist, proglumide, and found that it has the opposite effect following its intraventricular injection, i.e., LH levels are elevated (Vijayan and McCann, In preparation). Therefore, it appears that CCK may be a physiologically signficant inhibitor of LH release in the castrate. By contrast, vasoactive intestinal polypeptide (VIP) was a powerful stimulant of LH release in the castrate (24) and released LHRH from hypothalamic synaptasomes incubated in vitro suggesting that it may play a role in the stimulation of LH release in the castrate (25); however, studies with antisera directed against the peptide have yet to be carried out. Third ventricular injection of both pancreatic polypeptide (26) and neuropeptide Y (16) also inhibited LH release in the castrate, and this was the opposite of the action of the latter peptide at the pituitary level to stimulate LH and FSH release by direct action on the gland in the perifusion system as mentioned above (16). These inhibitory actions of the peptides have been confirmed by others (27). Neuropeptide Y is a good example of a peptide which has different actions depending on the hormonal state of the animal. In the estrogen-primed animal, instead of inhibiting LH release, neuropeptide Y stimulates it (27).

It is now generally accepted that opiates inhibit LH release under most circumstances (22). Beta-endorphin appears to play a physiological role in inhibition of LH release in the castrate female rat as determined by studies with intraventricular injection of beta-endorphin antibodies (28). The physiologic significance of the endorphins is born out by the ability of naloxone to elevate LH release under most conditions in the rat.

Alpha melanocyte-stimulating hormone (αMSH) appears to have a physiologically significant inhibitory action to decrease LH release in the castrate female rat since intraventricular injection of minute doses of αMSH lowers LH in the castrate (29) and a significant elevation in LH levels in plasma follows the intraventircular injection of the globulin fraction of a highly specific α-MSH antesera (30). Similarly, αMSH inhibits prolactin release and antisera directed against the peptide elevate prolactin. These inhibitory actions of αMSH on LH and prolactin are mediated by dopamine. Gastric inhibitory polypeptide has yet to be demonstrated in the brain, however, it has a unique ability to suppress FSH release following its intraventricular injection without altering LH release and, as already mentioned, has a stimulatory effect on both FSH and LH release from perifused pituitary cells in vitro (15). Its selective inhibitory effect on FSH release is shared by the gonadal peptide, inhibin, which also inhibits only FSH and not LH release following its intraventricular injection in the face of normal responsiveness of the pituitary to LHRH (31).

It has long been postulated that pituitary hormones have a short-loop feedback effect to suppress their own release by intrahypothalamic action (32). We have found suggestive evidence for this in the case of LH and prolactin clearly inhibits its own release rapidly by an intrahypothalamic action but, in these acute experiments, had no reproducible effect on LH release (33). So far, we have not been able to determine any reproducible effect of FSH to modify its own release or that of other hormones in the ovariectomized rat.

Ultrashort-loop feedback of peptides to alter their own release

Many years ago, Martini postulated ultrashort-loop feedback of releasing factors to suppress their own release (32). Several years ago, we discovered what appears to be an ultra short-loop feedback of somatostatin to suppress its own release. Following the intraventricular injection of small doses of the peptide in conscious ovariectomized rats, a paradoxical suppression of plasma growth hormone resulted which was accompanied by a lowering of FSH, LH and TSH levels. We postulated that the peptide was taken up from the third ventricle and then inhibited either its own release or stimulated the release of GRF or both and that this resulted in the suppression of GH. We postulated that the peptide also inhibited the release of LHRH and TRH to account for the lowering in the plasma concentrations of those three peptides (34).

Subsequently, we searched for similar ultrashort-loop feedback actions of other releasing factors to alter their own release. We found that injection of GRF into the third ventricle elevated plasma GH until the dose was drastically reduced at which point a lowering ensued. Presumably, following the higher doses of the peptide, the uptake was sufficient in portal vessels to deliver a quantity to the somatotrophs which evoked release of GH. When the dose was cut, the intrahypothalamic concentration was sufficient to exert its inhibitory action, but insufficient quantities reached the pituitary to directly effect the somatotrophs (35).

These experiments do not differentiate between an effect of the peptide on the secretion of GH-releasing factor and/or an effect on somatostatin release. Consequently, we have evaluated the effect of GRF on the release of somatostatin from small median eminence-stalk fragments of male rat hypothalami incubated in vitro (36). We have observed a highly potent effect of GRF to enhance somatostatin release with a minimal effective dose of 10^{-11}M. We have further determined the mechanism of this effect. Since it is not blocked by a variety of synaptic receptor blockers but is blocked by naloxone, which should block certain opioid receptors, consequently, it appears that the somatostatin-releasing action of GRF may be mediated via opioid peptides (37). There are many β-endorphin terminals within this small fragment which may mediate this effect.

We have evaluated a possible ultrashort-loop feedback of LHRH to inhibit its own release in the castrate rat and have found that, as in the case of GRF, administration of relatively high doses of LHRH intraventricularly resulted in an elevation of LH but if the dose was cut to 1 ng, a rapid and short-lived inhibition of LH release ensued. Presumably then, LHRH may have a physiologically significant ultrashort-loop negative feedback effect which may play a role in terminating the pulses of LH release which characterize the castrate condition (38).

Similarly, there appears to be an ultrashort-loop negative feedback of oxytocin to suppress its own release since intraventricular injection of the peptide resulted in inhibition of prolactin release, whereas its direct action on the pituitary is to stimulate prolactin release (1).

With the determination of structure and synthesis of CRF, we also evaluated this peptide for a possible ultrashort-loop feedback action. Relatively high doses of CRF injected into the third ventricle (greater than 1 nmole) elevated plasma ACTH in a dose-related manner which was attributed to uptake of the peptide from the ventricle, diffusion to the median eminence and transport via the portal vessels to the pituitary with stimulation of the corticotrophs. Consequently, we diminished the dose of CRF looking for a lowering of the already quite low plasma ACTH levels in these resting rats; however, no lowering was detected regardless of the dose injected. Therefore, we evaluated the effect of intraventricular injection of CRF in small doses on the release of ACTH during ether stress. To our surprise, employing doses 1,000 times less than those required to elevate ACTH by an action on the pituitary, we found an augmentation of the stress-induced ACTH-release (39). We therefore postulate that in stress, CRF exerts an ultrashort-loop positive feedback to augment its own release. This may be of physiologic significance since one would wish to amplify the release of CRF in stress and not to suppress it as might be the case under resting conditions. The situation is quite analogous to the stimulatory action of gonadal steroids around the time of the preovulatory release of LH which is the classical example of positive feedback on the hypothalamus (22).

Since the release of hormones other than ACTH is also altered by stress, we monitored the effects of the microinjection of ovine CRF into the third ventricle on the plasma levels of LH and GH. Lower doses of CRF had no effect on plasma LH; however, plasma LH levels were lowered by the highest dose of CRF injected (1 nmole) with the first significant lowering detectable at 30 min. Interestingly, the pattern of LH release following stress in the ovariectomized rat is an initial elevation followed by a lowering. This raises the possibility that release of CRF during stress could be responsible for the inhibition of LH release which supervenes. Microinjections of an even lower dose of 0.1 nmole of CRF lowered plasma GH levels within 15 min and the suppression persisted for the duration of the experiment. There was no additional effect of the higher 1 nmole dose. Stress is followed by a lowering of GH in the rat and it, therefore, seems reasonable to speculate that intrahypothalamic release of CRF following stress may mediate this lowering of GH (39). Similar results on both LH and GH have been obtained following the lateral ventricular injection of the peptide (40).

It is already known that vasopressin can potentiate the release of ACTH induced by CRF by actions at both pituitary and hypothalamic levels (41,42). To determine the physiologic significance of arginine vasopressin in stress-induced release of anterior pituitary hormones, vasopressin antisera or normal rabbit serum was microinjected into the third ventricle of freely-moving ovariectomized female rats. Because we had found this paradigm useful in prior experiments (30), a single 3 μl injection was given and 24 hours later the injection was repeated 30 min prior to application of ether stress for 1 min. Although arginine vasopressin antisera had no effect on basal plasma ACTH concentrations, the elevation of plasma ACTH induced by ether stress was lowered significantly (43). Plasma LH tended to increase following ether stress as expected; however, the level of LH following stress was significantly lower in the vasopressin antisera-treated group than in the group pre-treated with normal rabbit serum. Ether stress lowered plasma GH levels as expected in the rat and this lowering was slightly but significantly antagonized by vasopressin antisera. Ether stress also elevated plasma prolactin as expected, but these changes were not quite significantly modified by the antisera. As expected, vasopressin induced a dose-related release of ACTH in doses ranging from 10 ng (10 pMoles) to 10 ug per tube when incubated with dispersed anterior pituitary cells for 2 hrs. There was no effect of any of these doses of arginine vasopressin on the release of LH or other anterior pituitary hormones in vitro except for a significant stimulation of TSH release at a high dose which we have also found in other experiments.

These results indicate that vasopressin is involved in induction of ACTH and LH release during stress. The inhibitory action of the vasopressin antisera on ACTH release may be mediated intrahypothalamically by blocking the stimulatory action of vasopressin on CRF release. We have previously alluded to experiments indicating that microinjection of vasopressin into the hypothalamus could augment the action of CRF (42). Part of the effect of the vasopressin antisera could be related to direct blockade of the stimulatory action of vasopressin on the pituitary. Because of the minute amounts of antisera given, we are inclined to discount this possibility. The effects of vasopressin on LH release and stress are presumably brought about by blockade of a stimulatory action of vasopressin on the LHRH neural terminals since there is no action of the peptide directly on the pituitary to alter LH release (43).

We have evaluated the effects of intravenous injection of antibodies directed against CRF which are highly specific for this peptide. These antibodies nearly completely blocked the ACTH-releasing action of ether stress in agreement with previous findings of Vale's group (44). This is consistent with the statements made above which indicate that CRF is the most important ACTH-releasing agent. Intraventricular injection of antibodies directed against CRF, following the same protocol just mentioned for vasopressin antibodies, resulted in no alteration in the resting levels of any of the pituitary hormones measured, but largely blocked the ACTH release from ether anesthesia. In view of the minute amount of antibody given in these experiments, we speculate that the antibodies are taken up from the ventricle and block the ultrashort-loop positive feedback of CRF to augment its own release during stress (45). We cannot, however, completely exclude the possibility that some of the antibodies reached the pituitary and had a direct blocking action on the ACTH-releasing effects of CRF at the gland. The intraventricular injection of the antisera directed against CRF completely blocked the stress-induced lowering of GH which suggests that CRF induces this lowering following its release during stress by an intrahypothalamic action. This, of course, is consistent with the previously described ability of intraventricular CRF to lower GH (39). Surprisingly, the release of LH was unaffected by the CRF antibodies which could mean that the inhibition of LH release during stress does not involve CRF and/or that the passive neutralization was not complete permitting the unneutralized CRF to exert its effect. The fact that it takes a higher dose of CRF given intraventricularly to lower LH than GH mitigates against this possibility.

In conclusion, these studies reveal a complex interplay of peptides within the hypothalamus and directly on the pituitary to modulate the secretion of anterior pituitary hormones. Additional studies with antibodies or antagonists against other peptides will be necessary to unravel the situation completely. Furthermore, the interactions of the peptides with the various monoaminergic systems already shown to play an important role in control of pituitary hormone secretion must be further studied.

Abstract

A host of peptides act within the hypothalamus and also directly on the pituitary to control the secretion of anterior pituitary hormones. The classical releasing and inhibiting hormones act directly on the gland to inhibit or stimulate the release of anterior pituitary hormones. In addition to those which have been structurallty identified, strong evidence exists for the presence of an FSHRF and a pepetidic prolactin-inhibiting factor within the hypothalamus. The releasing and inhibiting hormones appear also to act within the brain to modulate their own release and that of other releasing factors. For example, somatostatin acts centrally to stimulate GH secretion by altering the release of somatostatin and/or GH-releasing factor. GH-releasing factor acts back to inhibit GH probably by inhibiting its own release and by activating the release of somatostatin. This has been demonstrated by incubating the peptide with median eminence fragments in vitro. LHRH appears to be able to inhibit its own release following its intraventricular injection into ovariectomized rats. Oxytocin, which has the capacity to stimulate prolactin release by direct action on the pituitary, appears to inhibit its own release following its injection into the third ventricle with resultant decline in prolactin release. CRF had no ability to suppress its own release in resting animals but appeared to exert an ultrashort-loop positive feedback to stimulate ACTH release to even higher levels following imposition of ether stress. At the same time, CRF is capable of inhibiting LH and GH secretion following its intraventricular injection which suggests that it induces the stress pattern of release of hypothalamic peptides which then brings about the stress pattern of pituitary hormone secretion. Similarly, somatostatin not only affects the release of GH but also suppresses the release of TSH and gonadotropins indicative of its inhibitory effects on the release of other hypothalamic peptides.

Many other peptides affect the release of pituitary hormones via hypothalamic action. The most clearly identified example is the opioid peptides. Beta-endorphin appears to have a tonic inhibitory effect on the release of LH and to be involved in the stress-induced release of prolactin as evidenced by studies with opioid receptor blockers and with antisera directed against β-endorphin. CCK is very potent to alter the secretion of pituitary hormones. This action may have physiologic significance in view of the opposite effects produced by the administration of the CCK antagonist, proglumide. MSH has powerful effects to suppress prolactin and LH secretion which may have physiological significance since they can be reversed by intraventricular injection of antisera directed against the peptide. These are but a few examples of the myriad interactions of peptides within the brain to alter pituitary hormone secretion.

REFERENCES

1. McCann, S.M., Lumpkin, M.D., Ono, N., Khorram, O., Ottlecz, A. and Samson, W.K., 1984, Interactions of brain peptides within the hypothalamus to alter anterior pituitary hormone secretion. In, Endocrinology, F. Labrie and L. Proulx, (Eds.) Excerpta Medica, Amsterdam, pp. 185-190.

2. Schally, A.V., Arimura, A., Kastin, A.J., Matsuo, H., Baba, R., Redding, T.W., Nair, R.M.G., Debeljuk, L. and White, W.F., 1971, The gonadotropin-releasing hormone: a single hypothalamic peptide regulates the secretion of both LH and FSH. Science 173, 1036.

3. Kalra, S.P., Ajika, K., Krulich, L., Fawcett, C.P., Quijada, M. and McCann, S.M., 1971, Effects of hypothalamic and preoptic electrochemical stimulation on gonadotropin and prolactin release in proestrous rats. Endocrinology 98, 927.

4. Chappel, S.C. and Barraclough, C.A., 1976, Hypothalamic regulation of pituitary FSH secretion. Endocrinology 98, 927.

5. Ojeda, S.R., Jameson, H.E. and McCann, S.M., 1977, Hypothalamic areas involved in prostaglandin (PG)-induced gonadotropin release. II. Effects of PGE_2 and PGF implants on follicle stimulating hormone release. Endocrinology 100, 1585.

6. Ojeda, S.R., Jameson, H.E. and McCann, S.M., 1977, Hypothalamic areas involved in prostaglandin (PG)-induced gonadotropin release. I. Effects of PGE_2 and PGF implants on LH release. Endocrinology 100, 1585.

7. Lumpkin, M.D. and McCann, S.M., 1984, Effect of destruction of the dorsal anterior hypothalamus on follicle-stimulating hormone secretion in the rat. Endocrinology 115, 2473-2480.

8. Lumpkin, M.D., Samson, W.K., McDonald, J.E. and McCann, S.M., 1983, Lesions of the paraventricular region suppress pulsatile FSH but not LH release. 13th Annual Mtg. Neuroscis., #208, p. 708.

9. Bishop, W., Kalra, P.S. , Fawcett, C.P., Krulich, L., and McCann, S.M., 1972, The effects of hypothalamic lesions on the release of gonadotropins and prolactin in response to estrogen and progesterone treatment in female rats. Endocrinology 91, 1404.

10. Culler, M.D. and Negro-Vilar, A., 1985, Passive LHRH immunization suppresses pulsatile LH but not FSH secretion in the castrate male rat. Program of the 67th Annual Mtg. of the Endocrine Society, Abstract No. 788.

11. Dees, L., Rettori, V., Kozlowski, J. and McCann, S.M., 1985, Ethanol and the pulsatile release of luteinizing hormone, follicle stimulating hormone and prolactin in ovariectomized rats. Alcohol (in press).

12. Mizunuma, H., Samson, W.K., Lumpkin, M.D. and McCann, S.M., 1983, Evidence for an FSH-releasing factor in the posterior portion of the rat median eminence. Life Sci. 33, 2003-2009.

13. Samson, W.K., Snyder, G.D., Fawcett, C.P. and McCann, S.M., 1980, Chromatographic and biological analysis of ME and OVLT LHRH. Peptides 1, 97-102.

14. Lumpkin, M.D., Moltz, J.H., Yu, W., Fawcett, C.P. and McCann, S.M., 1985, Purification of sheep FSH-releasing factor and detection of its activity by in vivo bioassay. (in preparation)

15. Ottlecz, A., Samson, W.K. and McCann, S.M., 1985, The effects of gastric inhibitory polypeptide (GIP) on the release of anterior pituitary hormones. Peptides 6, 115-119.

16. McDonald, J.K., Lumpkin, M.D., Samson, W.K. and McCann, S.M., 1985, Neuropeptide Y affects secretion of luteinizing hormone and growth hormone in ovariectomized rats. Proc. Natl. Acad. Sci. 82, 561-564.

17. McCann, S.M., Lumpkin, M.D., Mizunuma, H., Khorram, O., Ottlecz, A. and Samson, W.K., 1984, Peptidergic and dopaminergic control of prolactin release. Trends in Neuroscis. 7, 127-131.

18. Dhariwal, A.P.S., Antunes-Rodrigues, J., Grosvenor, C. and McCann, S.M., 1968, Purification of ovine prolactin-inhibiting factor (PIF). Endocrinology 82: 1236-1241.

19. Mizunuma, H., Khorram, O. and McCann, S.M., 1985, Purification of a non-dopaminergic and non-GABAergic prolactin release-inhibiting factor (PIF) in sheep stalk-median eminence. Proc. Soc. Experi. Biol. and Med. 178, 114-120.

20. Khorram, O., DePalatis, L.D., and McCann, S.M., 1984, Hypothalamic control of prolactin secretion during the perinatal period in the rat. Endocrinology 115: 1698-1704.

21. Adelman, J.P., Mason, A.J., Hayflick, J.S. and Seeburg, P.H., 1985, Isolation of gene and hypothalamic cDNA for common

precursor of gonadotropin releasing hormone (GnRH) and prolactin release inhibiting factor (PIF) in human and rat. Proc. Natl. Acad. Scis. (In press).

22. McCann, S.M., 1983, Present status of LHRH: its physiology and pharmacology. In Role of Peptides and Proteins in Control of Reproduction, S.M. McCann and D.S. Dhindsa (Eds.), pp. 3-26. Elsevier, New York.

23. Vijayan, E., Samson, W.K. and McCann, S.M., 1979, In vivo and in vitro effects of cholecystokinin on gonadotropin, prolactin, growth hormone and thyrotropin release in the rat. Brain Res. 172: 295-302.

24. Vijayan, E., Samson, W.K., Said, S.I. and McCann, S.M., 1979, Vasoactive intestinal peptide (VIP): evidence for a hypothalamic site of action to release growth hormone, luteinizing hormone and prolactin in conscious ovariectomized rats. Endocrinology 104, 53-57.

25. Samson, W.K., Burton, K.P., Reeves, J.P. and McCann, S.M., 1981, Vasoactive intestinal peptide stimulates LHRH release from median eminence synaptosomes. Regulatory Peptides 2, 253-264.

26. McDonald, J.K., Lumpkin, M.D., Samson, W.K. and McCann, S.M., 1985, Pancreatic polypeptides affect LH and growth hormone secretion in rats. Peptides 6: 79-84.

27. Kalra, S.P. and Crowley, W.R., 1984, Differential Effects of Pancreatic Polypeptide on Luteinizing Hormone Release in Female Rats. Neuroendocrinology 38: 511-513.

28. Bedran de Castro, J.C. and McCann, S.M. Role of opioid peptides in control of gonadotropin secretion. (Submitted).

29. Khorram, O., DePalatis, L.D. and McCann, S.M., 1984, The effect and possible mode of action of αMSH on gonadotropin release in the ovariectomized rat. Endocrinology 114: 227-233.

30. Khorram, O. and McCann, S.M., 1984, Physiological role of alpha melanocyte-stimulating hormone in modulating the secretion of prolactin and luteinizing hormone in the female rat. Proc. Natl. Acad. Sci. 81: 8004-8008.

31. Lumpkin, M.D., Negro-Villar, A., Franchimont, B. and McCann, S.M., 1981, Evidence for a hypothalamic site of action of inhibin to suppress FSH release. Endocrinology 108: 1101-1104.

32. Piva, F., Motta, M., and Martini, L., 1979, Regulation of hypothalamic and pituitary function: long, short and ultrashort-loop feedback. In: Endocrinology (Ed.) L.J. DeGroot, et. al., Grune and Stratton, New York.

33. Mangat, H.K. and McCann, S.M., 1983, Acute inhibition of prolactin and TSH secretion after intraventricular injection of prolactin in ovariectomized rats. Am. J. of Physiol. 244: E31-E36.

34. Lumpkin, M.D., Samson, W.K. and McCann, S.M. , 1981, Effects of paradoxical elevation of growth hormone by intraventricular somatostatin: possible ultrashort-loop feedback. Science 211: 1072-1074.

35. Lumpkin, M.D., Samson, W.K. and McCann, S.M., 1985, Effects of intraventricular growth hormone-releasing factor on growth hormone release: further evidence for ultrashort-loop feedback. Endocrinology 116: 2070-2074.

36. Aguila, M.C. and McCann, S.M., 1985, Stimulation of somatostatin release in vitro by synthetic human growth hormone-releasing factor by a nondopaminergic mechanism. Endocrinology 117 (2): 762-765.

37. Aguila, M.C. and McCann, S.M., 1986, Evidence that GRF stimulates SRIF release via an opioid mechanism in vitro. (In preparation).

38. Bedran de Castro, J.C., Khorram, O., and McCann, S.M., 1985, Possible negative ultrashort-loop feedback of luteinizing hormone releasing hormone (LHRH) in the ovariectomized rat. Proc. Soc. Exper. Biol. and Med. 179: 132-135.

39. Ono, N., Bedran de Castro, J.C. and McCann, S.M., 1985, Ultrashort-loop positive feedback of corticotropin-releasing factor (CRF) to enhance ACTH release in stress. Proc. Natl. Acad. Sci. 82: 3528-3531.

40. Rivier, C. and Vale, W., 1984, Corticotropin-releasing factor (CRF) acts centrally to inhibit growth hormone secretion in the rat. Endocrinology 114: 2409-2411.

41. Yates, F.E., Russel, S.M., Dallman, M.F., Hedge, G.A., McCann, S.M. and Dhariwal, A.P.S., 1971, Potentiation by vasopressin of corticotropin release induced by corticotropin-releasing factor. Endocrinology 88: 3-15.

42. Hedge, G.A., Yates, M.B., Marcus, R. and Yates, F.E., 1966, Site of action of vasopressin in causing corticotropin release. Endocrinology 79: 328-340.

43. Ono, N., Bedran de Castro, J., Khorram, O. and McCann, S.M., 1985, Role of arginine vasopressin in control of ACTH and LH release during stress. Life Sci. 36: 1779-1786.

44. Rivier, C. and Vale, W.W., 1985, Effects of corticotropin releasing factor, neurohypophyseal peptides and catecholamines on pituitary function. Fed. Proc. 44: 189-196.

45. Ono, N., Samson, W.K., McDonald, J.K., Lumpkin, M.D., Bedran de Castro, J.C. and McCann, S.M., 1985, The effects of intravenous and intraventricular injection of antisera directed against corticotropin releasing factor (CRF) on the secretion of anterior pituitary hormones. Proc. Natl. Acad. Scis. (In press).

PHOSPHOLIPID TURNOVER, Ca^{2+} MOBILIZATION AND PROTEIN
KINASE C ACTIVATION IN GnRH ACTION ON PITUITARY GONADOTROPHS

Zvi Naor, Rona Limor and Jacob Hermon

Department of Hormone Research
The Weizmann Institute of Science
Rehovot, 76100, Israel

INTRODUCTION

Gonadotropin releasing hormone (GnRH) binds to pituitary gonadotrophs (LH + FSH containing cells) and stimulates the biosynthesis and release of the gonadotropins. Peptide hormones in general exert their biological effects via the formation of 'second messengers'. The messenger molecule activates a respective protein kinase resulting in the phosphorylation and activation of key proteins and enzymes, and hence in the physiological response (Greengard, 1978). The classical 'second messengers' cAMP, cGMP and prostaglandins were ruled out as potential mediators of the acute actions of GnRH (for review see Naor, 1982). On the other hand, Ca^{2+} mobilization and phospholipid turnover were implicated in the hormone action (Pickering and Fink, 1979; Naor and Catt, 1981; Naor et al., 1980; 1985a; Conn et al., 1981; Snyder and Bleasdale, 1982; Raymond et al., 1984; Kiesel and Catt, 1984; Schrey, 1985). In this chapter we will review the recent findings, from several laboratories, concerning the mechanism of action of GnRH.

PHOSPHOINOSITIDE TURNOVER

The importance of phosphatidylinositol (PI) turnover in signal transduction was first introduced by Hokin and Hokin (1953) and later proposed as a more general mechanism for Ca^{2+}-mobilizing receptors (Michell, 1975) (Fig. 1). It is now thought that the activated receptor transfers the information via a GTP-binding protein analogous to Ni (Okajima et al., 1985), to a specific phospholipase C. Unlike other phospholipases, this particular enzyme can be activated even at very low Ca^{2+} concentrations (Wilson et al., 1984). The activated enzyme hydrolyses phosphatidylinositol-4,5-bisphosphate (PIP_2) to 1,2-diacylglycerol (DG) and myo-inositol-1,4,5-trisphosphate (IP_3). The two products are regarded as 'second messengers' since DG activates the Ca^{2+}/phospholipid-dependent protein kinase (C-kinase; Nishizuka, 1984), and IP_3 mobilizes intracellular Ca^{2+} (Berridge

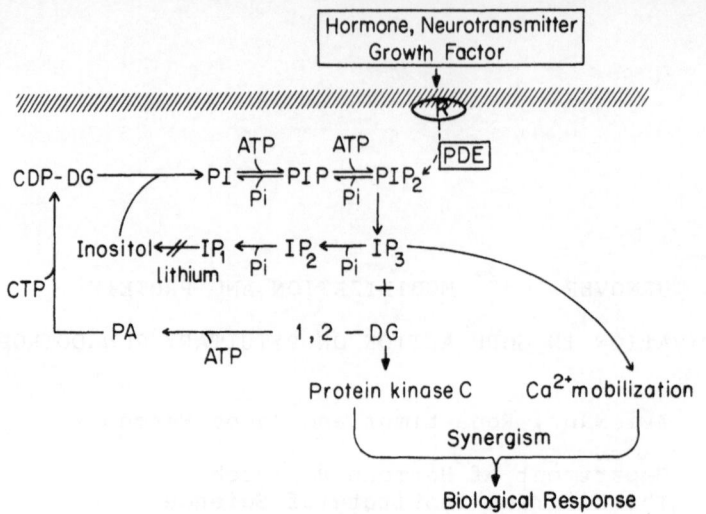

Fig. 1. Phosphoinositide turnover. Receptor (R) activation
results in breakdown of phosphatidylinositol 4,5
bisphosphate (PIP_2) to 1,2-diacylglycerol (DG) and
inositol trisphosphate (IP_3) by a specific phospholipase C (PDE). Phosphorylation by ATP and activation by diacylglycerol (DG) kinase forms phosphatidic acid (PA). Conjugation with CTP by
phosphatidic acid: CTP cytidyltransferase forms
CDP-diacylglycerol. An exchange of the activated
CDP with free inositol by CDP-diacylglycerol inositol phosphatidyltransferase forms PI. Multiple
phosphorylation produces phosphatidylinosi-
tol-4-phosphate (PIP) and PIP_2. A series of inositol phosphatases converts IP_3 to free inositol
which interacts with CDP-diacylglycerol to resyn-
thesize PI. (Modified from Nishizuka, 1984; Berridge and Irvine, 1984; and Wilson et al., 1985).

and Irvine, 1984). DG is then phosphorylated to phosphatidic acid and IP_3 is converted to inositol. The activated DG
and inositol will produce PI, and phosphorylation of PI will
result in the formation of phosphatidylinositol-4-phosphate
(PIP) and PIP_2. It was recently suggested that while IP_3 is
formed from PIP_2, most of the DG is produced from direct hydrolysis of PI rather than PIP_2 (Wilson et al., 1985). The
importance of the cycle is therefore the production of the
second messengers DG and IP_3 which are involved in signal
transduction mechanisms.

Several reports have appeared indicating that GnRH stimulates pituitary PI turnover (Snyder and Bleasdale, 1982;
Raymond et al., 1984; Kiesel and Catt, 1984; Naor et al.,
1985a). The effect is rapid, receptor-mediated and
Ca^{2+}-independent. More recently, it was suggested that GnRH
stimulates the hydrolysis of polyphosphoinositides in pituitary and ovarian granulosa cells (Schrey, 1985; Davis et
al., 1984). However, in the pituitary study (Schrey, 1985),
the effect observed was too slow to account for the produc-

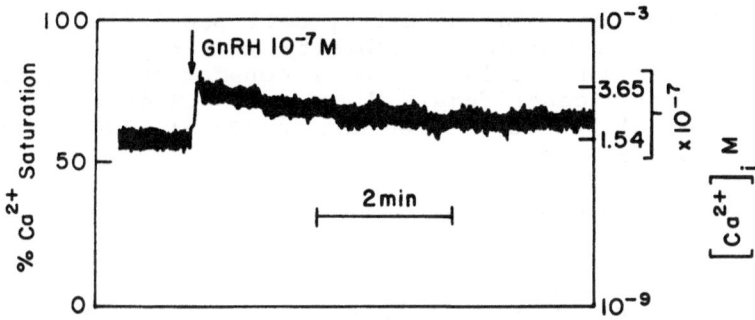

Fig. 2. Effect of GnRH on $[Ca^{2+}]_i$ in cultured gonadotrophs. The enriched cells were prepared by centrifugal elutriation as described by Hyde et al. (1982). Traces of Quin-2 fluorescence are shown. Note the early 'spike phase' of GnRH action which is followed by a sustained phase of elevated $[Ca^{2+}]_i$.

tion of a second messenger. Studies utilizing purified gonadotrophs should provide the clue whether GnRH exerts a rapid effect on IP_3 formation, and whether the early stimulation of pituitary PI turnover by GnRH is involved in raising free cytosolic Ca^{2+} levels.

Ca^{2+} MOBILIZATION

Several investigators have suggested that Ca^{2+} is a mediator of GnRH-induced gonadotropin release (Wakabayashi et al., 1969; Williams, 1976; Hopkins and Walker, 1978; Pickering and Fink, 1979; Naor et al., 1980; Conn et al., 1981). This was based on the observations that extracellular Ca^{2+} was required for GnRH-induced LH release; the Ca^{2+} ionophore A23187 mimicked GnRH action; Ca^{2+}-channel blockers such as verapamil and La^{3+} inhibited GnRH-induced LH release; calmodulin antagonists blocked GnRH action; and GnRH induced a transient rise in $^{45}Ca^{2+}$ efflux concomitant with elevated release of LH. Nevertheless, detailed quantitative analyses and direct measurements of changes in free cytosolic Ca^{2+} concentrations ($[Ca^{2+}]_i$) induced by GnRH in pituitary cells have only recently been performed. Such measurements are now possible with fluorescent probes such as Quin 2/AM, which can be trapped into cells and enable direct determination of $[Ca^{2+}]_i$ (Tsien et al., 1982). We have recently used the technique to directly quantitate the changes in $[Ca^{2+}]_i$ in response to GnRH in enriched gonadotrophs (Limor et al., submitted). Studies of this kind have been hampered so far by the fact that pituitary cells are heterogeneous and GnRH-responsive cells (gonadotrophs) constitute only about 10% of the total pituitary cell population (Naor et al., 1982a). The recent availability of a rapid and reliable method for the isolation of a gonadotroph-enriched cell population by centrifugal elutriation (Hyde et al., 1982; Childs et al., 1983; Thieulant and Duval, 1985) made this study possible.

GnRH induced a rapid spike (6-8 sec) in gonadotroph's $[Ca^{2+}]_i$ followed by a plateau phase of prolonged elevated $[Ca^{2+}]_i$ which lasted several min (Fig. 2 and Limor et al., submitted). GnRH elevated $[Ca^{2+}]_i$ only in gonadotroph-enriched cells, while thyrotropin-releasing hormone (TRH) and growth hormone-releasing factor (GHRF) elevated $[Ca^{2+}]_i$ in mammotrophs- and somatotrophs-enriched cells, respectively. The rapid spike phase in $[Ca^{2+}]_i$ induced by GnRH was also observed in Ca^{2+}-free medium containing EGTA but this was terminated within 2 min. Readdition of Ca^{2+} to the medium induced a second slower rise in $[Ca^{2+}]_i$ ('plateau phase'). The rise in $[Ca^{2+}]_i$ in GnRH-stimulated gonadotrophs originates partly from intracellular pools ('spike phase') and partly through influx across the cell membrane ('plateau phase'). This profile of elevation of $[Ca^{2+}]_i$ was also observed in clonal rat pituitary cells (GH$_3$) stimulated by TRH (Albert and Tashjian, 1984a, b; Gershengorn and Thaw, 1983).

The recent finding that GnRH stimulates the relese of inositol trisphosphate (IP$_3$) from hemipituitaries (Schrey, 1985) supports our finding that GnRH mobilizes Ca^{2+} also from intracellular stores. GnRH-induced gonadotropin secretion is biphasic (Naor et al., 1982b). The first phase of LH release is evident immediately upon GnRH stimulation and is followed by a second phase that is slower but more sustained. The kinetics of GnRH-induced elevation of $[Ca^{2+}]_i$ described here are in good agreement with the LH release profile. The two phases of GnRH-induced Ca^{2+} mobilization might therefore be involved in the biphasic release process of the gonadotropins.

Interestingly, the elevation of $[Ca^{2+}]_i$ induced by GnRH is relatively small and remained below the level needed generally for a $[Ca^{2+}]_i$ rise to elicit secretion by itself (Rink et al., 1982). It is therefore possible that both Ca^{2+} and diacylglycerol are needed to elicit the full response as second messengers for GnRH action, as we have recently demonstrated (Naor and Eli, 1985). We, therefore, suggest that the elevation of $[Ca^{2+}]_i$ by GnRH might establish a 'set point' enabling the full expression of protein kinase C activation.

PROTEIN KINASE C

The newly discovered Ca^{2+}-activated, phospholipid-dependent protein kinase (C-kinase) is activated by association with membrane phospholipids in particular phosphatidylserine (PS) in the presence of elevated Ca^{2+} (Nishizuka, 1984). Unsaturated diacylglycerol (DG) increases the apparent affinity of C-kinase for PS and decreases the Ca^{2+} concentration needed for maximal enzyme activity (Nishizuka, 1984).

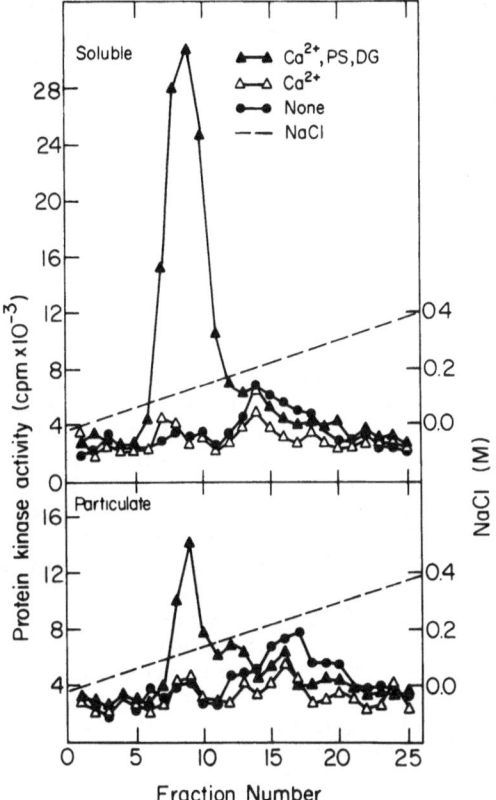

Fig. 3. DEAE-cellulose (DE-52) chromatography of pituitary protein kinase C (PKC).

Since DG is generated during phosphoinositide turnover, C-kinase is most likely involved in signal transduction mechanisms of Ca^{2+}-mobilizing receptors. We, and others, have recently reported the presence of C-kinase in the pituitary (Turgeon et al., 1984; Hirota et al., 1985; Naor et al., 1985b). Pituitary C-kinase is mostly soluble (70%) and partly particulate (30%; Fig. 3). However, while the soluble form is recovered in an inactive state, the particulate form is found in a cofactor-insensitive state (Naor et al., 1985b; Hermon et al., submitted). Therefore the soluble form of the enzyme can also be detected in a crude cytosolic preparation, while the particulate enzyme is detectable only after solubilization and anion exchange chromatography (Naor et al., 1985b; Hermon et al., submitted). Pituitary C-kinase is recruited from the cytosol to the membrane upon GnRH or TRH challenge (Hirota et al., 1985; Naor et al., 1985b; Drust and Martin, 1985; Fearon and Tashjian, 1985). Translocation of C-kinase to the membrane by GnRH or TRH following increased PI turnover will expose the inactive enzyme to DG and PS present in the membrane and permit activation even in the face of micromolar concentrations of Ca^{2+}.

The tumor promoters phorbol esters exert pleiotropic effects on a variety of cells and recent findings suggest that C-kinase is a major initial cellular binding site and transducer of their actions (Nishizuka, 1984). Thus, C-kinase might be involved in signal-transduction mechanisms responsible for secretion, proliferation and transformation. When cultured pituitary cells were stimulated with synthetic diacylglycerol such as 1-oleoyl-2-acetylglycerol (OAG), or with the potent tumor promoter 12-0-tetradecanoyl-phorbol-13-acetate (TPA), which are known stimulators of C-kinase, enhanced release of LH was observed (Naor and Eli, 1985). Similarly, LH release was also stimulated by the Ca^{2+}-ionophore, A23187. Simultaneous presence of A23187 and OAG or TPA resulted in a synergistic response that mimicked the full physiological response to $GnRH_2$. Thus simultaneous activation of the two branches of the Ca^{2+} messenger system, namely C-kinase activation and elevation of $[Ca^{2+}]_i$, are needed for full expression of GnRH action. GnRH induced LH release is therefore most likely optimally mediated by coordinated production of functionally different 'second messengers' (Ca^{2+} and DG) which are responsible for a different phase or part of the final exocytotic reaction mechanism. Since GnRH-induced LH release is biphasic (Naor et al., 1982b) it is possible that Ca^{2+} mediates the first fast phase while C-kinase mediates the more prolonged second phase of release as suggested by some investigators in another system (Kolesnick and Gershengorn, 1985). Alternatively, both Ca^{2+} and C-kinase are needed simultaneously for the full expression of the two release phases as also suggested for TRH-induced prolactin secretion (Albert and Tashjian, 1984a, b).

ARACHIDONIC ACID AND METABOLITES

The arachidonic acid (AA) needed for prostaglandin (PG) and leukotriene (LT) production can be derived from several sources. The diglyceride (DG) formed during PI turnover can be hydrolyzed by diglyceride lipase to yield AA. The second potential source is a phosphatidic acid (PA)-specific phospholipase A_2 that hydrolyzes PA and produces lysophosphatidic acid (LPA) and AA. Third, LPA can serve as a Ca^{2+} ionophore and activate phospholipase A_2 that acts on other phospholipids such as phosphatidylcholine and phosphatidylethanolamine to release more AA. Fourth, AA can also be liberated from cellular phospholipids by phospholipase A_2 independently of the PI cycle but in response to elevated cytosolic free Ca^{2+}. We have previously demonstrated that GnRH induces the release of free AA from cultured gonadotrophs (Naor and Catt, 1981). Prostaglandins are not involved in the pituitary actions of GnRH (Naor 1982). On the other hand, we and others have recently suggested that lipoxygenase or epoxygenase derivatives of AA might mediate GnRH action on gonadotropin release (Snyder et al., 1983; Naor et al., 1983; 1985c). We found that lipoxygenase inhibitors blocked GnRH-induced gonadotropin release while the prostaglandin synthesis inhibitor, indomethacin, potentiated GnRH action (Naor et al., 1983, 1985c). More recent studies reveal the presence of lipoxygenase activity in purified gonadotrophs (Vanderhoek et al., 1984), and that leukotriene C_4 stimulates LH release from anterior pituitary cells (Hulting et al., 1985).

CONCLUSIONS

GnRH binds to rat pituitary gonadotrophs and stimulates PI turnover in a Ca^{2+}-independent fashion (Fig. 4). Whether the first substrate of the activated phospholipase C is PI or PIP_2 is not clear yet, but preliminary studies suggest it is PIP_2. GnRH elevates cytosolic free $[Ca^{2+}]_i$ by mobilizing an intracellular pool and by influx across the cell membrane. It is suggested that the early activated PI turnover is involved in Ca^{2+} mobilization. GnRH translocates cytosolic C-kinase to the membrane and several endogenous substrate proteins for pituitary C-kinase were noticed. Activators of C-kinase such as diacylglycerol and phorbol ester stimulate LH release and act synergistically with Ca^{2+} ionophore to mimic the full response to GnRH. A possible site for Ca^{2+} action is the release of arachidonic acid and production of active metabolites such as leukotriene C_4 which is capable of stimulating LH release. The known biphasic response of pituitary gonadotrophs to GnRH might therefore be mediated by Ca^{2+} mobilization and C-kinase activation.

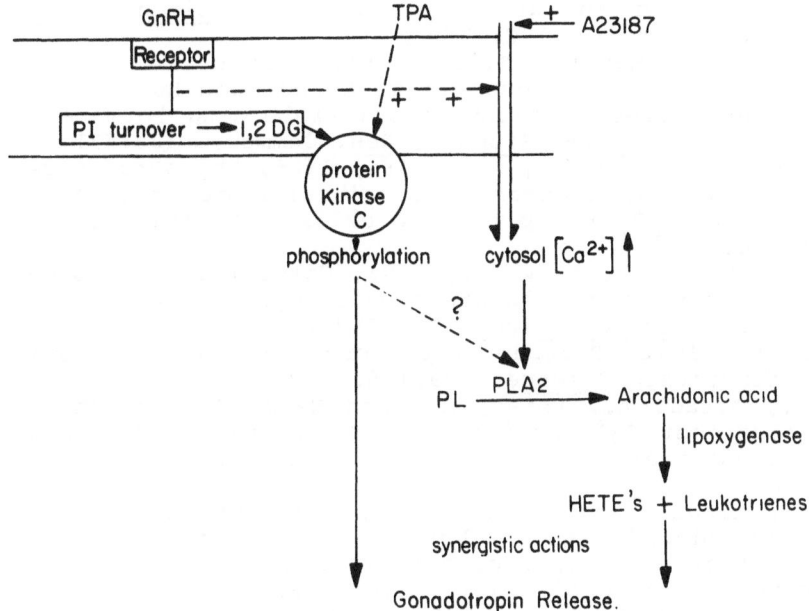

Fig. 4. Proposed mechanism of action of GnRH. PL, phospholipid; PLA_2, phospholipase A_2; HETE, hydroxyeicosatetraenoic acid.

Acknowledgements: Supported by NIH Grants HD-16279 and by the United States Israel Binational Science Foundation. The active collaboration with Drs. K. Catt, G. Childs and A.Capponi and the excellent technical assistance of A. Azrad, Y. Eli and T. Hannoch made this work possible. We also thank Mrs. M. Kopelowitz for typing the review.

REFERENCES

Albert, P.R., and Tashjian, A.H., Jr., 1984a, Thyrotropin-releasing hormone-induced spike and plateau in cytosolic free Ca^{2+} concentrations in pituitary cells. J. Biol. Chem., 259:5827.

Albert, P.R., and Tashjian, A.H., Jr., 1984b, Relationship of thyrotropin-releasing hormone-induced spike and plateau phases in cytosolic free Ca^{2+} concentrations to hormone secretion. J. Biol. Chem. 259:15350.

Berridge, M.J., and Irvine, R.F., 1984, Inositol trisphosphate, a novel second messenger in cellular signal transduction. Nature, 312:315.

Childs, G.V., Hyde, C., Naor, Z., and Catt, K.J., 1983, Heterogeneous luteinizing hormone and follicle stimulating hormone storage patterns in subtypes of gonadotrophs separated by centrifugal elutriation. Endocrinology, 113:2120.

Conn, P.M., Marian, J., McMillian, M., Stern, J., Rogers, D., Hamby, M., Penna, A., and Grant, E., 1981, Gonadotropin releasing hormone action in the pituitary: A three step mechanism, Endocr. Rev., 2:174.

Davis, J.S., West, L.A., and Farese, R.V., 1984, Gonadotropin-releasing hormone rapidly alters polyphosphoinositide metabolism in rat granulosa cells. Biochem. Biophys. Res. Commun., 122:1289.

Drust, D.S., and Martin, T.F.J., 1985, Protein kinase C translocates from cytosol to membrane upon hormone activation: Effects of thyrotropin-releasing hormone in GH_3 cells. Biochem. Biophys. Res. Commun. 128:531.

Fearon, C.W., and Tashjian, A.H., Jr., 1985, Thyrotropin-releasing hormone induces redistribution of protein-kinase C in GH_4C_1 rat pituitary cells. J. Biol. Chem., 260:8366.

Gershengorn, M.C., and Thaw, C., 1983, Calcium influx is not required for TRH to elevate free cytoplasmic calcium in GH_3 cells. Endocrinology, 113:1522.

Greengard, P., 1978, Phosphorylated proteins as physiological effectors. Science, 199:146.

Hermon, J., Azrad, A., Reiss, N., and Naor, Z., Phospholipid-dependent Ca^{2+}-activated protein kinase (C-kinase) in the pituitary: Further characterization and endogenous substrate. (Submitted).

Hirota, K., Hirota, T., Aguilera, G., and Catt, K.J., 1985, Hormone-induced redistribution of calcium-activated phospholipid-dependent protein kinase in pituitary gonadotrophs. J. Biol. Chem., 260:3243.

Hokin, M.R., and Hokin, L.E., 1953, Enzyme secretion and the incorporation of [^{32}P] into phospholipids of pancreas slices. J. Biol. Chem. 203:967.

Hopkins, C.R., and Walker, A.M., 1978, Calcium as a second messenger in the stimulation of luteinizing hormone secretion. Mol. Cell. Endocrinol., 12:189.

Hulting, A.L., Lindgren, J.A., Hokfelt, T., Eneroth, P., Werner, S., Patrono, C., and Samuelsson, B., 1985, Leukotriene C_4 as a mediator of luteinizing hormone release from rat anterior pituitary cells. Proc. Natl. Acad. Sci., U.S.A., 82:3834.

Hyde, C.L., Childs (Moriarty), G., Wahl, L.M., Naor, Z., and Catt, K.J., 1982, Preparation of gonadotroph enriched cell population from adult rat anterior pituitary cells by centrifugal elutriation. Endocrinology, 111:1421.

Kiesel, L., and Catt, K.J., 1984, Phosphatidic acid and the calcium-dependent actions of gonadotropin-releasing hormone in pituitary gonadotrophs. Arch. Biochem. Biophys., 231:202.

Kolesnick, R.N., and Gershengorn, M.C., 1985, Direct evidence that burst but not sustained secretion of prolactin stimulated by thyrotropin-releasing hormone is dependent on elevation of cytoplasmic calcium. J. Biol. Chem., 260:5217.

Limor, R., Ayalon, D., Capponi, A.M., and Naor, Z. Cytosolic free calcium levels in cultured pituitary cells separated by centrifugal elutriation: Effect of gonadotropin releasing hormone. (Submitted).

Michell, R.H., 1975, Inositol phospholipids and cell surface receptor function. Biochem. Biophys. Acta, 415:81.

Michell, R.H., Kirk, C.J., Jones, L.M., Downes, C.P., and Creba, J.A., 1981, The stimulation of inositol lipid metabolism that accompanies calcium mobilization in stimulated cells: defined characteristics and unanswered questions. Phil. Trans. R. Soc. Lond. B., 296:123.

Naor, Z., 1982, Cyclic nucleotide production and hormonal control of anterior pituitary cells. in: "Multihormonal Regulation in Neuroendocrine Cells", A. Tixier-Vidal, P. Richard, eds., Inserm 110:395.

Naor, Z., and Catt, K.J., 1981, Mechanism of action of gonadotropin-releasing-hormone: Involvement of phospholipid turnover in luteinizing hormone release. J. Biol. Chem., 256:2226.

Naor, Z., and Eli, Y., 1985, Synergistic stimulation of luteinizing hormone release by protein kinase C activators and Ca^{2+}-ionophore. Biochem. Biophys. Res. Commun., 130:848.

Naor, Z., Leifer, A.M., and Catt, K.J., 1980, Calcium-dependent actions of gonadotropin-releasing hormone on pituitary guanosine 3'5'-monophos- phate production and gonadotropin release. Endocrinology, 107:1438.

Naor, Z., Childs, G.V., Leifer, A.M., Clayton, R.N., Amsterdam, A., and Catt, K.J., 1982a, Gonadotropin releasing hormone binding and activation of enriched population of pituitary gonadotrophs. Mol. Cell. Endocrinol., 25:85.

Naor, Z., Katikineni, M., Loumaye, E., Garcia Vella, A., Dufau, M.L., and Catt, K.J., 1982b, Compartmentalization of luteinizing hormone pools: Dynamics of gonadotropin releasing hormone action in superfused pituitary cells. Mol. Cell. Endocrinol., 27:213.

Naor, Z., Vanderhoek, J.Y., Lindner, H.R., and Catt, K.J., 1983, Arachidonic acid products as possible mediators of the action of gonadotropin releasing hormone. in: "Advances in Prostaglandins Thromboxane and Leukotriene Research", B. Samuelsson, R. Paoletti, P. Ramwell, eds., Raven Press, 12:259.

Naor, Z., Molcho, J., Zakut, H., and Yavin, E., 1985a, Calcium-independent phosphatidylinositol response in gonadotropin-releasing hormone-stimulated pituitary cells. Biochem. J. 231:19.

Naor, Z., Zer, J., Zakut, H., and Hermon, J., 1985b, Characterization of pituitary calcium-activated phospholipid-dependent protein-kinase: redistribution by gonadotropin releasing hormone. Proc. Nat. Acad. Sci., (in press).

Naor, Z., Kiesel, L., Vanderhoek, J.Y., and Catt, K.J., 1985c, Mechanism of action of gonadotropin releasing hormone: Role of lipoxygenase products of arachidonic acid in leutinizing hormone release. J. Steroid Biochem. (in press).

Nishizuka, Y., 1984, The role of protein kinase C in cell surface signal transduction and tumor promotion. Nature, 308:693.

Okajima, F., Katada, T., and Ui, M., 1985, Coupling of the guanine nucleotide regulatory protein to chemotactic peptide receptors in neutrophil membranes and its uncoupling by islet-activating protein, pertusis toxin. J. Biol. Chem., 260:6761.

Pickering, A.J.M.C., and Fink, G. 1979, Priming effect of LHRH in vitro: role of protein synthesis, contractile elements, Ca^{2+} and cyclic AMP. J. Endocrinol., 81:223.

Raymond, V., Leung, P.C.K., Veilleux, R., Lefevre, G., and Labrie, F., 1984, LHRH rapidly stimulates phosphatidylinositol metabolism in enriched gonadotrophs. Mol. Cell. Endocrinol., 36:157.

Rink, T.J., Smith, S.W., and Tsien, R.Y., 1982, Cytoplasmic free Ca^{2+} in human platelets: Ca^{2+} thresholds and Ca-independent activation for shape-change and secretion. Febs. Letters. 148:21.

Schrey, M.P., 1985, Gonadotropin-releasing hormone stimulates the formation of inositol phosphates in rat anterior pituitary tissue, Biochem. J., 226:563.

Snyder, G.D., and Bleasdale, J.E., 1982, Effect of LHRH on incorporation of [^{32}P]-orthophosphate into phosphatidylinositol by dispersed anterior pituitary cells, Mol. Cell. Endocrinol. 28:55.

Snyder, G.D., Capdevila, J., Chacos, N., Manna, S., and Falck, J.R., 1983, Action of luteinizing hormone releasing hormone: Involvement of novel arachidonic acid metabolites. Proc. Natl. Acad. Sci. U.S.A., 80:3504.

Thieulant, M.L., and Duval, J., 1985, Differential distribution of androgen and estrogen receptors in rat pituitary cell populations separated by centrifugal elutriation, Endocrinology, 116:1299.

Tsien, R.Y., Pozzan, T., and Rink, T.J., 1982, Calcium homeostasis in intact lymphocytes: cytoplasmic free calcium monitored with a new intracellularly trapped fluorescent indicator. J. Cell Biol., 94:325.

Turgeon, J.L., Ashcroft, S.J.H., Waring, D.W., Milewski, M.A., and Walsh, D.A., 1984, Characteristics of adenohypophyseal Ca^{2+}-phospholipid dependent protein kinase. Mol. Cell. Endocrinol. 34:107.

Vanderhoek, J.Y., Kiesel, L., Naor, Z., Bailey, J.M., and Catt, K.J., 1984, Arachidonic acid metabolism in gonadotroph enriched pituitary cells, Prostag. Leukotriene Med., 15:375.

Wakabayashi, K., Kamberi, I.A., and McCann, S.M., 1969, In vitro response of the rat pituitary to gonadotropin releasing factors and to ions. Endocrinology, 85:1046.

Williams, J.A., 1976, Stimulation of $^{45}Ca^{++}$ efflux from rat pituitary by LHRH and other pituitary stimulants. J. Physiol., 260:105.

Wilson, D.B., Bross, T.E., Hofman, S.L., and Majerus, P.W., 1984, Hydrolysis of polyphosphoinositides by purified sheep seminal vesicles phospholipase C enzymes. J. Biol. Chem., 259:11718.

Wilson, D.B., Neufeld, E.J., and Majerus, P.W., 1985, Phosphoinositide interconversion in thrombin stimulated human platelets. J. Biol. Chem., 260:1046.

HORMONAL REGULATION OF PITUITARY

GONADOTROPIN GENE EXPRESSION

R. Counis, M. Corbani, A. Starzec and M. Jutisz

Laboratoire des Hormones Polypeptidiques, CNRS
91190 Gif sur Yvette, France

INTRODUCTION

Lutropin (LH) and follitropin (FSH) are two members of a family of structurally related polypeptide hormones, that also includes pituitary thyrotropin (TSH) and placental human chorionic gonadotropin (hCG). Each hormone of this family contains two non identical, non covalently linked subunits, α and β. The primary structure of the α subunits is identical among hormones within a species, whereas the primary structures of the β subunits differ greatly and confer to the hormones their specific biological activities (Pierce and Parsons, 1981). Although they are not identical, the β subunits show enough homology in their primary structures to suggest that they arose by duplications and mutations of a single ancestral gene (Fontaine and Burzawa-Gérard, 1977).

These structural properties of gonadotropic hormones were extensively exploited during the past few years to study the mechanisms of their biosynthesis. Using translation of rat, ovine and bovine pituitary poly (A^+)mRNA in heterologous cell free protein synthesising systems, we and others showed that gonadotropin subunits α, LHβ and FSHβ are synthesised independently, as precursors encoded by separate messengers (Landefeld, 1979; Landefeld and Kepa, 1979; Keller et al., 1980; Counis et al., 1981, 1982a; Alexander and Miller, 1981).

The production of pituitary gonadotropins is subject to a complex regulatory process involving a number of factors including the hypothalamic gonadotropin-releasing hormone (GnRH) which is a potent stimulator of gonadotropin release, gonadal steroid hormones, which exert complex, negative and positive feedback effects on gonadotropin release and/or synthesis, and probably other regulatory substances such as thyroid hormones, neurotransmitter amines and neuropeptides. Therefore LH and FSH provide an interesting model for studying the multihormonal regulation of gene expression. With this objective, different approaches were developped in our laboratory. Cell-free translation of pituitary mRNAs in an heterologous protein synthesising system and molecular hybridization of specific mRNAs with complementary DNA sequences allowed us to assess changes in messenger activity and/or in the number of messages after in vivo treatment of rats or in vitro incubation of pituitary cells in culture. Complementary to these techniques, influences on the biosynthesis of the polypeptide chains of the gonadotropin subunits was explored using the incorporation of labeled amino acids by cultured pituitary cells.

REGULATION OF THE SYNTHESIS OF GONADOTROPIN SUBUNITS
BY GONADAL STEROIDS

First we examined the effect of gonadectomy of male and female rats on the specific mRNA content of their pituitary glands (Counis et al., 1982b; 1982c; Corbani et al., 1984). Using the cell-free translation of pituitary mRNA, we have observed that the translational capacity of the specific messages encoding precursors to α, LHβ and FSHβ increases with time during the 4 weeks following gonadectomy. The rate of synthesis of all 3 precursors rose rapidly with a significant increase 4 days after gonadectomy in both males and females, and a plateau was reached after 21 days. At that time, the translational capacity of each mRNA was enhanced by 10-15 fold as compared to normal rats. As already known (McCann and Ramirez, 1964; Schally et al., 1972), serum LH rapidly increases during the 4 weeks following castration, but differently in male and female rats (Corbani et al., 1984). Thus, although no direct correlation appears to exist between serum levels of LH, and pituitary levels of LH subunit mRNAs, it is evident that the two phenomena are affected by gonadal steroids.

In order to investigate the mechanisms by which gonadal steroids control the synthesis of gonadotropin subunits, we compared (Fig.1) mRNA preparations from normal and castrated rats and ewes (Counis et al., 1984) for their content in specific mRNAs using hybridization techniques. Oligodeoxynucleotides (ODN), 15- to 16-base long, corresponding to a short portion in the DNA strand of the genes coding for each of the gonadotropin subunit mRNAs, were chemically synthesized. Their nucleotide sequence was deduced by exploiting complete homologies in amino acid sequence of highly conserved regions present in subunits of pituitary gonadotropins, and the subunits of hCG, the cDNA of which was already obtained by recombinant technology by Fiddes and Goodman (1979, 1980).

The 3 synthetic ODNs were 5'-end labeled and their hybridization to rat and ovine pituitary poly(A$^+$)RNAs, previously separated by agarose gel electrophoresis and transferred onto solid support (gene screen, NEN), was tested. As a result of this experiment, the 15-mer ODNs devised for α- and FSHβ-mRNA, each hybridized with a single complementary sequence derived from rat pituitary, while the 16-mer ODN devised for LHβ hybridized to a single ovine mRNA. These results illustrate the high specificity in each RNA-ODN interaction; only one or two mismatches in a short sequence of 15- to 16-base ODN may render the hybrid unstable (Wallace et al., 1979) and therefore compromise detection of mRNA. Furthermore, our data reveal the presence of cryptic mutations in the selected regions of the genes, encoding highly conserved amino acid sequences of the glycoprotein hormones.

According to their species specificity, ODNs were used as probes to compare α- and FSHβ-mRNA in normal and ovariectomized rat pituitaries, and LHβ-mRNA in normal and ovariectomized ewe pituitaries. It is evident from Fig.1, which shows the autoradiograph of filters after hybridization, that castration results in an increase in the number of each of the 3 specific RNA messages encoding precursors to α, LHβ and FSHβ. This autoradiograph confirms these precursors are encoded by distinct mRNAs, of 820, 640 and 680 nucleotides, respectively.

Further studies (Counis et al., 1983) showed that supplementation, 4 weeks after ovariectomy, of female rats with a single injection (s.c.) of 25 µg estradiol benzoate resulted in a rapid decrease of the synthesis of all 3 gonadotropin precursors α, LHβ and FSHβ, produced in cell-free

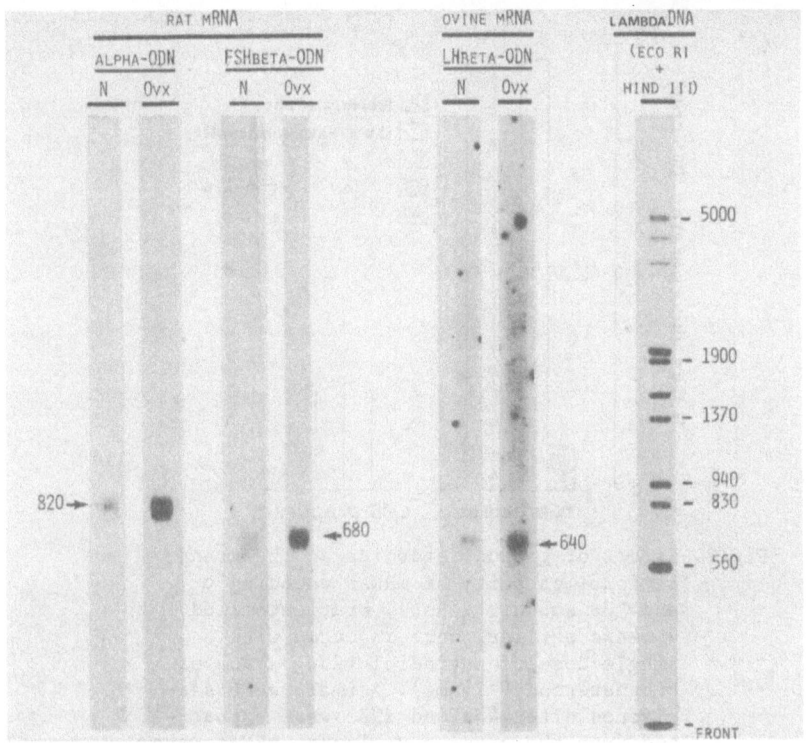

Fig.1. Autoradiograph of the filters after transfer of
pituitary poly(A)$^+$RNAs derived from normal (N)
and ovariectomized (ovx) adult rats and ewes and
hybridization with ^{32}P-labeled, synthetic
oligodeoxynucleotides (ODNs). ODNs were specific
to a part of the mRNA sequences encoding α, LHβ
and FSHβ (see the text). Messenger RNAs were glyo-
xal-denatured, separated by agarose gel electrophoresis and transfered to a solid support (gene
screen, NEN). On the right are reference-DNAs with
their nucleotide numbers. (From Counis et al.,
1984; Courtesy of Plenum Press).

conditions by translation of pituitary mRNA, to values near those observed
in normal rats. Fig.2 illustrates that a dramatic decrease of 75% in the
translational capacity of α- and LHβ-mRNA, is already observed 48 hours
after steroid injection. Similar results in the inhibition of FSHβ
(Alexander and Miller, 1982), LHβ and α subunits (Landefeld et al., 1983;
Landefeld and Kepa, 1984) have been reported with estradiol injected into
castrated sheep. Recent data from our laboratory indicate that testosterone
and dihydrotestosterone, also participate in this regulation in the male
rat (unpublished results). Progesterone, alone or in combination with
estradiol appeared, in the conditions tested, to have no effect. However,
progesterone receptors are estradiol-inducible and the possibility that
they do not work after 48 hours estradiol treatment is now under
investigation.

As was observed for FSHβ by Alexander and Miller (1982) in the sheep,
we obtained evidence that steroids act directly, at least in part, at the
pituitary level, using pituitary cells cultured in the presence of gonadal
steroids. That estradiol and testosterone negatively control the synthesis

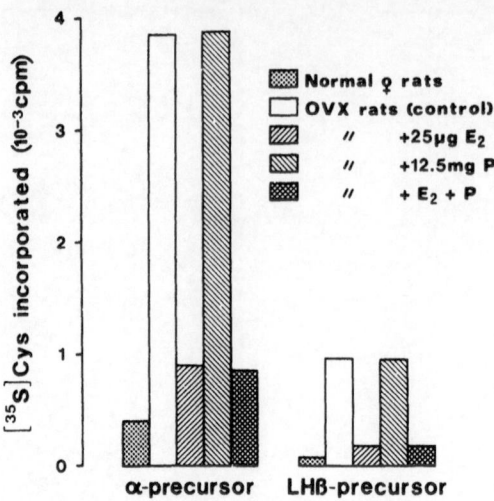

Fig.2. Effect of gonadal steroids on the translational capacity of mRNAs encoding α
and LHβ subunits. Rats, ovariectomized
4 weeks earlier, were injected with a
single dose of estradiol (25μg), and of
progesterone (12,5mg). Animals were sacrificed after 48h and RNAs were extracted from the pituitary glands. mRNAs were
translated in heterologous cell-free media
in the presence of ^{35}S-Met, in parallel with those derived from normal rats
and castrated non treated rats. The subunit precursors were immunoprecipitated
and their radioactivity was counted after
SDS-polyacrylamide gel electrophoresis.

of the polypeptide chains of the pituitary gonadopropin subunits was
further substantiated using our specific oligonucleotides as probes. Using
cloned cDNA, comparable results were obtained for α- and LHβ-mRNA in the
sheep (Nilson et al., 1983; Landefeld et al., 1984) and in the rat (Chin
and Habener, 1984). Our data demonstrate FSHβ-mRNA is regulated as well,
indicating α, LHβ and FSHβ are co-ordinately regulated by gonadal steroids.

More detailed information will be obtained by the use of long DNA
probes complementary to RNA messages. Therefore, we employed our specific,
synthetic ODNs to detect in cDNA libraries recombinant clones having such
inserts. We have already obtained cDNA sequences for α (fully
characterized) and LHβ (nucleotide sequencing in progress). We are now
currently screening large cDNA libraries for FSHβ.

REGULATION OF THE SYNTHESIS OF GONADOTROPIN SUBUNITS BY GnRH

It was repeatedly reported that GnRH stimulates, in addition to
release, the synthesis of pituitary gonadotropins (see the review by Khar
and Jutisz, 1980). However, a controversy existed as to the mechanisms

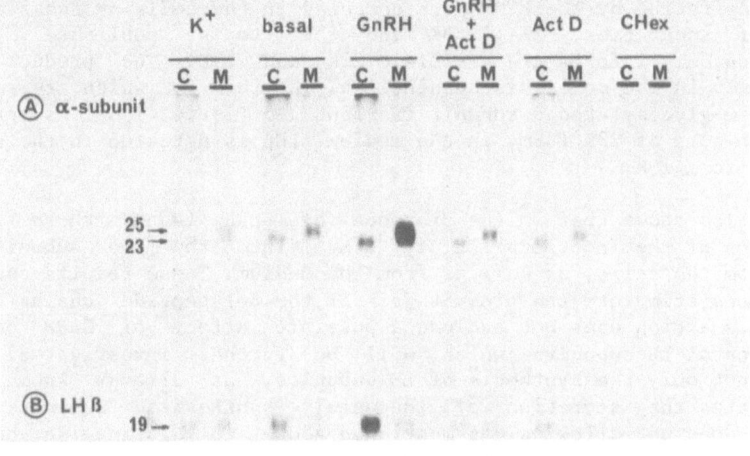

Fig.3. Fluorography of radiolabeled polypeptides immunologically related to α- and LHβ-subunits,
isolated by immunoprecipitation with specific
antisera, from cells (C) and media (M), and
electrophoresed on SDS-polyacrylamide gel.
Anterior pituitary cells in culture from castrated adult male rats were incubated for 5h
in the presence of ^{35}S-Met and in the
absence or presence of GnRH (40 nM), 59 mM
K$^+$, cycloheximide (CHex, 1mM) and actinomycin D (Act D, 150nM).

involved in the GnRH action on the synthesis of gonadotropins. Most of the
investigators involved in this research field supported the idea that GnRH
stimulates the glycosylation of gonadotropin subunits rather than the
synthesis of their polypeptide chains (Liu et al., 1976 ; Liu and Jackson,
1978, 1985 ; Azhar et al., 1978; Vogel et al., 1983). In contrast to these
data, early results from our laboratory (Khar et al., 1978 ; Khar and
Jutisz, 1980) suggested GnRH stimulates the incorporation of labeled amino
acids into the polypeptide chains of gonadotropins and consequently
stimulates the synthesis of the polypeptide chains of LH and FSH. As the
method of immunoprecipitation used in our early studies for isolating
radiolabeled gonadotropins was not selective enough to get ride of all
contaminants, we have recently re-evaluated this effect of GnRH using a
more reliable and more precise methodology.

Our new methodology involved the culture of dispersed anterior
pituitary cells from normal or castrated adult male rats. Cultured cells
were incubated for 5 hours in the presence of ^{35}S-methionine and in the
presence or absence of GnRH (40nM). At the end of the incubation period,
media and cells were processed separately. After immunoprecipitation using
specific antisera, labeled polypeptides were separated and characterized by
SDS-polyacrylamide slab gel electrophoresis (SDS-PAGE). The bands
corresponding to polypeptides immunologically related to α and LHβ
subunits, revealed by fluorography, were excised and counted for their
radioactivity. In the conditions used for SDS-PAGE, the associated form α-β
of LH cannot be detected, but only separated subunits, as well as
radiolabeled impurities (if any).

Fig.3 shows a fluorograph of radiolabeled polypeptides immunologically
related to subunits α and LHβ, isolated by SDS-PAGE from the cells (C) and
media (M) following a 5 h-incubation in the presence of ^{35}S-Met. Three

forms of α differing by their M_r were detected in the cells as function of time of incubation, 21K, 23K and 25K (to be published). After 5h-incubation only 2 forms are visible, 23K and 25K. The product 23K, present mainly in the cells, is authentic α. The form 25K which represents probably an O-glycosylated α-subunit (Parsons and Pierce, 1984) is present mainly, with some of 23K form, in the medium. LHβ is detected in the cells, only in the form 19K.

Fig.3 also shows that in the presence of GnRH (40nM) there is an amplification of the incorporation of ^{35}S-Met into the two subunits isolated from the cells, as well as from the medium. These results suggest that GnRH does stimulate the biosynthesis of the polypeptide chains of α and LHβ. This action does not exclude a possible effect of GnRH on the glycosylation of LH subunits which will be further investigated. GnRH stimulates not only the synthesis of LH subunits, as already known, but also stimulates the secretion of the newly synthesized subunits. Our methodology does not allow us, as mentioned abose, to distinguish between associated and dissociated forms of LH, but as α is present in a large excess over LHβ, it could be presumed that α is mainly released in a free state. The release of the free α subunit has been already reported by different authors (Franchimont and Pasteels, 1972; Hagen and McNeilly, 1975; Grotjan, Jr. et al., 1984; Parsons and Pierce, 1984).

In order to investigate the mechanism by which GnRH acts on the synthesis of LH subunits, we checked whether 59mM K^+, which releases LH in a non specific manner as well as other pituitary hormones, is able or not to mimick the action of GnRH. Fig.3 shows that a non specific release of LH with a high potassium medium is unable to induce the synthesis of the hormone. Cycloheximide which inhibits the synthesis of proteins, also inhibits incorporation of ^{35}S-Met into LH subunits. Actinomycin D alone, has no significant effect on the synthesis of LH subunits during a 5h-period of incubation but it inhibits completely the stimulatory effect of GnRH. These later results would be in favor of an effect of GnRH on the transcription of genes encoding LH subunits. However, we recently found that incubation of the pituitary cells in the presence of GnRH did not increase the translational capacity of mRNAs prepared from those cells (to be published). Thus, the effect of GnRH seems not to follow the same mechanisms as those involved in the translation of pituitary mRNAs from castrated rats (see above).

These results would suggest that GnRH acts neither directly on the genes encoding LH subunits, nor on the messenger activity of specific mRNAs coding for LH subunits. We postulate that GnRH acts on a gene encoding a protein which participates in the regulation of the expression of specific messengers for LH subunits. We are currently pursuing our investigation on this mechanism which is not very classical in the case of Eucaryotes.

In order to study thoroughly the mechanism by which GnRH acts, we recently checked the effect of an analog of cAMP, the 8-Br-cAMP (Br-cAMP), on the biosynthesis of LH subunits. Fig.4 shows that 1mM Br-cAMP mimicks the stimulatory action of GnRH on the incorporation of ^{35}S-Met into subunits α and LHβ. These results suggest that cAMP is a good candidate as an intracellular mediator of GnRH action on the synthesis of gonadotropin subunits. It should be noted that the involvement of cAMP in the mechanism by which GnRH acts on LH and FSH release, controversial for many years, has been definitively discarded (Bérault et al., 1980). Nevertheless, our hypothesis concerning the cAMP mediation of GnRH action is based only on preliminary results which should be substantiated. In particular, it should be shown that GnRH action leads to an intracellular accumulation of cAMP and that a correlation exists between both cAMP accumulation and the

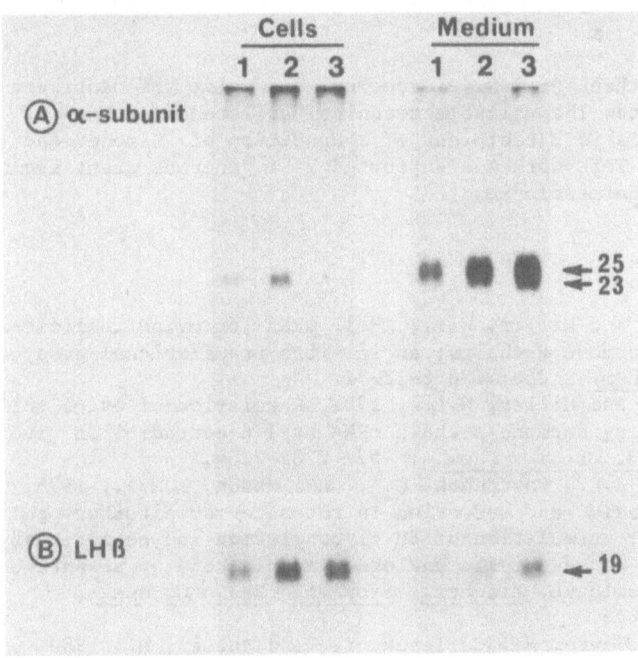

Fig.4. Fluorography of radiolabeled polypeptides immunologically related to α- and
LHβ-subunits isolated by immunoprecipitation with specific antisera from
cells (C) and media (M) and electrophoresed on SDS-polyacrylamide gel. Anterior pituitary cells in culture from
castrated adult male rats were incubated for 5h in the presence of ^{35}S-Met
and in the absence (1) or presence of
40 nM GnRH (2) or 1 mM 8-Br-cAMP (3).

incorporation of ^{35}S-Met into LH subunits. It should also be
demonstrated that substances capable of stimulating intracellular
accumulation of cAMP, such as cholera toxin or forskolin, are able to
mimick the action of GnRH. These investigations are now in progress in our
laboratory.

CONCLUSIONS

Recent data from our laboratory and those reported by others suggest
the synthesis of gonadotropin hormones is under a double hormonal control,
by gonadal steroids and by GnRH. According to our hypothesis, the negative
action of steroids and the stimulatory effect of GnRH proceed through
different mechanisms. The gonadal steroids act by inhibiting transcription
of specific genes coding for gonadotropin subunits, while GnRH proceeds
through an stimulatory effect, within the pituitary cell, on the expression
of specific mRNAs encoding the synthesis of the polypeptide chains of
gonadotropins.

ACKNOWLEDGEMENTS

We wish to thank Dr J.G. Pierce for his gift of antisera to RCMX bovine LH subunits. The valuable technical assistance of Mme G. Ribot is acknowledged. A.S. is a recipient of a Fondation of Simone and Cino del Duca fellowship. This work was supported by a generous grant from Fondation de Recherche en Hormonologie.

REFERENCES

Alexander, D.C., and Miller, W.L., 1981, mRNA for ovine follicle-stimu-
        lating hormone β-chains; an in vitro translational assay, J.
        Biol. Chem. , 256: 12628-12631.
Alexander, D.C., and Miller, W.L.., 1982, Regulation of ovine follicle-
        stimulating hormone β-chain mRNA by 17β-estradiol in vivo and in
        vitro, J. Biol. Chem. , 257: 2282-2286.
Azhar, S., Reel, J.R., Pastushok, C.A., and Menon, K.M.J., 1978, LH
        biosynthesis and secretion in rat anterior pituitary cell
        cultures: stimulation of LH glycosylation and secretion by
        GnRH and an agonistic analogue and blockade by an antago-
        nistic analogue, Biochem. Biophys. Res. Commun. ,
        80: 659-666.
Bérault, A., Théoleyre, M., Colleaux, Y., and Jutisz, M., 1980,
        Further evidence that cyclic AMP is not the mediator of
        the releasing action of GnRH, Neuroendocr. Letters , 2: 31-37.
Chin, W.W. and Habener, J.F., 1984, Organization and expression of
        lutropin (LH) subunit genes, 7th Int. Congress of Endocri-
        nology, Quebec City, Canada, Abstr. no S 47.
Corbani, M., Counis, R., Starzec, A., and Jutisz, M., 1980,
        Effect of gonadectomy on pituitary levels of mRNA encoding
        gonadotropin subunits and secretion of luteinizing
        hormone, Mol. Cell. Endocr. , 35: 83-87.
Counis, R., Ribot, G., Corbani, M., Poissonnier, M., and Jutisz, M.,
        1981, Cell-free translation of the rat pituitary messenger
        RNA coding for the precursors of α- and β-subunits of
        lutropin, FEBS Letters , 123: 151-155.
Counis, R., Corbani, M., Ribot, G., and Jutisz, M., 1982a,
        Characterization of the precursors of α- and β-subunits
        of follitropin following cell-free translation of rat and
        ovine pituitary mRNAs, Biochem Biophys. Res.
        Commun. , 107: 998-1005.
Counis, R., Corbani, M.,and Jutisz, M., 1982b, Studies on cell-
        free biosynthesis of lutropin (LH) and characterization
        of its subunit precursors, in : "Pituitary hormones
        and related peptides ", M. Motta, M. Zanisi, and
        F. Piva, eds., Academic Press, London and New-York,
        pp 49-61.
Counis, R., Corbani, M., and Jutisz, M., 1982c, Régulation de la
        biosynthèse des gonadotropines hypophysaires, in : "Régulations
        cellulaires multihormonales en Neuroendocrinologie",
        A. Tixier-Vidal and Ph. Richard, eds., Colloques de l'INSERM,
        Paris, 110: 509 -524.
Counis, R., Corbani, M.,and Jutisz, M., 1983, Estradiol regulates
        mRNAs encoding precursors to rat lutropin (LH) and
        follitropin (FSH) subunits, Biochem. Biophys. Res.
        Commun. , 114: 65 -72.
Counis, R., Corbani, M., and Jutisz, M., 1984, In vivo regulation
        by estradiol of the messenger RNAs encoding LH and FSH
        subunits and the secretion of gonadotropins, in :
        "Hormonal control of the hypothalamo-pituitary-gonadal

axis", K.W. McKerns and Z. Naor, eds., Plenum Press, pp. 397-410.

Fiddes, J.C., and Goodman, H.M., 1979, Isolation, cloning and sequence analysis of the cDNA for the α-subunit of human chorionic gonadotropin, Nature , 281: 351-356.

Fiddes, J.C., and Goodman, H.M., 1980, The cDNA for the β-subunit of human chorionic gonadotropin suggests evolution of a gene by readthrough into the 3'-untranslated region, Nature , 286: 684-687.

Fontaine, Y.A., and Burzawa-Gérard, E., 1977, Esquisse de l'évolution des hormones gonadotropes et thyréotropes des vertébrés, Gen. Comp. Endocr. , 32: 341-347.

Franchimont, P., and Pasteels, J.L., 1972, Sécrétion indépendante des hormones gonadotropes et de leurs sous-unités, C.R. Acad. Sci. Paris , Série D, 275: 1799-1802.

Grotjan, Jr., H.E., Leveque, N.W., Berkowitz, A.S., and Keel, B.A., 1984, Quantitation of LH subunits released by rat anterior pituitary cells in primary culture, Mol. Cell. Endocr. , 35: 121-129.

Hagen, C., and McNeilly, A.S., 1975, Changes in circulating levels of LH, FSH, LHβ and α-subunit after gonadotropin-releasing hormone, and of TSH, LHβ and α-subunits after thyrotropin-releasing hormone, J. Clin. Endocrinol. Metab. , 41: 466-470.

Keller, D., Fetherson, J., and Boime, I., 1980, Isolation of mRNA from bovine pituitary: the cell free synthesis of the α- and β-subunits of luteinizing hormone, Eur. J. Biochem. , 108: 367-372.

Khar, A., and Jutisz, M., 1980, Role of gonadotropin releasing hormone in the biosynthesis of LH and FSH by rat anterior primary cells in culture, in : "Synthesis and release of adenohypophyseal hormones", M. Jutisz and K.W. McKerns, eds., Plenum Press, New York and London, pp. 217-235.

Khar, A., Debeljuk, L., and Jutisz M., 1978, Biosynthesis of gonado-tropin by rat pituitary cells in culture and in pituitary homogenates: effect of gonadotropin-releasing hormone, Mol. Cell. Endocr. , 12: 53-65.

Landefeld, T.D., 1979, Identification of in vitro synthesized pituitary glycoprotein α subunit. Translation of a possible precursor, J. Biol. Chem. , 254: 3585-3688.

Landefeld, T.D., and Kepa, J., 1979, The cell free synthesis of bovine lutropin β-subunit, Biochem. Biophys. Res. Commun. , 90: 1111-1118.

Landefeld, T.D., Kepa, J., and Karsh, F.J., 1983, Regulation of α-subunit synthesis by gonadal steroid feedback in the sheep anterior pituitary, J. Biol. Chem. , 258: 2390-2393.

Landefeld, T.D., Kepa, J., and Karsh, F., 1984, Estradiol feedback effects on the α-subunit mRNA in the sheep pituitary gland: correlation with serum and pituitary hormone concentrations, Proc. Natl. Acad. Sci. USA, 81: 1322-1326.

Liu, T.C., and Jackson, G.L., 1978, Modifications of luteinizing hormone biosynthesis and release by GnRH, cycloheximide and antinomycin D, Endocrinology , 103: 1253-1263.

Liu, T.C., and Jackson, G.L., 1985, Synthesis and release of LH in vitro: effects of gallopamil hydrochloride (D600) and pimozide, 67th Annual Meeting of the Endocrine Society, Baltimore, Abstr. no 423.

Liu, T.C., Jackson, G.L., and Gorski, J., 1976, Effects of synthetic gonadotropin-releasing hormone on incorpora-tion of radioactive glucosamine and amino acids into luteinizing hormone and total protein by rat pituitaries in vitro, Endocrinology , 98: 151-163.

McCann, S.M., and Ramirez, V.D., 1964, The neuroendocrine regulation of hypophyseal luteinizing hormone secretion; Recent Progr. Horm. Res. , 20: 131-181.

Nilson, J.H., Nejedlik, M.T., Virgin, J.B., Crowder, M.E., and Nett, T.M., 1983, Expression of α-subunit and luteinizing hormone β-genes in the ovine anterior pituitary ; estradiol supresses accumulation of mRNAs for both α-subunit and luteinizing hormone-β, J. Biol. Chem. , 258: 12087-12090.

Parsons, T.C., and Pierce, J.G., 1984, Free α-like material from bovine pituitaries, J. Biol. Chem. , 259: 2662-2666.

Pierce, J.G., and Parsons, T.F., 1981, Glycoprotein hormones: structure and function, Ann. Rev. Biochem. , 50: 465-495.

Schally, A.V., Kastin, A.J., and Arimura, A., 1972, FSH-releasing hormone and LH-releasing hormone, Vitamins and Hormones , 30: 83-164.

Vogel, D.L., Mayner, J.A., and Sherins, R.J., 1983, Biosynthesis of LH-alpha and -beta subunits in intact and orchiectomized rats, 65th Annual Meeting of the Endocrine Society, San Antonio, Abstr. no 144.

Wallace, R.B., Shaffer, J., Murphy, R.F., Bonner, J., Hirose, T., and Itakura, K., 1979, Hybridization of synthetic oligodeoxynucleotides to Øx 174 DNA: the effect of single base pair mismatch, Nucleic Acids Res. , 6: 3543-3557.

THE MAMMALIAN TACHYKININS AND THEIR RECEPTORS: STRUCTURE-ACTIVITY

RELATIONSHIP OF THE MAMMALIAN TACHYKININS

Sadao Kimura, Katsutoshi Goto*, Yuka Shigematsu,
Yoshiki Sugita, and Ichiro Kanazawa**

Department of Biochemistry, *Department of Pharmacology
Institute of Basic Medical Sciences
**Department of Neurology, Institute of Clinical Medicine
University of Tsukuba, Ibaraki 305, Japan

INTRODUCTION

The tachykinins are a family of naturally occuring bioactive peptides, which share the common C-terminal amino acid sequence, Phe-X-Gly-Leu-Met-NH$_2$ (Fig.1) and exhibit a wide spectrum of biological actions such as smooth muscle contraction, sialogogic action and hypotension (Ersparmer, 1981). Until recently, substance P (SP)(X=Phe) has been the only known tachykinin shown to be present in the mammalian tissues. The accumulating evidence suggest that SP is a neurotransmitter or neuromodulator in the central and peripheral nervous systems (Otsuka and Konishi, 1975; Pernow, 1983). We have recently determined the amino acid sequences of two additional and novel tachykinins (X=Val), neurokinin A (NKA) and neurokinin B (NKB), which are remarkably similar to kassinin (Kimura et al., 1983). The identical peptides were independently reported and named differently; NKA as neuromedin L or substance K (Maggio et·al., 1983; Minamino et al., 1984; Nawa et al., 1983) and NKB as neuromedin K (Kangawa et al., 1983). Pharmacologically, it was revealed that both neurokinins (NKs) possess the typical tachykinin activities and SP-E type ligand characteristics using peripheral tissues (Hunter and Maggio, 1984; Kimura et al., 1984; Nawa et al., 1984) and central nervous tissues (Buch et al., 1984; Mantyh et al., 1984; Quirion and Pilapil, 1984; Torrens et al., 1984). Electrophysiologically, it was found that both NKA and NKB have potent excitatory actions on neurons in the isolated spinal cord of neuborn rat (Matsuto et al., 1984). Moreover, we and others have shown using radioimmunoassay system that NKA and NKB are distributed unevenly in rat central nervous system (Kanazawa et al., 1984; Minamino et al., 1984). Especially, considerable amounts of NKs were found in the spinal cord and localized predominantly in the dorsal horn (Ogawa et al., 1985). Furthermore, the nucleotide sequences of two different SP precursors isolated from bovine striatum have been reported; one contains each copy of SP and NKA (substance K) while the other contains only substance P (Nawa et al., 1983). It was suggested that NKA (SK) may be coreleased with SP from the common precursor and serves as a second type of the endogenous tachykinin in mammalian tissue. These lines of evidence strongly suggested that NKs may act as neurotransmitters or neuromodulators like SP in mammalian central nervous system, especially, in the spinal cord.
    In this paper, we describe the structure-activity relationship study using SP, NKA, NKB and their chemically related peptides on smooth muscle

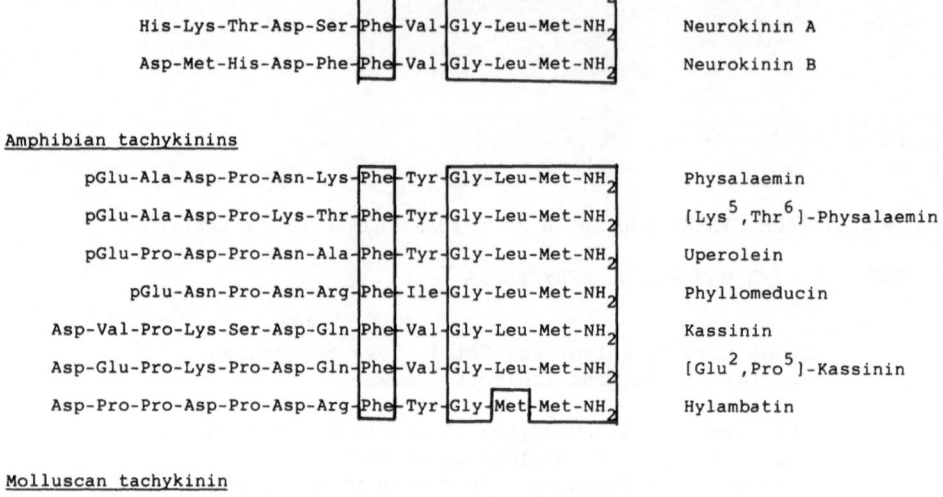

Mammalian tachykinins

Arg-Pro-Lys-Pro-Gln-Gln-Phe-Phe-Gly-Leu-Met-NH$_2$ Substance P

His-Lys-Thr-Asp-Ser-Phe-Val-Gly-Leu-Met-NH$_2$ Neurokinin A

Asp-Met-His-Asp-Phe-Phe-Val-Gly-Leu-Met-NH$_2$ Neurokinin B

Amphibian tachykinins

pGlu-Ala-Asp-Pro-Asn-Lys-Phe-Tyr-Gly-Leu-Met-NH$_2$ Physalaemin

pGlu-Ala-Asp-Pro-Lys-Thr-Phe-Tyr-Gly-Leu-Met-NH$_2$ [Lys^5,Thr^6]-Physalaemin

pGlu-Pro-Asp-Pro-Asn-Ala-Phe-Tyr-Gly-Leu-Met-NH$_2$ Uperolein

pGlu-Asn-Pro-Asn-Arg-Phe-Ile-Gly-Leu-Met-NH$_2$ Phyllomeducin

Asp-Val-Pro-Lys-Ser-Asp-Gln-Phe-Val-Gly-Leu-Met-NH$_2$ Kassinin

Asp-Glu-Pro-Lys-Pro-Asp-Gln-Phe-Val-Gly-Leu-Met-NH$_2$ [Glu^2,Pro^5]-Kassinin

Asp-Pro-Pro-Asp-Pro-Asp-Arg-Phe-Tyr-Gly-Met-Met-NH$_2$ Hylambatin

Molluscan tachykinin

pGlu-Pro-Ser-Lys-Asp-Ala-Phe-Ile-Gly-Leu-Met-NH$_2$ Eledoisin

Fig.1 Amino acid sequences of tachykinin family

preparations to obtain the further information about the nature of the
tachykinin receptors and the signalling mechanisms induced by such receptors.

STRUCTURE-ACTIVITY RELATIONSHIP OF THE TACHYKININS ON THE SMOOTH MUSCLE
PREPARATIONS

SP and other nonmammalian tachykinins have been found to possess the
similar spectrum of activity of tachykinins although their rank order of
potencies in various pharmacological tests are considerably different.
This strongly indicates that there exists more than one class of tachykinin
receptor. For instance, kassinin and eledoisin are approximately 2 orders
of magnitude more potent than SP on some peripheral tissues, while SP and
kassinin are almost equipotent on some other tissues. To explain these
differences, the existence of two subtypes of tachykinin receptor has been
proposed, namely, SP-P receptor to SP and physalaemin and SP-E receptor to
eledoisin and kassinin (Iversen et al., 1982; Lee et al., 1982; Watson et
al., 1983).

NKA and NKB have a common C-terminal structure (X=Val) identical to
that of kassinin, and are expected to have their own functional roles
distinct from that of SP. In our previous study, we have shown that NKA
and NKB have SP-E ligand characteristics on the gut smooth muscles (Kimura
et al., 1984). In addition, we suggested that an amino acid at position X
in the C-terminal structure is especially important for recognition to
tachykinin receptors (Kimura et al., 1985a; Kimura et al., 1985b). To
clarify further the recognition site in the tachykinin molecule for
tachykinin receptor subtypes and to distinguish further the pharmacological
characteristics of two types of tachykinin receptors known in the gut smooth
muscles, structure-activity relationship studies were performed on three
smooth muscle preparations, guinea-pig ileum, rat vas deferens and rat
rectum. A series of SP, NKA and NKB analogues with different chain length
and/or an amino acid replacement at position X in the C-terminal region
were examined.

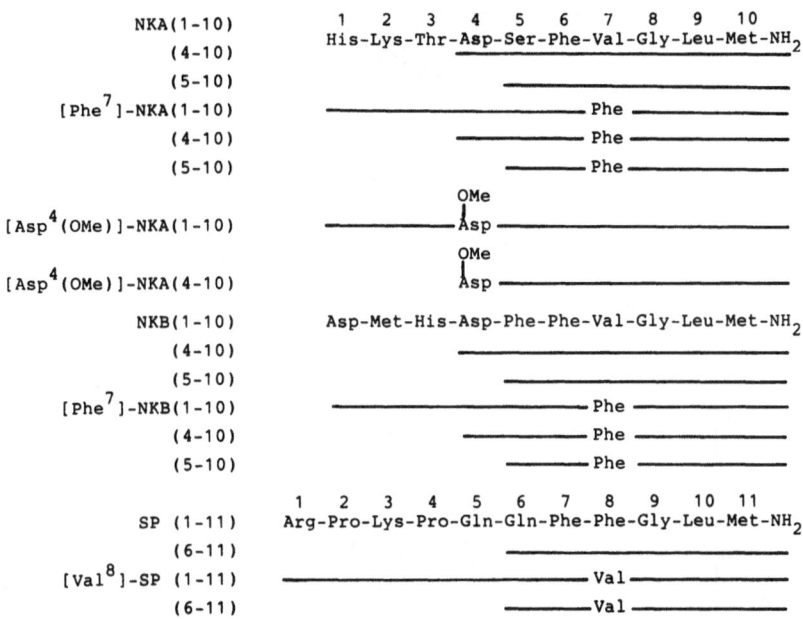

Fig.2   Amino acid sequences of tachykinins and their analogues

1. Effects of replacement of an amino acid at position X in the common C-terminal sequence on biological activity (Role of the C-terminal region)

In order to evaluate the importance of an amino acid residue at position X in the tachykinin common C-terminal sequence, Phe-X-Gly-Leu-Met-NH$_2$, the tachykinin analogues listed in Fig.2 were synthesized and submitted to guineapig ileum (GPI) and rat vas deferens (RVD) assays together with naturally occuring tachykinins (Fig.3, Table 1). In the GPI assay, SP, NKA and NKB were almost equipotent although NKA and NKB were less potent than SP. When Val at position 7 in NKs was replaced by Phe, [Phe[7]]-NKA and [Phe[7]]-NKB became more potent than NKA and NKB by 8.6 and 2.4 times, respectively, and more comparable with SP. Similarly, [Val[8]]-SP did not change the agonistic potency compared with SP. Replacements of position X in the C-terminal region gave various effects, both increasing and decreasing the agonistic potency but not great effects on the potency in the GPI assay. It can be, therefore, said that the tachykinins, whatever the amino acid at position X in their molecules is an aromatic or aliphatic amino acid, have almost equipotent in SP-P tissue (the GPI), which probably contains both SP-P and SP-E receptors, moreover, that two types of tachykinins (X=Phe and X=Val) exert the equipotent activity through the corresponding SP-P and SP-E type receptors, respectively. In the RVD assay, NKA and NKB are more than 2 orders of magnitude active than SP and physalaemin. Similarly, eledoisin (X=Ile) was 57.4 times more active than SP. On the contrary, the potencies of [Phe[7]]-NKA and [Phe[7]]-NKB, in which Val at position X is replaced by Phe, decreased about 50 - 100 fold compared with those of NKA and NKB, and were almost comparable with that of SP. These results strongly suggest that the RVD, classified as one of SP-E type tissue, contains at least one NK specific receptor (possibly SP-E receptor), recognizing an aliphatic amino acid (especially, Val) not an aromatic amino acid (both Phe and Tyr) at position X in the C-terminal sequence of the tachykinins because an amino acid replasement of Val[7] in NKs by Phe greatly reduced the potencies. These results on the strict ligand selectivity of SP-E type tachykinin receptor are in good accordance with the results of the recent binding study (Buch et al., 1984). It should be noticed that [Val[8]]-SP was only 3.8

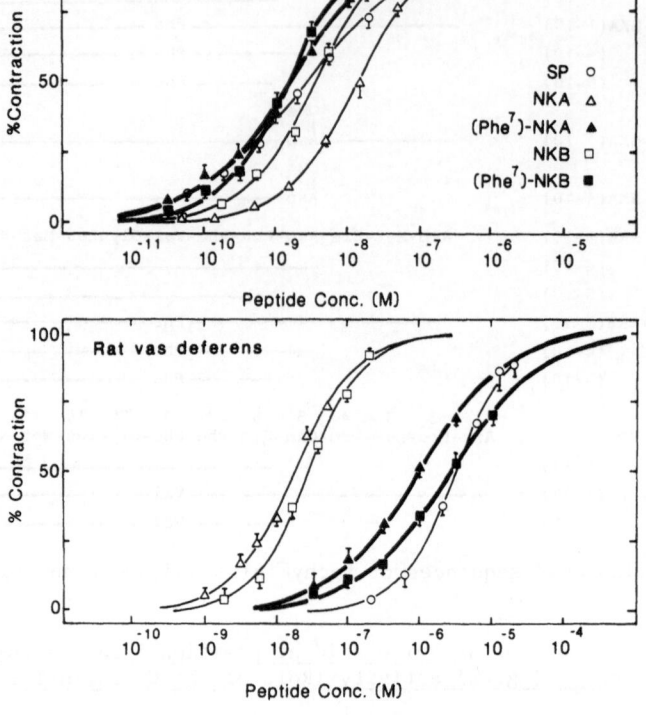

Fig.3   Dose-reponse curves of tachykinins and their analogues.

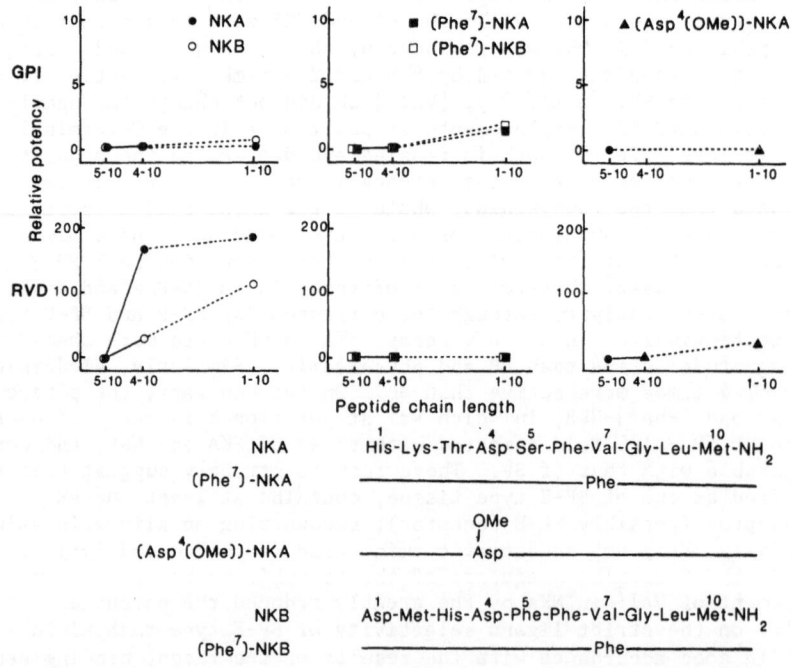

Fig.4   Effect of N-terminal chain length on contractile activity (SP=1.0)

Peptide	Guinea - pig ileum			Rat vas deferens		
	EC_{50}	R.P.	n	EC_{50}	R.P.	n
Substance P (SP)	2.9×10^{-9}	1.00	10	3.1×10^{-6}	1.0	3
Physalamin	4.6×10^{-9}	0.63	12	1.4×10^{-6}	2.2	6
Eledoisin	2.7×10^{-9}	1.07	6	5.4×10^{-8}	57.4	7
SP (6-11)	8.0×10^{-9}	0.36	9	1.2×10^{-5}	0.3	2
[Val^{8}]-SP	1.7×10^{-9}	1.71	6	8.1×10^{-7}	3.8	9
[Val^{8}]-SP (6-11)	1.6×10^{-7}	0.02	12	1.6×10^{-6}	2.0	4
Neurokinin A (NKA)	1.4×10^{-8}	0.21	4	1.7×10^{-8}	186.0	5
NKA (4-10)	1.2×10^{-8}	0.24	6	1.9×10^{-8}	166.0	6
NKA (5-10)	6.2×10^{-8}	0.05	6	2.3×10^{-6}	1.3	6
[Phe^{7}]-NKA	1.6×10^{-9}	1.81	10	9.0×10^{-7}	3.4	6
[Phe^{7}]-NKA (4-10)	1.9×10^{-8}	0.51	6	1.1×10^{-6}	2.7	6
[Phe^{7}]-NKA (5-10)	4.1×10^{-8}	0.07	6	3.8×10^{-6}	0.8	6
[Asp^{4}(OMe)]-NKA	2.8×10^{-8}	0.10	8	9.9×10^{-8}	31.3	6
[Asp^{4}(OMe)]-NKA (4-10)				7.5×10^{-7}	4.1	6
Neurokinin B (NKB)	3.4×10^{-9}	0.85	4	2.6×10^{-8}	119.0	5
NKB (4-10)	2.3×10^{-8}	0.13	6	1.2×10^{-7}	26.6	6
NKB (5-10)	2.0×10^{-8}	0.15	6	1.1×10^{-6}	2.8	6
[Phe^{7}]-NKB	1.4×10^{-9}	2.07	6	2.5×10^{-6}	1.2	6
[Phe^{7}]-NKB (4-10)	7.4×10^{-9}	0.39	6	6.5×10^{-6}	0.5	6
[Phe^{7}]-NKB (5-10)	5.0×10^{-8}	0.06	6	6.6×10^{-5}	0.05	6

times more potent than SP on the RVD although it contains a Val residue at position X. The reason why it elicits so low activity is due to lacking an Asp residue at position 5, which enhances greatly the affinity to the SP-E type receptor (discussed in the next section).

2. Properties of C-terminal NK peptides with different chain length (Role of N-terminal region)

The structural information of the tachykinins is mainly restricted to the respective C-terminal hexapeptide sequence, which is regarded as the minimal part necessary for biological activity. This is in accordance with the principle that the C-terminal region indispensable for function remain stable during evolution (Ersparmer, 1981; Niedrich et al., 1981). In order to understand the role of N-terminal region of the tachykinins, C-terminal SP and NK peptide analogues with different chain length were synthesized and submitted to the GPI and the RVD assays (Fig.4, Table 1). It was found that the C-terminal hexapeptides, NKA(5-10) and NKB(5-10), are the minimal fragments for biological activity, although both peptides exhibited very little activity on the GPI; however, on the RVD, they were almost equipotent as SP. Furthermore, it was found that NKA(4-10) and NKB(4-10), which contain Asp residue at position 4, are 166 and 22.6 times more potent than SP, respectively. On the contrary, [Phe^{7}]-NKA(5-10), (4-10) and [Phe^{7}]-NKB(5-10), (4-10) did not have such an increase and were almost equipotent or less than SP on the rat vas deferens. These results clearly indicated that Phe and Val type tachykinins possess the characteristics of SP-P and SP-E type ligands, respectively, and N-terminal portions of NKA and NKB, especially the Asp^{4} residue appeared to have a role for enhaning the activity of the C-terminal hexapeptides of NKA and NKB. Now, we can estimate the reason why [Val^{8}]-SP did not elicit the potency comparable to NKA and NKB despite that it contains a Val residue at position X. It seems likely that the lack of the corresponding Asp residue in it makes the

139

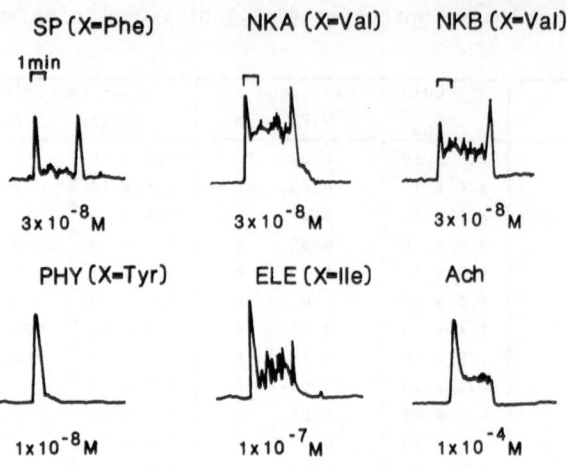

Fig.5   Contractile patterns of tachykinins on the rat rectum

potency low as same as that of SP. It can be, therefore, concluded that
the N-terminal part of NKs, especially Asp residue, is also a major
determinant of tha relative potency of the tachykinins with SP-E type
ligand characteristics in SP-E tissue. To confirm this interpretation on
the role of Asp^4 residue, a methyester derivative, $[Asp^4(OMe)]$-NKA, was
synthesized and tested on the gut smooth muscle assays. As shown in Fig.4,
on the RVD, the lack of the negative charge by methylesterification,
decrease the affinity to SP-E type tachykinin receptor. These results
suggest that SP-E receptor contains a positively charged group (possibly a
basic amino acid) at the ligand binding site, while SP-P receptor might not
have such a group crucial for ligand binding.

3. Contractile properties of the rat rectum to the mammalian tachykinins

       As described above, we have discriminated between SP and NKS based
on their potencies on SP-E type smooth muscles. On the contrary, we have
distinguished them by their characteristic contractile actions on the rat
rectum, although they showed SP-P type potency characteristics apparently.
Interestingly, both NKs showed biphasic contractle actions on the rat
rectum longitudinal muscle, while SP exhibited almost monophasic action
(Fig.5). Similarly, physalaemin (X=Tyr) and eledoisin (X=Ile) showed
monophasic and biphasic actions, respectively. These results indicate that
1st phase seems to be generated by aromatic type (X=Phe or Tyr) and non-
aromatic hydrophobic type (X=Val or Ile) C-terminal structures, which is
apparently nonspecific, and however, 2nd phase seems to be specific to the
latter one. Acetylcholine also exhibited biphasic actions although 2nd
phase of its contractile pattern is somewhat different from those of the
tachykinins tested. As shown in Fig.6, 1st phase of the tachykinins showed
almost equipotent patterns. On the contrary, 2nd phases of NKA and NKB
showed similar dose-response curves to those of 1st phases of NKs. Second
phase of SP was observed only at higher concentration (Fig.7). EC_{50} values
for the tachykinin are listed in Table 2. As seen in Fig.7 and Table 2,
the potency differencies among the tachykinins tested indicated that 2nd
phase on rat rectum has SP-E type characteristics, because that EC_{50} of 2nd
phase of SP and physalaemin are almost 2 orders of magnitude less potent

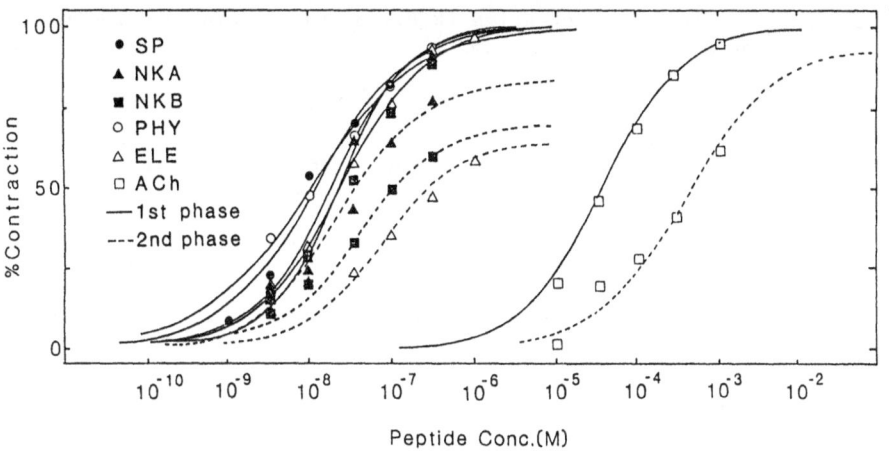

Fig.6   Dose-response curves of tachykinins and their analogues on the rat
        rectum.

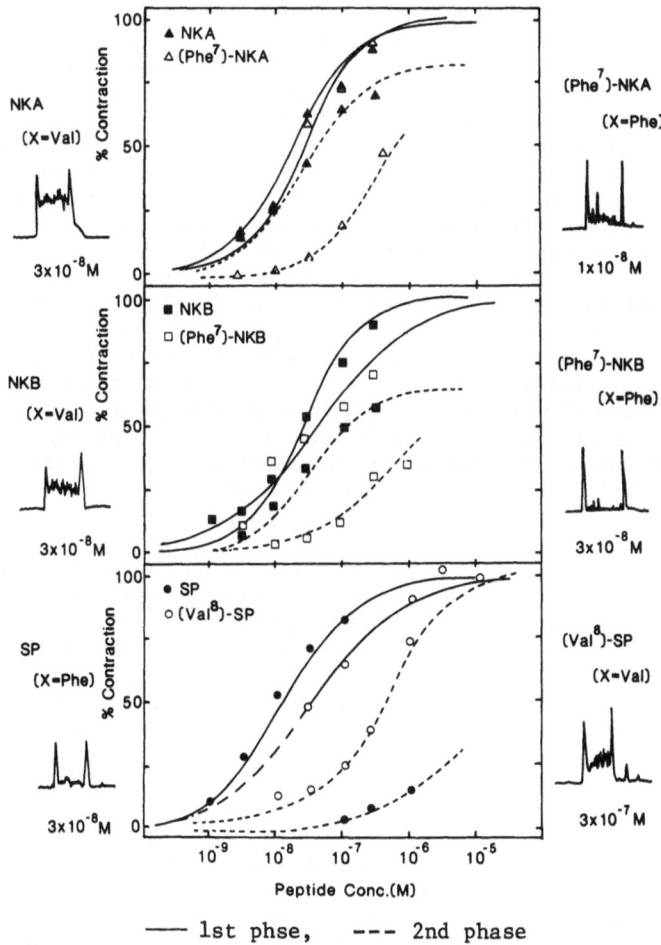

Fig.7   Dose-response curves and contractile patterns of tachykinins and
        their analogues on the rat rectum.

Table 2. EC$_{50}$ and relative potencies (R.P.) of tachykinins and their analogues on the rat rectum.

	1st phase		2nd phase		$EC_{50}^{2nd}/EC_{50}^{1st}$	n
	EC_{50}	R.P. a)	EC_{50}	R.P. b)		
Substance P (SP)	1.3×10^{-8}	1.00	$> 10^{-6}$	< 0.02	> 77.1	6
[Val^8]-SP	3.4×10^{-8}	0.37	4.7×10^{-7}	0.05	13.8	4
Physalaemin	1.1×10^{-8}	1.18	$> 10^{-6}$	< 0.02	> 90.1	6
Eledoisin	2.6×10^{-8}	0.50	7.7×10^{-8}	0.33	3.0	6
Neurokinin A (NKA)	1.9×10^{-8}	0.68	2.5×10^{-8}	1.00	1.3	6
[Phe^7]-NKA	2.6×10^{-8}	0.50	$>6.0 \times 10^{-7}$	< 0.04	> 23.1	6
[Asp^4(OMe)]-NKA	4.0×10^{-8}	0.33	-	-	-	6
Neurokinin B (NKB)	2.5×10^{-8}	0.52	2.5×10^{-8}	1.00	1.0	6
[Phe^7]-NKB	4.2×10^{-8}	0.31	$> 10^{-6}$	< 0.02	> 23.8	6
Acetylcholine	3.9×10^{-5}	3.2×10^{-4}	3.2×10^{-4}	7.0×10^{-5}	6.2	7

a) SP=1.0 b) NKA=NKB=1.0

than those of NKs. When we compare the potency differencies between 1st and 2nd phases of tachykinins, the differences in the cases of both NKA and NKB are very small. On the contrary, the differences in the case of SP and physalaemin are approximately 2 orders of magnitude. Interestingly, the similar difference in the case of acetylcholin was found although the difference is approximately one order. In the case of eledoisin, the difference is fairly great and seems to be between of NKs and SP. These results indicated that the rat rectum smooth muscle has mixed charecteristics of both SP-P (1st phase) and SP-E (2nd phase) type tissues, which have never been found so far. In order to characterize the receptor system present in the rat rectum, the substituted derivatives of SP and NKs were tested on the rat rectum. As seen in Fig.7, when Val at position 7 in NKs are replaced by Phe, EC$_{50}$ of 2nd phase decreased almost 2 orders of magnituede, which is comparable with that of SP. On the other hand, 1st phase of [Val^8]-SP, in which Phe in SP was replaced by Val, was similar to that of SP but the 2nd phase became more potent. This suggests that [Val^8]-SP turned to be recognized by SP-E type receptor. However, the potency difference between 1st and 2nd phases of [Val^8]-SP was found to be one order of magnitude. The difference can be attributed to the lack of Asp residue because tachykinins without Asp have the week affinity to SP-E recorder. The above results found on the rat rectum raised us an interesting question whether an SP-E type receptor activated by NKs produces both 1st and 2nd phases of the contraction. In the case of SP, an SP-P type receptor produces only monophasic contraction. Whatever one or two receptor are responsible to produce biphasic contraction, we can expect that there exists two contractile mechanisms after binding, in contrast to one contractile mechanism in the cases of SP and physalaemin. If a single receptor is responsible for the biphasic contraction on the rat rectum, it should have two coupling sites which link an SP-E receptor to the two intracellular signalling systems. Alternatively if there are two receptors in the rat rectum, SP-E receptors should be subdivided into SP-E1 and SP-E2 receptors. Further studies are needed to clarify the next two question; One is how many receptors are involved and the other is how many intracellular (contractile) mechanisms are coupled to produce biological activity.
(Fig.8)

CONCLUSIONS

    In summary, the existence of at least two subtypes of tachykinin receptors is demonstrated from structure-activity study of the mammalian tachykinins on the gut smooth muscles. Furthermore, our results have extablished the importance of the specific amino acid sequence in the chemical structures of the tachykinins that a non-aromatic hydrophobic residue (especially Val) at position 7, and an acidic amino acid at position 4, are important for the expression of the SP-E type of activities.

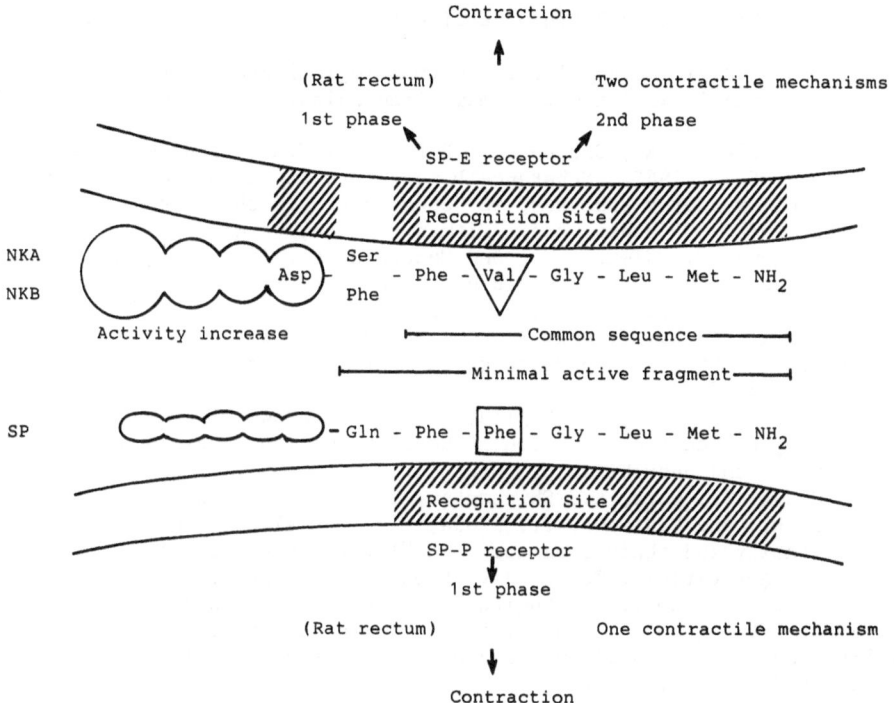

Fig.8   The summary of structure-activity relationship study of tachykinins.

Based on comparable potencies between NKA and NKA(5-10), NKB and NKB(4-10), it can be said that the amino acid sequence containing Asp^4 and the characteristic Phe-X-Gly-Leu-Met-NH_2 (X= non-aromatic hydrophobic residue, Val or Ile) are enough for the expression of the nearly full activity of neurokinins, especially NKA.

In addition, we have proposed the existence of two intracellular signalling systems to produce neurokinin type activity. However, little is known about the intracellular chemical signals generated by the tachykinin receptors. Recently, the possibility has been suggested that inositol phospholipid hydrolysis system, not cAMP production system, may be the second messenger system for tachykinins, but no conclusive evidence for the second messenger has been established. Thus, more detailed studies are needed concerning the nature of at least two types of tachykinin receptors and the biochemical phathway that link receptor activation to physiological response, including the second messenger associated with tachykinin receptor activities. Fig.8 summarizes our structure-activity relationship study.

Acknowledgements

        We thank Prof. M. Otsuka, Tokyo Medical and Dental University for his encouragement and valuable advice. This work was partly supported by grants from the Ministry of Education, Science and Culture of Japan, University of Tsukuba Project Research and Protein Research Foundation.

REFERENCES

Buck, S.H., Burcher, E., Shults, C.W., Lovenberg, W., and O'Donohue, T.L., 1984, Novel pharmacology of substance K-binding sites: a third type of tachykinin receptor, Science, 226:987.

Ersparmer, V., 1981, The tachykinin peptide family, Trends neurosci., 4:267.

Hunter, J.C., and Maggio, J.E., 1984, Pharmacological characterization of a novel tachykinin isolated from mammalian spinal cord, Eur. J. Pharmacol. 97: 159.

Iversen, L.L., Hanley, M.R., Sandberg, B.E.B., Lee, C.M., Pinnock, R.D., and Watson, S.P., 1982, Substance P receptors in the nervous system and possible receptor subtypes, in "Substance P in the nervous system" R. Porter and M. O'Conner, eds., Pitman, London, pp186.

Kanazawa, I., Ogawa, T., Kimura, S., and Munekata, E., 1984, Regionaldistribution of substance P, neurokinin α, and neurokinin β in rat central nervous system, Neurosci. Res., 2:111.

Kangawa, K., Minamino, K., Fukuda, A., and Matsuo, H., 1983, A novel tachykinin identified in porcine spinal cord, Biochem. Biophys. Res. Commun., 114:533.

Kimura, S., Goto, K., Ogawa, T., Sugita, Y., and Kanazawa, I., 1984, Pharmacological characterization of novel mammalian tachykinins, neurokinin α and neurokinin β, Neurosci. Res., 2:97.

Kimura, S., Ogawa, T., Goto, K., Shigematsu, Y., Yamashita, H., Sugita, Y., Munekata, E., and Kanazawa, I., 1985a, The mammalian tachykinin peptide family and their receptors, in "Natural products and Biological Activities", Tokyo Univ. Press, Tokyo, in press.

Kimura, S., Ogawa, T., Goto, K., Sugita, Y., Munekata, E., and Kanazawa, I., 1985b, Endogenous ligands for tachykinin receptors in mammals, in "Substance P --Metabolism and Biological Actions--", Talor & Francis Ltd, London, in press.

Kimura, S., Okada, M., Sugita, Y., Kanazawa, I., and Munekata, E., 1983, Novel neuropeptides, neurokinin α and β, isolated from porcine spinal cord, Proc. Japan Acad., 59B:101.

Lee, C.-M., Iversen, L.L., Hanley, M.R., and Sandberg, B.E.B., 1982, The possible existence of multiple receptors for substance P, Naunyn-Schmiedeberg's Arch. Pharmacol., 318:281.

Maggio, J.E., Sandberg, B.E.B., Bradley, C.V., Iversen, L.L., Santikarn, S., Williams, B.H., Hunter, J.C., and Hanley, M.R., 1983, Substance K: a novel tachykinin in mammalian spinal cord, in "Substance P --Dublin 1983", P. Skarabanek, ed., Boole press, Dublin, pp20.

Mantyh, P.W., Maggio, J.E., and Hunt, S.P., 1984, The autographic distribution of kassinin and substance K binding sites is different from the distribution of substance P binding sites in rat brain, Eur. J. Pharmacol., 102:361.

Matsuto, T., Yanagisawa, M., Otsuka, M., Kanazawa, I., and Munekata, E., 1984, The exitatory action of the newly-discovered mammalian tachykinins, neurokinin α and neurokinin β, on neurons of the isolated spinal cord of the newborn rat, Neurosci. Res., 2:105.

Minamino, N., Matsuda, H., Kangawa, K., and Matsuo, H., 1984, Regional distribution of neuromedin K and neuromedin L in rat brain and spinal cord, Biochem. Biophys. Res. Commun., 124:731.

Nawa, H., Doteuchi, M., Igano, K., Inouye, K., and Nakanishi, S., 1984, Substance K: a novel mammalian tachykinin that differs from substance P in its pharmacological profile, Life Sci., 34:1153.

Nawa, H., Hirose, T., Takashima, H., Inayama, S., and Nakanishi, S., 1983, nucleotide sequences of cloned cDNAs for two types of bovine brain substance P precorsors, Nature, 306:32.

Ogawa, T., Kanazawa, I., and Kimura, S., 1985, Regional distribution of substance P, neurokinin α and neurokinin β in rat spinal cord, nerve roots and dorsal root ganglia, and the effects of dorsal root section on spinal transection, Brain Res., in press.

Otsuka, M., and Konishi, S., 1975, Substance P and excitatory transmitter of primary sensory neurons, Cold Spring Harbor Symp. Quant. Biol., 40:135.

Pernow, B., 1983, Substance P, Pharmacol. Res., 35:135.

Quirion, R., and Pilapil, C., 1984, Conparative potencies of substance P, substance K and neuromedin K on brain substance P receptors, Neuropeptides, 4:325.

Torrens, Y., Lavielle, S., Chassaing, G., Marguet, A., Glowinski, J., and Beaujouan, J.C., 1984, Neuromedin K, a tool to further distinguish two central tachykinin binding sites, Eur. J. Pharmacol., 102:381.

Watson, S.P., Sandberg, B.E.B., Hanley, M.R., and Iversen, L.L., 1983, Tissue selectivity of substance P alkyl esters: suggesting multiple receptors, Eur. J. Pharmacol., 87:77.

Watson, S.P., 1984, Are the proposed substance P receptor subtypes, substance P receptors? Life Sci., 25:797.

BIOSYNTHESIS OF THE TACHYKININS AND SOMATOSTATIN

Anthony J. Harmar, Adrian R. Pierotti, and *Peter Keen

MRC Brain Metabolism Unit, Royal Edinburgh Hospital
Edinburgh, EH10 5HF, and *University Department of
Pharmacology, University Walk, Bristol BS8 1TD

INTRODUCTION

There are two mechanisms by which the expression of a single
neuropeptide gene may, in different tissues, give rise to alternative
patterns of biologically active peptides:

1) Tissue-specific RNA splicing of a single gene transcript may result in
the generation of messenger RNA (mRNA) species encoding different
polypeptide precursors, which may be processed into different products.
For example, transcription of the calcitonin gene in thyroid tissue
results in the production of a mRNA encoding the calcitonin precursor,
whereas in nervous tissue a mRNA encoding the neuropeptide calcitonin
gene-related peptide (CGRP) is generated (Rosenfeld et al., 1983; Craig
et al., this volume).

2) Tissue-specific post-translational modifications of a single
polypeptide precursor may generate different polypeptide products.   The
best known example is pro-opiomelanocortin, the common precursor to
adrenocorticotrophic hormone (ACTH), the melanocyte-stimulating hormone
(α, β- and γ-MSH) and the endorphin family of opioid peptides.   In the
anterior pituitary gland, the predominant products of POMC processing are
ACTH and β-endorphin, whereas in the pars intermedia αMSH,
corticotrophin-like intermediate lobe peptide (CLIP) and acetylated,
biologically inactive forms of endorphin are produced (Krieger & Liotta,
1979).

We here describe recent studies of the biosynthesis and
post-translational processing of the tachykinins and of somatostatin.
The tachykinins substance P and neurokinin A are synthesised from two
polypeptide precursors, α- and β- preprotachykinin, which are, in turn,
derived from a single gene by tissue-specific RNA splicing.   In contrast,
multiple molecular forms of somatostatin are present in tissues, in
proportions determined by tissue-specific patterns of post-translational
processing of a single gene product.

Morphological studies have demonstrated the existence of two populations of sensory neurones in dorsal root ganglia (DRG): cells with 'large, light' perikarya and myelinated axons and cells with 'small, dark' perikarya and largely unmyelinated axons.   A number of small peptides including substance P (SP: Hokfelt et. al., 1976), somatostatin (SS: Hokfelt et. al., 1976), vasoactive intestinal polypeptide (Lundberg et. al.,1978), calcitonin gene-related peptide (Rosenfeld et. al., 1983) and cholecystokinin (Larsson and Rehfeld, 1979) have been described in sub-populations of the 'small, dark' cells, where they may function in the transmission of nociceptive information.   When incubated in a simple salts medium at 37°C, isolated DRG continue to incorporate radiolabelled amino acids into protein for many hours and thus provide a valuable system for the study of neuropeptide biosynthesis in vitro.

Biosynthesis of the tachykinins

In preliminary studies on the biosynthesis of SP in DRG (Harmar & Keen, 1982), we incubated isolated DRG in vitro for periods of up to 24h and measured their content of SP-like immunoreactivity by radioimmunoassay.   We found that the SP content of ganglia increased linearly with time and that this increase was inhibited by inclusion of the protein synthesis inhibitors cycloheximide (10^{-4})M or anisomycin (10^{-5})M in the incubation medium.

In subsequent studies, we used a more direct approach to demonstrate the biosynthesis of SP in DRG: ganglia were incubated with [^{35}S]-methionine for periods of up to 9 hours, SP-related peptides immunoprecipitated with SP antibody and characterised by high performance liquid chromatography (HPLC). Two major peaks of [^{35}S]-labelled material were observed (Fig. 1), one of which coincided with authentic carrier SP. There was a lag phase of about 1.5h before the incorporation of [^{35}S]-methionine into this peak became apparent; subsequently incorporation of radiolabel increased linearly with time.   This lag phase probably represents the time taken for a polypeptide precursor to SP, which may not be identified by our immunoprecipitaion and chromatographic procedures, to be processed into SP.   Incorporation of radiolabel into SP was completely blocked by the inclusion of cycloheximide (10^{-5}M) in the incubation medium, suggesting that the peptide is synthesised by a conventional ribosomal mechanism.   In ganglia from rats which had been treated neonatally with capsaicin, incorporation of [^{35}S]-methionine into SP was reduced to 14% of that in normal animals.   This treatment causes degeneration of small-diameter chemosensitive neurones, including those which are thought to use SP as their transmitter. Confirmation of the identity of the major radiolabelled peak which co-eluted with SP was obtained from an experiment in which hydrogen peroxide was used to oxidise this peak to its sulphoxide. The oxidised material was then re-run on HPLC and a parallel conversion of biosynthetically labelled peptide and of synthetic SP to SP sulphoxide was observed.

Further confirmation of the identity of this peak with authentic SP, and information regarding the nature of the other major [^{35}S]-labelled peak (P2) was obtained in studies using a variety of amino acids for radiolabelling and a number of SP antisera of different specificity for immunoprecipitation (Fig. 1). Using C-terminally directed SP antisera, it could be shown that both [^{3}H]-proline and [^{35}S]-methionine were incorporated in vitro into SP, but that only methionine was incorporated into P2. When an antiserum directed towards the N-terminus of the SP sequence (Lee et. al., 1980) was used for immunoprecipitation, only a single major peak of radiolabel, corresponding to SP, was observed on HPLC

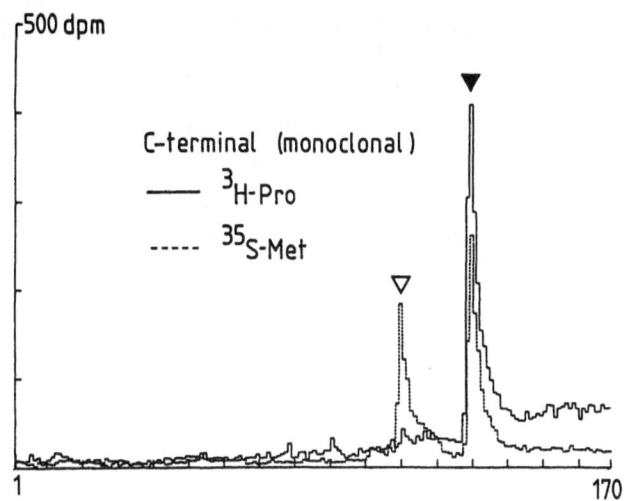

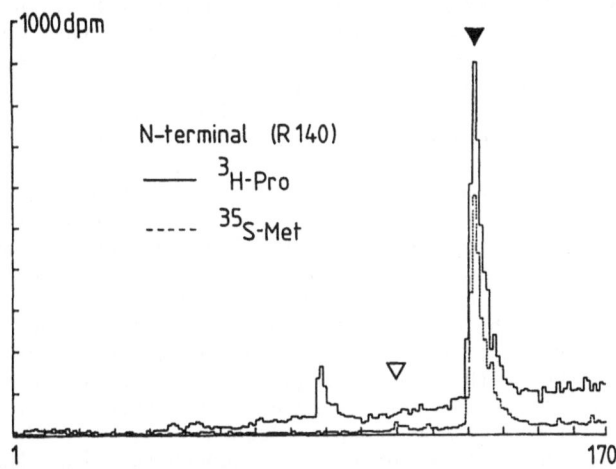

Figure 1 HPLC of SP-related radiolabelled peptides from dorsal root ganglia which had been incubated for 9h with a mixture of [^{35}S]-methionine and [^{3}H]-proline. The antisera used for immunoprecipitation are indicated: monoclonal SP antiserum was kindly supplied by Dr. A.C. Cuello and polyclonal N-terminally-directed antiserum by Drs. C.-M. Lee and P.C. Emson. (————): [^{3}H] dpm. (-------): [^{35}S] dpm.   The elution positions of SP (▼) and a second SP-like peptide (▽) are indicated. From Harmar et. al.,1981, with permission.

analysis. When ganglia were incubated simultaneously with [^{3}H]-phenylalanine and [^{35}S]-methionine both amino acids were incorporated into both peaks; however the ratio [^{3}H]-phenylalanine: [^{35}S]-methionine in P2 was almost exactly half that in SP implying – since SP contains two phenylalanine residues – that P2 might contain only one such residue. The inability of N-terminally directed SP antisera to immuno-precipitate P2 and the lack of incorporation of [^{3}H]-proline into the peptide suggested that it had a sequence similar to SP in the C-terminal region but different at the N-terminus.

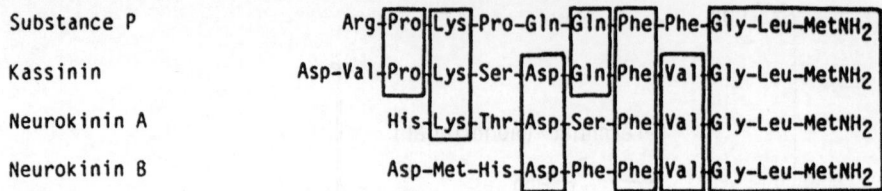

Substance P						Arg-Pro-Lys-Pro-Gln-Gln-Phe-Phe-Gly-Leu-MetNH₂

Figure 2 Structures of the tachykinins referred to in this chapter, showing sequence homologies.

Substance P was, until recently, the only known representative in mammals of the tachykinins, a family of peptides characterised by the common amino acid sequence -Phe-X-Gly-Leu-Met-NH_2 at the C-terminus, where X is a hydrophobic or aromatic residue (Erspamer, 1981; Fig. 2). Our results suggested that P2 might represent a novel tachykinin, and were consistent with reports (Iversen et. al., 1982) that more than one type of tachykinin receptor, and hence more than one tachykinin, might be present in mammalian tissues. In 1983, Kimura and his colleagues (see Kimura et. al., 1983 and this volume) reported the isolation of two new tachykinins, Neurokinins A and B (NKA and NKB), from porcine spinal cord (see Fig. 2). The amino acid sequences of both of these peptides exhibited striking sequence homology with the amphibian tachykinin kassinin and were consistent with our knowledge of the properties of P2. At about the same time, Nawa et. al. (1983) reported the nucleotide sequences of two precursors for SP from the bovine striatum (α- and β- preprotachykinin; α- and β- PPT). The structures of α- and β- PPT are identical, except that a 54-nucleotide segment of the cDNA sequence of β- PPT is missing in α-PPT. The 19-amino acid sequence corresponding to this region contains the sequence of NKA together with flanking sequences appropriate for the liberation of the NKA sequence from its putative precursor and for the formation of its C-terminal amide group.   Accordingly, we conducted experiments to determine whether our P2 corresponded to either NKA or NKB.

We were able to establish that P2, radiolabelled with [^{35}S]-methionine, had an identical retention time to synthetic NKA. Although NKA had only 1/70 of the affinity of SP for the antiserum used in these studies, under immunoprecipitation conditions the antiserum is used at high concentration and is therefore likely to precipitate both SP and NKA quantitatively. [^{3}H]-Histidine, [^{3}H]-serine and [^{3}H]-valine, amino acids which are present in the sequence of NKA but not of SP, were also incorporated into material co-eluting with NKA but not into SP (Fig. 3). No radiolabel was incorporated into material with the chromatographic properties of NKB. Consistent with previous observations, and in accord with the lack of homology of NKA with SP in the N-terminal region, no NKA-like material was immunoprecipitated by N-terminally directed SP antisera.

In summary, the evidence that the peptide synthesised by DRG is authentic NKA is as follows: 1). The peptide and its sulphoxide co-elute with NKA and NKA sulphoxide respectively. 2). The peptide contains methionine, phenylalanine, histidine, valine and serine. 3). It contains phenylalanine and methionine in a 1:1 ratio. 4). It does not contain proline. 5). It is immunoprecipitated by C-terminally, but not N-terminally directed SP antisera.

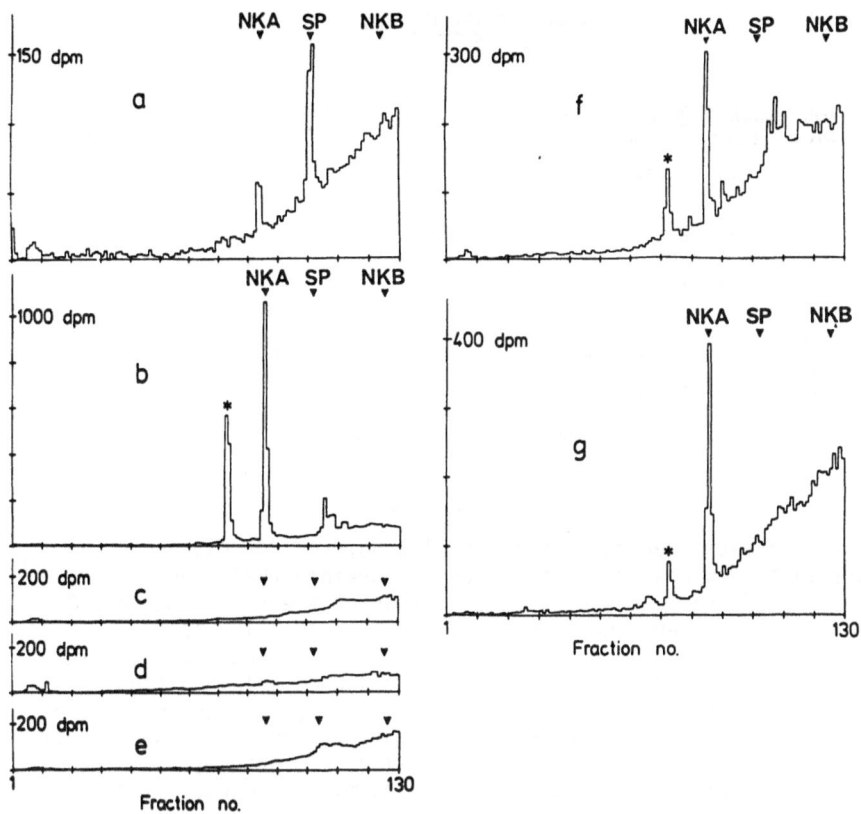

Figure 3 HPLC of radiolabelled peptides isolated from rat dorsal root ganglia which had been incubated for 9h with various radiolabelled amino acids: a, [^{3}H]-phenylalanine; b-e, [^{3}H]-histidine; f, [^{3}H]-serine; g, [^{3}H]-valine. All immunoprecipitated with a C-terminal-directed SP antiserum except: c, N-terminal-directed antiserum; d, pre-immune serum. In e, immunoprecipitation was carried out in the presence of 100µg added synthetic SP. Arrows denote elution positions of SP, NKA and NKB. From Harmar and Keen (1984), with permission.

Recent studies from Nakanishi's laboratory (Nawa et. al.,1984) indicate that the two bovine tachykinin precursors (α- and β-PPT) are derived from a single gene by alternative patterns of RNA splicing (Fig. 4). There are striking differences between tissues in the proportions of α- and β-PPT synthesised. If rat dorsal root ganglia contain a similar pair of precursors, our radiolabelling data suggest that α- and β-PPT are present in a ratio of 1.6:1, a value close to that determined in bovine sensory ganglia by S_1 nuclease protection analysis (Nawa et. al., 1984). Our studies suggest that, as predicted by recombinant DNA studies, SP and NKA may be synthesised in sensory neurones from a common precursor.

Biosynthesis of somatostatin (SS)

It is now clear that, in both rat and human tissues, the primary product of the translation of SS messenger RNA is a polypeptide (preproSS) of 116 amino acids (Figure 5: Goodman et al., 1983). A sequence of 24 amino acids at the N-terminus of preproSS constitutes the signal peptide, a hydrophobic region present in all secreted hormones and proteins, which promotes the attachment of the nascent prohormone to the membrane of the

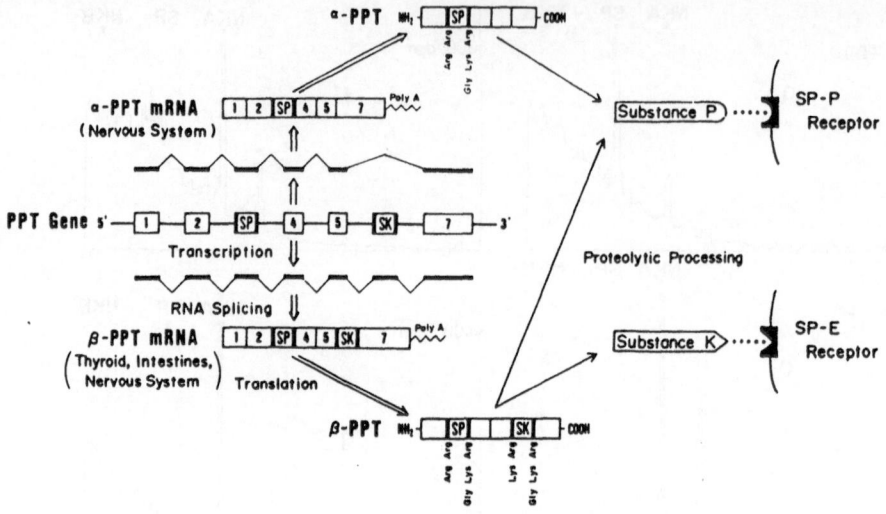

Figure 4 Schematic representation of alternative RNA splicing
pathways in the expression of the preprotachykinin (PPT) gene. The
gene contains 7 exons (numbered boxes) of which the third encodes SP
and the sixth NKA (here shown as substance K, SK). Alternative
splicing events give rise to two messenger RNAs (α- and β-PPT), both
of which contain a region encoding the SP sequence. In addition,
β-PPT contains a region encoding NKA (substance K). SP and NKA may be
the preferred ligands, respectively, for the SP-P and SP-E subtypes of
tachykinin receptor. Reprinted by permission from Nature, vol 312,
no.5996, pp 729-734.   Copyright (c) 1984 Macmillan Journals Ltd.

rough endoplasmic reticulum and its translocation through the membrane
into the intracisternal space.   The signal peptide is cleaved from the
prohormone before the completion of translation to generate proSS (92
amino acids), the largest form of the peptide present in significant
quantities in tissues.   As the prohormone passes from the endoplasmic
reticulum into the Golgi apparatus and is packaged into secretory
vesicles, a number of proteolytic cleavages take place, generating a
spectrum of oligopeptide products, only some of which contain the SS14
sequence.   The work of Benoit and colleagues (1982b, 1984) suggests that
there are three cleavage sites within the proSS sequence, and that at
least seven different peptides are generated from the precursor (Figure
5).   Four of these peptides contain the SS14 sequence: the
tetradecapeptide itself, somatostatin 28 (SS28:   Pradayrol et al., 1980),
a 6,000-7,500 molecular weight form and the intact proSS molecule.

    We have investigated the biosynthesis of SS in dorsal root ganglia,
using techniques similar to those described for the tachykinins. In these
studies ganglia were radiolabelled with [^{3}H]-phenylalanine, an amino
acid that is present three times in the sequence of SS14..   We observed a
time-dependent incorporation of [^{3}H]-phenylalanine into a peptide which
co-eluted with synthetic SS14 on HPLC (Fig. 6).   Inclusion of anisomycin
in the incubation medium totally inhibited incorporation of radiolabel
into SS14 indicating a ribosomal mechanism of synthesis.
[^{3}H]-phenylalanine was also incorporated into a second immunoreactive
peptide eluting with an identical retention time to synthetic SS28.
Incorporation of radiolabel into this peptide was apparent after 1h and
appeared not to increase thereafter.   These studies suggested that, in
sensory neurones, SS28 may be an intermediate in the biosynthesis of
SS14.   In the amino acid sequence of SS28, the C-terminal SS14 sequence
is preceded by a pair of basic amino acids;   in other precursor molecules

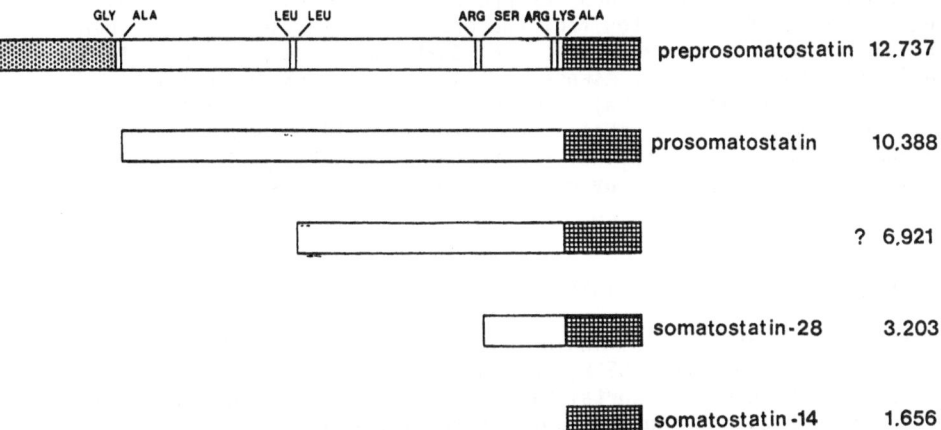

GLY ALA   LEU LEU   ARG SER ARG LYS ALA

preprosomatostatin 12,737

prosomatostatin 10,388

? 6,921

somatostatin-28 3,203

somatostatin-14 1,656

Figure 5 Peptides derived from the preprosomatostatin sequence.
Molecular weights of the multiple molecular forms are shown and
regions corresponding to the signal peptide (dotted area), to the SS14
sequence (hatched area) and amino acid residues at sites of
proteolytic processing are indicated, Tissues contain four peptides
which can be detected with antisera directed against the C-terminus of
SS14 and a further 3 peptides which crossreact with SS28 (1-12)
antisera. Based on the work of Benoit et al. (1982a, b; 1984) and
Goodman et al. (1983).

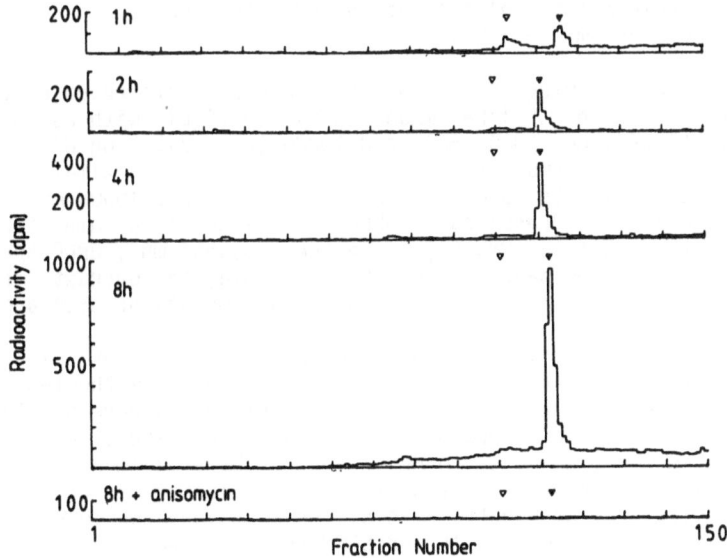

Figure 6 HPLC of somatostatin-related radiolabelled peptides
immunoprecipitated from dorsal root ganglia which had been incubated
for 1, 2, 4 and 8h with [^{3}H]-phenylalanine. Bottom trace
demonstrates inhibition of somatostatin biosynthesis when ganglia were
incubated for 8h in the presence of anisomycin (10^{-5}M). Arrows
denote elution positions of authentic SS28 ($\triangledown$) and SS14 ($\blacktriangledown$).
From Harmar et. al. (1982), with permission.

such sequences appear to be sites recognised by the trypsin-like endopeptidases which process peptide precursors into their functional end products. If SS14 originates from the enzymatic cleavage of SS28 then the remaining amino-terminal fragment of SS28 should be present in tissues where SS14 is found. The presence of this fragment ($SS28_{1-12}$) in tissues (Benoit et. al., 1982a) is consistent with the hypothesis that SS28 is the immediate biosynthetic precursor of SS14. However, the occurrence in brain of N-terminally extended forms of $SS28_{1-12}$ indicates that, in some tissues, SS14 may be generated by a pathway independent of SS28.

TISSUE-SPECIFIC POST-TRANSLATIONAL PROCESSING OF PREPROSOMATOSTATIN

Although in most tissues, including sensory ganglia, SS14 is present in much larger amounts than SS28, median eminence synaptosomes (Kewley et. al., 1981) and hypophysial portal blood (Millar et. al., 1983) have been shown to contain almost equal amounts of the two peptides. There is also evidence that SS14 and SS28 may play separate roles in the regulation of endocrine function. SS28 is more potent than SS14 in inhibiting the secretion of GH from the pituitary gland (Brazeau et. al., 1981); in the pancreas, SS28 inhibits insulin release with a potency some ten times greater than that of SS14 (Mandarino et. al., 1981), whereas SS14 is more potent in the inhibition of glucagon release. There is evidence that pituitary SS receptors may be of a type exhibiting a selective affinity for SS28, whereas those in the rest of the hypothalamus are more sensitive to SS14 (Srikant & Patel, 1981). These findings suggest that the processing of preproSS in the median eminence may differ from that in other tissues; accordingly, we have investigated the patterns of molecular forms of SS contained in, and released from, the rat hypothalamus and the isolated median eminence by gel filtration chromatography and high performance liquid chromatography.

Gel filtration chromatography of hypothalamic extracts revealed the presence of four forms of SS-like immunoreactivity (SSLI) with estimated molecular weights of 1500, 3000, 6000 and 10000 (Fig. 7). On HPLC, the 1500 and 3000 mol. wt. forms of SSLI eluted with retention times identical, respectively, to SS14 and SS28. The 6000 and 10000 mol. wt. forms were not completely resolved on HPLC, and eluted as a composite peak of high molecular weight SS (HMW-SS). In the hypothalamus, HPLC analysis indicated that the predominant form of SSLI was SS14; in contrast, median eminence tissue contained approximately equimolar amounts of SS14 and SS28.

To determine whether this difference in the proportions of SS14 and SS28 was reflected in the composition of SSLI released from the two tissues in vitro, hypothalamic slices or median eminences were perifused with modified Krebs medium, exposed to a depolarising pulse of KCl and the peptides released were subjected to HPLC. Under the experimental conditions employed, the release of SS from these tissues was calcium-dependent. Perifusates of hypothalamic slices contained two molecular forms of SSLI corresponding to SS14 and SS28 (Fig. 8, a-c). K^+ depolarisation increased by 12 fold the rate of release of SS14, which represented 76% of the total SSLI released. Small amounts of SS28 were also detected, but depolarisation had no significant influence on the rate of release of this peptide. In contrast, perifusates of isolated median eminences contained SS14, SS28 and HMW-SS (58%, 25% and 17% of total SSLI, respectively: Fig. 8, d-f) and K^+ depolarisation increased the rates of release of all three peptides.

These results provide supportive evidence for the existence of two populations of SS-containing neurons in the hypothalamus, already suggested by neuroanatomical (Kawano et. al., 1982) and ontogenic (Daikoku

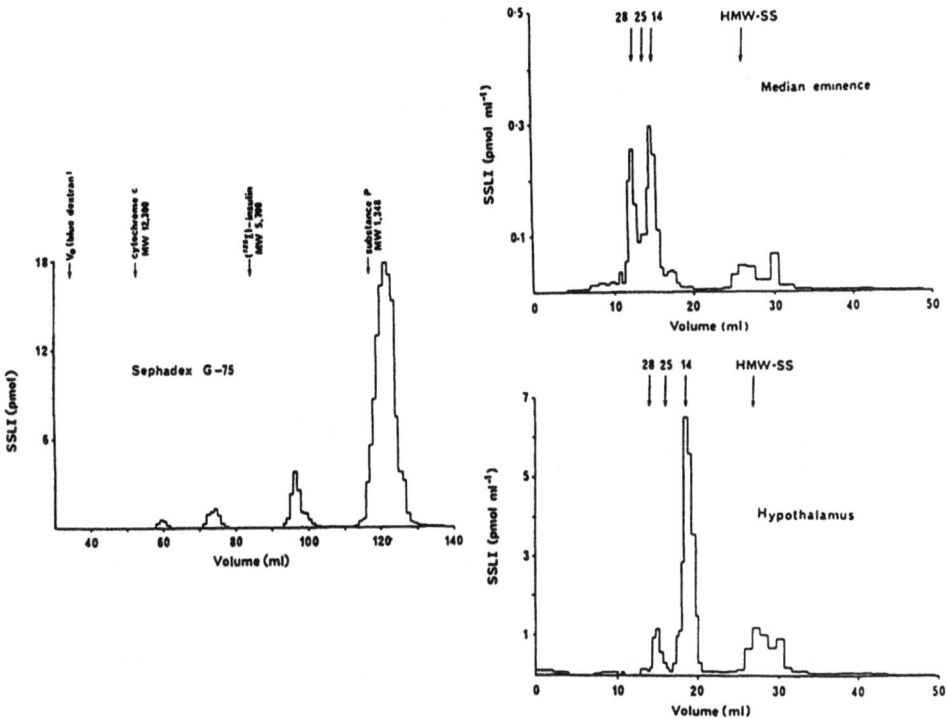

Figure 7 Elution profile of SS-like immunoractivity (SSLI) extracted
from the rat hypothalamus and median eminence. Left panel shows
elution profile following gel filtratiion chromatography of
hypothalamic extracts on Sephadex G-75.   Four molecular forms of SS,
with approximate molecular weights of 1500 (SS14), 3000 (SS28), 6000
and 10,000, are present. The elution positions of markers used to
calibrate the column are indicated. Right panels show elution
profiles following high performance liquid chromatography of extracts
of hypothalamus and median eminence. Arrows indicate the elution
positions of somatostatins 14, 25 and 28 and of high molecular weight
somatostatin (HMW-SS: consists of 6,000 Mol. Wt. and 10,000 Mol. Wt.
forms). Re-drawn from Pierotti and Harmar (1985).

et. al., 1983) studies.   One population, with cell bodies in the rostral
periventricular area which send projections to the median eminence and
possibly to the neurohypophysis, synthesises and releases both SS14 and
SS28 as hormonal regulators of pituitary growth hormone secretion.   The
second population of SS-producing cells are interneurones in the arcuate,
ventromedial and other nuclei, which synthesise and release predominantly
SS14.

      Evidence is thus accumulating to suggest that, in different tissues,
proSS may be processed to generate different patterns of biologically
active products.   In most regions, SS14 is the predominant end-product of
biosynthesis, but in some tissues SS28 is synthesised in comparable or
greater amounts.   The submucosa and muscle layer of the small intestine
contain almost exclusively SS14 of neuronal origin, whereas in the
endocrine D cells of the mucosa SS28 predominates (Baskin &
Ensinck,1984).   SS14- and SS28- producing endocrine cells may be
distinguished at the ultrastructural level by immunohistochemical
techniques (Ravazzola et. al., 1983).   In D cells from the pancreas and
stomach, which synthesise predominantly SS14, SS28 immunoreactivity is
confined to vesicular structures close to the Golgi apparatus, consistent

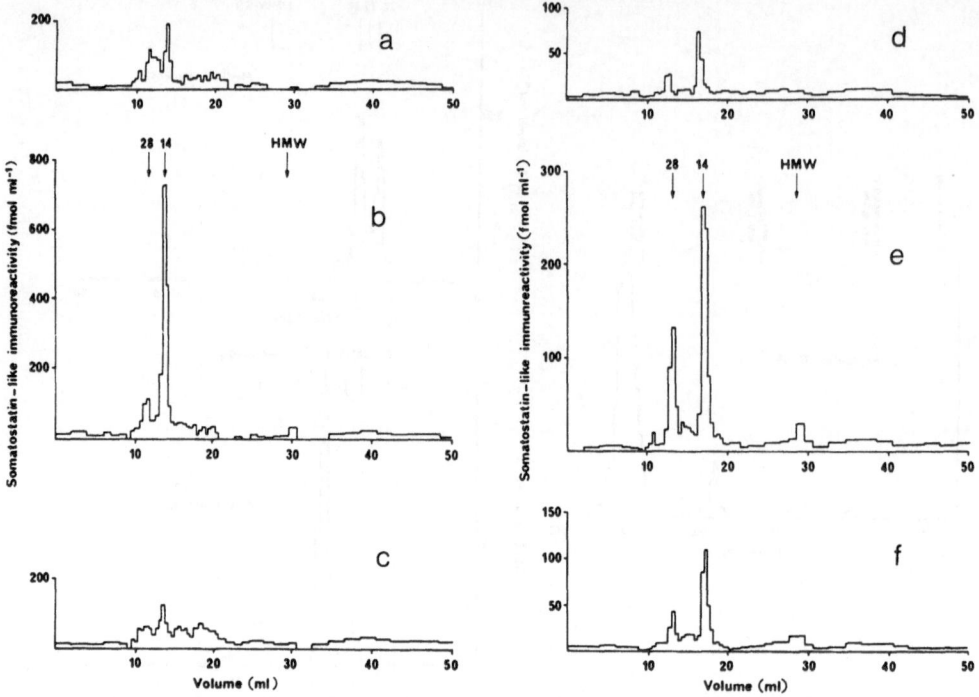

Figure 8 High performance liquid chromatography of SSLI released from (a–c) hypothalamic slices and (d–f) median eminences in vitro. Perifusate samples were collected for 9 minute periods before (a,d), during (b,e) and after (c,f) exposure of tissues to a depolarising pulse of KCl. The elution positions of synthetic SS28 and SS14, and of HMW-SS, are indicated. From Pierotti et. al. (1985), with permission.

with a role for this peptide as an intermediate in the biosynthesis of SS14. Mature secretory granules contain only SS14 and SS28(1-12). In intestinal D cells, which secrete predominantly SS28, SS28-like immunoreactivity is seen both in the Golgi apparatus and in mature secretory granules.

It is likely that two different processing enzymes are responsible for the generation of SS28 and of SS14 from the proSS precursor. One of these (I) cleaves the precursor following a pair of basic amino acids (Arg-Lys) to generate SS14. An enzyme possessing such activity has recently been described (Cohen et. al., this volume) and enzymes of similar specificity have been described in many hormone-secreting tissues. The site at which SS28 is cleaved from the proSS molecule (following the sequence Leu-Gln-Arg-) may be the substrate for a processing enzyme (II) of different specificity. In some tissues (e.g. pancreas, hypothalamic interneurones) enzyme I may be abundant and SS14 is the major end product of proSS processing, whereas in SS28-producing tissues (e.g. intestinal D cells, bovine retina) enzyme I may be absent or reduced in activity. The presence of roughly equimolar proportions of SS14 and SS28 in the median eminence and neurohypophysis may be due to the existence of separate populations of SS14 and SS28-synthesising neurones, or to limited activity of enzyme I within a single population of neurones. For a complete understanding of the tissue-specific processing of proSS in neurones and endocrine cells, further studies of the distribution and properties of this enzyme will be required.

ACKNOWLEDGEMENTS

We thank Norma Brearley for expert preparation of this manuscript.
ARP is an MRC Research Student.

REFERENCES

Baskin, D.G. and Ensinck, J.W., 1984, Somatostatin in epithelial cells
    of intestinal mucosa is present primarily as somatostatin 28,
    Peptides, 5:615.
Benoît, R., Bohlen, P., Esch, F. and Ling, M., 1984, Neuropeptides derived
    from prosomatostatin that do not contain the somatostatin-14 sequence,
    Brain Res. 311:23.
Benoît, R., Bohlen, P., Ling, N., Briskin, A., Esch, F. Brazeau, P.,
    Ying, S-Y and Guillemin, R., 1982, Presence of somatostatin 28 (1-12)
    in hypothalamus and pancreas, Proc. Natn. Acad. Sci. U.S.A. 79:917.
Benoit, R., Ling, N., Alford, B. and Guillemin, R., 1982, Seven peptides
    derived from pro-somatostatin in rat brain, Biochem. Biophys. Res.
    Commun. 107:944.
Brazeau, P., Ling, N., Esch, F., Bohlen, P., Benoit, R. and Guillemin, R.,
    1981, High biological activity of the synthetic replicates of
    somatostatin-28 and somatostatin-25, Reg. Peptides 1:255.
Daikoku, S., Hisano, S., Kawano, H., Okamura, Y. and Tsuruo, Y., 1983,
    Ontogenetic studies on the topographical heterogeneity of
    somatostatin-containing neurones in rat hypothalamus, Cell Tissue Res.
    233:347.
Erspamer, V., 1981, The tachykinin peptide family, Trends Neurosci. 4:267.
Goodman,R.H., Aron, D.C. and Roos, B.A., 1983, Rat prepro-somatostatin,
    structure and processing by microsomal membranes, J. Biol. Chem.
    258:5570.
Harmar, A.J. and Keen, P., 1982, Synthesis, and central and peripheral
    axonal transport of substance P in a dorsal root ganglion-nerve
    preparation in vitro, Brain Res. 231:379.
Harmar, A.J. and Keen, P., 1984, Rat sensory ganglia incorporate
    radiolabelled amino acids into substance K (neurokinin α) in vitro,
    Neurosci. Letts. 51:387.
Harmar, A.J., Ivell, R. and Keen, P., 1982, The de novo biosynthesis of
    somatostatin and a related peptide in isolated rat dorsal root
    ganglia, Brain Res. 242:365.
Harmar, A.J., Schofield, J.G. and Keen, P., 1981, Substance P
    biosynthesis in dorsal root ganglia: An immunochemical study of
    [35S]-methionine and [3H]-proline incorporation in vitro, Neuroscience
    6:19172.
Hokfelt, T., Elde, R., Johansson, O., Luft, R., Nilsson, G. and
    Arimura, A., 1976, Immunohistochemical evidence for separate
    populations of somatostatin-containing and substance P-containing
    primary afferent neurons in the rat, Neuroscience 1:131.
Iversen, L.L., Hanley, M.R., Sandberg, B.E.B., Lee, C.M., Pinnock, R.D.
    and Watson, S.P., 1982, Substance P receptors in the nervous system
    and possible receptor subtypes, in: "Substance P in the nervous
    system," R. Porter and M. O'Connor, eds., Pitman, London.
Kawano, H., Diakoku, S. and Saito, S., 1982, Immunohistochemical studies
    of intrahypothalamic somatostatin-containing neurones in rat,
    Brain Res. 242:227.
Kewley, C.F., Millar, R.P., Berman, M.C. and Schally, A.V., 1981,
    Depolarization- and ionophore-induced release of octacosa somatostatin
    from stalk median eminence synaptosomes, Science 213:913.
Kimura, S., Oada, M., Sugita, Y., Kanazawa, I. and Munekata, E., 1983,
    Novel neuropeptides, neurokinin α and β, isolated from porcine spinal
    cord, Proc. Japan. Acad. Ser. B., 59:101.

Krieger, D.T. and Liotta, A.F. 1979, Pituitary hormones in brain: where, how and why?, Science, 205:366.

Larsson, L.I. and Rehfeld, J.F., 1979, Localization and molecular heterogeneity of cholecystokinin in the central and peripheral nervous system, Brain Res. 165: 201.

Lee, C-M., Emson, P.C. and Iversen, L.L., 1980, The development and application of a novel N-terminal directed substance P antiserum, Life Sci. 27:535.

Lundberg, J.M., Hokfelt, T., Nilsson, G., Terenius, L. and Rehfeld, J. R. Elde and S. Said, (1978), Acta physiol. Scand. 104, 499-

Mandarino, L., Stenner, D., Blanchard, W., Nissen, S., Gerich, J., Ling, N., Brazeau, P., Bohlen, P., Esch, F. and Guillemin, R., 1981, Selective effects of somatostatin-14,25,28 on in vitro insulin and glucagon secretion, Nature 291:767.

Millar, R.P., Sheward, W.J., Wegener, I. and Fink, G., 1983, Somatostatin-28 is a hormonally active peptide released into hypophysial portal vessel blood, Brain Res. 260:334.

Nawa, H., Hirose, T., Takashima, H., Inayama S & Nakanishi, S., 1983, Nucleotide sequence of cloned cDNAs for two types of bovine brain substance P precursor, Nature 306:32.

Nawa, H., Kotani, H. and Nakaniski, S., 1984, Tissue-specific generation of two preprotachykinin mRNAs from one gene by alternative RNA splicing, Nature 312:729.

Pierotti, A.R. and Harmar, A.J., 1985, Multiple forms of somatostatin-like immunoreactivity in the hypothalamus and amygdala of the rat: selective localization of somatostatin-28 in the median eminence, J. Endocrinol. 105:383.

Pierotti, A.R., Harmar, A.J., Tannahill, L. and Arbuthnott, G.W., 1985, Different patterns of molecular forms of somatostatin are released by the rat median eminence and hypothalamus, Neurosci. Letts. 57:215.

Pradayrol, L., Jornvall, H., Mutt, V. and Ribet, A., 1980, N-terminally extended somatostatin: the primary structure of somatostatin-28, FEBS Lett. 109:55.

Ravazzola, M., Benoit, R., Ling, N., Guillemin, R. and Orci, L., 1983, Immunocytochemical localization of prosomatostatin fragments in maturing and mature secretory granules of pancreatic and gastrointestinal D-cells, Proc. Natn. Acad. Sci. U.S.A. 80:215.

Rosenfeld, M.G., Mermod, J-J., Amara, S.G., Swanson, L.W., Sawchenko, P.E. Rivier, J., Vale, W.W. and Evans, R.M., 1983, Production of a novel neuropeptide encoded by the calcitonin gene via tissue-specific RNA processing, Nature 304:129.

Srikant, C.B. and Patel, Y.C., 1981, Receptor binding of somatostatin-28 is tissue specific, Nature 294:259.

BIOSYNTHESIS OF SOMATOSTATIN, VASOACTIVE INTESTINAL POLYPEPTIDE, AND THYROTROPIN RELEASING HORMONE

R.H. Goodman, M.R. Montminy, M.J. Low, T. Tsukada, S. Fink, R.M. Lechan, P. Wu*, I.M.D. Jackson*, and G. Mandel

Division of Endocrinology, Dept. of Medicine, New England Medical Center, Boston, Massachusetts and *Division of Endocrinology, Providence Hospital, Providence, Rhode Island

The purpose of this review is to describe the recent advances in our understanding of the biosynthesis of three brain/gut peptides -- somatostatin, vasoactive intestinal polypeptide, and thyrotropin releasing hormone. Although each of these hormones is found in both the brain and gastrointestinal tract, their physiology is quite distinct. Furthermore, each peptide poses a unique set of problems for isolating and characterizing its gene and gene products. This review focuses on the different molecular approaches that we have used to characterize the precursors of these three peptides. It further describes the more general techniques that we are currently using to examine the expression of the three genes.

SOMATOSTATIN

Somatostatin is a tetradecapeptide initially named for its ability to inhibit pituitary growth hormone release (Brazeau et al., 1973). It was subsequently found to inhibit the secretion of a number of other pituitary, gastrointestinal, and pancreatic islet hormones. Like other regulatory peptides, somatostatin is synthesized as a high molecular weight precursor. The first evidence for a larger form of somatostatin was provided by Arimura et al. (1976), using antisera directed toward the somatostatin tetradecapeptide. The fact that certain antisera raised against somatostatin-14 also detect high molecular weight forms of the hormone has greatly facilitated the elucidation of the precursor, pre-prosomatostatin. Another factor that has aided the characterization of pre-prosomatostatin is the existence of a rich source of the hormone precursor -- the pancreatic islets (Brockmann bodies) of the anglerfish, Lophius americanus. Noe and co-workers have made extensive use of these islets (which contain levels of somatostatin exceeding 1 ug/mg) to study the conversion of high molecular weight precursor forms to the bioactive somatostatin-14 molecule (Noe and Spiess, 1983). An additional feature that has been helpful to molecular biological approaches is the low level of ribonuclease in anglerfish islets. Because of the abundance of somatostatin in the anglerfish islets, this tissue represented an ideal

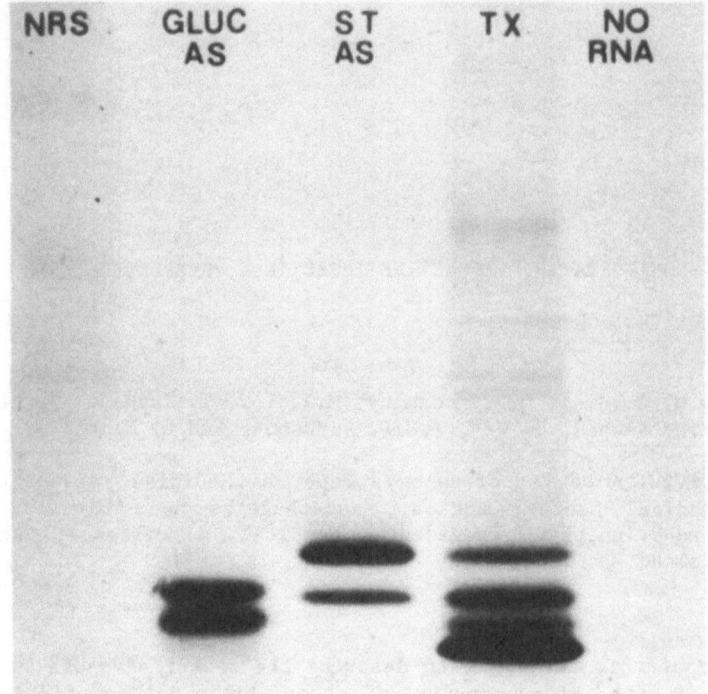

Fig. 1.   Immunoprecipitations of islet cell-free translation products displayed on an SDS-polyacrylamide
gel.   NRS designates immunoprecipitation using
normal rabbit serum.   Gluc AS and ST AS designate
immunoprecipitations using antisera to glucagon
and somatostatin.   TX designates total translation
products.   No RNA lane shows translation in the
absence of exogenous mRNA.   Size of the translation products range from 16,000 to 11,000 daltons.

starting-point for studies aimed at elucidating the structure of the
somatostatin precursor.

Translation of islet mRNA in a cell-free system derived from wheat
germ extract generates four major protein products when analyzed on an
SDS-polyacrylamide gel (Fig. 1).   These products range in size from
12,000 to 16,000 daltons.   When the cell-free translations are
immunoprecipitated using antisera to synthetic somatostatin, two
discrete proteins are observed.   Although these studies suggested that
there were two forms of somatostatin precursor in anglerfish (Goodman et
al., 1980), this analysis proved to be an oversimplification.
Two-dimensional gel electrophoresis revealed that each of the protein
bands in Fig. 1 is composed of multiple proteins with separate
isoelectric focusing characteristics.   Warren and Shields (1982) have
reported that several of these separate translation products react with
antiserum to somatostatin, and have suggested that there may be as many
as nine somatostatin precursors.   These precursor forms fall into two
classes -- migrating on gel electrophoresis with apparent molecular
weights of 16,000 and 14,000 daltons.   Within each class, individual
precursors seem to vary by one or a few amino acids (Warren and Shields,

SP   ---28---

                                     -14-

Fig. 2. Structure of rat pre-prosomatostatin. SP refers
to signal peptide. 28 refers to somatostatin-28.
14 refers to somatostatin-14. The entire molecule
is 116 amino acids in length. A pro-region of 92
amino acids separates the signal peptide from the
somatostatin-28 sequence.

1984). The significance of these variations, which occur within the
pro-regions of the precursors, is uncertain. Subsequent studies by
Hobart and colleagues (1980) established the nucleotide and
corresponding amino acid sequences of two cDNAs representing precursors
of anglerfish somatostatin. These precursors have been designated
pre-prosomatostatin I and II and appear to represent the 16,000 and
14,000 M_r forms, respectively.

Our laboratory used the technique of hybridization arrest to
identify a cDNA clone encoding pre-prosomatostatin I (Goodman et al.,
1980). This approach, which involves the identification of test cDNAs
by hybridization to total mRNA and subsequent cell-free translation, was
successful only because of the extraordinarily high levels of
somatostatin mRNA in the anglerfish islets. We subsequently used the
anglerfish somatostatin cDNA to screen a cDNA library derived from a rat
medullary thyroid carcinoma (Goodman et al., 1983). Because the amino
acid structures of the anglerfish and mammalian somatostatin-14
molecules are so highly conserved, we reasoned that the nucleotide
sequence homology between the two genes might be high enough to allow us
to detect the mammalian precursor.

The rat medullary thryoid carcinoma cDNA library was screened using
the procedure of Hanahan and Meselson (1980). The somatostatin cDNA
that was identified encodes a protein of 12,737 daltons (Fig. 2),
consistent in size with estimates of the precursor determined using
pulse-labeling techniques (Patzelt et al., 1980). Like the anglerfish
precursors, the sequences encoding the biologically active forms of
somatostatin are located at the carboxy-terminus of the prohormone. At
the amino-terminus of the precursor is a sequence of hydrophobic amino
acids characteristic of a signal region. Cell-free translations of
medullary thyroid carcinoma mRNA supplemented with microsomal membranes
subsequently established that the cleavage site of the signal region
from the prohormone occurs at the glutamine residue at amino acid
position 24. Cleavage at this site generates a prohormone of 92 amino
acids (10,389 daltons). A cDNA encoding human pre-prosomatostatin,
subsequently reported by Shen et al. (1982), is remarkably similar to

the rat precursor. Only four amino acids within the two 116 amino acid precursors differ. Two of these differences occur within the signal regions and preserve the hydrophobic character of this domain.

The somatostatin-14 sequence is flanked in the precursor by the amino acids arginine and lysine. These amino acids are frequently found at sites of post-translational proteolytic cleavage. In general, however, Lys-Lys or Lys-Arg sequences are more usual cleavage sites than the Arg-Lys sequence found in the somatostatin precursors. A single arginine residue separates the 28 amino acids at the carboxy-terminus of the molecule from the remainder of the prohormone. Single arginine residues have been found to be sites of cleavage in a number of other hormone precursors including prorelaxin (Hudson et al., 1983), propressophysin (Land et al., 1982), and propancreatic polypeptide (Leiter et al., 1984). Cleavage at this site would generate the icosapeptide somatostatin-28, which has been identified in brain, retina, and gastrointestinal tract (Pradayrol et al., 1980; Schally et al., 1980; Esch et al., 1980). Somatostatin-28, like somatostatin-14, is biologically active. Several reports have suggested that the biological activity of somatostatin-28 may differ from that of somatostatin-14 (Mandarino et al., 1981; Brown et al., 1980; Konturek, 1981). Somatostatin-28 (1-12) also appears to be generated in vivo from the precursor (Benoit et al., 1982). This fragment represents the amino-terminal 12 residues of somatostatin-28. The biological function of this peptide is unknown.

The function of the pro-region of the somatostatin in precursor is also unknown. This region contains a glycosylation signal (Asn-Gln-Thr), but the evidence that prosomatostatin is glycosylated is contradictory (Patzelt et al., 1980). Our laboratory has found that prosomatostatin cannot be glycosylated by microsomal membranes in vitro (Goodman et al., 1983). It is possible that the pro-region of the precursor may be required as a spacer -- to produce a precursor large enough to span the membranes involved in proteolytic processing and intracellular trafficking. This possibility is supported by the observation that all presecretory proteins are at least 65 amino acids long. Furthermore, deletion mutants of the somatostatin gene in which the pro-region has been removed apppear to produce a protein that is secreted inefficiently (Sevarino, unpublished observations). These findings suggest that there might be a length requirement for efficient processing of hormonal precursors.

If the pro-region only satisfied a length requirment, however, it would not be necessary to conserve the sequence of this portion of the precursor. The nearly complete conservation of the entire pre-prosomatostatin molecule between rodents and humans suggests that other features of the pro-region may also be important. Although not quite as striking as the similarity between the rodent and human sequences, a relatively high level of conservation also exists between anglerfish pre-prosomatostatin I and the mammalian precursors. Immunohistochemical studies using antisera directed against the prohormone have suggested that at least a portion of the somatostatin pro-region is transported to nerve terminals (Lechan et al., 1983). Whether peptides derived from the pro-region sequence have independent biological activities is unknown. Benoit et al. (1984) and Aron et al. (1984) have both found evidence for proteolytic cleavage of prosomatostatin within the pro-region. Low et al. (1983) in our laboratory have made similar observations. It is also possible that conservation of the pro-region sequence is essential for proper post-translational processing.

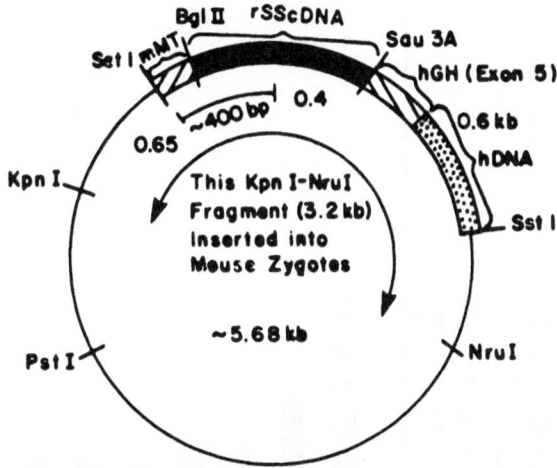

Fig. 3. Structure of the metallothionein-somatostatin-
growth hormone recombinant.   mMT refers to
mouse metallothionein promotor.   rSS refers to
rat somatostatin structural gene.   hGH and
hDNA refers to human growth hormone 3'
untranslated region and flanking region.

The cellular factors that control the processing of hormone
precursors to biologically active peptides are clearly of great
interest.   It is not known, for example, whether prohormonal processing
enzymes are specific for an individual precursor or whether a generic
protease recognizes a homologous sequence in a number of unrelated
prohormones.   Studies by Moore et al. (1983) have suggested that
prohormonal processing enzymes may not be very specific.   These
investigators introduced the rat insulin gene into AtT-20 cells, a
pituitary cell line which normally makes ACTH.   The transfected
pituitary cells synthesized pre-proinsulin and furthermore had the
ability to efficiently process the precursor to mature insulin.
Similarly, Hellerman et al. (1984) showed that GH_4 cells, a growth
hormone-producing pituitary cell line, had the ability to process
pre-proparathyroid hormone to the mature form.   These studies all
utilized transformed cells however, whose processing abilities might not
reflect those of normal cells.

By examining somatostatin production in transgenic mice, we sought
to determine whether normal cells shared this ability to correctly
process heterologous hormonal precursors (Low et al., 1985).   For these
studies, a fusion gene was constructed that contained the mouse
metallothionein promoter, the rat pre-prosomatostatin structural gene,
and the human growth hormone 3' flanking region (Fig. 3).   The
metallothionein promoter in this fusion gene allows expression in a wide
variety of cell types.   The growth hormone portion provides a

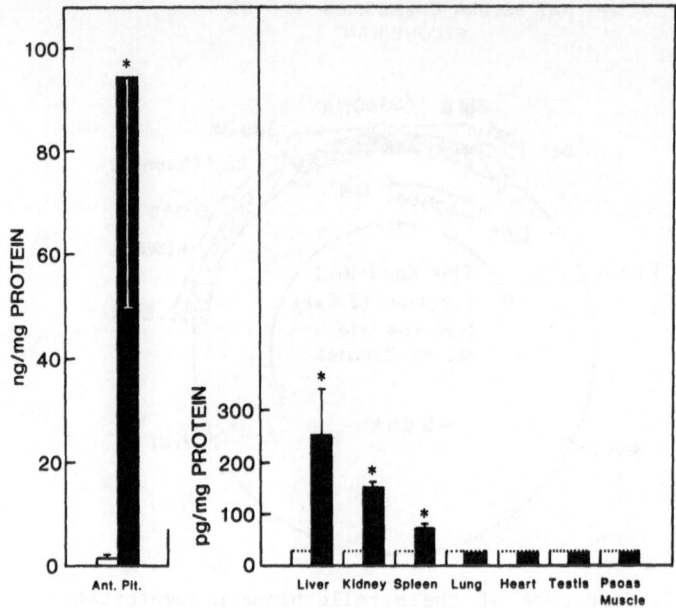

Fig. 4. Tissue concentration of immunoreactive somatostatin
in control and transgenic mice. Open bars give
values for controls, black bars refer to transgenic
animals.

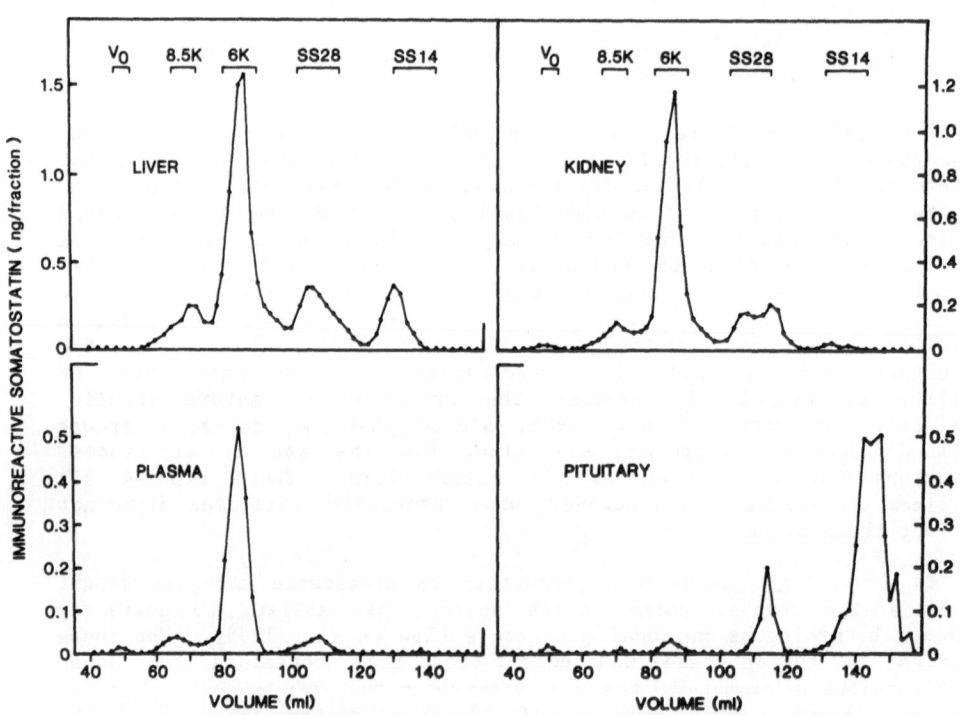

Fig. 5. Sephadex G-50 SF chromatography of immunoreactive
somatostatin in tissue extracts from transgenic
mice.

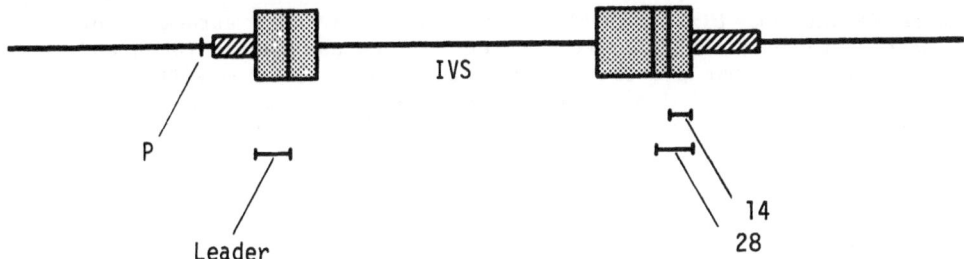

Fig. 6.   Structure of the rat somatostatin gene. P refers to promotor,
          Leader designates signal region, IVS designates intervening
          sequence. Somatostatin-14 and -28 peptides are indicated. 5'
          and 3' untranslated regions are designated by slashed boxes.

transcriptional termination signal and makes it possible to distinguish
somatostatin mRNA transcribed from the fusion gene from the endogenous
message. The translational start and stop signals are entirely provided
by the somatostatin cDNA. Consequently, the only protein product
translated from the fusion gene is pre-prosomatostatin. By
microinjecting the fusion gene into the pronuclei of fertilized mouse
eggs, it was possible to generate several lines of mice which express
pre-prosomatostatin ectopically in a large number of tissues.

Several tissues in the transgenic mice produced immunoreactive
somatostatin, including liver, kidney, and spleen (Fig. 4). By far the
highest concentration of somatostatin was found in the anterior
pituitary, which, like liver, kidney, and spleen, does not normally
produce somatostatin. The anterior pituitary cells, furthermore,
processed the precursor to somatostatin-14 and somatostatin-28 (Fig.
5). These experiments confirm the impression that the cellular factors
involved in prohormonal processing are not very specific. Pituitary
cells, never normally called upon to process pre-prosomatostatin,
nonetheless have the ability to synthesize, transport, and process the
precursor in an apparently normal fashion.

Unlike the anglerfish, mammals appear to have only one somatostatin
gene. As a first step toward characterizing the factors that control
the expression of the somatostatin gene, we have isolated the gene from
a rat genomic library (Montminy et al., 1984). The structure of the
gene is depicted in Fig. 6. The gene is interrupted by a single intron,
630 bases in length, which occurs near the middle of the pro-region.
The position of this intron does not coincide with any of the known
cleavage sites of the prohormone. Other characteristic features of the
gene include a somewhat atypical promotor sequence, TTTAAA, 31 bases
upstream from the transcriptional initiation site. A repetitive
sequence of alternating guanine and thymine residues occurs
approximately 650 bases upstream from the promotor. The significance of
this sequence is its potential to adopt a Z-DNA configuration. This
alternative structure of the DNA helix has been proposed to influence
transcriptional activity of other eukarotic genes (Wang et al.,
1979). By S-1 nuclease analysis, Dixon and co-workers, who have also
sequenced the rat somatostatin gene (Tavianini et al., 1985), have
determined that the repetitive sequences do in fact adopt a Z-DNA

structure (Hayes and Dixon, 1985). Although the overall structure of the somatostatin gene is conserved between rodents and humans (Shen and Rutter, 1984), the Z-DNA portion of the human gene is located within the intron as opposed to the 5' flanking region. Whether the Z-DNA regions within the somatostatin genes have any biological significance has not been determined.

VASOACTIVE INTESTINAL POLYPEPTIDE

Vasoactive intestinal polypeptide was initially characterized as a vasodilatory hormone in the small intestine (Said and Mutt, 1970). More recent studies have indicated that VIP additionally functions as a neurotransmitter, neuromodulator, and neuroendocrine releasing factor (Kato et al., 1978; Marley and Emson, 1982). VIP appears to have a physiological role in regulating pituitary prolactin secretion (Abe et al., 1985). Additionally, though its stimulatory effects on adenyl cyclase, VIP may influence the biosynthesis of other neuropeptides such as somatostatin (Montminy et al., submitted for publication).

The amino acid sequence of VIP is homologous to that of a number of other hormones including glucagon, secretin, gastric inhibitory peptide, and growth hormone releasing factor. Itoh et al. (1983) have determined the nucleotide and corresponding amino acid sequence of the human VIP precursor using mRNA isolated from a neuroblastoma cell line. These studies indicated that VIP is derived from a precursor that generates, in addition to VIP, a 27 amino acid peptide designated PHM (Peptide Histidine Methionine). PHM is structurally very similar to PHI (Peptide Histidine Isoleucine), a peptide isolated from porcine intestine (Tatemoto and Mutt, 1981). Our laboratory has confirmed the sequence of the human VIP precursor, using a cDNA derived from a human islet cell tumor (Tsukada et al., 1985). The structure of the human VIP precursor is depicted in Fig. 7. Both the VIP and PHM sequences within the precursor are flanked at their carboxy-termini by the sequences Gly-Lys-Arg. These sequences generally predict sites of amidation and proteolytic processing. A single arginine residue flanks the PHM sequence at its amino terminus. As discussed above, single basic residues can also serve as proteolytic cleavage sites. The VIP sequence is flanked by a more typical Lys-Arg sequence.

We have used the human VIP cDNA to isolate a clone encoding the rat precursor from a hypothalamic lambda gt11 bacteriophage library. The rat hypothalamic precursor is strikingly similar to that of the human. The amino acid sequence of VIP itself is completely conserved between the two species. The PHI/PHM region contains four amino acid substitutions, all conservative in nature. One of these substitutions is an isoleucine for a methionine at the carboxy-terminus of the molecule, similar to the structure of PHI found in porcine intestine. The peptide separating PHI and VIP as well as the carboxy-terminal peptide is highly conserved bewteen the rat and human precursors. This high level of conservation suggests that the non-VIP peptides within the prohormone may also have some specific biological function. Nishizawa et al. (1985) have recently published an identical rat pre-proVIP cDNA structure.

As a first step toward understanding the molecular basis of pre-proVIP gene expression, we have isolated the human pre-proVIP gene and have determined its structure (Tsukada et al., 1985). The gene, also depicted in Fig. 7, is approximately 9 kilobases in length and is interupted by six introns. These introns divide the precursor into domains that correspond very closely to the known products of the

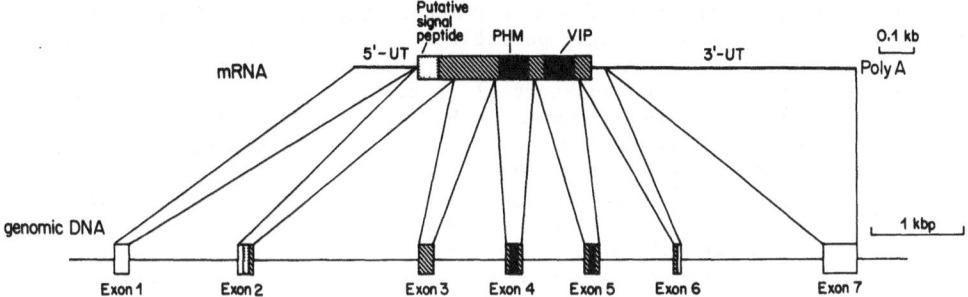

Fig. 7.   Structure of the human vasoactive intestinal polypeptide cDNA
          and gene. 5' UT and 3' UT refer to untranslated regions. PHM
          and VIP sequences are indicated.   In the lower portion of the
          figure, the positions of introns within the gene are depicted.

precursor.   Exon 1 encodes the 5' untranslated region, exon 2 encodes
the signal region, exon 3 encodes the amino-terminal peptide, exon 2
encodes PHM, exon 5 encodes VIP, exon 6 encodes the carboxy-terminal
peptide, and exon 7 encodes the 3' untranslated region. The presence of
an intron within the 3' untranslated region is an especially unusual
feature of the pre-proVIP gene.   These studies confirm the partial
characterization of the human pre-proVIP gene reported by Bodner et al.
(1985).

     It is likely that the tandem arrangement of the highly homologous
VIP and PHM sequences within the precursor resulted from a gene
duplication.   The presence of an intron between the VIP and PHM
sequences supports this possibility. The biological role of PHM/PHI has
not been determined, however.

     Hayakawa et al. (1984) have demonstrated that VIP biosynthesis is
regulated by cyclic AMP.   To further characterize this aspect of VIP
gene regulation, we have constructed a number of fusion genes containing
the VIP promoter linked to the bacterial chloramphenical acetyl
transferase (CAT) gene. We have introduced these fusion genes into PC12
cells, a cell line derived from a rat pheochromocytoma (Greene and
Tischler, 1976), and measured CAT activity in a transient assay system.
PC12 cells were used because they were believed likely to contain the
cellular factors necessary for regulation of VIP expression.   These
assays have indicated that a region that confers activation by cyclic
AMP is located in the 5' flanking region of the VIP gene, not very
distant from the promoter. Further deletional analysis is necessary to
precisely   characterize   this   region.   These   studies   support   the
hypothesis of Hayakawa et al. (1984) that cyclic AMP regulates VIP
biosynthesis at the transcriptional level.

THYROTROPIN RELEASING HORMONE

     Thyrotropin releasing hormone (TRH) has a central role in the
regulation of the hypothalamic-pituitary-thyroid axis. Although TRH was
the first hypophysiotropic peptide to be isolated, the structure of the
mammalian precursor has remained elusive.   In part, this is due to the
special problems presented by the TRH molecule itself. TRH consists of

only three amino acids, two of which are post-translationally modified. Consequently, it has not been possible to use antisera to TRH to identify the precursor. This problem also precludes the approaches that we used to identify the somatostatin precursor. A cDNA molecule encoding the amphibian TRH prohormone has been isolated by using an oligonucleotide primer to screen a frog skin cDNA library, where TRH sequences are very abundant (Richter et al., 1984). This approach has not proven useful in characterizing the mammalian precursor, however, presumably due to the low concentration of TRH mRNA in mammalian hypothalamus. Finally, there have been no examples of a TRH-producing tumor, which would facilitate the identification of pre-proTRH cDNA clones.

To circumvent these problems, we developed an antiserum directed toward a synthetic peptide (Cys-Lys-Arg-Gln-His-Pro-Gly-Lys-Arg-Cys) presumed to represent a portion of the TRH prohormone (Jackson, et al., in press). The Gln-His-Pro within the peptide corresponds to the amino acid sequence of TRH, with the glycine necessary for carboxy-terminal amidation and the paired flanking basic residues representing putative sites of post-translational cleavage. The amino- and carboxy-terminal cysteines were added to allow reductive cyclization of the peptide in order to increase the probability of generating a mid-portion directed antiserum. We assumed that authentic pro-TRH antisera would recognize a product concentrated in the paraventricular nucleus and lateral hypothalamus. This criterion was used to identify proTRH-specific antisera.

We used the proTRH antiserum to screen a lambda gt11 bacteriophage library derived from rat hypothalamus (Lechan et al., submitted for publication). In the lambda gt11 cloning vector, cDNAs are inserted adjacent to the ß-galactosidase promoter (Young and Davis, 1983). In the correct reading frame and orientation, the cDNAs are translated to produce fusion proteins linked to the ß-galactosidase enzyme. These fusion proteins can be detected immunologically, using antisera to the proteins of interest. Because the library is packaged in bacteriophage, large numbers of clones can be screened simultaneously.

Our expression library was prepared from 65 adult rat hypothalami and contained approximately 3×10^7 recombinant phage. 750,000 plaques were screened with the proTRH antiserum. Of these, eight immunopositive clones were identified and purified by low density screening. One of these encoded a protein that contained five copies of the sequence Gln-His-Pro-Gly. In each instance, this sequence was flanked by pairs of basic amino acids, suggesting that five TRH molecules could be generated from the precursor. The structure of the putative pre-proTRH molecule is depicted in Fig. 8.

Because of the possibility that the sequence Lys-Arg-Gln-His-Pro-Gly-Lys-Arg could theoretically be part of a protein that does not actually generate TRH, we used the cDNA to perform in situ hybridization histochemistry assays. The cDNA detected neurons in the same regions of

SP

Fig. 8. Structure of the rat thyrotropin releasing hormone precursor. SP designates signal peptide. The five TRH coding regions within the precursor are shown as shaded boxes.

the hypothalamus as those shown previously to produce TRH (Lechan and Jackson, 1982). This analysis substantiates our hypothesis that the molecule depicted in Fig. 8 represents the TRH precursor.

Like the precursor to enkephalin (Gubler et al., 1982), processing of pro-TRH could yield multiple copies of the biologically active peptide. In fact, the presence of repeated TRH coding units dispersed throughout the precursor is the only feature maintained between the amphibian and mammalian prohormones. The biological advantage achieved by amplifying the sequences of a regulatory peptide within a precursor molecule has yet to be determined.

SUMMARY AND CONCLUSIONS

We have outlined the various strategies used to characterize the precursors of three brain-gut hormones -- somatostatin, VIP, and TRH. In the case of somatostatin and VIP, isolation of the cDNA clones was relatively straightforward and was greatly facilitated by the use of tissues that were extremely rich in specific mRNA. The isolation of the cDNA encoding TRH was more circuitous and required the production of a synthetic peptide, the generation of an antiserum to the peptide, the screening of the antiserum by immunohistochemistry, the isolation of a cDNA from a bacteriophage expression library, and the ultimate confirmation of the identity of the clone by in situ hybridization. Characterization of the structures of the precursors is clearly only the first step in understanding the regulation of neuropeptide biosynthesis. As discussed in the sections on somatostatin and VIP biosynthesis, the cDNAs can be used to address questions related to both gene expression and post-translational processing. Hopefully, these investigations will help to unify our concepts of brain-gut hormone synthesis.

ACKNOWLEDGEMENTS

We thank R. Hammer, R. Palmiter, R. Brinster, and J. Habener for help with the experiments utilizing transgenic mice. This work was supported by NIH grants AM31400, CA37370, P30AM39428.

REFERENCES

Abe, H., Engler, D., Molitch, M. E., Bollinger-Gruber, J., Reichlin, S., 1985, Vasoactive intestinal peptide is a physiological mediator of prolactin release in the rat. Endocrinology, 116:1383.

Aron, D. C., Andrews, P. C., Dixon, J. E., Roos, B. A., 1984, Identification of cellular prosomatostatin and nonsomatostatin peptides derived from its amino terminus. Biochem. Biophys. Res. Commun., 124:450.

Arimura, A., Smith , W., Schally, A. V., 1976, Blockade of the stress-induced decrease in blood GH by anti-somatostatin serum in rats. Endocrinology, 98:450.

Benoit, R., Bohlen, P., Ling, N., Briskin, A., Esch, F., Brazeau, P., Ying, S.Y., Guillemin, R., 1982, Presence of somatostatin-28 (1-12) in hypothalamus and pancreas. Proc. Natl. Acad. Sci. USA, 79:917.

Benoit, R., Bohlen, R., Esch, R., Ling, N., 1984, Neuropeptides derived from prosomatostatin that do not contain the somatostatin-14 sequence. Brain Res., 311:23.

Bodner, M., Fridkin, M., Gozes, I., 1985, Coding sequences for vasoactive intestinal peptide and PHM-27 peptide are located on two adjacent exons in the human genome. Proc. Natl. Acad. Sci. USA, 82:3548.

Brazeau, P., Vale, W., Burgus, R., Ling, N., Butcher, M., Rivier, J., Guillemin, R., 1973, Hypothalamic peptide that inhibits the secretion of immunoreactive pituitary growth hormone. Science 129:77.

Brown, M., Rivier, J., Vale, W., 1980, Somatostatin-28: selective action on the pancreatic ß cell and brain. Endocrinology, 108:2391.

Esch, F., Bohlen, P., Ling, N., Benoit, R., Brazeau, P., Guillemin, R., 1980, Primary structure of ovine hypothalamic somatostatin-28 and somatostatin-25. Proc. Natl. Acad. Sci. USA, 77:6827.

Goodman, R. H., Lund, P. K., Jacobs, J. W., Habener, J. F., 1980, Pre-prosomatostatins: products of cell-free translations of messenger RNAs from anglerfish islets. J. Biol. Chem., 255:6549.

Goodman, R. H., Jacobs, J. W., Chin, W. W., Lund, P. K., Dee, P. C., Habener, J. F., 1980, Nucleotide sequence of a cloned structural gene coding for a precursor of pancreatic somatostatin. Proc. Natl. Acad. Sci. USA, 77:5869.

Goodman, R. H., Aron, D. C., Roos, B. A., 1983, Rat pre-prosomatostatin: structure and processing by microsomal membranes. J. Biol. Chem., 257:1156.

Greene, L. A., Tischler, A. S., 1976, Establishment of a noradrenergic clonal cell line of rat pheochromocytoma cells which respond to nerve growth factor. Proc. Natl. Acad. Sci. USA, 73:2424.

Gubler, U., Seeburg, P., Hoffman, B. J., Gage, L. P., Udenfriend, S., 1982, Molecular cloning establishes proenkephalin as a precursor of enkephalin containing peptides. Nature, 295:206.

Hanahan, D., Meselson, M., 1980, Plasmid screening at high colony density. Gene, 10:63.

Hayakawa, Y., Obata, K., Itoh, N., Yanaihara, N., Okamoto, H., 1984, Cyclic AMP regulation of provasoactive intestinal polypeptide/PHM-27 synthesis in human neuroblastoma cells. J. Biol. Chem., 259:9207.

Hayes, T. E., Dixon, J. E., 1985, Z-DNA in the rat somatostatin gene. J. Biol. Chem., 260:8145.

Hellerman, S. G., Cone, R. C., Potts, J. T., Rich, A., Mulligan, R. C., Kronenberg, H. M., 1984, Secretion of human parathyroid hormone from rat pituitary cells infected with a recombinant retrovirus encoding preproparathyroid hormone. Proc. Natl. Acad. Sci. USA, 81:5340.

Hobart, P., Crawford, R., Shen, L., Pictet, R., Rutter, W., 1980, Cloning and sequence analysis of cDNAs encoding two distinct somatostatin precursors found in endocrine pancreas of anglerfish. Nature, 288:137.

Hudson, P., Haley, J., John, M., Cronk, M., Crawford, R., Haralambides, J., Tregear, G., Shine, J., Niall, H., 1983, Structure of a genomic clone encoding biologically active human relaxin. Nature, 301:628.

Itoh, N., Obata, K., Yanaihara, N., Okamoto, H., 1983, Human prepro-vasoactive intestinal polypeptide contains a novel PHI-27-like peptide, PHM-27. Nature, 304:547.

Jackson, I. M. D., Wu, P., Lechan, R. M., 1985, Immunohistochemical localization in the rat brain of the precursor for thyrotropin releasing hormone. Science, (in press).

Kato, Y., Iwasaki, Y., Iwasaki, J., Abe, H., Yanaihara, N., Imura, H., 1978, Prolactin release by vasoactive intestinal polypeptide in rats. Endocrinology, 103:554.

Konturek, S. J., Tasler, J., Jaworek, J., Pawlik, W., Walus, K. M., Schusdziarra, V., Meyers, C. A., Coy, D. H., Schally, A. V., 1981, Gastrointestinal, secretory, motor, circulatory and metabolic effects of prosomatostatin. Proc. Natl. Acad. Sci. USA, 78:1967.

Land, H., Schultz, G., Schmale, H., Richter, D., 1982, Nucleotide sequence of a cloned cDNA encoding bovine arginine vasopressin-neurophysin II precursor. Nature, 295:299.

Lechan, R. M., Jackson, I. M. D., 1982, Immunohistochemical localization of throtropin-releasing hormone in rat hypothalamus and pituitary. Endocrinology, 111:55.

Leiter, A. B., Keutmann, H. T., Goodman, R. H., 1984, Structure of a precursor to human pancreatic polypeptide. J. Biol. Chem., 259:14702.

Low, M. J., Lechan, R. M., Rosenblatt, M., Goodman, R. H., 1983, Distribution and content of prosomatostain-specific antigen in rat brain. Program of the 65th Annual Meeting of the Endocrine Society, San Antonio, 1983 Abs. 281.

Low, M. J., Hammer, R. E., Goodman, R. H., Habener, J. F., Palmiter, R. D., Brinster, R. L., 1985, Tissue-specific postranslational processing of pre-prosomatostatin encoded by a metallothionein-somatostatin fusion gene in transgenic mice. Cell, 41:211.

Mandarino, L., Stenner, D., Blanchard, W., Nissen, S., Gerich, J., Ling., N., Brazeau, P., Bohlen, P., Esch, F., Guillemin, R., 1981, Selective effects of somatostatin-14, -25, and -28 on in vitro insulin and glucagon secretion. Nature, 291:76.

Marley, P., Emson, P., 1982, in: "Vasoactive Intestinal Peptide," S. I. Said, ed. (Raven Press, New York) p. 341.

Montminy, M. R., Goodman, R. H., Horovitch, S., Habener, J. F., 1984, Primary sturucture of the gene encoding rat pre-prosomatostatin. Proc. Natl. Acad. Sci. USA, 81:3337.

Moore, H. P. H., Walker, M. D., Lee, F., Kelly, R. B., 1983, Expressing a human proinsulin cDNA in a mouse ACTH-secreting cell. Intracellular storage, proteolytic processing, and secretion on stimulation. Cell, 35:531.

Nishizawa, M., Hayakawa, Y., Yanahihara, N., Okamoto, H., 1985, Nucleotide sequence divergence and functional constraint in VIP precursor mRNA evolution between human and rat. FEBS Lett., 183:55.

Noe, B. D., Spiess, J., 1983, Evidence for biosynthesis and differential post-translational proteolytic processing of different pre-prosomatostatins. J. Biol. Chem., 258:1121.

Patzelt, C., Tager, H. S., Carroll, R. J., Steiner, D. F., 1980, Identification of prosomatostatin in pancreatic islets. Proc. Natl. Acad. Sci. USA, 77:2410.

Pradayrol, L., Jornvall, H., Mutt, V., Ribet, A., 1980, N-terminally extended somatostatin: the primary structure of somatostatin-28. FEBS Lett., 109:55.

Richter, K., Kawashima, R., Egger, G., Krell, G., 1984, Biosynthesis of thyrotropin releasing hormone in the skin of Xenopus laevis. partial sequence of the precursor deduced from a cloned cDNA. EMBO J., 3:617.

Said, S. I., Mutt, V., 1970, Polypeptide with broad biological activity: isolation from small intestine. Science, 169:1217.

Schally, A. V., Huang, W., Chang, C. C., Arimura, A., Redding, T. W., Millar, R. P., Hunkapillar, M. W., Hood, L. E., 1980, Isolation and structure of pro-somatostatin: a putative somatostatin precursor from pig hypothalamus. Proc. Natl. Acad. Sci. USA, 77:4489.

Shen, L., Pictet, R. L., Rutter, W. J., 1982, Human somatostatin I: sequence of the cDNA. Proc. Natl. Acad. Sci. USA, 79:4575.

Shen, L. P., Rutter, R., 1984, Sequence of the human somatostatin I gene. Science, 224:168.

Tatemoto, K., Mutt, V., 1981, Isolation and characterization of the intestinal peptide porcine PHI (PHI-27), a new member of the glucagon-secretin family. Proc. Natl. Acad. Sci. USA, 78:6603.

Tavianini, M. A., Hayes, T. E., Magazin, M. D., Minth, C. D., Dixon, J. E., 1985, Isolation, characterization, and DNA sequence of the somatostatin gene. J. Biol. Chem., 259:11798.

Tsukada, T., Horovitch, S. J., Montminy, M. R., Mandel, G., Goodman, R. H., 1985, Structure of the human vasoactive intestinal polypeptide gene. DNA, (in press).

Wang, A., Quigley, G. J., Kuplack, F. J., Crawford, J. L., VanBoon, J. H., VanderMarel, G., Rich, A., 1979, Molecular structure of a left-handed double helical DNA fragment at atomic resolution. Nature, 282:680.

Warren, T. G., Shields, D., 1982, Cell-free biosynthesis of somatostatin precursors: evidence for multiple forms of pre-prosomatostatin. Proc. Natl. Acad. Sci. USA, 77:4074.

Warren, T. G., Shields, D., 1984, Cell-free biosynthesis of multiple pre-prosomatostatins: characterization by hybrid selection and amino-terminal sequencing. Biochemistry, 23:2684.

Young, R. A., Davis, R. W., 1983, Efficient isolation of genes by using antibody probes. Science, 222:778.

ENZYMES PROCESSING SOMATOSTATIN-PRECURSORS, THE SOMATOSTATIN-28 CONVERTASE
OF RAT BRAIN, A SYSTEM CONVERTING SOMATOSTATIN-28 TO BOTH SOMATOSTATINS-14
AND -28(1-12)

Paul Cohen, Pablo Gluschankof, Sophie Gomez, Alain Morel,
Hamadi Boussetta, Christine Clamagirand and Pierre Nicolas

Groupe de Neurobiochimie Cellulaire et Moléculaire
Unité Associée au CNRS 554, Université P. et M. Curie
96 bld Raspail, 75006 PARIS, France

INTRODUCTION

In the biosynthetic pathways of neuropeptides both co- and post-
translational events are essential in confering to these messengers their
biological activity. Regulation of these mechanisms may be critical in giving rise to diversity in the products derived from the processing of a
single, but multipotential, biosynthetic precursor (Eipper and Mains,
1980 ; Benoît et al., 1982a ; Kimura et al., 1986). Of particular importance is the proteolysis which allows the active fragments to be released from their larger proforms. Cleavage of peptide bonds appears to
occurs generally at loci consisting in basic amino acids (Arg or Lys)
arranged within the precursor sequence as doublets (Nakanishi et al.,
1979), sometime as triplets (Craig et al., 1982) or even as quadruplets
(Nakanishi et al., 1979 ; Furutani et al., 1983). Attempts to characterize the enzyme(s) possibly involved in these cleavages were hampered by
difficulties in obtaining sufficient amounts of purified hormone precursors. The biosynthetic system of somatostatin-14 (S-14), a tetradecapeptide found both in the central nervous system and in the gastro-intestinal tract, is particularly well-suited for such studies since, in mammals,
the peptide hormone seems to derive from a single precursor (Morel et al.,
1983). Its structure was determined in various systems through the sequencing of cDNA, to the corresponding mRNAs (Hobart et al., 1980 ;
Goodman et al., 1983). In all cases, the S-14 molecule occupies the COOH-
terminal end of the precursor and is preceded at positions -1 and -2 by
a basic doublet (Lys^{-1}, Arg^{-2}). Amino terminal extension of this particular structure leads to an octaeicosapeptide, somatostatin-28 (S-28),
which includes in its sequence both the NH_2-terminal fragment, called
S-28(1-12) and S-14. Therefore, in theory, selective excision of the
Arg Lys doublet should release stoichiometric amounts of both S-14 and
S-28(1-12). The corollary is that the octaeicosapeptide S-28 was considered as a possible common precursor to both the NH_2- and COOH- terminal
dodeca- and tetradeca- peptides respectively (see for a discussion
Patel Y.C., 1986).

We have attempted the characterization of enzymes possibly involved
in the conversion of somatostatin precursors by using a family of synthetic peptide substrates either reproducing or mimicking the S-28 amino
acid sequence around the Arg Lys dibasic stretch. The results are discus-

SOMATOSTATIN-14

ALA-GLY-CYS-LYS-ASN-PHE-PHE-TRP-LYS-THR-PHE-THR-SER-CYS

SOMATOSTATIN-28 (1-12)

SER-ALA-ASN-SER-ASN-PRO-ALA-MET-ALA-PRO-ARG-GLU

SOMATOSTATIN-28

1   5   10
SER-ALA-ASN-SER-ASN-PRO-ALA-MET-ALA-PRO-ARG-GLU-ARG-LYS-
15   20   25
ALA-GLY-CYS-LYS-ASN-PHE-PHE-TRP-LYS-THR-PHE-THR-SER-CYS

8K SOMATOSTATIN-28 (1-12) OR PRE-PROSOMATOSTATIN (25-100)

ALA-PRO-SER-ASP-PRO-ARG-LEU-ARG-GLN-PHE-LEU-GLN-LYS-SER-
LEU-ALA-ALA-ALA-THR-GLY-LYS-GLN-GLU-LEU-ALA-LYS-TYR-PHE-
LEU-ALA-GLU-LEU-LEU-SER-GLU-PRO-ASN-GLN-THR-GLU-ASN-ASP-
ALA-LEU-GLU-PRO-GLU-ASP-LEU-PRO-GLN-ALA-ALA-GLU-GLN-ASP-
GLU-MET-ARG-LEU-GLU-LEU-GLN-ARG-SER-ALA-ASN-SER-ASN-PRO-
ALA-MET-ALA-PRO-ARG-GLU

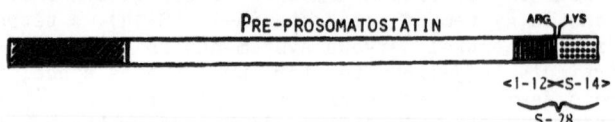

PRE-PROSOMATOSTATIN

ARG LYS

<1-12><S-14>

S-28

Figure 1 : Peptides derived from pre-prosomatostatin
and present in the central nervous system.

sed in connection with the sub-cellular localization and possible _in vivo_ involvment of the convertase.

MATERIALS AND METHODS

Peptide synthesis was performed using the solide phase method modified (Merrifield, 1963 ; Gluschankof et al., 1984). Purification of the compounds was done successively by ion-exchange chromatography and HPLC and their structure was verified both by microsequencing, amino acid composition analysis, and chromatography under various conditions. The following peptides were synthesized :

Peptide I (Ala^{17}, Tyr^{20}) somatostatin-28(10-20) NH_2

Peptide II (Ala^{17}, Tyr^{20}) S-28(15-20) NH_2

Peptide III (Ala^{17}, Tyr^{20}) S-28(6-20) NH_2

Peptide IV (Ala^{17}, Tyr^{20}) S-28(14-20) NH_2

Peptide V   S-28(6-13)

Peptide VI   des Arg^{13} peptide V, i.e.
                                       S-28(6-12)

Somatostatins-14 and -28 were from CRB (Cambridge, G.B.). The antibodies directed against S-28(1-12) were kindly donated by Dr R. Benoit (Montreal, Canada) (Benoit et al., 1982b ; Gluschankof et al., 1985) and the antiserum 36-38 (from C. Rougeot, URIA, Pasteur Institute, Paris) was continuously used in the S-14 radioimmunoassay (Morel et al.,1983 ; Gluschankof et al., 1984). Either cortices, or hypothalami, were dissected from rat brains after decapitation. The tissue extracts were prepared after homogeneization as reported (Gluschankof et al., 1984). Neurosecretory granules were extracted as described by Masse et al., 1982) and purified both by sucrose and Percoll density gradients (Masse et al., 1982 ; Gomez et al., 1985). For the routine convertase assay conversion of (^{125}I) peptide I into (^{125}I) peptide II was monitored after ion exchange separation of the product from the unmodified substrate (Gluschankof et al., 1984). Determination of both the amino-peptidase like and carboxypeptidase-like activities were made either by monitoring conversion of Peptide IV into Peptide II or else of Peptide V into Peptide VI respectively. The reaction products were, in all cases, unequivocally identified by HPLC, NH_2-terminal analysis and amino acid composition determination.

RESULTS

The convertase activity was measured and quantitatively expressed as amounts of (^{125}I) – Peptide II produced by conversion of (^{125}I) – Peptide I. Non specific cleavage occuring at the single Lys^{10} residue of Peptide I was monitored and evaluated through the quantity of $Asn(^{125}I)$-Tyr NH_2 produced during the reaction. Preliminary experiments indicated that, in the cortex, the production of Peptide II was maximal with respect to non-specific cleavage giving the dipeptide. Therefore the brain cortex extracts were choosen as primary source of enzymes and submitted to successive fractionations both on Sephadex G-150 and on a cationic resin (Mono Q) by FPLC procedure. The convertase activity behaved as 80-90 Kda species and, in this enriched fraction, essentially no-contaminating trypsin-like activity was detected (Figure 2).

Conversion of either Peptide III into Peptide II or of S-28 into S-14 by this preparation was quantitative and neither further degradation of the product nor secondary cleavage of the substrate were observed. Furthermore HPLC analysis coupled with RIA performed using anti S-28(1-12)

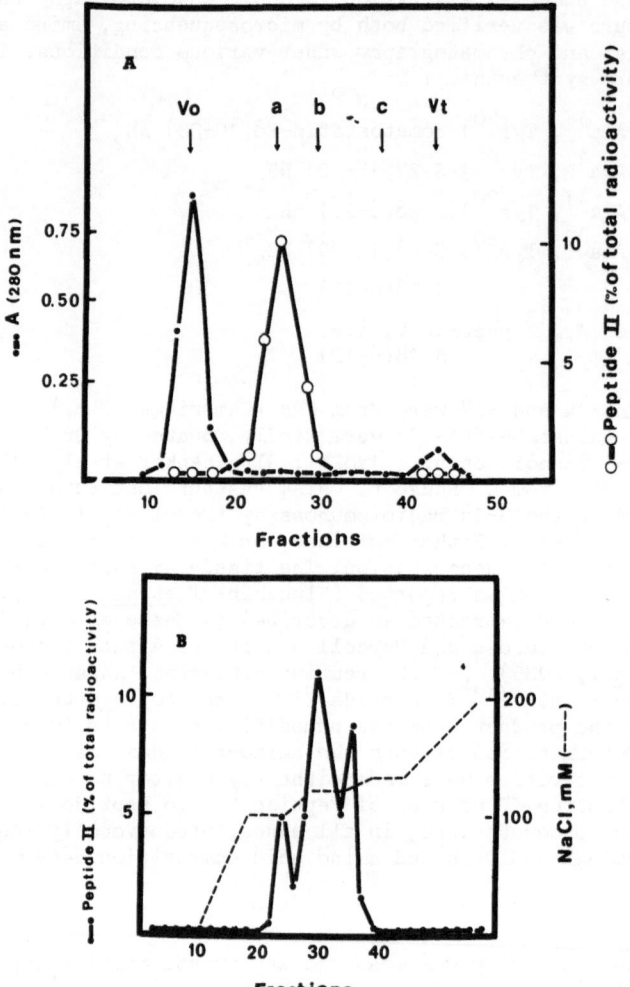

Figure 2 : Fractionation of the convertase activity extracted
from secretory granules.

Top : Elution of the convertase on a Sephadex G-150
column.

Bottom : Behaviour of the convertase activity on
ion-exchange chromatography using a cationic resin
(adapted from Gomez et al., 1985).

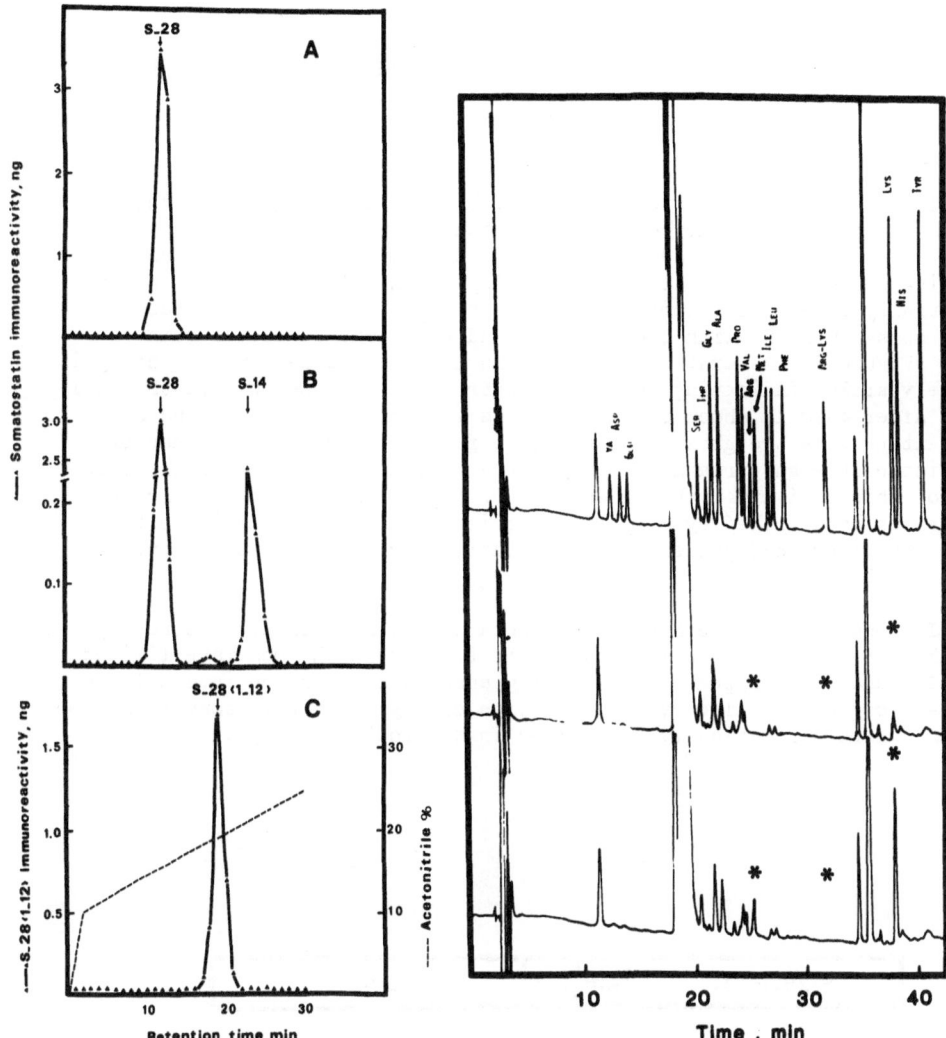

Figure 3 : Processing of S-28 into S-14 and S-28(1-12) by the convertase activity (from Gomez et al., 1985). RIA/HPLC identification of the products.

Figure 4 : HPLC analysis of the amino acids released in the conversion of S-28 into S-14 and S-28(1-12). The results show that no Arg Lys doublet is detected (middle) while free Arg and Lys are produced in this processing (from Gluschankof et al., 1985).

antibodies indicated that the latter dodecapeptide fragment was produced in stoichiometric amounts with S-14 (Figure 3). Therefore it was interesting to determine if the Arg Lys doublet was excised from the substrate as a dipeptide or as free Arg and Lys. That the latter is the case was established using HPLC analysis of the enzyme reaction product which showed no Arg Lys dipeptide but stoichiometric amounts of free Arg and Lys (Figure 4).

The results indicate that three peptide bonds are cleaved in the reaction catalyzed by the convertase and that stable amino- and carboxy-terminals fragments are produced. They suggest that the system may act as an Arg Lys esteropeptidase.

In order to investigate further the biological relevance of this converting activity, its sub-cellular localization was examined. Neurosecretory granules were prepared by successive density gradient fraction either by using sucrose or Percoll gradients. An enriched fraction of granules essentially free of lysosomial contaminants was used. The convertase was found co-sedimenting with the granules and with immunoreactive somatostatin (Figure 5). A burst of activity was produced subsequent to the lysis of the vesicles indicating that the convertase is trapped within the vesicles. After osmotic shock of the granules the convertase was found co-sedimenting with the ghosts but could be further solubilized either by mild ionic strenght treatment or by ultra-sonication (Table I). The identity of this activity with the enzyme purified from the whole tissue extracts was established by comparison of its physico-chemical properties (i.e. behaviour on Sephadex G-150 filtration, or on a Mono-Q cationic resin ; selectivity toward the substrates).

In conclusion the brain cortex contains, within its neurosecretory granules, a convertase activity which is trapped in the membrane-limited vesicles and which appears to be weakly associated with the internal face of these organelles. The overall reaction catalyzed by this converting system can be schematically summarized in the following way :

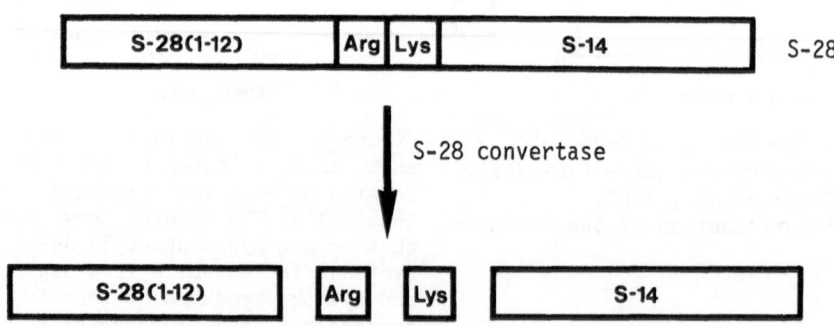

Table I

Quantitative analysis of S-14 produced from S-28 by the converting
activity obtained at different fractionation steps.

Source of enzyme	S-14 produced
	ng
Purified granules	53
Granule membranes	41
Granule lysate	8
Solubilized enzyme from granule membranes	43
Membranes after treatment with either sonication or 200 mM KCl, 50 mM phosphate buffer	6

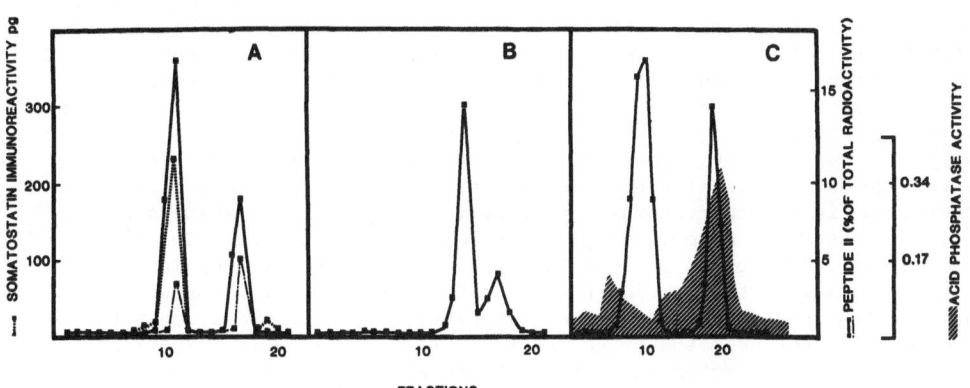

Figure 5 : Association of convertase activity with granules
membranes. A/ Banding of granules on sucrose gradient.
B/ Activity associated with ghosts after osmotic shock of
granules. C/ Banding of granules on Percoll gradient. Acid
phosphatase is in the hatched region of the gradient.

Since three peptide bonds were hydrolyzed in this catalytic phenomenon a sequential process can be hypothesized. Interestingly an aminopeptidase-like and a carboxypeptidase-like activity were detected in the convertase using the peptide substrates and the assays described in the Experimental Section. However their possible relevance to the overall catalytic phenomenon remains to be established unambiguously. Altogether, the above reported data support the idea that S-28 may act, in certain tissues, as a precursor for both S-28(1-12) and S-14. Of interest is the Brockman organ of the Teleostean fish Lophius piscatorius which provides a system giving substance to this hypothesis. In these endocrine pancreatic tissues two S-28 species (named respectively S-28 I and S-28 II) and a single corresponding S-14 I can be detected. This was taken as evidence for an incomplete processing of pro-somatostatin II (Hobart et al., 1980) blocked at the S-28 II level in vivo (Morel et al., 1984a and b). Since conversion of this octaeicosapeptide could be achieved in vitro by the brain convertase above described (Morel et al., 1984 a), the failure to process S-28 II into S-14 II observed in vivo can possibly be attributed to an incomplete enzyme machinery in the producing cells. Together these data support a metabolic pathway of prosomatostatin in which conversion of S-28 to S-14 and S-28(1-12) may alternate with direct production of these peptides from the larger precursor as proposed by Benoit et al., 1982b (Figure 6).

DISCUSSION

In the last few years both the techniques of recombinant DNA and of cDNA sequencing have allowed the structure determination of a great number of pro-neuropeptide hormones. Comparative analysis of these structural data together with quantitative evaluation of the tissue - specific, differential processing of these precursors have provided a basis for the understanding, at the post-translational level, of their enzymatic maturation. Although of somewhat spectacular achievement these genetic engineering techniques together with those allowing the study of gene expression in heterologous cells (Shields et al., 1986) could not provide direct information on the enzyme systems involved in the processing. Furthermore, they only suggested indirectly the involvment of regulatory mechanisms on which little is known at the molecular level. The purpose of our approach was to characterize directly enzyme systems using peptide substrates carefully selected for their possible relevance to the proteolytic process. At the present time, the convertase we have characterized in the brain cortex, and in the corresponding neurosecretory granules, exhibits a set of properties which make it of possible relevance to the expected biological function. However it remains to be seen how the activities involved in the breakage of the peptide bonds are together connected. Extensive purification and separation of the components together with reconstitution experiments might help answering these questions. Moreover, comparison of this particular system with other proteases purified from other regions of the brain (Clamagirand et al., 1986) and possibly involved in the post-translational processing of other pro-hormones should provide further insight on the question.

Experiments performed on the hypothalamo-neurohypophysis tract on the pro-neurophysin/vasopressin system suggested that processing of the precursor might occur in granules during axonal transport (Gainer et al., 1977). A set of experiments conducted on similar systems showed that, if axoplasmic transport is impaired by pharmacological agents like colchicine (Camier et al., 1985) processing of secretory material is completed at the cell bodies level. Therefore it may be inferred that the finding of the convertase system in the vesicles simply indicates that at least some of the processing machinery is trapped within these granules. This does not allow, however, to exclude that processing may already be active at

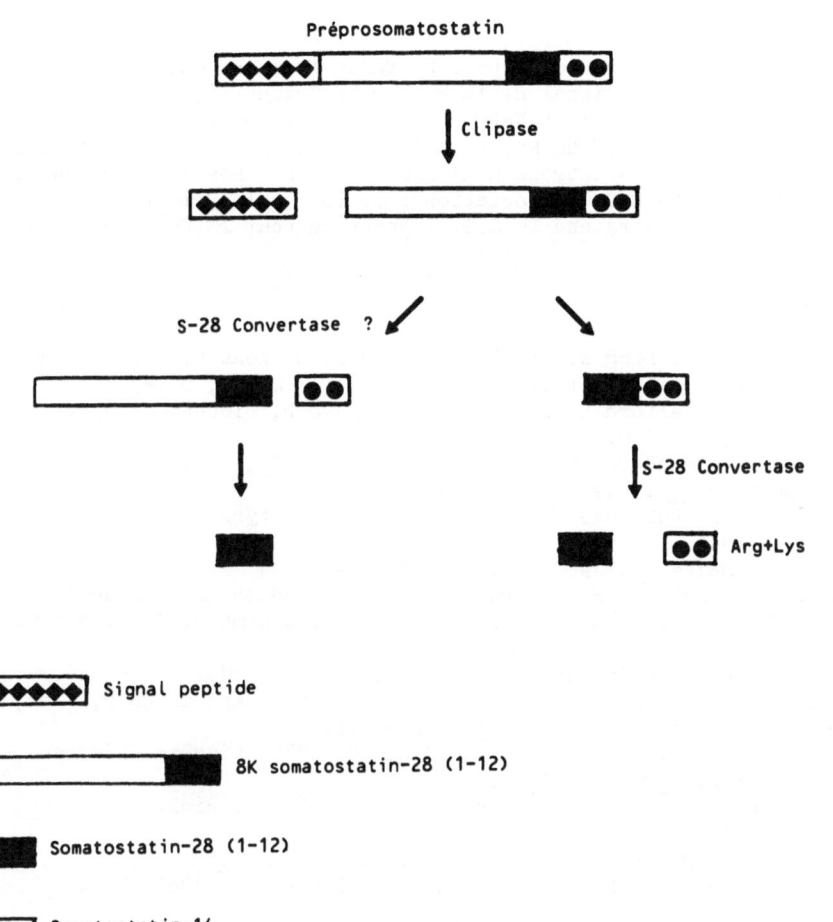

Figure 6 : Processing of pre-prosomatostatin in mammalian hypothalamus.

the level of the Golgi apparatus. Particularly relevant to this question is the problem of the internal pH of vesicles prepared from storage organs (neurohypophysis for example) at the nerve terminal ends. Comparison of these pH values (Russel and Holz, 1981 ; Scherman and Nordmann, 1982) with those obtained for the optimal pH, measured _in vitro_, for the putative processing enzymes should be taken with some caution. The existence of selective systems (Scherman _et al._, 1982) located at the membrane level indicates that either ionic strenght or else pH (or both) can be modified inside the vesicles and that this may possibly play a role in the regulation of enzyme mechanisms occuring inside the granules.

ACKNOWLEDGEMENTS

This work was supported in part by funds from the Université P. et M. Curie, the Centre National de la Recherche Scientifique (Unité Associée n°554), the Institut National de la Santé et de la Recherche Médicale (CRE n°834006), the Mutuelle Générale de l'Education Nationale, the Fondation pour la Recherche Médicale Française. Research fellowships were granted both to P.G. (Association pour le Développement de la Recherche sur le Cancer) and to S.G. (Fondation pour la Recherche Médicale Française).

REFERENCES

Benoit R, Böhlen P, Ling N, Briskin A, Esch F, Brazeau P, Ying SY and Guillemin R, 1982a, Proc. Natl. Acad. Sci. USA 79, 917-921.
Benoit R, Ling N, Alford B and Guillemin R, 1982b, Biochem. Biophys. Res. Comm. 107, 944-950.
Camier M, Barre N and Cohen P, 1985, Brain Res. 334, 1-8.
Clamagirand Ch et al., 1986, in preparation.
Craig RK, Hall L, Edbrooke NMR, Allison J and McIntyre I, 1982, Nature 295, 345-348.
Eipper BA and Mains RE, 1980, Endocrinol. Rev. 1, 1
Furutani Y, Morimoto Y, Shibahara S, Noda M, Takahashi H, Hirose T, Asai M, Inayama S, Hayashida H, Miyata T and Numa S, 1983, Nature 301, 537-540.
Gainer H, Sarne Y and Brownstein MJ, 1977, Science 195, 1354-1356.
Gluschankof P, Morel A, Gomez S, Nicolas P, Fahy Ch and Cohen P, 1984, Proc. Natl. Acad. Sci. USA 81, 6662-6666.
Gluschankof P, Morel A, Benoit R and Cohen P, 1985, Biochem. Biophys. Res. Commun. 128, 1051-1057.
Gomez S, Gluschankof P, Morel A and Cohen P, 1985, J. Biol. Chem. 260, 10541-10545.
Goodman RH, Aron DC and Ross B, 1983, J. Biol. Chem. 258, 5570-5573.
Hobart P, Crawford R, Shen LP, Pictet R and Rutter WJ, 1980, Nature (London) 288, 137-141.
Kimura S, Goto K, Shigematsu Y, Sugita Y, Ogawa T and Kanazawa I, 1986, see in this Book.
Masse MJO, Desbois-Perrichon P and Cohen P, 1982, Eur. J. Biochem. 127, 609-617.
Merrifield RB, 1963, J. Am. Chem. Soc. 85, 2149-2154.
Morel A, Nicolas P and Cohen P, 1983, J. Biol. Chem. 258, 8273-8276.
Morel A, Gluschankof P, Gomez S, Fafeur V and Cohen P, 1984a, Proc. Natl. Acad. Sci. USA 81, 7003-7006.
Morel A, Chang JY and Cohen P, 1984b, FEBS Lett. 175, 21-24.
Nakanishi S, Inoue A, Kita T, Nakamura M, Chang AYC, Cohen SN and Numa S, 1979, Nature 278, 423-427.
Patel YC, 1986, see in this Book.
Russel JT and Holz RW, 1981, J. Biol. Chem. 256, 5950-5953.
Sherman D and Nordmann JJ, 1982, Proc. Natl. Acad. Sci.USA 79, 476-479.
Sherman D, Nordmann JJ and Henry JP, 1982, Biochem. 21, 687-694.
Shields D et al., 1986, see in this Book.

OXYTOCIN AND VASOPRESSIN SECRETION: NEW PERSPECTIVES

Dennis W. Lincoln and John A. Russell

MRC Reproductive Biology Unit
Centre for Reproductive Biology and
Department of Physiology
University of Edinburgh, U.K.

INTRODUCTION

   Six naturally occurring oxytocin-like and three vasopressin-like
posterior pituitary hormones have been chemically characterized;   all have
nine amino-acid residues and six of these are consistent throughout.   The
posterior pituitary hormones of about 50 species have been fully characterized.   This is a small sample, representing about 0.1% of all vertebrate species, but it has allowed a number of principles to be established
(Acher, 1980).   Most species have two posterior pituitary hormones of the
oxytocin-vasopressin family;   only the cyclostomes (lampreys) have one.
This suggests that a gene duplication must have occurred before the evolution of fishes, some 450-500 million years ago.   Point mutations then
occurred separately within these two genes to give rise to two families of
posterior pituitary hormones.   The functions of these two hormones also
underwent progressive change.   The oxytocin-like peptides became linked
primarily to reproduction and the vasopressin-like peptides to water and
mineral metabolism.   However, it remains impossible to determine the
posterior pituitary peptides of extinct ancestral forms.   A modern day
North American opossum (Didelphis virginiana), often referred to as a
living fossil, has been exposed to just as many years of evolution as any
other mammalian species.   One possible sequence for the derivation of the
posterior pituitary hormones is presented in Figure 1.   This is based on
vasotocin, a hormone found in birds, reptiles, amphibia, fishes and cyclostomes.   Single codon substitutions in the genome could account for all
the changes illustrated.   The switch from mesotocin and vasotocin to
oxytocin and arginine vasopressin appears to coincide with the evolution
of mammals.   The substitution of isoleucine in position 3 of vasotocin
for phenylalanine, producing arginine vasopressin, results in a marked
increase in antidiuretic activity and loss of oxytocic activity, as
measured in rats.   The change from mesotocin to oxytocin seems to have
been of less significance in that both are very effective stimulators of
milk ejection, and experimentally one can be used to substitute for the
other.   The Prototherian (egg laying) and Eutherian mammals are relatively
uniform in having oxytocin and arginine vasopressin (Chauvet et al., 1985);
the only notable exception being the substitution of lysine for arginine
in position 8 in the pig and related species.   In the rat, at least, these
two hormones are encoded by single genes (Richter, this volume).   By comparison, the Metatherian mammals (the marsupials) display great diversity.
The Phalangids of Australia have mesotocin and arginine vasopressin,

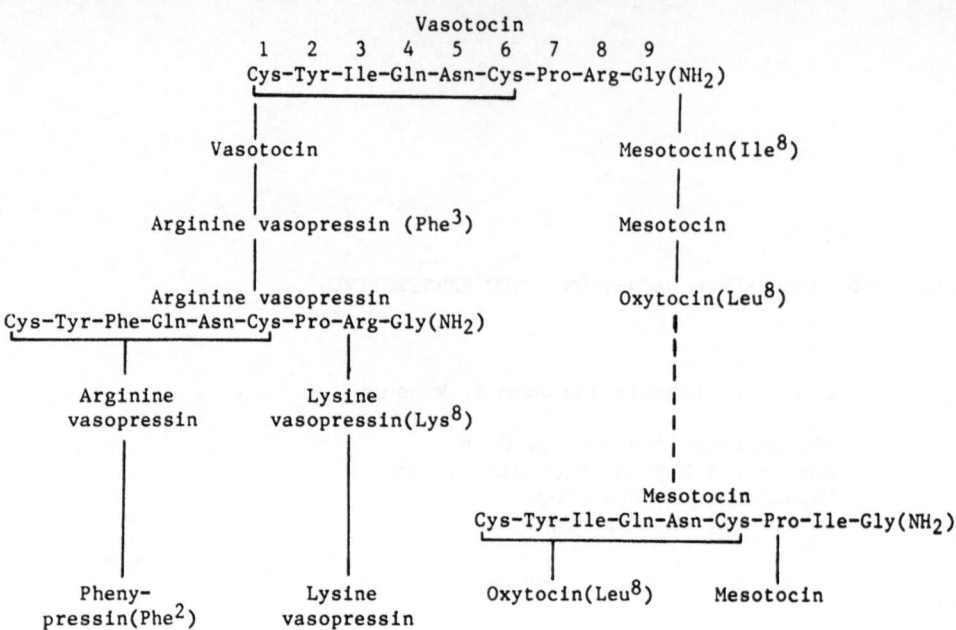

Fig. 1. A family tree depicting the evolution and structure
of the posterior pituitary peptides of mammals.

suggesting a possible reverse mutation.   All other groups have two vasopressin-like peptides, indicative of a second gene duplication.   Thus
all kangaroos and wallabies that have been studied have mesotocin, lysine
vasopressin and phenypressin (Chauvet et al., 1983).   Whilst the North
American opossum (Didelphis virginiana) is the most complex of all, and
produces oxytocin, mesotocin, lysine vasopressin and arginine vasopressin
(Chauvet et al., 1984).

The evolution of these two peptide lineages will undoubtedly be
advanced further by the analysis of the much more extensively substituted
precursors and the analysis of the determining gene sequences (Ivell and
Richter, 1984).   Both the oxytocin and vasopressin precursor encoding
sequences consist of three exons and two introns.   These occur at
identical positions with respect to the reading frame, and very marked
homology exists between the neurophysin regions of the two precursors
(Ruppert et al., 1984).

Whilst oxytocin and vasopressin production is thought to be a consequence of gene duplication, the two genes do not appear to be expressed
together;   thus a neurone produces either oxytocin or vasopressin (and
their related precursor peptides) but never both.   This lack of co-
existence, demonstrated initially with immuno-cytochemical techniques
(Dierickx 1980), has been supported by in situ hybridization studies
using appropriate cDNA probes (Coghlan et al., 1984).   It is plausible
that an oxytocin cell could switch to the production of vasopressin, and
vice versa, but this seems unlikely given the many other differences
expressed by the two cell types.

The oxytocin- and vasopressin-containing magnocellular neurones of
the supraoptic and paraventricular nuclei project to the posterior

pituitary gland (Figs. 2-3), at least in the rat, guinea pig, monkey and man (Swanson and Sawchenko, 1983). Neurones in the anterior and medial parvocellular portions of the paraventricular nucleus also produce vaso-pressin, and a smaller number produce oxytocin; these project to the external zone of the median eminence and there terminate on the primary capillaries of the hypothalamo-hypophysial portal plexus (Vandesande et al., 1977; Zimmerman and Silverman, 1983). Vasopressin and oxytocin cells in the posterior parts of the paraventricular nucleus project to the brain stem and spinal cord, and particularly to those areas involved in the regulation of autonomic functions (Swanson and Sawchenko, 1983).

We are therefore presented with two related gene families, only one of which is expressed by an individual neurone. However, the expression of these two genes is very closely associated with the expression of other specific aspects of the genome. Oxytocin and vasopressin cells display profound differences in the co-production of other regulatory peptides, electrophysiological properties, connectivities and morphological specializations.

CO-EXISTENCE AND SECRETION OF OXYTOCIN AND VASOPRESSIN WITH OTHER REGULATORY PEPTIDES

The co-existence and secretion of oxytocin and vasopressin with three types of regulatory peptide will be considered, namely CRF, CCK and opioids. It is important, however, to appreciate that these peptides co-exist in vastly different concentrations. The posterior pituitary gland of the rat contains about 1,000 nmole of vasopressin/g tissue, and a similar concentration of oxytocin. The three groups of regulatory pep-tides to be discussed exist at concentrations of 0.1-1.0 nmole/g tissue: this suggests local rather than systemic action of these co-peptides.

Corticotrophin-Releasing Factor (CRF)

CRF^{1-41} (Vale et al., 1983), hereinafter referred to as CRF, has been demonstrated within terminals in the median eminence that arise from neurones situated in the anterior and medial parvocellular regions of the paraventricular nucleus (Swanson et al., 1983; Bruhn et al., 1984a); 1-2% of these also contain vasopressin (Sawchenko et al., 1984), co-localized in the same small neurosecretory granules as CRF (Whitnall et al., 1985). Adrenalectomy increases the density of vasopressin stain-ing in the median eminence without any change in the posterior pituitary gland, and the effect is reversed by glucocorticoid therapy (Sawchenko and Swanson, 1985). These changes, a week after adrenalectomy, reflect a dramatic increase to more than 70% in the proportion of CRF neurones containing vasopressin and prepropressophysin mRNA (Tramu et al., 1983; Sawchenko et al., 1984; Kiss et al., 1984a; Wolfson et al., 1985). By contrast, up to one third of magnocellular neurones projecting to the posterior pituitary gland produce both oxytocin and CRF (Swanson et al., 1983; Dreyfuss et al., 1984). The perikarya of these CRF/oxytocin cells are somewhat smaller than the other magnocellular neurones, and it is plausible that such cells have different afferent connections and electrical properties. Pituitary stalk section results in a 90% loss of CRF, and thus CRF is probably released when the CRF/oxytocin cells are electrically excited (Saavedra et al., 1984). Conversely, no vaso-pressinergic magnocellular neurones appear to synthesise CRF (Burlet et al., 1983). These distributions are illustrated in Figure 2. Co-existence of CRF with oxytocin or vasopressin within neurosecretory granules (Dreyfuss et al., 1984; Whitnall et al., 1985) should result in their co-release, but this has not been fully established. The potential release patterns should be as follows: separate release of vasopressin, and possibly oxytocin, from the median eminence and of both hormones from the posterior pituitary gland, and co-release of oxytocin and CRF from the posterior pituitary gland and of vasopressin and CRF (or CRF alone) from the median eminence. Oxytocin and vasopressin are present in the pituitary portal blood at concentrations 15-50 times those in peripheral plasma, and there is some evidence for differential release between the median eminence and posterior pituitary (Gibbs, 1984a; Bruhn et al., 1984b; Plotsky et al., 1985). CRF is also present in portal blood (Gibbs, 1985). Both vasopressin and CRF are released from the median eminence in vitro in response to depolarizing stimuli (Gillies et al., 1984). Both hormones may also be released centrally to influence other neural systems or modulate their own secretion. The intracerebro-ventricular injection of CRF has been reported to reduce the secretion of vasopressin and oxytocin into pituitary portal blood by 80% (Plotsky et al., 1985), whilst a similar injection of vasopressin depresses the secretion of CRF (Plotsky et al., 1984). Oxytocin, vasopressin or even CRF could, after release from the posterior pituitary gland, reach the anterior pituitary gland in high concentrations via local blood vessels (Baertschi et al., 1983). In terms of the peripheral circulation, the intraventricular injection of CRF increases the levels of oxytocin without changing the levels of vasopressin (Bruhn et al., 1984b). Where CRF acts to elicit these effects is unknown, since CRF receptors have not been identified in either the paraventricular nucleus or the median eminence (Aguilera et al., 1984).

ACTH secretion is provoked by a diversity of stressful stimuli, and this is accompanied by, and in part caused by, the secretion of CRF, vasopressin and oxytocin. Vasopressin is secreted in response to osmotic stimulation induced by water deprivation or salt loading, haemorrhage involving hypovolemia or hypotension, and arterial chemoreceptor stimulation. It is not secreted in response to environmental cold, noise or supine immobilization (except in female rats) (Lang et al., 1983; Gibbs, 1984b; 1985). Oxytocin is secreted in response to suckling and during parturition and, like vasopressin, in response to haemorrhage and osmotic stimuli. Indeed, it appears to be released by some stresses that do not release vasopressin, e.g. supine immobilization and enforced swimming (Lang et al., 1983; Williams et al., 1985). Changes in CRF, vasopressin and oxytocin content have been reported in relation to these stimuli, but such results are difficult to interpret given the existence of large storage pools and changes in biosynthesis designed to compensate for changes in secretion.

Vasopressin stimulates ACTH release but is less potent than CRF (Gibbs et al., 1984b). Secondly, vasopressin augments the ACTH releasing action of CRF some 2- to 4-fold (Knepel et al., 1984a). These two actions could involve separate receptors (Buckingham, 1985). The binding of (3)H-vasopressin to anterior pituitary cell membranes and competition studies with various analogues (including oxytocin) indicate an action through a specific receptor to which rat-CRF (and other peptides) does not bind (Spinedi and Negro-Vilar, 1984). Studies with V1 (pressor) and V2 (anti-diuretic) vasopressin-receptor analogues and antagonists indicate that a new type of vasopressin receptor is involved in the ACTH-releasing actions of vasopressin (Antoni, 1984; Baertschi and Friedli, 1984; Knepel et al., 1984b; Koch and Lutz-Bucher, this volume), though this may only apply to the CRF-augmenting action (Knepel et al., 1984a). Others have antagonized both vasopressin effects with a V1 antagonist (Rivier et al., 1984). Oxytocin has no ACTH-releasing activity on its own, but potentiates the effect of CRF, doubling the release of ACTH (Antoni et al., 1983; Baertschi and Friedli, 1984). Qualitatively and quantitatively different combinations of CRF, vasopressin and oxytocin secretion induced by the activation of subsets of neurones could serve to modulate ACTH secretion in subtle ways.   It is interesting to note from an evolutionary standpoint that the CRF and vasopressin precursors display a considerable homology in their peptide sequences (Furutani et al., 1983).

Several studies have attempted to establish the relative roles of these and other releasers of ACTH using immunoneutralization and pharmacological blockade.   Passive immunization with anti-CRF serum reduces ACTH release evoked by the stress of restraint or ether or formaldehyde vapour by 70-80% (Rivier and Vale, 1983; Linton et al., 1985), demonstrating a key role for CRF.   Vasopressin antiserum reduces the ACTH response by 50% (Linton et al., 1985).   After CRF release has been blocked by lesions of the paraventricular nucleus, stress-induced or vasopressin-induced ACTH release can be blocked by V1 antagonists (Bruhn et al., 1984a; Rivier et al., 1984).   This indicates a role for vasopressin in ACTH release, as well as an action via V1 receptors.   The role of oxytocin has not been investigated in this way.   No effects of CRF in the posterior pituitary gland have been reported.

Cholecystokinin

Cholecystokinin-8 (sulphated, CCK-8s) is the most abundant regulatory peptide in the brain, and the posterior pituitary gland of the rat contains the highest known concentration (220 pg/gland, 264 pmol/g tissue) (Deschepper et al., 1983; Rehfeld, 1985).   CCK-8s, as measured by RIA, disappears from the posterior pituitary gland after stalk section (Palkovits et al., 1984; Marley et al., 1984) and the content falls by 60-80% after lesions of the paraventricular nuclei (Beinfeld et al., 1980; Palkovits et al., 1984).   Both the paraventricular and supraoptic nuclei contain CCK-8s, based on immunohistochemistry, and the distribution of these cells in the magnocellular group follows that of oxytocin (Kiss et al., 1984b), and co-existence has been confirmed with double staining (Vanderhaeghen et al., 1981).   Furthermore, both oxytocin and CCK-8s appear to be present in the same neurosecretory granules (Martin et al., 1983a).   Not all magnocellular oxytocin neurones or their terminals contain CCK-8s or vice versa, but CCK-8s has never been observed in vasopressin neurones.   Considering the proportions of CRF/oxytocin and CCK-8s/oxytocin neurones it follows that there should be a subset of CRF/CCK-8s/oxytocin cells.   CCK-8s-containing terminals have also been identified in the external zone of the median eminence (Kiss et al., 1984b), and the content of CCK-8s in this region falls by 80% after lesions of the paraventricular nuclei (Palkovits et al., 1984).   These CCK-8s-containing fibres arise from CCK-8s immunoreactive parvocellular neurones in the antero-medial paraventricular nucleus, since these neurones react by accumulating CCK-8s immunoreactive material after section of the lateral retrochiasmatic paraventriculo-infundibular pathway.   Co-existence of CCK-8 and oxytocin in the parvocellular division of the paraventricular nucleus has not been described, but about 60% of the paraventricular CCK-8s neurones are located in the parvocellular regions (Kiss et al., 1984b).   CCK-8s perikarya are not found among the centrally-projecting neurones in the posterior paraventricular nucleus (Kiss et al., 1984b).

The sequence of rat CCK precursor has been determined via tumour-derived mRNA and cloning of cDNA.   Rat brain mRNA hybridizes with this cDNA, suggesting that the brain produces a similar CCK precursor (Deschenes et al., 1984).   The structure of the CCK preprohormone gene and its promoter have also been determined (Deschenes et al., 1985). However the sequential cleavage of CCK-58 to yield CCK-8s from the C-terminal (Rehfeld, 1985) has not yet been described for the magnocellular-neurohypophysial system.

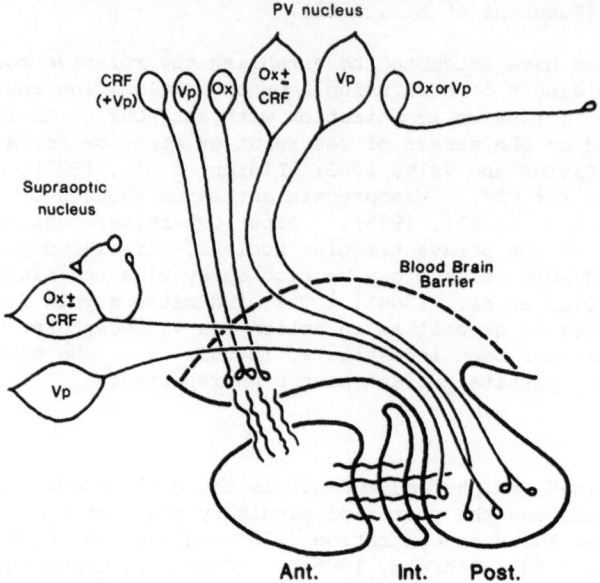

Fig. 2. A diagrammatic illustration depicting the co-existence of
oxytocin (Ox), vasopressin (Vp) and corticotrophin releasing factor (CRF) in the supraoptic and paraventricular (Pv)
nuclei of the hypothalamus of the rat.

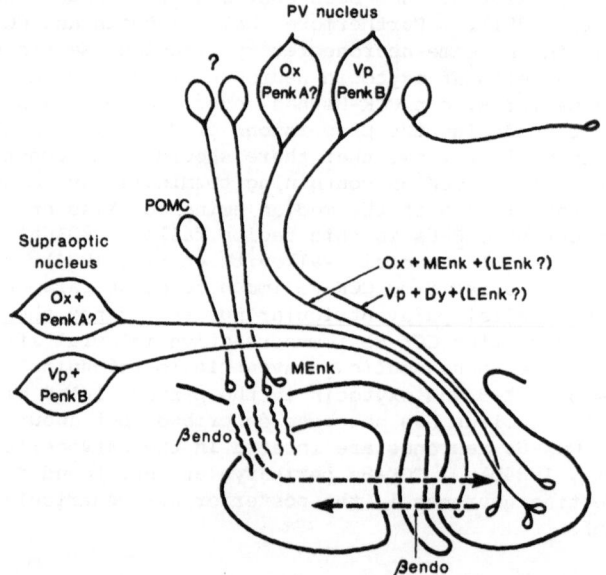

Fig. 3. A diagrammatic illustration depicting the co-existence of
oxytocin (Ox), vasopressin (Vp) and opioid peptides in
the supraoptic and paraventricular (Pv) nuclei of the
hypothalamus of the rat.
PenkA and B, Pro-enkephalin A and B; MEnk, Met-enkephalin;
LEnk, Leu-enkephalin; Dy, Dynorphin; POMC, pro-opiomelanocortin; ßendo, beta-endorphin.

Depolarization of the posterior pituitary gland in vitro with potassium increases CCK-8s release (Marley et al., 1984). Similarly, CCK-8s release from the median eminence is stimulated in a calcium-dependent fashion in vitro (Anhut et al., 1983). The evidence for release in vivo is based on chronic changes in posterior pituitary gland content, which could reflect decreased synthesis rather than increased release. The content of CCK-8 is decreased by 80% in male and female rats given 2% sodium chloride to drink for 5 days or water-deprived for 2 days (Beinfeld et al., 1980; Deschepper et al., 1983). Magnocellular oxytocin neurones are activated in both these situations (Russell, 1983; Lincoln and Russell, 1985), and since only oxytocin terminals in the neuro-hypophysis contain CCK-8s the changes in content may indicate co-release of CCK-8s and oxytocin. The posterior pituitary glands of male rats contain twice the content of CCK-8 found in females. This difference disappears after orchidectomy or oestrogen treatment of males (Deschepper et al., 1983).

The amount of CCK-8s in the median eminence is not altered by water deprivation, but is decreased by 25% after adrenalectomy, and restored by dexamethasone. Adrenalectomy does not affect the content of the posterior pituitary gland (Anhut et al., 1983). These changes suggest a possible role for median eminence CCK-8 in the control of ACTH secretion, but the changes are the opposite of those for vasopressin previously described.

In vitro, CCK-8 does not affect potassium-stimulated vasopressin secretion (Pittman et al., 1983) nor electrically-stimulated vasopressin and oxytocin secretion (Marley et al., 1984). However, these studies are confounded by the possibility of stimulation releasing CCK-8s as well as vasopressin and oxytocin. The use of a CCK-8 antagonist might be more fruitful. Specific, high-affinity CCK-8 receptors have been demonstrated in brain membrane preparations (Wennogle et al., 1985) and in hypothalamus (Saito et al., 1980), but have not been reported in the posterior pituitary gland. If secreted into the pituitary portal system, CCK-8s could act as a regulator of anterior pituitary secretion. In vitro, CCK-8s stimulates the release of growth hormone and ACTH (Matsumura et al., 1984).

CCK-8s has anti-opioid actions in vivo. Centrally administered CCK-8s reduces the analgesic effects of some exogenous and endogenously released opioids, and the CCK-8 antagonist, proglumide, and CCK-8s antiserum have the opposite action (Faris et al., 1983; Itoh et al., 1985; Watkins et al., 1985). Speculatively, it is possible that CCK-8s released from oxytocin terminals could oppose inhibitory actions of opioids released from vaso-pressin or oxytocin terminals.

Endogenous Opioid Peptides

Met-enkephalin is present in the posterior pituitary gland, as judged by immunocytochemistry (Martin et al., 1983a) and radioimmunoassay (Fig. 3) (Morris and Livingston, 1983; Zamir et al., 1985), and appears to be present in oxytocin-containing, but not vasopressin-containing, neuro-secretory granules (Martin et al., 1983a,b). Thus co-release of met-enkephalin and oxytocin is predicted. With immunocytochemistry, met-enkephalin-Arg^6-Gly^7-Leu^8 is undetectable in the supraoptic and para-ventricular magnocellular neurones (Weber et al., 1983a) and, relative to met-enkephalin, there is little of this or other extended met-enkephalin products of pro-enkephalin A in the posterior pituitary gland or neuro-intermediate lobe (Giraud et al., 1983; Martin et al., 1983b; Zamir et al., 1985). However, with immunocytochemistry the 72 amino acid sequence of pro-enkephalin A has been shown in bovine supraoptic and paraventricular magnocellular neurones (Vanderhaeghen et al., 1983). This suggests that pro-enkephalin A could be processed to yield met-enkephalin before it leaves the magnocellular nuclei.

Leu-enkephalin is present in the posterior pituitary gland in such large amounts relative to met-enkephalin that it is unlikely to be produced entirely from pro-enkephalin A. Some may be derived from pro-enkephalin B, or an unknown precursor (Giraud et al., 1983; Zamir et al., 1985). The antisera to leu-enkephalin used in many immunocytochemical studies have not necessarily distinguished met- from leu-enkephalin, nor pro-enkephalin A from B. Immunoreactivity with leu-enkephalin antisera has been shown with light microscopy in magnocellular perikarya in the supraoptic and paraventricular nuclei and with electron microscopy in nerve terminals in the posterior pituitary gland. Such immunoreactivity is lost after lesions of the nuclei or stalk section (Rossier et al., 1979). There is evidence that some enkephalin terminals, devoid of neurosecretory granules, abut onto pituicytes (van Leeuwen et al., 1983); but others claim that all enkephalin-immunoreactive terminals contain oxytocin or vasopressin (Martin et al., 1983a).

The products of pro-enkephalin B, i.e. dynorphin and related peptides (Matsuo, 1984), co-exist with vasopressin but not oxytocin in magnocellular neurones (Fig. 3) (Watson et al., 1982; Weber and Barchas, 1983). Furthermore, dynorphin A (1-17), A (1-8), dynorphin B (rimorphin), and α- and β-neoendorphin are present in approximately equimolar amounts (Weber et al., 1983b; Seizinger et al., 1984; Zamir et al., 1985). There is no precursor in the posterior pituitary gland which suggests that complete cleavage must occur before or during axonal transport (Cone et al., 1983; Seizinger et al., 1984). Bilateral lesions of the supraoptic nuclei result after 10 days in the loss of about one-third of the dynorphin and α-neo-endorphin (along with vasopressin) from the posterior pituitary gland (Millan et al., 1983), and lesions of the paraventricular nuclei have similar effects (Palkovits et al., 1983; Millan et al., 1984). Parallel changes in the content of vasopressin and dynorphin-related peptides have now been observed in several contexts and provide supportive evidence for their co-production and release (Millan and Herz, 1985). However, normal synthesis of vasopressin is not required for the production of dynorphin, since homozygous Brattleboro' rats, genetically incapable of synthesizing prepropressophysin (Schmale and Richter, 1984), produce dynorphin and related peptides (at normal or reduced levels) in their putative vasopressinergic neurones (Watson et al., 1982; Weber et al., 1983b).

The third family of opioid peptides, derived from pro-opiomelanocortin (POMC) has not been found to co-exist with oxytocin or vasopressin. β-Endorphin is secreted into the pituitary portal vessels in considerable amounts from neurones located in the arcuate nucleus of the hypothalamus (Wardlaw et al., 1982; Wehrenberg et al., 1982). This, and β-endorphin derived from the neurointermediate lobe, could reach the posterior pituitary gland via local vascular bridges.

It can be expected from the above that met-enkephalin should be released with oxytocin, and dynorphin (and related peptides) should be released with vasopressin from the posterior pituitary gland. Such selective release is difficult to demonstrate since stimulation of the posterior pituitary gland in vivo or in vitro usually releases oxytocin and some vasopressin or vice versa. Depolarization releases dynorphins from the posterior pituitary gland in vitro in a calcium-dependent fashion (Seizinger et al., 1982; Maysinger et al., 1984). With chronic stimulation, dynorphin-related peptides are secreted pari passu with vasopressin (Zamir et al., 1985), but a similar correlation of met-enkephalin release with oxytocin or vasopressin secretion has not been found (Morris and Livingston, 1983; Zamir et al., 1985).

Extensive studies have been conducted with opioid agonists and antagonists. Morphine acts centrally to inhibit vasopressin secretion in a naloxone reversible manner (Millan and Herz, 1985) but, paradoxically, morphine also causes antidiuresis due to renal haemodynamic changes which are independent of vasopressin (Grell et al., 1985). Morphine, leu-enkephalin, β-endorphin and K-agonists diminish vasopressin secretion in response to hyperosmotic or hypovolemic stimuli, and in response to excitatory transmitters (Summy-Long et al., 1981; 1983; Szligi and Ludens, 1982). However, studies with antagonists suggest little inhibitory activity of endogenous opioid peptides on vasopressin secretion. Naltrexone does not alter plasma vasopressin concentrations during chronic water deprivation or acute haemorrhage (Summy-Long et al., 1984), while the intravenous infusion of naloxone attenuates vasopressin secretion and antidiuretic responses to acute hyperosmotic or hypovolemic stimulation (Ishikawa and Schrier, 1982). A minor role for central morphine-sensitive opioid receptors is indicated by the finding that, in morphine-dependent rats, naloxone provokes an increase in the electrical and secretory activity of vasopressin neurones but the effect is small relative to the changes observed in oxytocin neurones (Bicknell et al., 1985a).

Opioids have been shown to inhibit oxytocin secretion during suckling (Clarke et al., 1979; Clarke and Wright, 1984; Russell and Spears, 1984), at parturition (Cutting et al., 1985) and after excitation by neurotransmitters (Haldar et al., 1982; Keil et al., 1984). Naloxone has also been shown to increase oxytocin secretion, especially during anaesthesia, immobilization (Bicknell et al., 1984b; Samson et al., 1985) and parturition (Hartman et al., 1984), when it can speed up delivery (Leng et al., 1985a). Naloxone does not increase the plasma oxytocin concentration during suckling (Summy-Long et al., 1984), which is the only state investigated in which vasopressin (and presumably dynorphin) release is not also stimulated. The sensitivity of oxytocin neurones to opioid peptides is well demonstrated by the marked excitation and massive oxytocin release observed in morphine-dependent rats injected with naloxone (Russell, 1984; Bicknell et al., 1984). However, there is much confusion regarding the sites at which opioid peptides interact to regulate oxytocin and vasopressin secretion, though studies on the isolated posterior pituitary gland have identified powerful effects involving the magnocellular nerve terminals and/or their interaction with neuroglial elements.

Morphine inhibits the release of vasopressin from the isolated posterior pituitary gland stimulated electrically (Iversen et al., 1980) but inhibition of stimulated release is not fully reversed by naloxone (Clarke and Patrick, 1983). Chronic morphine treatment does not induce any sign of dependence in the posterior pituitary gland (Bicknell et al., 1985b). There are conflicting results regarding the effects of naloxone on vasopressin secretion in vitro, with both augmentation of secretion and lack of effect being reported (Bicknell and Leng, 1982b; Maysinger et al., 1984; Bicknell et al., 1985c). These differences could be attributed, in part, to the pattern or method of stimulation applied. Vasopressin release by phasic stimulation is calcium-dependent and tetrodotoxin-sensitive, whereas release by continuous stimulation is only calcium-dependent (Racké et al., 1982). D-ala-D-leu-enkephalin (DADLE) at 5×10^{-6}M inhibits phasically-stimulated vasopressin secretion by 70% but not if calcium concentration is raised (Iversen et al., 1980; Lightman et al., 1982). DADLE fails to inhibit vasopressin secretion evoked by continuous stimulation (Bicknell and Leng, 1982a; Bicknell et al., 1985c), unless the calcium concentration is reduced (Maysinger et al., 1984). Thus it appears that leu-enkephalin, released either with met-enkephalin from oxytocin terminals or with dynorphin from vasopressin terminals, could inhibit further vasopressin secretion. However, stimulatory effects have also been described. At 5×10^{-6}M, DADLE has been found to enhance phasically-stimulated vasopressin secretion (Knepel et al., 1983) and similar effects are observed with dynorphin at 10^{-6} or 10^{-7}M (Maysinger et al., 1984). These findings suggest an interesting possibility. In the absence of depolarization, the basal release of vasopressin would be reduced by auto-inhibition via co-released dynorphins. Yet these same dynorphins could facilitate release under conditions of enhanced secretion. Interestingly the stimulatory effect of dynorphin under these conditions is not diminished by naloxone (Maysinger et al., 1984), but at a high concentration a k-receptor antagonist inhibits stimulated release (Bicknell et al., 1985c).

The pulsatile secretion of oxytocin during suckling or in response to electrical stimulation of the posterior pituitary gland can be monitored by the electrophysiological recording of magnocellular neurones, by radio-immunoassay of circulating hormone, and by the measurement of intramammary pressure. Such a model provides a unique opportunity to distinguish the effects of opioid peptides on the terminals from more central actions. The pulsatile secretion of oxytocin in response to suckling is inhibited by morphine given intravenously or into the cerebral ventricles without noticeably affecting the electrical activation of the oxytocin neurones in the magnocellular nuclei (Clarke et al., 1979). Furthermore, morphine inhibits oxytocin secretion from the posterior pituitary gland in vivo and in vitro (Clarke and Patrick, 1983). Naloxone has a clear enhancing action on oxytocin secretion in vivo, without, except in morphine-dependent rats, affecting the electrical activity of oxytocin neurones (Bicknell et al., 1984b). In vitro naloxone also markedly increases the amount of oxytocin released in response to both phasic and continuous electrical stimulation, and such stimulation, of course, should release the products of all nerve terminals (Bicknell and Leng, 1982b; Maysinger et al., 1984). A lack of a naloxone effect has also been reported (Pitzel and König, 1984). Together, these data demonstrate that an opioid ligand is released within the posterior pituitary gland and this acts to inhibit oxytocin secretion. The ligand has not been identified with certainty but it is likely to be an opioid peptide, rather than a dopamine-derived morphine-like product, since a monoclonal antibody against the aminoterminal tetrapeptide of the common opioid sequence produces a similar enhancing effect to naloxone (Maysinger et al., 1984). The opioid control of oxytocin secretion could be governed by autoregulation through the co-secretion of met-enkephalin and CCK-8s. Alternatively, there may be cross-talk, with the dynorphin released from vasopressin terminals regulating oxytocin secretion. In this context it is interesting to note that leu-enkephalin (10^{-6}M) and dynorphin 1-17 (10^{-7} or 10^{-6}M) enhance, rather than inhibit, oxytocin secretion evoked by submaximal electrical stimulation (Maysinger et al., 1984) and a k-receptor antagonist at high concentration inhibits stimulated secretion (Bicknell et al., 1985c). Thus high levels of vasopressin secretion might augment the amount of oxytocin secretion in response to a fixed electrical input to the oxytocin terminals.

Radioreceptor assays and autoradiography have been used in recent studies with (3)H-etorphine and ligands relatively specific for the opioid receptor sub-types (μ, δ and k). These show both high and low affinity sites for etorphine in bovine and rat posterior pituitary glands. Neither of these sites bind μ-ligands, but the low affinity site is displaced by δ-ligands. Only k-ligands (including dynorphin 1-8) displace (3)H-etorphine from the high affinity site (Gerstberger and Barden, 1984). These effects are consistent with the observed actions of morphine, enkephalins and dynorphins (Zukin and Zukin, 1984). A selective k-antagonist, but not a δ-antagonist has enhancing effects on stimulated oxytocin secretion in vitro (Bicknell et al., 1985c). The localization of these receptors within the posterior pituitary gland has not been achieved with high resolution, but the persistence of binding sites two and four weeks after stalk section leads to the conclusion that opioid receptors are present on pituicytes (Lightman et al., 1983). Thus opioid peptides released from nerve terminals could influence pituicytes through these receptors, and perhaps alter the pituicyte encapsulation of the magnocellular nerve terminals in relation to the vascular bed (Tweedle, 1983).

ELECTROPHYSIOLOGICAL SPECIALIZATIONS

A striking feature of the oxytocin and vasopressin neurones, in the rat at least, is their contrasting patterns of electrical activity. These relate in part to differences in the intrinsic properties of the cell membranes and in part to differences in their connections.

The explosive activation of the continuously-firing supraoptic and paraventricular neurones for a brief period immediately before milk ejection in the rat has identified these cells as putatively oxytocinergic (Lincoln and Wakerley, 1974). By exclusion, the phasically-firing neurones that were not activated by suckling were classified as putatively-vasopressinergic. This classification based on patterns of electrophysiological activity has stood the test of time. Only rarely has a phasic neurone been observed to display explosive activation in relation to milk ejection (Poulain and Wakerley, 1982), and the activation of the continuously firing cells has also been recorded during suckling in the unanaesthetized rat (Summerlee and Lincoln, 1981). No one has managed to record from and then label by intracellular dye injection neurones electrophysiologically classified in vivo in the rat, so as to permit the identification of the peptide they contain. However, some of the patterns of electrical activity are expressed in vitro (Hatton, 1984; Bourque and Renaud, 1985a), but for obvious reasons the association between suckling and the activation of the continuously firing magnocellular neurones cannot be made in an isolated slice of hypothalamic tissue. Intracellular recording and dye injection in vitro, with subsequent immunohistochemistry, indicates that phasically-firing neurones contain vasopressin (Yamashita et al., 1983; Cobbett et al., 1984) and, by exclusion, continuously active neurones are designated as oxytocinergic.

In vivo, these two cell types have a second distinguishing feature. The phasic bursts of the vasopressin neurones occur asynchronously, whereas the explosive bursts of activity displayed by the oxytocin cells during suckling are tightly coupled between cells within all four magnocellular nuclei (Belin et al., 1985). One result of asynchronous phasic activity is to generate, in theory, a continuous profile of vasopressin secretion, whose amplitude is a function of the ratio of the mean "on-phase" to the "off-phase" and the rate of firing during the "on-phase". The continuous background firing of the oxytocin cell would also appear to promote a continuous background of secretion, but superimposed upon this are pulses of oxytocin secretion designed to activate the mammary gland or uterus at discrete intervals during suckling or labour.

Phasic firing of vasopressin neurones

A neurone in order to produce a self-sustaining burst of action potentials must, after an initiation spike, display a depolarizing plateau potential of sufficient amplitude to sustain the generation of further spikes (Andrew and Dudek, 1984a). The phasic firing of the vasopressin cell is very much an intrinsic property of the individual cell and its initiation involves a gradual depolarization which can be augmented by synaptic activation, changes in the osmolarity of the bathing medium (Mason, 1980; Leng et al., 1982), stimulation by antidromic potentials (Dreifuss et al., 1976), and by current injection (Andrew and Dudek, 1984a; Bourque

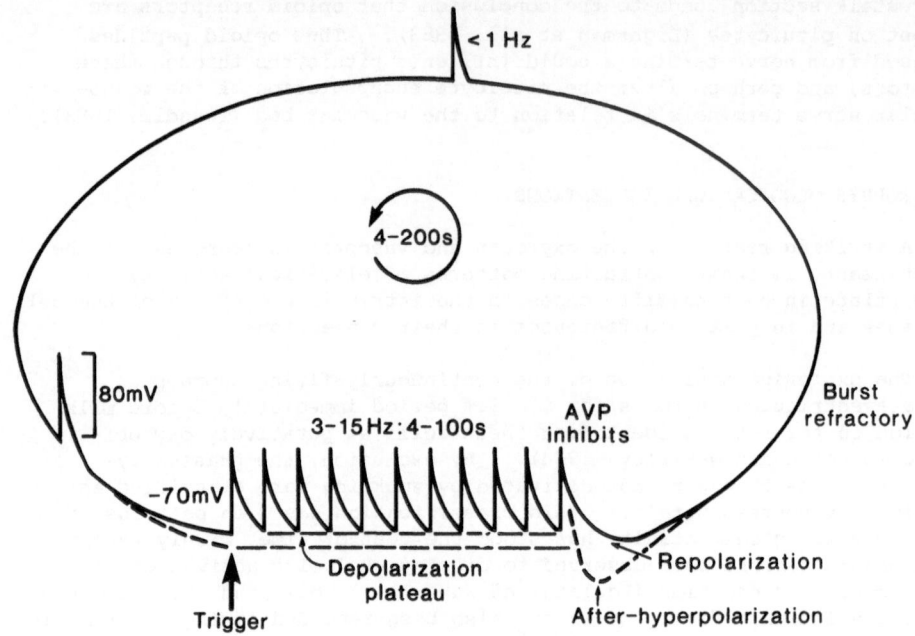

Fig. 4. A schematic illustration summarising the cycle of events associated with the phasic discharge of the vasopressin neurone.

and Renaud, 1985a). The maintenance of spike activity does not depend on the summation of epsps; it is governed by a positive feedback mechanism involving a repetitive sequence of events that serves to maintain a depolarization plateau (Andrew and Dudek, 1984a). Each action potential in a burst is a composite of an initial low-threshold Na^+ conductance and a high-threshold Ca^{++} conductance (Andrew and Dudek, 1984a; Bourque and Renaud, 1985a). Individual action potentials last longer with increasing frequency, increasing from about 1.2 ms during continuous low frequency firing to more than double this duration at rates of activity observed during the "on-phase" of a vasopressinergic neurone (Bourque and Renaud, 1985b). This phenomenon is calcium dependent and reflects a prolongation of the calcium conductance, and can be recorded as a shoulder on the falling phase of the spike (Mason and Leng, 1984). These changes consequent on the clustering of spikes within a burst have the effect of prolonging the duration of depolarization and increasing the Ca^{++} influx. If the terminals of magnocellular neurones have similar ionic properties this could account for why high frequency stimulation is a more efficient way of releasing oxytocin or vasopressin for a given number of action potentials (Harris et al., 1969; Lincoln, 1974; Dutton and Dyball, 1979; Bicknell and Leng, 1981). Phasic discharge also avoids the fatigue that accompanies continuous stimulation (Bicknell et al., 1984a). Firing rate within a burst is limited by the refractoriness of the cell such that the minimum interspike interval is around 40 msec (Leng, 1981) and rarely less than 30 msec (Thomson, 1984). The after-spike refractoriness can be attributed to a hyperpolarization evoked by a Ca^{++}-activated K^+ conductance, and is not synaptically mediated. Such a phenomenon is seen in an exaggerated form at the end of a period of accelerated activity (Andrew and Dudek, 1984b). After the last spike in a burst the cell slowly repolarizes, potentially aided by the Ca^{++}-dependent conductance that evokes hyperpolarization. This can last for a long period and account for the post-burst silence displayed by both oxytocin and vasopressin cells (Andrew and Dudek, 1984b).

There is some evidence that vasopressin could contribute to the cessation of each phase of accelerated firing. Repetitive antidromic constant-collision stimulation does not affect the intra-burst firing of phasic neurones but it shortens the burst without changing the inter-burst interval (Leng, 1981). This is not seen in Brattleboro' rats which suggests that centrally released vasopressin could be involved. Indeed, vasopressin inhibits phasic firing when applied to isolated hypothalamic slices (Leng and Mason, 1982). Curiously, it generates a reversible depolarization so that, if the cell is hyperpolarized, spike activation can result (Abe et al., 1983). Vasopressin does not appear to influence the firing of phasic cells when given into the cerebral ventricles (Freund-Mercier and Richard, 1984). The burst cycle of the vasopressin cell is depicted in Figure 4.

Synchronized bursting of oxytocin neurones

The bursts of electrical activity displayed by oxytocin neurones in the rat are shorter and involve a higher frequency of firing than is displayed by phasically-active vasopressin neurones, and these cells do not burst "spontaneously". The only physiological trigger known to drive an oxytocin neurone into a burst is the gated input derived from the suckling stimulus (Poulain and Wakerley, 1982; Lincoln and Russell, 1985) or from the dilatation of the cervix during labour (Summerlee, 1981). The suckling stimulus is distributed to all four magnocellular nuclei and this promotes all the oxytocin neurones within these nuclei to burst within 400 ms of each other (Belin et al., 1985). Quite amazingly this input is so specific as not to influence the vasopressin neurones which are intermingled with the oxytocin cells. Such specificity could be achieved anatomically through synaptic connectivity. Alternatively, the oxytocin cells could respond by virtue of a specific chemo-sensitivity. Various other stimuli, including changes in osmotic pressure, increase the firing of the oxytocin cell, but they do so by increasing the continuous pattern of discharge (Poulain et al., 1977).

Reciprocal connections between the magnocellular nuclei (Silverman et al., 1981; Saphier and Feldman, 1985) could be involved in the synchronous activation of oxytocin neurones, as could intranuclear connections. Many nerve terminals in the supraoptic nuclei arise from within the nuclei or close by, but few appear to contain dense-cored vesicles (Léranth et al., 1975). In immunocytochemically prepared EM-sections, a few oxytocin-containing terminals, of unknown origin, have been found to synapse on oxytocin-containing magnocellular neurones (Theodosis, 1985). During lactation major structural changes have also been observed to occur within the magnocellular nuclei brought about by the withdrawal of neuroglial processes. This increases the area of direct apposition between the somata and dendrites of adjacent oxytocinergic neurones (but not vasopressinergic neurones), and increases the incidence with which terminals synapse onto two adjacent neurones (Theodosis and Poulain, 1983; Theodosis et al., 1984). Such plasticity may be important for synchronized firing. However, similar changes occur during chronic osmotic stimulation - a situation that does not involve the synchronized bursting of oxytocin neurones (Tweedle and Hatton, 1984). There is also evidence for chemical specificity. Depolarization increases the amount of oxytocin (and vasopressin) released from the supraoptic region in hypothalamic slices (Chapman et al., 1983). Secondly, oxytocin but not vasopressin increases the release of oxytocin from isolated paraventricular and supraoptic nuclei in a calcium-dependent fashion (Moos et al., 1984). Effects consistent with this are observed in vivo. An oxytocin antagonist given by intraventricular injection during suckling reduces the number of spikes in each explosive burst and increases the inter-burst interval (Freund-Mercier and Richard, 1984). Oxytocin given by the same route has the opposite action and augments the bursting phenomenon, but only in the presence of

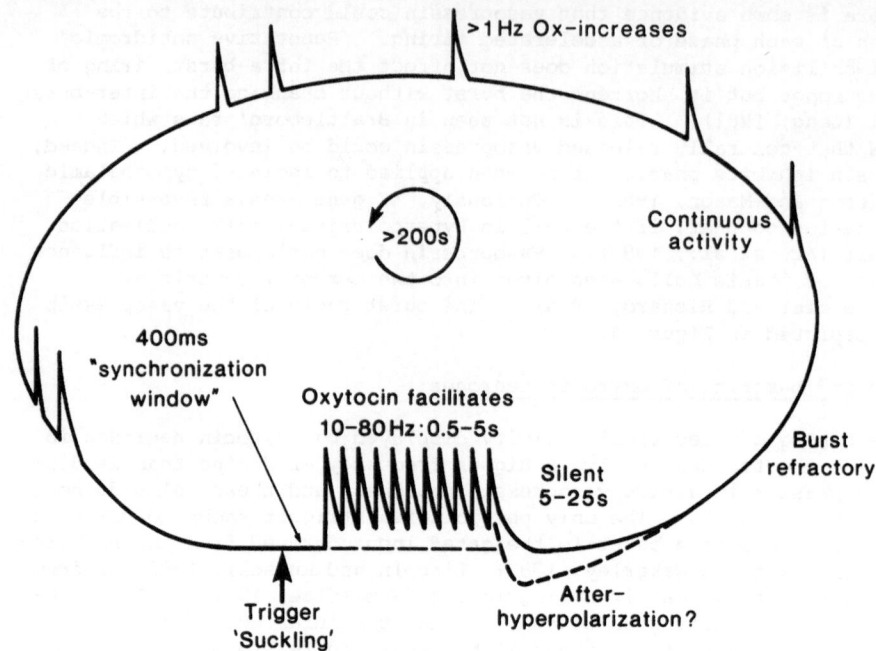

>1Hz Ox-increases

>200s

Continuous
activity

400ms
"synchronization
window"

Oxytocin facilitates
10–80Hz:0.5–5s

Silent
5–25s

Burst
refractory

Trigger
'Suckling'

After-
hyperpolarization?

Fig. 5.   A schematic illustration summarising the cycle of events
associated with the synchronized bursting of the oxytocin
neurone during suckling.

the suckling stimulus.   Constant collision stimulation applied anti-
dromically to the pituitary stalk evokes mini-bursts of activity in some
putative oxytocinergic neurones (Leng, 1981).   This effect could be
mediated by other oxytocin cells releasing oxytocin within the magnocellular nuclei or via neurones synaptically activated close to the nuclei
(Leng,   1982;   Blount and Leng, 1985).   There is little sign of magnocellular neurones influencing one another synaptically (Akaishi and
Ellendorff, 1983) and, in vivo, adjacent magnocellular neurones show no
signs of mutual interaction in a time domain that could indicate orthodox
synaptic activation (Leng and Dyball, 1983).   There is accumulating
evidence, therefore, to indicate that oxytocin can, by a central action,
augment its own secretion, and enhance the efficacy of the suckling
stimulus.   The burst cycle of the oxytocin neurone is depicted in
Figure 5.

Control of inter-burst intervals in oxytocin and vasopressin neurones

The firing patterns of oxytocin and vasopressin neurones differ
markedly, but so far they have been found to have similar underlying
electrophysiological properties.   From in vitro studies, the wave form of
their action potentials, membrane resistances and osmosensitivity, and
monovalent and divalent ionic conductances appear to be similar.   Both
types of cell can be provoked to burst in a similar manner by current
injection, yet the current-elicited burst in both types of cell resembles
the phasic discharge of the vasopressin neurone in vivo (Bourque and
Renaud, 1985a).   However, such bursting differs fundamentally from the
bursting of an oxytocin cell in vivo - the milk ejection burst is faster
and shorter.   The underlying ionic mechanisms have not been elucidated,

but it would appear that the calcium conductance must be suppressed to enable the cell to discharge so rapidly.

It is clear that a fundamental difference lies in the fact that the oxytocin cells are triggered synchronously by a gated modality-specific afferent input, whereas the vasopressin cells appear to be triggered independently of each other in response to several humoral or synaptic signals. For both types of neurone, the character of the burst or phase of activity is determined by factors intrinsic to the magnocellular complex. The dichotomy in the electrophysiological behaviour of these two cell types is clear in the rat, as just described, but is considerably less marked in primates, rabbits, pigs and guinea-pigs. Indeed, in these species it is often difficult to distinguish between the firing patterns of the two cells, though there have not been exhaustive investigations.

The factors determining in the rat the interval between one oxytocin burst and the next, and the length of the silent periods displayed by vasopressin cells are less well understood, but in both situations the cells appear to be relatively refractory to immediate restimulation. Both cell types receive extensive synaptic inputs and some of these appear to arise from neurones nearby. Electrophysiological and immunocytochemical evidence indicates that neurones just dorsolateral to the supraoptic nuclei receive axon collaterals from oxytocin and vasopressin neurones (Leng, 1982; Mason et al., 1983; 1984). Some of these neurones are inhibited and others excited when supraoptic neurones are activated (Blount and Leng, 1985). In return, there are about 100 cholinergic neurones in the area dorsolateral to the supraoptic nucleus that send axons, widely branching, into the nucleus. These neurones excite vasopressin neurones via nicotinic receptors but, like acetylcholine, they do not excite oxytocin neurones (Hatton et al., 1983). These neurones could be involved in osmotic activation and provide the epsps important in triggering the phasic activity of the vasopressin cell (Leng et al., 1982). Oxytocin neurones would, by contrast, appear not to have a cholinergic-innervation, and presumably the nicotinic acetylcholine receptor gene is not expressed.

The intermittent activation of the oxytocin neurones of the rat during suckling appears to involve a gating of the afferent input at the level of the brain stem or spinal cord (Lincoln and Russell, 1985). Double milk ejections, involving explosive bursts of firing in all oxytocin cells at intervals as short as 30s have been observed following the placement of lesions in the ventral tegmentum of the midbrain (Juss and Wakerley, 1981) and following the administration of 5-HT antagonists (Moos and Richard, 1983). Such observations suggest that the 5-HT projection from the raphe nuclei, to either the magnocellular nuclei or spinal cord, could be involved in determining the refractory interval between one milk ejection and the next. Clearly the refractory interval is not a function of the magnocellular neurones themselves because, using electrical stimulation, pulsatile oxytocin secretion can be evoked at intervals much shorter than normally observed during suckling. A hiatus in spike activity is observed following each explosive burst, presumably due to hyperpolarization, but this is relatively short (5-25s) compared with the interval between bursts (120-600s) (Lincoln and Wakerley, 1974). Not all afferent inputs to oxytocin neurones lead to explosive activation; they respond very effectively to haemorrhage (Poulain et al., 1977) and to the intraperitoneal injection of hypertonic saline, an osmotic stimulus without marked cardiovascular effects. In these situations the activity of the cells changes from slow and irregular to faster and more continuous; bursting has never been observed. Some of these responses could involve angiotensin II and a projection from the subfornical organ to the magnocellular nuclei (Lang et al., 1981; Miselis, 1981; Sgro et al., 1984; Lincoln and Russell, 1985; Leng et al., 1985b). The interval between one phase of activity

in a vasopressin cell and the next differs from that discussed above in
that it involves sustained inactivity, and probably relates more to a protracted hyperpolarization of the cell than to a lack of synaptic drive.
Without question, these vasopressin cells are driven by synaptic inputs
from distant sites within the brain, including the projection from the
subfornical organ just discussed. Indeed, an increase in the magnitude
of this afferent input could serve to shorten the interval between one
phase of activity and the next. Conversely, in the absence of effective
stimulation, vasopressin cells cease to fire phasically and assume a slow
intermittent pattern of activity difficult to distinguish from that of an
unstimulated oxytocin cell.

CONCLUDING COMMENTS

     Five hundred million years of evolution have, with respect to oxytocin
and vasopressin, effectively generated two neuroendocrine systems whose
parallels are still apparent but whose functional organization is now quite
divergent. As a result of gene duplication, oxytocin- and vasopressin-
related genes co-exist within an individual neurone, but it seems that only
one of these two genes is expressed. An oxytocin cell should not, however,
be viewed as simply a cell that produces and secretes oxytocin - it is an
integrated neuroendocrine unit of great complexity, and vasopressin cells
should be viewed likewise. The genome of these two cell types is so
selectively regulated that each expresses a remarkable divergence in terms
of 1) the co-production and secretion of presumptive regulatory peptides,
2) synaptic- and chemo-sensitivity, 3) patterns of spike generation and
stimulus-secretion coupling, and 4) morphological specializations related
to afferent and efferent connections and relationships to other neural
elements. As Mozart could have told us, the selection of the key is only
one step in the composition of a symphony.

REFERENCES

Abe H., Inoue, M., Matsuo, T., and Ōgata, N., 1983, The effects of vaso-
     pressin on electrical activity in the guinea-pig supraoptic nucleus
     in vitro, J. Physiol. (Lond.), 337:665.
Acher, R., 1980, Molecular evolution of biologically active polypeptides,
     Proc. R. Soc. Lond. B., 210:21.
Aguilera, G., Wynn, P.C., Haugher, R.L., Holmes, M.C., Millan, M.A.,
     Mendelsohn, F.A.O., Crewe, C., and Catt, K.J., 1984, Receptors and
     actions of angiotensin II (AII) and corticotropin-releasing factor
     (CRF) in specific regions of the central nervous system, Excerpta
     Medica (Int. Congress Series 655):609, Elsevier, Amsterdam.
Akaishi, T., and Ellendorff, F., 1983, Electrical properties of para-
     ventricular neurosecretory neurones with and without recurrent
     inhibition, Brain Res., 262:151.
Andrew, R.D., and Dudek, F.E., 1984a, Analysis of intracellularly recorded
     phasic bursting by mammalian neuroendocrine cells, J. Neurophysiol.,
     51:552.
Andrew, R.D., and Dudek, F.E., 1984b, Intrinsic inhibition in magnocellular
     neuroendocrine cells of rat hypothalamus, J. Physiol. (Lond.), 353:171.
Anhut, M., Meyer, D.K., and Knepel, W., 1983, Cholecystokinin-like immuno-
     reactivity of rat medial basal hypothalamus: investigations on a
     possible hypophysiotropic function, Neuroendocrinology, 36:119.
Antoni, F.A., 1984, Novel ligand specificity of pituitary vasopressin
     receptors in the rat, Neuroendocrinology, 39:186.
Antoni, F.A., Holmes, M.C., and Jones, M.T., 1983, Oxytocin as well as
     vasopressin potentiate ovine CRF in vitro, Peptides, 4:411.

Baertschi, A.J., and Friedli, M., 1984, Novel anterior pituitary
receptor(s) for neurohypophysial hormones, J. Steroid Biochem.,
20(6B):1504.

Baertschi, A.J., Bény, J-L., Gähwiler, B.H., and Kolodziejczyk, E., 1983,
Vasopressin, corticoliberins and the central control of ACTH
secretion, Prog. Brain Res., 60:505.

Beinfeld, M.C., Meyer, D.K., and Brownstein, M.J., 1980, Cholecystokinin
octapeptide in the rat hypothalamo - neurohypophysial system, Nature,
288:376.

Belin, V., Moos, F., and Richard, Ph., 1985, Synchronization of oxytocin
cells in the hypothalamic and supraoptic nuclei in suckled rats:
direct proof with paired extracellular recordings, Ex. Brain Res.,
57:201.

Bicknell, R.J., and Leng, G., 1981, Relative efficiency of neural firing
patterns for vasopressin release in vitro, Neuroendocrinology,
33:295.

Bicknell, R.J., and Leng, G., 1982a, Enkephalin analogue inhibits oxytocin
secretion from the rat neurohypophysis, J. Physiol. (Lond.), 332:87P.

Bicknell, R.J., and Leng, G., 1982b, Endogenous opiates regulate oxytocin
but not vasopressin secretion from the neurohypophysis, Nature,
298:161.

Bicknell, R.J., Brown, D., Chapman, C., Hancock, P.D., and Leng, G., 1984a,
Reversible fatigue of stimulus secretion coupling in the rat neurohypophysis, J. Physiol. (Lond.), 348:601.

Bicknell, R.J., Leng, G., Lincoln, D.W., and Russell, J.A., 1984b,
Activation of oxytocin neurones following naloxone administration to
rats treated chronically with intracerebroventricular (i.c.v.)
morphine., J. Physiol. (Lond.), 357:97P.

Bicknell, R.J., Chapman, C., Leng, G., and Russell, J.A., 1985a, Vasopressin
release following naloxone administration to rats treated chronically
with intracerebroventricular (i.c.v.) morphine, J. Physiol. (Lond.), 364:61P.

Bicknell, R.J., Chapman, C., Leng, G., and Russell, J.A., 1985b, Chronic
morphine exposure and oxytocin neurones in rats: lack of both morphine
dependence and cross-tolerance to endogenous opioids in the neurohypophysis, J. Physiol. (Lond.), 361:32P.

Bicknell, R.J., Chapman, C., and Leng, G., 1985c, Effects of opioid
agonists and antagonists on oxytocin and vasopressin release in vitro,
Neuroendocrinology, 41:142.

Blount, C.A., and Leng, G., 1985, Synaptic excitation and inhibition in the
lateral hypothalamus following stimulation of the neural stalk in the
rat, J. Physiol. (Lond.), 361:30P.

Bourque, C.W., and Renaud, L.P., 1985a, Calcium-dependent action potentials
in rat supraoptic neurosecretory neurones recorded in vitro,
J. Physiol. (Lond.), 363:419.

Bourque, C.W., and Renaud, L.P., 1985b, Activity dependence of action
potential duration in rat supraoptic neurosecretory neurones recorded
in vitro, J. Physiol. (Lond.), 363:429.

Bruhn, T.O., Plotsky, P.M., and Vale, W.W., 1984a, Effect of paraventricular lesions on corticotropin-releasing factor (CRF)-like immunoreactivity in the stalk-median eminence: studies on the adrenocorticotropin response to ether stress and exogenous CRF,
Endocrinology, 114:57.

Bruhn, T.O., Sutton, S.W., and Vale, W.W., 1984b, Corticotropin-releasing
factor (CRF) stimulates oxytocin secretion into systemic circulation,
Excerpta Medica (Int. Congress Series 652):464, Elsevier, Amsterdam.

Buckingham, J.C., 1985, Two distinct corticotrophin releasing activities
of vasopressin, Br. J. Pharmacol., 84:213.

Burlet, A., Tonon, M-C., Tankosic, P., Coy, D., and Vaudry, H., 1983,
Comparative immunocytochemical localization of corticotropin releasing
factor (CRF-41) and neurohypophysial peptides in the brain of
Brattleboro and Long-Evans rats, Neuroendocrinology, 37:64.

Chapman, C., Hatton, G.I., Ho, Y.W., Mason, W.T., and Robinson, I.C.A.F., 1983, Release of oxytocin (OXT) and vasopressin (AVP) from slices of guinea-pig hypothalamus containing supraoptic or paraventricular nucleus, J. Physiol. (Lond.), 343:40P.

Chauvet, M.-T., Colne, T., Hurpet, D., Chauvet, J., and Acher, R., 1983, A multigene family for the vasopressin-like hormones? Identification of mesotocin, lysipressin and phenypressin in Australian macropods. Biochem. Biophys. Res. Comm., 116:258.

Chauvet, J., Hurpet, D., Michel, G., Chauvet, M.-T., and Acher, R., 1984, Two multigene families for marsupial neurohypophysial hormones? Identification of oxytocin, mesotocin, lysipressin and arginine vaso-pressin in the North American opossum (Didelphis virginiana), Biochem. Biophys. Res. Comm., 123:306.

Chauvet, J., Hurpet, D., Michel, G., Chauvet, M.-T., Carrick, F.M., and Acher, R., 1985, The neurohypophysial hormones of the egg-laying mammals: identification of arginine vasopressin in the platypus (Ornithorhynchus anatinus), Biochem. Biophys, Res. Comm., 127:277.

Clarke, G., and Patrick, G., 1983, Differential inhibitory action by morphine on the release of oxytocin and vasopressin from the isolated neural lobe, Neurosci. Lett., 39:175.

Clarke, G., and Wright, D.M., 1984, A comparison of analgesia and suppression of oxytocin release by opiates, Br. J. Pharmacol., 83:799.

Clarke, G., Wood, P., Merrick, L., and Lincoln, D.W., 1979, Opiate inhibition of peptide release from the neurohumoral terminals of hypothalamic neurones, Nature, 282:746.

Cobbett, P., Smithson, K.G., and Hatton, G.I., 1984, Phasic neurons of rat hypothalamic paraventricular nucleus are immunoreactive to vasopressin but not oxytocin-associated neurophysin antiserum, Soc. Neurosci. Abstr., 10:609.

Coghlan, J.P., Aldred, P., Butkus, A., Crawford, R.J., Darby, I.A., Fernley, R.T., Haralambidis, J., Hudson, P.J., Mitri, R., Niall, H.D., Penschow, J.D., Roche, P.J., Scanlon, D.B., and Tregear, G.W., 1984, Hybridization histochemistry, Excerpta Medica (Int. Congress Series 655):18, Elsevier, Amsterdam.

Cone, R.I., Weber, E., Barchas, J.D., and Goldstein, A., 1983, Regional distribution of dynorphin and neo-endorphin peptides in rat brain, spinal cord, and pituitary, J. Neurosci., 3:2146.

Cutting, R., Fitzsimons, N., Gosden, R.G., Humphreys, E.M., and Russell, J.A., 1985, Evidence that morphine interrupts parturition in rats by inhibiting oxytocin secretion, J. Physiol. (Lond.), Proceedings, September, 1985:In press.

Deschenes, R.J., Lorenz, L.J., Haun, R.S., Roos, B.A., Collier, K.J., and Dixon, J.E., 1984, Cloning and sequence analysis of a cDNA encoding rat preprocholecystokinin, Proc. Natl. Acad. Sci. USA., 81:726.

Deschenes, R.J., Haun, R.S., Funckes, C.L. and Dixon, J.E., 1985, A gene encoding rat cholecystokinin. Isolation, nucleotide sequence, and promoter activity, J. Biol. Chem., 260:1280.

Deschepper, C., Lotstra, F., Vandesande, F., and Vanderhaeghen, J-J., 1983, Cholecystokinin varies in the posterior pituitary and external median eminence of the rat according to factors affecting vasopressin and oxytocin, Life Sci., 32:2571.

Dierickx, K., 1980, Immunocytochemical localization of the vertebrate cyclic nonapeptide neurohypophyseal hormones and neurophysins, Int. Rev. Cytol., 62:119.

Dreifuss, J.J., Tribollet, E., Baertschi, A.J., and Lincoln, D.W., 1976, Mammalian endocrine neurones: control of phasic activity by antidromic action potentials, Neurosci. Lett., 3:281.

Dreyfuss, F., Burlet, A., Tonon, M.C., and Vaudry, H., 1984, Comparative immunoelectron microscopic localization of corticotropin-releasing factor (CRF-41) and oxytocin in the rat median eminence, Neuroendocrinology, 39:284.

Dutton, A., and Dyball, R.E.J., 1979, Phasic firing enhances vasopressin release from the rat neurohypophysis, J. Physiol., 290:433.

Faris, P.L., Komisaruk, B.R., Watkins, L.R., and Mayer, D.J., 1983, Evidence for the neuropeptide cholecystokinin as an antagonist of opiate analgesia, Science, 219:310.

Freund-Mercier, M.J., and Richard, Ph., 1984, Electrophysiological evidence for the facilitatory control of oxytocin neurones by oxytocin during suckling in the rat, J. Physiol., 352:447.

Furutani, Y., Morimoto, Y., Shibahara, S., Noda, M., Takahashi, H., Hirose, T., Asai, M., Inayama, S., Hayashida, H., Miyata, T., and Numa, S., 1983, Cloning and sequence analysis of cDNA for ovine corticotropin-releasing factor precursor, Nature, 301:537.

Gerstberger, R., and Barden, N., 1984, Dynorphin (1-8) binds to opiate-kappa-receptors in the neurohypophysis and is involved in the regulation of vasopressin release, Excerpta Medica (Int. Congress Series 652):631, Elsevier, Amsterdam.

Gibbs, D.M., 1984a, High concentrations of oxytocin in hypophysial portal plasma, Endocrinology, 114:1216.

Gibbs, D.M., 1984b, Dissociation of oxytocin, vasopressin and corticotropin secretion during different types of stress, Life Sci., 35:487.

Gibbs, D.M., 1985, Measurement of hypothalamic corticotrophin-releasing - factors in hypophyseal portal blood, Fed. Proc., 44:203.

Gibbs, D.M., Vale, W., Rivier, J., and Yen, S.S.C., 1984, Oxytocin potentiates the ACTH-releasing activity of CRF(41) but not vasopressin, Life Sci., 34:2245.

Gillies, G., Puri, A., Hodgkinson, S., and Lowry, P.J., 1984, Involvement of rat corticotrophin-releasing factor-41-related peptide and vasopressin adrenocorticotrophin-releasing activity from superfused rat hypothalami in vitro, J. Endocrinol., 103:25.

Giraud, P., Castanas, E., Patey, G., Oliver, C., Rossier, J., 1983, Regional distribution of methionine-enkephalin-Arg[6]-Phe[7] in the rat brain: comparative study with the distribution of other opioid peptides, J. Neurochem., 41:154.

Grell, S., Christensen, J.D., and Fjalland, B., 1985, Morphine anti-diuresis in conscious rats: contribution of vasopressin and blood pressure, Acta Pharmacol. Toxicol. (Copenh.), 56:38.

Haldar, J., Hoffman, D.L., and Zimmerman, E.A., 1982, Morphine, beta-endorphin and D-Ala[2] met-enkephalin inhibit oxytocin release by acetylcholine and suckling, Peptides, 3:663.

Harris, G.W., Manabe, Y., and Ruf, K.B., 1969, A study of the parameters of electrical stimulation of unmyelinated fibres in the pituitary stalk, J. Physiol. (Lond.), 203:67.

Hartman, R., Miller, D., Rosella-Dampman, L., Emmert, S., and Summy-Long, J., 1984, Role for endogenous opioid peptides in regulating oxytocin release at parturition, J. Steroid Biochem., 20(6B):1503.

Hatton, G.I., 1984, Hypothalamic neurobiology, in: "Brain Slices", R. Dingledine, ed., Plenum, New York.

Hatton, G.I., Ho, Y.W., and Mason, W.T., 1983, Synaptic activation of phasic bursting in rat supraoptic nucleus neurones recorded in hypothalamic slices, J. Physiol. (Lond.), 345:297.

Ishikawa, S., and Schrier, R.W., 1982, Evidence for a role of opioid peptides in the release of arginine vasopressin in the conscious rat, J. Clin. Invest., 69:666.

Itoh, S., Katsuura, G., Yoshikawa, K., and Rehfeld, J.F., 1985, Potentiation of β-endorphin effects by cholecystokinin antiserum in rats, Can. J. Physiol. Pharmacol., 63:81.

Ivell, R., and Richter, D., 1984, Structure and comparison of the oxytocin and vasopressin genes from rat, Proc. Natl. Acad. Sci. USA, 81:2006.

Iversen, L.L., Iversen, S.D., and Bloom, F.E., 1980, Opiate receptors influence vasopressin release from nerve terminals in rat neuro-hypophysis, Nature, 284:350.

Juss, T.S., and Wakerley, J.B., 1981, Mesencephalic areas controlling pulsatile oxytocin release in the suckled rat, J. Endocrinol., 91:233.

Keil, L.C., Rosella-Dampman, L.M., Emmert, S., Chee, O., and Summy-Long, J.Y., 1984, Enkephalin inhibition of angiotensin-stimulated release of oxytocin and vasopressin, Brain Res., 297:329.

Kiss, J.Z., Mezey, E., and Skirboll, L., 1984a, Corticotropin-releasing factor-immunoreactive neurones of the paraventricular nucleus become vasopressin positive after adrenalectomy, Proc. Natl. Acad. Sci. USA, 81:1854.

Kiss, J.Z., Williams, T.H., and Palkovits, M., 1984b, Distribution and projections of cholecystokinin-immunoreactive neurons in the hypo-thalamic paraventricular nucleus of rat, J. Comp. Neurol., 227:173.

Knepel, W., Nutto, D., and Meyer, D.K., 1983, Naloxone increases vaso-pressin secretion from the neurointermediate lobe of the hypophysis of the rat: search for the endogenous agonist, Life Sci., 33, Suppl. 1:499.

Knepel, W., Homolka, L., Vlaskovska, M., and Nutto, D., 1984a, Stimulation of adrenocorticotropin/beta-endorphin release by synthetic ovine corticotropin-releasing factor in vitro. Enhancement by various vasopressin analogs, Neuroendocrinology, 38:344.

Knepel, W., Homolka, L., Vlaskovska, M., and Nutto, D., 1984b, In vitro adrenocorticotropin/β-endorphin-releasing activity of vasopressin analogs is related neither to pressor nor to antidiuretic activity, Endocrinology, 114:1797.

Lang, R.E., Rascher, W., Heil, J., Unger, Th., Wiedemann, G., and Ganten, D., 1981, Angiotensin stimulates oxytocin release, Life Sci., 29:1425.

Lang, R.E., Heil, J.W.E., Ganten, D., Hermann, K., Unger, T., and Rascher, W., 1983, Oxytocin unlike vasopressin is a stress hormone in the rat, Neuroendocrinology, 37:314.

Leng, G., 1981, The effects of neural stalk stimulation upon firing patterns in rat supraoptic neurones, Exp. Brain Res., 41:135.

Leng, G., 1982, Lateral hypothalamic neurones: osmosensitivity and the influence of activating magnocellular neurosecretory neurones, J. Physiol., 326:35.

Leng, G., and Dyball, R.E.J., 1983, Intercommunication in the rat supraoptic nucleus, Q. J. Exp. Physiol., 68:493.

Leng, G., and Mason, W.T., 1982, Influence of vasopressin upon firing patterns of supraoptic neurons: a comparison of normal and Brattleboro rats, Ann. N.Y. Acad. Sci., 394:153.

Leng, G., Mason, W.T., and Dyer, R.G., 1982, The supraoptic nucleus as an osmoreceptor, Neuroendocrinology, 34:75.

Leng, G., Dyball, R.E.J., and Mason, W.T., 1985a, Electrophysiology of osmoreceptors, in: "Vasopressin", R.W. Schrier, ed., Raven Press, New York.

Leng, G., Mansfield, S., Bicknell, R.J., Dean, A.D.P., Ingram, C.D., Marsh, M.I.C., Yates, J.O., and Dyer, R.G., 1985b, Central opioids: a possible role in parturition?. J. Endocrinol., 106:219.

Léránth, C.S., Záborsky, L., Marton, J., and Palkovits, M., 1975, Quanti-tative studies on the supraoptic nucleus in the rat. I. Synaptic organization, Exp. Brain Res., 22:509.

Lightman, S.L., Iversen, L.L., and Forsling, M.L., 1982, Dopamine and [D-Ala^2,D-Leu^5]enkephalin inhibit the electrically stimulated neuro-hypophyseal release of vasopressin in vitro: evidence for calcium-dependent opiate action, J. Neurosci., 2:78.

Lightman, S.L., Ninkovic, M., Hunt, S.P., and Iversen, L.L., 1983, Evidence for opiate receptors on pituicytes, Nature (Lond.), 305:235.

Lincoln, D.W., 1974, Dynamics of oxytocin secretion, in: "Neurosecretion - the final neuroendocrine pathway", F. Knowles and L. Vollrath, eds., Springer-Verlag, Heidelberg.

Lincoln, D.W., and Wakerley, J.B., 1974, Electrophysiological evidence for the activation of supraoptic neurones during the release of oxytocin, J. Physiol. (Lond.), 242:533.

Lincoln, D.W., and Russell, J.A., 1985, The electrophysiology of magnocellular oxytocin neurons, in: "Oxytocin: Clinical and Laboratory Studies", J.A. Amico and A.G. Robinson, eds., Elsevier, New York.

Linton, E.A., Tilders, F.J.H., Hodgkinson, S., Berkenbosch, F., Vermes, I., and Lowry, P.J., 1985, Stress-induced secretion of adrenocorticotropin in rats is inhibited by administration of antisera to ovine corticotropin-releasing factor and vasopressin, Endocrinology, 116:966.

Marley, P.D., Lightman, S.L., Forsling, M.L., Todd, K., Goedert, M., Rehfeld, J.F., and Emson, P.C., 1984, Localization and actions of cholecystokinin in the rat pituitary neurointermediate lobe, Endocrinology, 114:1902.

Martin, R., Geis, R., Holl, R., Schäfer, M., and Voigt, K.H., 1983a, Co-existence of unrelated peptides in oxytocin and vasopressin terminals of rat neurohypophyses: immunoreactive methionine[5]-enkephalin, leucine[5]-enkephalin and cholecystokinin-like substances, Neuroscience, 8:213.

Martin, R., Moll, U., and Voigt, K.H., 1983b, An attempt to characterize by immunocytochemical methods the enkephalin-like material in oxytocin endings of the rat neurohypophysis, Life Sci., 33, Suppl. I:69.

Mason, W.T., 1980, Supraoptic neurones of rat hypothalamus are osmosensitive, Nature, 287:154.

Mason, W.T., and Leng, G., 1984, Complex action potential waveform recorded from supraoptic and paraventricular neurones of the rat: evidence for sodium and calcium spike components at different membrane sites, Exp. Brain Res., 56:135.

Mason, W.T., Ho, Y.W., Eckenstein, F., and Hatton, G.I., 1983, Mapping of cholinergic neurones associated with rat supraoptic nucleus: combined immunocytochemical and histochemical identification, Brain Res. Bull., 11:617.

Mason, W.T., Ho, Y.W., and Hatton, G.I., 1984, Axon collaterals of supraoptic neurones: anatomical and electrophysiological evidence for their existence in the lateral hypothalamus, Neuroscience, 11:169.

Matsumura, M., Yamanoi, A., Yamamoto, S., Mori, H., and Saito, S., 1984, In vivo and in vitro effects of cholecystokinin octapeptide on the release of growth hormone in rats, Horm. Metab. Res., 16:626.

Matsuo, H., 1984, Neo-endorphins and dynorphins processed out of proenkephalin B, Excerpta Medica (Int. Congress Series 655):637, Elsevier, Amsterdam.

Maysinger, D., Vermes, I., Tilders, F., Seizinger, B.R., Gramsch, C., Höllt, V., and Herz, A., 1984, Differential effects of various opioid peptides on vasopressin and oxytocin release from the rat pituitary in vitro, Naunyn Schmiedebergs Arch. Pharmacol., 328:191.

Millan, M.J., and Herz, A., 1985, The endocrinology of the opioids, Int. Rev. Neurobiol., 26:1.

Millan, M.J., Millan, M.H., and Herz, A., 1983, Contribution of the supraoptic nucleus to brain and pituitary pools of immunoreactive vasopressin and particular opioid peptides, and the interrelationships between these, in the rat, Neuroendocrinology, 36:310.

Millan, M.H., Millan, M.J., and Herz, A., 1984, The hypothalamic paraventricular nucleus: relationship to brain and pituitary pools of vasopressin and oxytocin as compared to dynorphin, β-endorphin and related opioid peptides in the rat, Neuroendocrinology, 38:108.

Miselis, R.R., 1981, The efferent projections of the subfornical organ of the rat: a circumventricular organ within a neural network subserving water balance, Brain Res., 230:1.

Moos, F., and Richard, P., 1983, Serotonergic control of oxytocin release during suckling in the rat: opposite effects in conscious and anesthetized rats, Neuroendocrinology, 36:300.

Moos, F., Freund-Mercier, M.J., Guerné, Y., Guerné, J.M., Stoeckel, M.E., and Richard, Ph., 1984, Release of oxytocin and vasopressin by magnocellular nuclei in vitro: specific facilitatory effect of oxytocin on its own release, J. Endocrinol., 102:63.

Morris, B., and Livingston, A., 1983, Hormone-releasing stimuli do not alter met-enkephalin levels in the rat neurohypophysis, Life Sci., 33, Suppl. I:511.

Palkovits, M., Brownstein, M.J., and Zamir, N., 1983, Immunoreactive dynorphin and α-neo-endorphin in rat hypothalamo-neurohypophyseal system, Brain Res., 278:258.

Palkovits, M., Kiss, J.Z., Beinfeld, M.C., and Brownstein, M.J., 1984, Cholecystokinin in the hypothalamo-hypophyseal system, Brain Res., 299:186.

Pittman, Q.J., Lawrence, D., and Lederis, K., 1983, Presynaptic interactions in the neurohypophysis: endogenous modulators of release, Prog. Brain Res., 60:319.

Pitzel, L., and König, A., 1984, Lack of response in the release of oxytocin and vasopressin from isolated neurohypophyses to dopamine, met-enkephalin and leu-enkephalin, Exp. Brain Res., 56:221.

Plotsky, P.M., Bruhn, T.O. and Vale, W., 1984, Central modulation of immunoreactive corticotropin-releasing factor secretion by arginine vasopressin, Endocrinology, 115:1639.

Plotsky, P.M., Bruhn, T.O., and Otto, S., 1985, Central modulation of immunoreactive arginine vasopressin and oxytocin secretion into the hypophysial-portal circulation by corticotropin-releasing factor, Endocrinology, 116, 1669.

Poulain, D.A., and Wakerley, J.B., 1982, Electrophysiology of hypothalamic magnocellular neurones secreting oxytocin and vasopressin, Neuroscience, 7:773.

Poulain, D.A., Wakerley, J.B., Dyball, R.E.J., 1977, Electrophysiological differentiation of oxytocin- and vasopressin-secreting neurones. Proc. R. Soc. Lond. B, 196:367.

Racké, K., Ritzel, H., Trapp, B., and Muscholl, E., 1982, Dopaminergic modulation of vasopressin release from isolated neurohypophysis of the rat. Possible involvement of endogenous opioids, Naunyn-Schmiedebergs Arch. Pharmacol., 319:56.

Rehfeld, J.F., 1985, Neuronal cholecystokinin: one or multiple transmitters, J. Neurochem., 44:1.

Rivier, C., and Vale, W., 1983, Modulation of stress-induced ACTH release by corticotropin-releasing factor, catecholamines and vasopressin, Nature, 305:325.

Rivier, C., Rivier, J., Mormede, P., and Vale, W., 1984, Studies of the nature of the interaction between vasopressin and corticotropin-releasing factor or adrenocorticotropin release in the rat, Endocrinology, 115:882.

Rossier, J., Battenberg, E., Pittman, Q., Bayon, A., Koda, L., Miller, R., Guillemin, R., and Bloom, F., 1979, Hypothalamic enkephalin neurones may regulate the neurohypophysis, Nature, 277:653.

Ruppert, S., Scherer, G., and Schütz, G., 1984, Recent gene conversion involving bovine vasopressin and oxytocin precursor genes suggested by nucleotide sequence, Nature, 308:554.

Russell, J.A., 1983, Combined morphometric and immunocytochemical evidence that in the paraventricular nucleus of the rat oxytocin but not vasopressin neurones respond to the suckling stimulus, Prog. Brain Res., 60:31.

Russell, J.A., 1984, Naloxone provokes protracted secretion of oxytocin in morphine-dependent lactating rats anaesthetized with urethane, J. Physiol. (Lond.), 355:34P.

Russell, J.A., and Spears, N., 1984, Morphine inhibits suckling-induced oxytocin secretion in conscious lactating rats but also disrupts maternal behaviour, J. Physiol. (Lond.), 346:133P.

Saavedra, J.M., Rougeot, C., Culman, J., Israel, A., Niwa, M., Tonon, M.C., Dray, F., and Vaudry, H., 1984, Decrease in corticotropin-releasing factor (CRF)-like immunoreactivity in rat intermediate and posterior lobes after stalk section, Excerpta Medica (Int. Congress Series 652):1290, Elsevier, Amsterdam.

Saito, A., Sankaran, H., Goldfine, I.D., Williams, J.A., 1980, Cholecysto-kinin receptors in the brain: characterization and distribution, Science, 208:1155.

Samson, W.K., McDonald, J.K., and Lumpkin, M.D., 1985, Naloxone-induced dissociation of oxytocin and prolactin releases, Neuroendocrinology, 40:68.

Saphier, D., and Feldman, S., 1985, Electrophysiologic evidence for neural connections between the paraventricular nucleus and neurons of the supraoptic nucleus in the rat, Exp. Neurol., 89:289.

Sawchenko, P.E., and Swanson, L.W., 1985, Localization, colocalization, and plasticity of corticotropin-releasing factor immunoreactivity in rat brain, Fed. Proc. , 44:221.

Sawchenko, P.E., Swanson, L.W., and Vale, W., 1984, Co-expression of corticotropin-releasing factor and vasopressin immunoreactivity in parvocellular neurons of the adrenalectomized rat, Proc. Natl. Acad. Sci. USA, 81:1883.

Schmale, M., and Richter, D., 1984, Single base deletion in the vaso-pressin gene is the cause of diabetes insipidus in Brattleboro rats, Nature, 308:705.

Seizinger, B.R., Maysinger, D., Höllt, V., Grimm, C., and Herz, A., 1982, Concomitant neonatal development and in vitro release of dynorphin and α-neo-endorphin, Life Sci., 31:1757.

Seizinger, B.R., Höllt, V., and Herz, A., 1984, Proenkephalin B (Prodynorphin)-derived opioid peptides: evidence for a differential processing in lobes of the pituitary, Endocrinology, 115:662.

Sgro, S., Ferguson, A.V., and Renaud, L.P., 1984, Subfornical organ-supraoptic nucleus connections: an electrophysiologic study in the rat, Brain Res., 303:7.

Silverman, A.J., Hoffman, D.L., Zimmerman, E.A., 1981, The descending afferent connections of the paraventricular nucleus of the hypotha-lamus, Brain Res. Bull., 6:47.

Spinedi, E., and Negro-Vilar, A., 1984, Arginine vasopressin and ACTH release: correlation between binding characteristics and bioactivity in anterior pituitary dispersed cells, Excerpta Medica (Int. Congress Series 652):1321, Elsevier, Amsterdam.

Summerlee, A.J.S., 1981, Extracellular recordings from oxytocin neurones during the expulsive phase of birth in unanaesthetized rats, J. Physiol. (Lond.), 321:1.

Summerlee, A.J.S., and Lincoln, D.W., 1981, Electrophysiological recordings from oxytocinergic neurones during suckling in the unanaesthetized lactating rat, J. Endocrinol., 90:255.

Summy-Long, J.Y., Rosella, L.M., and Keil, L.C., 1981, Effects of centrally administered endogenous opioid peptides on drinking behavior, increased plasma vasopressin and pressor response to hypertonic sodium chloride, Brain Res., 221:343.

Summy-Long, J.Y., Keil, L.C., Sells, G., Kirby, A., Chee, O., and Severs, W.B., 1983, Cerebroventricular sites for enkephalin inhibition of the central actions of angiotensin, Am. J. Physiol., 244:R522.

Summy-Long, J.Y., Miller, D.S., Rosella-Dampman, L.M., Hartman, R.D., and Emmert, S.E., 1984, A functional role for opioid peptides in the differential secretion of vasopressin and oxytocin, Brain Res., 309:362.

Swanson, L.W., and Sawchenko, P.E., 1983, Hypothalamic integration: organization of the paraventricular and supraoptic nuclei, Ann. Rev. Neurosci., 6:269.

Swanson, L.W., Sawchenko, P.E., Rivier, J., and Vale, W.W., 1983, Organization of ovine corticotropin-releasing factor immunoreactive cells and fibers in the rat brain: an immunohistochemical study, Neuroendocrinology, 36:165.

Szligi, G.R., and Ludens, J.H., 1982, Studies on the nature and mechanism of the diuretic action of the opioid analgesic ethylketocyclazocine, J. Pharmacol. Exp. Ther., 220:585.

Theodosis, D.T., 1985, Oxytocin-immunoreactive terminals synapse on oxytocin neurones in the supraoptic nucleus, Nature, 313:682.

Theodosis, D.T., and Poulain, D.A., 1983, Evidence for structural plasticity in the supraoptic nucleus of the rat hypothalamus in relation to gestation and lactation, Neuroscience, 11:183.

Theodosis, D.T., Chapman, D., Montagnese, C., Morris, J., Poulain, D.A., and Vincent, J.D., 1984, Structural plasticity in the hypothalamic supraoptic nucleus involves mainly oxytocin-secreting neurons, J. Steroid Biochem., 20(6B):1499.

Thomson, A.M., 1984, Supraoptic neurons sustain high frequency firing when extracellular Ca^{2+} is replaced with other divalent cations in rat brain slices, Neuroscience, 12:495.

Tramu, G., Croix, C., and Villez, A., 1983, Ability of the CRF immunoreactive neurons of the paraventricular nucleus to produce a vasopressin-like material, Neuroendocrinology, 37:467.

Tweedle, C.D., 1983, Ultrastructural manifestations of increased hormone release in the neurohypophysis, Prog. Brain Res., 60:259.

Tweedle, C.D., and Hatton, G.I., 1984, Synapse formation and disappearance in adult rat supraoptic nucleus during different hydration states, Brain Res., 309:373.

Vale, W., Rivier, C., Brown, M.R., Spiess, J., Koob, G., Swanson, L., Bilezikjian, L., Bloom, F., and Rivier, J., 1983, Chemical and biological characterization of corticotropin releasing factor., Rec. Prog. Horm. Res., 39:245.

Vanderhaeghen, J.J., Lotstra, F., Vandesande, F., and Dierickx, K., 1981, Coexistence of cholecystokinin and oxytocin-neurophysin in some magnocellular hypothalamo-hypophyseal neurons, Cell Tissue Res., 221:227.

Vanderhaeghen, J.J., Lotstra, F., Liston, D.R., and Rossier, J., 1983, Proenkephalin, [Met]enkephalin, and oxytocin immunoreactivities are colocalized in bovine hypothalamic magnocellular neurones, Proc. Natl. Acad. Sci. USA, 80:5139.

Vandesande, F., Dierickx, K., and De Mey, J., 1977, The origin of the vasopressinergic and oxytocinergic fibres of the external region of the median eminence of the rat hypophysis, Cell Tissue Res., 180:443.

Van Leeuwen, F.W., Pool, C.W., and Sluiter, A.A., 1983, Enkephalin immunoreactivity in synaptoid elements on glial cells in the rat neural lobe, Neuroscience, 8:229.

Wardlaw, S.L., Wehrenberg, W.B., Ferin, M., Antunes, J.L., and Frantz, A.G., 1982, Effect of sex steroids on β-endorphin in hypophyseal portal blood, J. clin. Endocrinol Metab., 55:877.

Watkins, L.R., Kinscheck, I.B., and Mayer, D.J., 1985, Prolongation of morphine analgesia by the cholecystokinin antagonist proglumide (BRE 10528), Brain Res., 327:169.

Watson, S.J., Akil, H., Fischli, W., Goldstein, A., Zimmerman, E.A., Nilaver, G., and Van Wimersma Greidanus, Tj.B., 1982, Dynorphin and vasopressin: common localization in magnocellular neurons, Science, 216:85.

Weber, E., and Barchas, J.D., 1983, Immunohistochemical distribution of dynorphin B in rat brain: relation of dynorphin A and α-neo-endorphin systems, Proc. Natl. Acad. Sci. USA, 80:1125.

Weber, E., Roth, K.A., Evans, C.J., Chang, J-K., and Barchas, J.D., 1983a, Immunohistochemical localization of dynorphin (1-8) in hypothalamic magnocellular neurons: evidence for absence of proenkephalin, Life Sci., 31:1761.

Weber, E., Geis, R., Voigt, K.H., and Barchas, J.D., 1983b, Levels of
pro-neo-endorphin/dynorphin-derived peptides in the hypothalamo-
posterior pituitary system of male and female Brattleboro rats,
Brain Res., 260:166.

Wehrenberg, W.B., Wardlaw, S.L., Frantz, A.G., and Ferin, M., 1982,
β-endorphin in hypophyseal portal blood; variations throughout the
menstrual cycle, Endocrinology, 111:879.

Wennogle, L.P., Steel, D.J., and Petrack, B., 1985, Characterization of
central cholecystokinin receptors using a radioiodinated octapeptide
probe, Life Sci., 36:1485.

Whitnall, M.H., Mezey, E., and Gainer, H., 1985, Co-localization of
corticotropin-releasing factor and vasopressin in median eminence
neurosecretory vesicles, Nature, 317:248.

Williams, T.D.M., Carter, D.A., and Lightman, S.L., 1985, Sexual dimorphism
in the posterior pituitary response to stress in the rat,
Endocrinology, 116:738.

Wolfson, B., Manning, R.W., Davis, L.G., Arentzen, R., and Baldino, F. Jr.,
1985, Co-localization of corticotropin releasing factor and vasopressin mRNA in neurones after adrenalectomy, Nature, 315:59.

Yamashita, H., Inenaga, K., Kawata, M., and Sano, Y., 1983, Phasically
firing neurons in the supraoptic nucleus of the rat hypothalamus:
immunocytochemical and electrophysiological studies, Neurosci. Lett.,
37:87.

Zamir, N., Zamir, D., Eiden, L.E., Palkovits, M., Brownstein, M.J.,
Eskay, R.L., Weber, E., Faden, A.I., and Feuerstein, G., 1985,
Methionine and leucine enkephalin in rat neurohypophysis: different
responses to osmotic stimuli and T_2 toxin, Science, 228:606.

Zimmerman, E.A., and Silverman, A.J., 1983, Vasopressin and adrenal
cortical interactions, Prog. Brain Res., 60:493.

Zukin, R.S., and Zukin, S.R., 1984, The case for multiple opiate receptors,
Trends in Neurosciences, 7(5):160.

ORGANIZATION AND EXPRESSION OF THE VASOPRESSIN
AND OXYTOCIN GENES

Dietmar Richter and Richard Ivell

Institut fuer Zellbiochemie und
Klinische Neurobiologie, Universitaet Hamburg
Martinistr. 52, 2 Hamburg 20, FRG

INTRODUCTION

Vasopressin is known to control water resorption in the distal kidney tubuli, oxytocin milk ejection and the contraction of the uterus during birth. In addition the two hormones may modulate other processes in the body (de Wied, 1983).

The precursors to the hormones vasopressin and oxytocin are typical examples of cellular polyproteins being composed of the respective hormone, and its functionally linked carrier, called neurophysin. Only the vasopressin precursor includes a third moiety, a glycoprotein at its C-terminus. Both hormones are synthesized in the magnocellular neurones of the hypothalamus and transported axonally to the posterior pituitary. During transport, maturation of the precursors into the biologically active peptides is assumed to take place.

GENE STRUCTURE AND REGULATION

So far the oxytocin and vasopressin genes have been determined from rat (Schmale et al., 1983; Ivell and Richter, 1984a) and cow (Ruppert et al., 1984). Each gene comprises three exons separated by two intervening sequences (Fig. 1). Each exon encodes one of the principal functional domains of the polyprotein - hormone, carrier protein, glycoprotein; in the case of the oxytocin gene the third exon comprises only the C terminal variable part of the neurophysin.

Regulation of the two genes has been studied by quantifying the mRNA encoding the hormone precursor by liquid or blot hybridization assays. Rats placed under osmotic stress by drinking 2% saline respond with an elevated plasma osmolality. To maintain the water balance the stores of vasopressin in the posterior pituitary are depleted while the hormone level in the plasma rises (Robertson, 1977). In this situation there is a significant increase of the mRNA encoding vasopressin (Burbach et al., 1984). Interestingly, the increase is greatest in the nucleus supraopticus, less in the nucleus

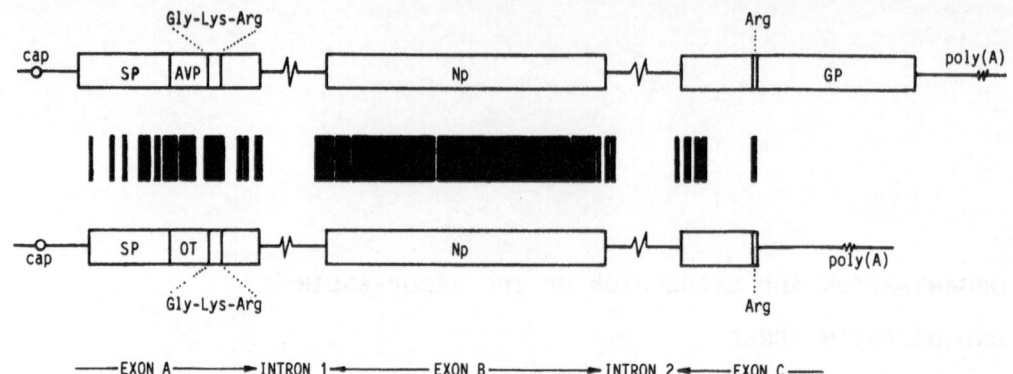

Figure 1. Comparison of the Vasopressin and Oxytocin Genes from Rat.

paraventricularis and absent in the suprachiasmatic nucleus. It is known that vasopressin-producing neurones of the SON and PVN project towards the posterior pituitary, those of the SCN towards other brain areas. Hence, it has been speculated that these data point to a specific regulation in the expression of the vasopressin gene in different hypothalamic regions (Burbach et al., 1984).

Whether the different levels of vasopressin-encoding mRNA in the three hypothalamic areas assayed reflect indeed differential regulation of the vasopressin gene has to be examined more rigorously, for instance by transcription run-on experiments measuring the levels of nascent vasopressin-mRNA in the cell nucleus. Also other physiological parameters are needed which may affect specifically the vasopressin mRNA level in the SCN but not in the other two areas.

A way to test the expression of the hormones more directly makes use of thin layer sections of the hypothalamus which are then hybridized in situ to specific radiolabeled DNA probes. This method combines a number of advantages; it allows not only the cells expressing the respective gene to be identified but also the mRNA in these cells to be quantified as well as a study of the effect of endogenously induced factors on the anatomy and number of cells expressing the respective gene. These experiments are outlined in greater detail by the collaborative studies presented by J. McCabe. (McCabe et al., this volume).

EXPRESSION OF VASOPRESSIN AND OXYTOCIN IN PERIPHERAL ORGANS

The refinement of antibody detection techniques has recently enabled vasopressin- and oxytocin-like immunoreactive material also to be identified in several non-neural, steroidogenic tissues such as the adrenal gland, ovary, testis and placenta. Because the bovine corpus luteum is particularly rich in oxytocin, most information is available for this tissue. For the remaining organs, the apparent levels of the neurophyseal hormones are so low as to preclude at this time detailed physiological and chemical studies.

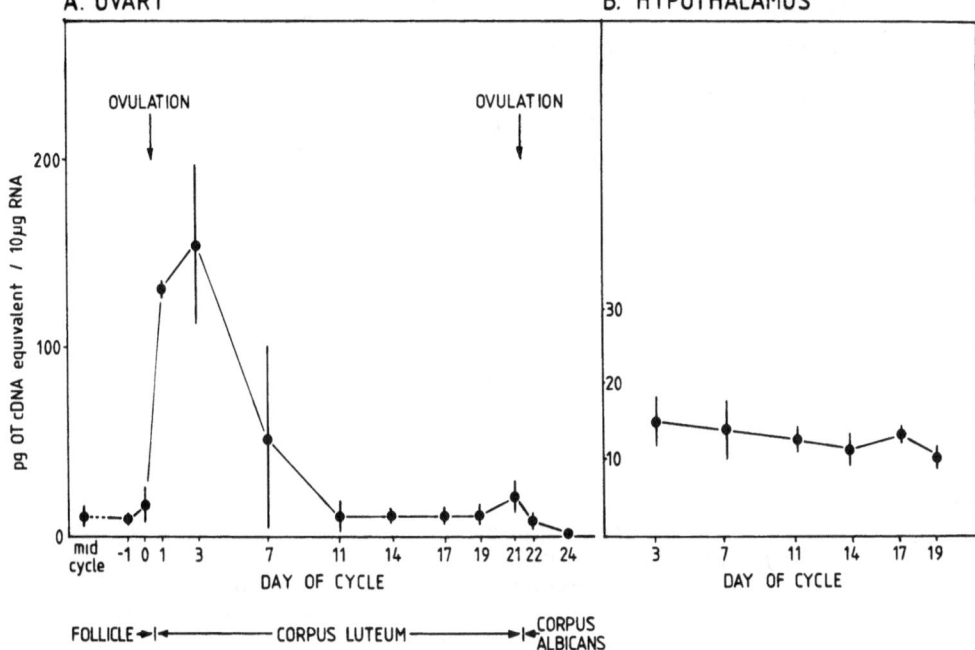

Figure 2. mRNA from Ovary and Hypothalamus Encoding the Oxytocin Precursor Taken from ref. Ivell et al., 1985 in press.

In the bovine corpus luteum it could be shown using a combination of DNA/RNA hybridization, cDNA cloning and in vivo labelling, that the single bovine oxytocin gene is expressed in this tissue via a similar mRNA and precursor polyprotein as in the hypothalamus (Ivell & Richter, 1984b; Swann et al., 1984). Significantly, though, the luteal oxytocin mRNA is shorter by some 60 adenosine residues of the polyadenylated 3' terminus. The meaning of this tissue-specific differential polyadenylation is not clear: in a cell-free system both hypothalamic and luteal oxytocin mRNAs appear to be equally translatable (Ivell & Richter, 1984b).

DNA/RNA hybridization analysis of total RNA extracted quantitatively from follicles and corpora lutea of non-pregnant cows shows that although the oxytocin gene is expressed already in the follicle, slightly above background level, a dramatic increase in transcription occurs accompanying ovulation and the luteinizing of the granulosa cells to form the corpus luteum (Fig. 2; Ivell et al., 1985a). This increase is such that total levels of oxytocin mRNA in a single corpus luteum may exceed those in the hypothalamus by a hundredfold. Transcription of the oxytocin gene evidently stops early in the cycle, such that oxytocin mRNA levels decline rapidly at mid-cycle and are very low towards luteolysis. Oxytocin mRNA in the pregnant corpus luteum, in contrast, is virtually undetectable in spite of maintained progesterone production.

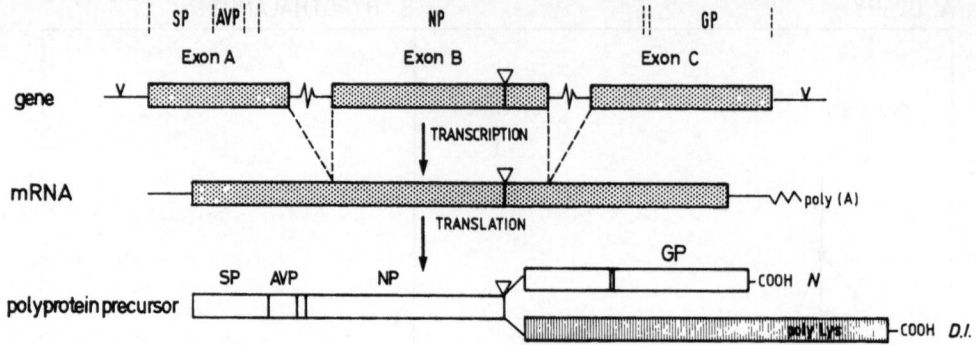

Figure 3. Comparison of the Vasopressin Gene Precursor from Wild-type (N) and Diabetes Insipidus (D.I.) Rats Deduced from their Gene Structures. For details see ref. Schmale and Richter, 1984.

Although the ovarian peptide appears to be released mostly in the mid- to luteolytic phase of the cycle, it is still quite unclear what function this luteal oxytocin is having. Parallel analyses of the bovine hypothalamus through the oestrous cycle show no evidence of a correlation, pointing to the independent regulation of luteal and hypothalamic oxytocin expression (Ivell et al., 1985a).

HEREDITARY HYPOTHALAMIC DIABETES INSIPIDUS

Hypothalamic disorder in the production of vasopressin leads to excessive thirst and to the excretion of large amounts of a very dilute urine. The defect can be reversed by applying arginine vasopressin or its analogues. In man, X-linked and autosomally dominant forms of diabetes insipidus exist (Green et al., 1967). An appropriate model for studying the genetic defect is the so-called Brattleboro rat with an autosomally recessive form of diabetes insipidus (Sokol and Valtin, 1982). These animals lack vasopressin as well as its neurophysin while oxytocin and its associated carrier are not affected.

Figure 4. (A) Antibodies Raised Against Vasopressin or the CP-14 of the Vasopressin-Like Precursor from Brattleboro Rats. The arrow head points to the deletion site. SP, signal peptide; AVP, arginine vasopressin; NP, neurophysin; GP, glycoprotein; shaded bars, peptides used for immunization. (B) Immunohistochemical Staining of the Supraoptic Nucleus of Hypothalami from Brattleboro (a-c) and Normal Rats (d). Staining procedure: peroxidase-anti peroxidase method using either antisera raised against the C-peptide of the mutated vasopressin precursor (a-c) or against vasopressin (d). In c the antibodies were inactivated by preincubation with the antigen. a, x 220; b, x 560; c, x 340; d, x 220; OT, optic tract (Richter et al., 1985).

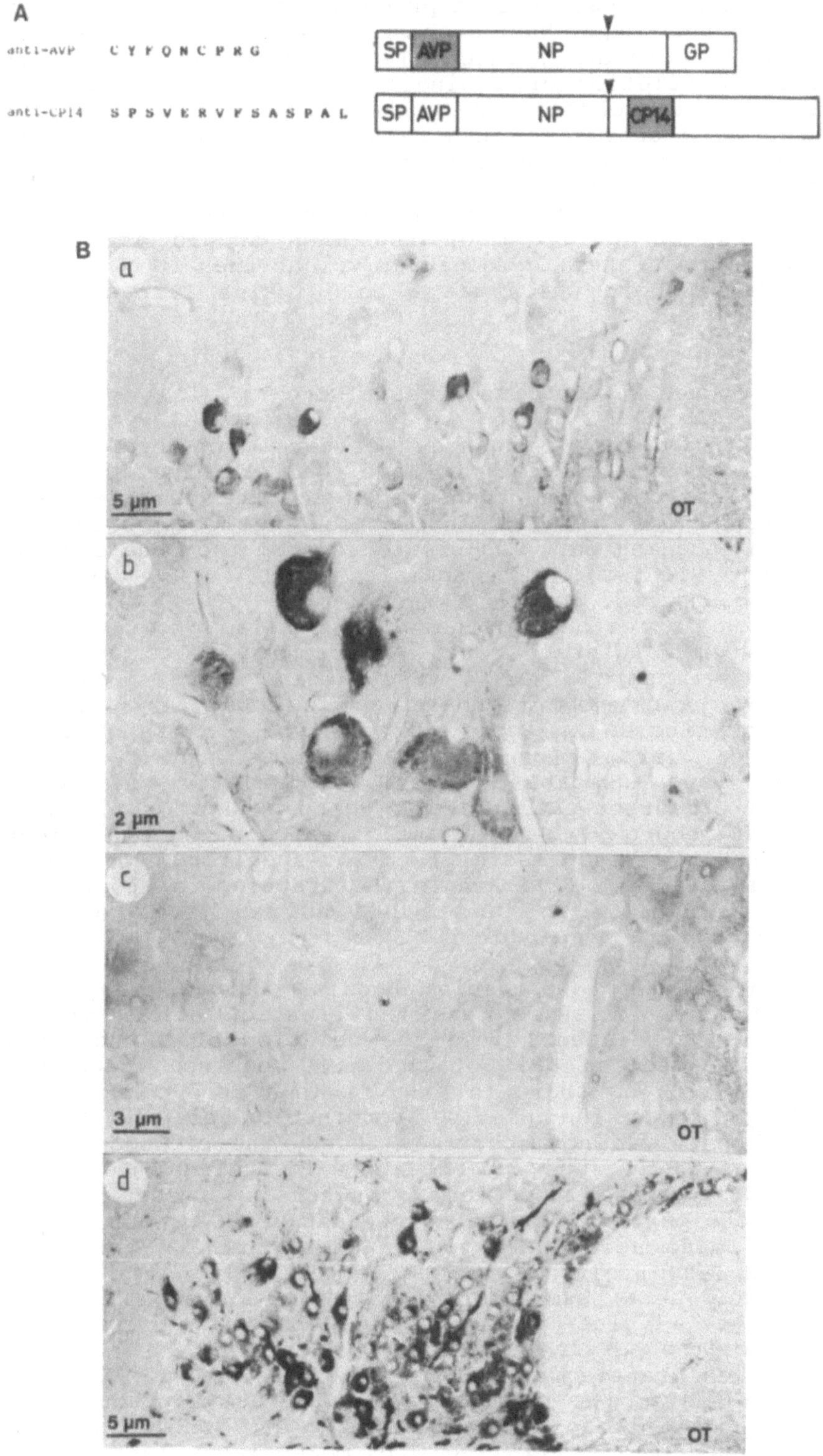

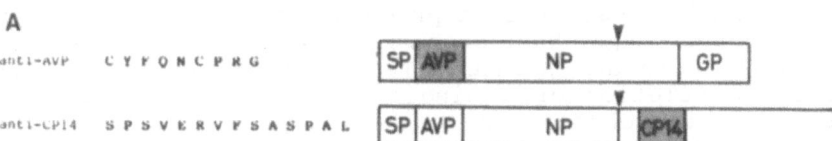

A

anti-AVP   C Y F Q N C P R G

anti-CP14   S P S V E R V F S A S P A L

B

a

5 µm   OT

b

2 µm

c

3 µm   OT

d

5 µm   OT

The gene for the vasopressin precursor from the mutant rats as well as the mRNA has been isolated and its sequence determined (Schmale and Richter, 1984; Schmale et al., 1984). It contains a deletion of a single G residue in the part of the molecule which encodes the conserved part of neurophysin. This mutation gives rise to an open reading frame predicting a hormone precursor with a different C terminus having lost its glycosylation site (Fig. 3). Because of the open reading frame there is no stop codon for terminating the translation process. Hence translation of the mRNA should lead to a product which is higher in molecular weight than the precursor from normal rats and which has a poly lysine tail at the C terminus.

Expression of the Mutated Vasopressin Gene

Transcription and mRNA Sequence. Northern blot analysis and cloning of hypothalamic cDNA have demonstrated that the mutated gene is correctly transcribed and that the resulting mRNA contains a single nucleotide deletion in the neurophysin-encoding part, as predicted from the gene sequence (Schmale and Richter, 1984; Schmale et al., 1984; Ivell et al., 1985, manuscript in preparation).

Inspection of Northern blots of hypothalamic poly(A)+ RNA from wild-type (wistar) and Brattleboro rats suggests comparable amounts of vasopressin and oxytocin precursor-specific mRNA (Schmale and Richter, 1984). More quantitative studies by dot blot analysis of hypothalamic poly(A)+ RNA from Long-Evans, Wistar and homozygous Brattleboro rats indicate the following relative amounts of vasopressin precursor-specific mRNA: Long-Evans 100%, Wistar 66% and Brattleboro 27%. Information from cloning experiments suggests that the vasopressin precursor-specific mRNA amounts to 0.04% of total poly(A)+ RNA in the Brattleboro, to 0.1% in the Wistar and 0.15% in the Long-Evans rat (Ivell et al., 1985, manuscript in preparation).

Translation. Though vasopressin encoding mRNA is present in hypothalami of Brattleboro rats the question arises whether this mRNA is translated. Since the deletion of a nucleotide residue in the second half of the mRNA sequence leads to a shift in the reading frame in a way that a stop codon signal is no longer read the precise termination of the nascent peptide chain should not occur. In theory the ribosome should read through the 3'-end of the mRNA including the poly(A) sequence giving rise to a larger precursor product. If one calculates the maximal size of such a potential precursor than roughly 70 amino acids have to be added to the normal sized protein (MW 19000) giving rise to a molecular weight between 26 and 28000. The C terminus of this precursor should consist of ca. fifty lysine residues corresponding to the 150 adenosines of the poly (A) tail.

Indeed when Brattleboro hypothalamic mRNA is translated in a cell-free system the normal-sized vasopressin precursor is not observed, instead small amounts of vasopressin-like precursors have been identified with molecular weights in the range from 19000 to 26000 (Ivell et al., 1985, manuscript in preparation). The heterogeneously-sized translation products may reflect how far these ribosomes have been able to translate the mRNA. The products have been identified by using antisera raised either against vasopressin, neurophysin or a 14mer peptide (CP-14) predicted from the new reading frame and specific for the mutated Brattleboro vasopressin precursor (Schmale et al., 1984).

Immunocytochemical Studies. Using antibodies against the 14mer peptide (CP-14 antibodies) in immunocytochemical studies at the light microscopic level, it can be demonstrated that the enlarged cells of the supraoptic nucleus can be specifically stained in Brattleboro but not in wild-type rat hypothalami (Richter et al., 1985; Guldenaar et al., manuscript in preparation)(Fig.4). Typically the stain is located in the vicinity of the Nissl bodies. Staining is restricted to the cell body and cannot be found in axons or in the pituitary. Very few cells are stained in the paraventricular nucleus, none in the suprachiasmatic nucleus. Electron microscopic studies show that this stain is situated in the rough endoplasmic reticulum, in the inner cisternae of the Golgi apparatus and in small lysosomes (Krisch et al., manuscript in preparation). Since the mutated vasopressin precursor is highly charged at its C terminus (ca. 50 lysine residues!) it is not too surprising that most of the synthesized material is fixed at the membrane.

CONCLUSION

The Brattleboro rat represents the rare case where a single nucleotide deletion leads to an open reading frame predicting a protein precursor which in theory should end in a poly lysine tail at its C terminus. Another case with a frameshift mutation has been reported for hemoglobin Wayne which is an alpha chain variant and which has been explained by assuming a minus one frameshift. This leads to the extension of the hemoglobin chain by five amino acids to be followed then by a stop codon (Seid-Akhavan et al., 1976). In that example only a few nucleotide residues of the 3' untranslated region and none of the poly(A) sequence are copied into protein.

Though the vasopressin gene is expressed in the hypothalamus of Brattleboro rats, relatively few cells in the supraoptic nucleus can be specifically stained with antibodies. This could be due to the modified precursor structure which because of its highly charged C terminus might be folded in a way not easily accessible to the antibodies used. The reduced vasopressin precursor synthesis however may be limited just because of the poly(A) sequence exhausting the supply of lysine, its tRNA or the corresponding aa-tRNA synthetase.

Alternatively translation of mRNA and packaging of the synthesized product into neurosecretory granules may be a more coordinated and feed-back controlled process than generally expected. The electron microscopic studies so far show that most of the mutated vasopressin precursor is found in the rough endoplasmic reticulum, in terminal Golgi vesicles and in lysosomes indicating that the 'false' precursor is recognized and discarded in an unknown way by the cell (Krisch et al., manuscript in preparation).

References

Burbach, J.P.H., De Hoop, M.J., Schmale, H., Richter, D., De Kloet, E. R., Ten Haar, J. A. & De Wied, D. J. Neuroendocrinology 39, 582-584 (1984).

De Wied, D. Progress in Brain Res. 60, 155-167 (1983).

Green, J.R., Buchan, G.C., Alvord, E.C., jr. & Swanson, A.G. Brain 90, 707-714 (1967).

Ivell, R. & Richter, D. Proc. Natl. Acad. Sci. USA 81, 2006-2010 (1984a).

Ivell, R. & Richter, D. EMBO J. 3, 2351-2354 (1984b).

Ivell, R., Brackett, K.H., Fields, M.J. & Richter, D. FEBS Letters, (1985) in press.

Richter, D., Ivell, R., Schmale, H., Nahke, P. & Krisch, B. in : Selected Topics of Neurobiochemistry (eds. Hamprecht, B. & Neuhoff, V.) Springer Verlag, Heidelberg, 1985 in press.

Robertson, G.L. Recent Progress Horm. Res. 33, 333-385 (1977).

Ruppert, D., Scherer, G. & Schutz, G. Nature 308, 554-557 (1984).

Schmale, H., Heinsohn, S. & Richter, D. EMBO J. 2, 763-767 (1983).

Schmale, H., Ivell, R., Breindl, M., Darmer, D. & Richter, D. EMBO J. 3, 3289-3293 (1984).

Schmale, H. & Richter, D. Nature 308, 705-709 (1984).

Seid-Akhavan, M., Winter, W.P., Abramson, R.K. & Rucknagel, D.L. Proc. Natl. Acad. Sci. USA 73, 882-886 (1976).

Sokol, H.W. & Valtin, H. (eds.) The Brattleboro Rat. Ann. N.Y. Acad. Sci. 394, 1-828 (1982).

Swann, R.W., O'Shaughnessy, P.J., Birkett, S.D., Wathes, D.C., Porter, D.G. & Pickering, B.T. FEBS Letters 174, 262-266 (1984).

MEASUREMENT OF EXPRESSION OF THE VASOPRESSIN AND OXYTOCIN GENES IN SINGLE

NEURONS BY IN SITU HYBRIDIZATION

Joseph T. McCabe, Joan I. Morrell, and Donald W. Pfaff

Laboratory of Neurobiology & Behavior
The Rockefeller University
New York, New York 10021, U.S.A.

Methods required for optimization of the in situ hybridization technique vary according to the message, the probe, the tissue used, and the regulatory question asked. There are independent technical dimensions which must be considered for discovering how to optimize each methodological step for a given experiment. Here, we offer a general outline of technical parameters that should be considered to optimize signal. This review is offered as an overview of relevant procedural principles in light of the extensive parametric reports of in situ hybridization (Bauman, et al., 1984; Brahic & Haase, 1978; Cox, et al., 1984; Gee & Roberts, 1983; Godard & Jones, 1979; Haase, et al., 1984; Henderson, 1982; Lawrence & Singer, 1985; McCabe, et al., 1985b,c; Pardue & Gall, 1975; Schachter, et al., 1985). Various investigators often report seemingly contradictory findings: these may have their basis in what tissue and probe are under study.

In the latter section of this chapter, we discuss some of our results and our attempts to develop in situ hybridization as a means to monitor gene expression quantitatively at the single-cell level. Based upon our observations that particular magnocellular neurons within the neurohypophyseal system appear to be intensely responsive to physiological challenges, we propose that in situ hybridization may provide a method for quantifying differential changes in gene expression within a population of neurons, all of which synthesize a specific protein. These observations would be in congruity with neuroanatomical and neurophysiological observations demonstrating neuronal subsystems and the complex "microheterogeneity" of the nervous system (Gilbert, 1983; Morrell, et al., 1985; Rakic, 1976).

TECHNICAL DIMENSIONS OF IN SITU HYBRIDIZATION

The ultimate goal of in situ hybridization is to obtain quality tissue sections; good cellular morphology with labelled cells surrounded by unlabelled cells and neuropil. To achieve this, one must consider the abundance of a specific mRNA/cell, the probe's accessibility to the message, the possible loss of message by diffusion or by RNase-caused degradation, preservation of tissue morphology, and minimization of background noise.

Abundance of Message

At present, _in situ_ hybridization is most easily performed in cases
where mRNA concentration is at least moderately abundant (> 10 copies/
cell). For example, earlier studies were concerned with abundant messages
such as satellite sequences, ribosomal nucleic acids, and poly-A+ RNA
(Angerer & Angerer, 1981; Gall & Pardue, 1969; McCabe, et al., 1985a;
Pardue & Gall, 1970). In these cases technical success was clear.
Studies with relatively abundant brain messenger RNAs that encode oxytocin,
proopiomelanocortin, somatostatin, and vasopressin are also reported
(e.g., Gee, et al., 1983; McCabe, et al., 1985b,c; Nojiri, et al., 1985;
Sherman, et al., 1984; Uhl, et al., 1985; Wolfson, et al., 1985). For
rarer messages, in terms of copies per cell, careful consideration must
be given to how the tissue is processed. Optimal tissue fixation will
help in this instance; however one must not fix the tissue to the extent
that protein cross-linking by the fixative masks what few hybridization
sites, that is mRNAs, are available. In addition to optimal fixation
procedures, one must also apply sufficient concentration of the probe to
ensure that the hybridization reaction occurs under saturation conditions
to allow all available mRNAs to be hybridized. Hybridization conditions
with high probe concentration, in turn, then require sufficiently
stringent hybridization and washing conditions so that background does not
render the signal undetectable. The use of single-stranded probes can
improve the situation when copy number is low since the complementary
strand is not available as a competitor which permits the probe to
re-anneal with itself. With double-stranded probes, even under saturation
conditions, it has been shown that the percentage of hybrids formed is
only 10% the amount of message estimated to be available in a tissue
(Cox, et al., 1984). If message levels are extremely low (0.1-10 copies/
cell), presently available techniques may not enable detection.

Probe Access

A second consideration for improving the _in situ_ technique is whether
the probe has sufficient access to the RNA of interest. Steps must be
taken to permeabilize the tissue to allow the probe to diffuse into the
section and anneal with the message. In addition to increasing penetration,
these steps also must remove some of the proteins and cytoskeletal frame-
work surrounding the message so that sufficient lengths of the message are
exposed to allow the probe to hybridize. Procedures at this step usually
entail denaturation by exposure to alkalai, acidic, or high-temperature
solutions, and deproteination steps such as acidic treatments or treatment
with enzymes such as pepsin or proteinase-K (Gee & Roberts, 1983; Haase,
et al., 1985; McCabe, et al., 1985b). These treatments may be in direct
opposition to the fixation procedures discussed above, which were
conducted to ensure mRNA is not diffusing out of the tissue. Systematic
experimentation with fixation and denaturation/deproteination parameters
will determine the best set of conditions for a particular tissue and
message.

Message Degradation by RNase

A third consideration when hybridization results are weaker than
expected is whether endogenous RNA has been degraded as a result of RNase
activity in the tissue, and/or from RNase due to the experimental
procedure. Endogenous as well as exogenous contamination with RNase will
potentially reduce substrate message, and this will be of crucial importance in neural tissue where many investigations involve low message
levels. Conditions that incorporate compounds to reduce RNase activity
(dithiothreitol, diethylpyrocarbonate, RNasin, vanadyl-ribonucleoside

complexes) can improve signal strength. In addition, wearing gloves when handling tissue, using RNase-free reagents, and sterile solutions and glassware will reduce RNase contamination.

Tissue Morphology

Throughout these procedures, it is necessary to maintain adequate tissue and cellular morphology. If tissue quality is compromised by deproteination/denaturation steps, this procedure must be shortened, or at least counterbalanced by a preceding stronger fixation procedure. In general, tissue from animals that were perfused with a fixative ensures excellent morphology, but this may produce a reduced signal in comparison to post-fixed tissue (McCabe, et al., 1985b,c).

Problems with Background

Instances may arise where background is so high that signal cannot be detected. This may result mostly from the probe binding non-specifically to tissue proteins and/or nucleic acids. In this case more stringent salt washes after hybridization may reduce non-specific annealing. In addition, including excess heterologous nucleic acids in the hybridization buffer (such as t-RNA, salmon sperm DNA, yeast total RNA) will improve specificity of results because these components act to reduce the nonspecific binding of probe to tissue proteins and heterologous nucleic acids.

Tissue artifact can also arise from autoradiographic procedures (Rogers, 1973). For example, background, which will not increase in severity as exposure time is lengthened, can result from warming the photographic emulsion to such high temperatures that grains are reduced. Also, high background within the emulsion above the tissue section may result from positive chemography: an interaction between reagents deposited in the tissue during processing which subsequently induce the reduction of silver grains in the emulsion. Finally, one may witness a loss of signal in some tissue sections if emulsion-coated slides are exposed to moisture during storage. Humidity can cause the latent image to fade.

Controls

All of these potential difficulties emphasize the need for several control conditions. First, some sections should be prepared by the standard procedure except that no radiolabelled probe should be applied. Second, one can treat the tissue with RNase to eliminate endogenous ribonucleic acids before applying the probe to demonstrate that the probe is annealing to endogenous ribonucleic acids. Third, if single-stranded probes are used, excess "cold" complementary-strand nucleic acids (analogous to a pre-absorption step in immunocytochemistry) can be added to the hybridization buffer before applying the probe to the tissue. Fourth, for quantitative experiments, one should conduct hybridization with several different probe concentrations to determine saturation. Fifth, a heterologous radiolabelled probe or a probe to another specific nucleic acid not present in the cells of interest can be tested to demonstrate specificity. Sixth, appropriate cellular specificity can be verified by conducting in situ hybridization in conjunction with immuno-cytochemistry to localize a particular mRNA and its protein product (Brahic, et al., 1984; Griffin, et al., 1983; Shivers, et al., 1986). Finally, one can conduct hybridizations and the post-hybridization rinses at increasingly higher temperatures and lower salt rinses to also demonstrate specificity.

METHODS

The in situ hybridization procedure has been outlined in detail
elsewhere (McCabe, et al., 1985b,c). Briefly, rat brain tissue was
rapidly removed from the skull, blocked, and frozen in liquid nitrogen.
Cryostat-cut sections were mounted on chrome-alum subbed slides and dried
on a slide warmer (39°C). The mounted tissue was then fixed with ethanol:
acetic acid, denatured at 70°C in acidic buffer, deproteinated with pepsin,
washed, and dried. Prehybridization buffer was applied to each section
for 2 hours in order to reduce non-specific binding of the probe. The
prehybridization buffer was then removed from the section and hybridization
buffer containing tritium- or ^{35}S-labelled probes specific for oxytocin or
vasopressin (Ivell & Richter, 1984; Schmale, et al., 1983) were applied to
the section and allowed to hybridize overnight. Sections were then
extensively rinsed, dehydrated, and coated with autoradiographic emulsion.
The slides were stored in dry, light-tight containers for various exposure
periods and were processed by standard autoradiographic development
procedures, counterstained with cresyl-violet, coverslipped, and analyzed
with the aid of a microscope equipped with a camera lucida. The number
of grains per cell were counted by eye and cellular areas were determined
using a digitizing pad and the BioquantII computer program. Data were
summarized both in terms of grains per cell and grains per unit cell area.

RESULTS

Results from two studies currently in progress are summarized. We
have examined how daily ingestion of 2% salt-water for 14 days affects
magnocellular vasopressin (VP) mRNA concentration, and second, whether
pregnancy alters cellular oxytocin (OT) mRNA concentration. Figure 1
shows labelled magnocellular neurons in the hypothalamic paraventricular
nucleus after in situ hybridization with an OT probe.

Vasopressin

Figure 2 summarizes grain-counts over individual hypothalamic magnocellular neurons in the supraoptic nucleus. The sampled sections are
from two animals processed at the same time. The sections received
identical amounts of VP probe. (See McCabe, et al. (1985b) for details:
VP and OT probes were generously donated by Dr. Richter and colleagues
(cf., Ivell & Richter, 1984; Schmale, et al. 1983).)

Comparison of the distributions of grains/cell from a control rat
(upper left panel), maintained on tap water, to a salt-loaded rat (lower
left panel) shows that VP mRNA levels were significantly increased in the
rat maintained on salt-water, with the mean grains/cell increased by 2.7
fold (means provided in figures). As can be seen in Figure 2 (lower
right), salt-treatment greatly increased the upper tail of the histogram,
indicating the increased proportion of highly-responsive cells
(Kolmogorov-Smirnov (K-S) test, $p < 0.001$). The salt treatment also significantly increased cell area. The upper and lower panels on the right
show grains per unit area for the control and the salt-loaded rat,
respectively. Statistical tests indicate the average grains/cell area is
significantly greater in the salt-loaded rat and again it is evident that
salt treatment increased the upper limits of grain distribution (K-S test;
$p < 0.001$). This finding concurs with reports of increased incorporation
of tritium-labelled cytidine in supraoptic nucleus magnocellular neurons
following salt-loading (George, 1973), and with work from others that
VP mRNA levels are increased when rats chronically drink salt-water
(Burbach, et al., 1984; Majzoub, et al., 1983; McCabe, et al., 1985a;
Sherman & McKelvy, 1984).

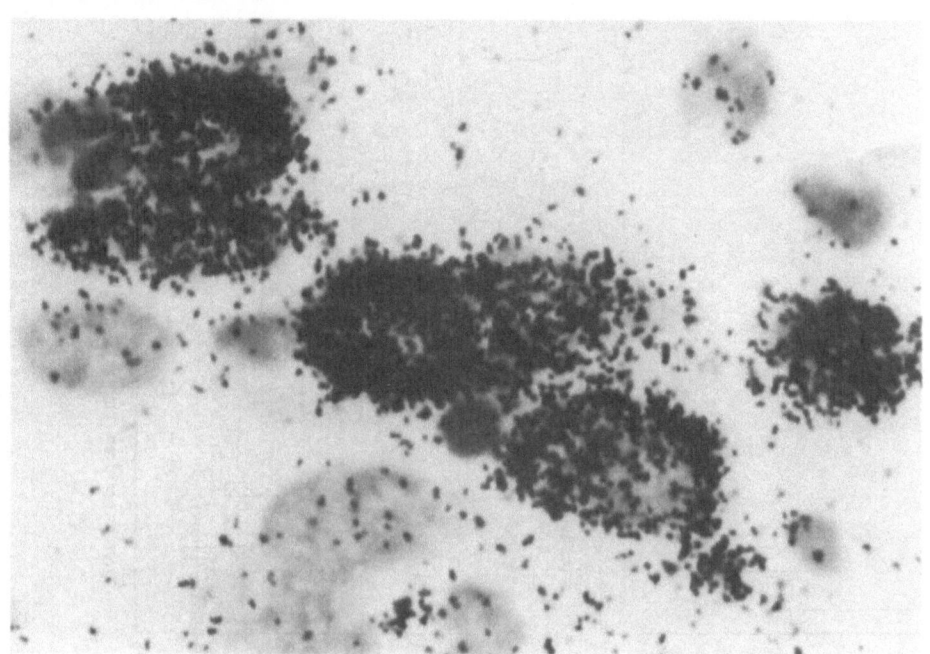

Figure 1. Photomicrograph of magnocellular neurons in
the hypothalamic paraventricular nucleus of the rat.
In situ hybridization was performed as described
(McCabe, et al., 1985b,c), in this case with an oxytocin-
specific probe that was prepared by nick-translation
with tritium-labelled nucleotides. Heavily-labelled
, oxytocinergic cells are observed and are surrounded by
unlabelled, presumably vasopressinergic, cells with few
overlying grains. Note that in cases where the tissue
section passed through the nucleus of an oxytocin cell,
the predicted cytoplasmic location of autoradiographic
grains are clearly observed.

Oxytocin

   A second study examines how pregnancy affects OT mRNA levels in
supraoptic nucleus OT-containing neurons. These results compare non-
pregnant, hormonally-intact females and 15-day pregnant females. The
upper and lower panels on the left in Figure 3 depict the grains/cell
distributions for a non-pregnant and a pregnant rat, respectively. The
histogram summarizing grains/cell from the pregnant rat (lower left
panel) shows a significant shift in the distribution toward the right
(K-S test; p<0.001). Likewise, pregnancy increased cellular area (data
not shown), and OT mRNA levels, expressed as grains/area (right panels),
significantly increased in the pregnant rat (lower right panel) compared
to the non-pregnant rat (upper right panel). Pregnancy also caused a
significant increase in the number of highly-responsive cells (K-S test;
p<0.001). Note that the grain density range was much higher with the

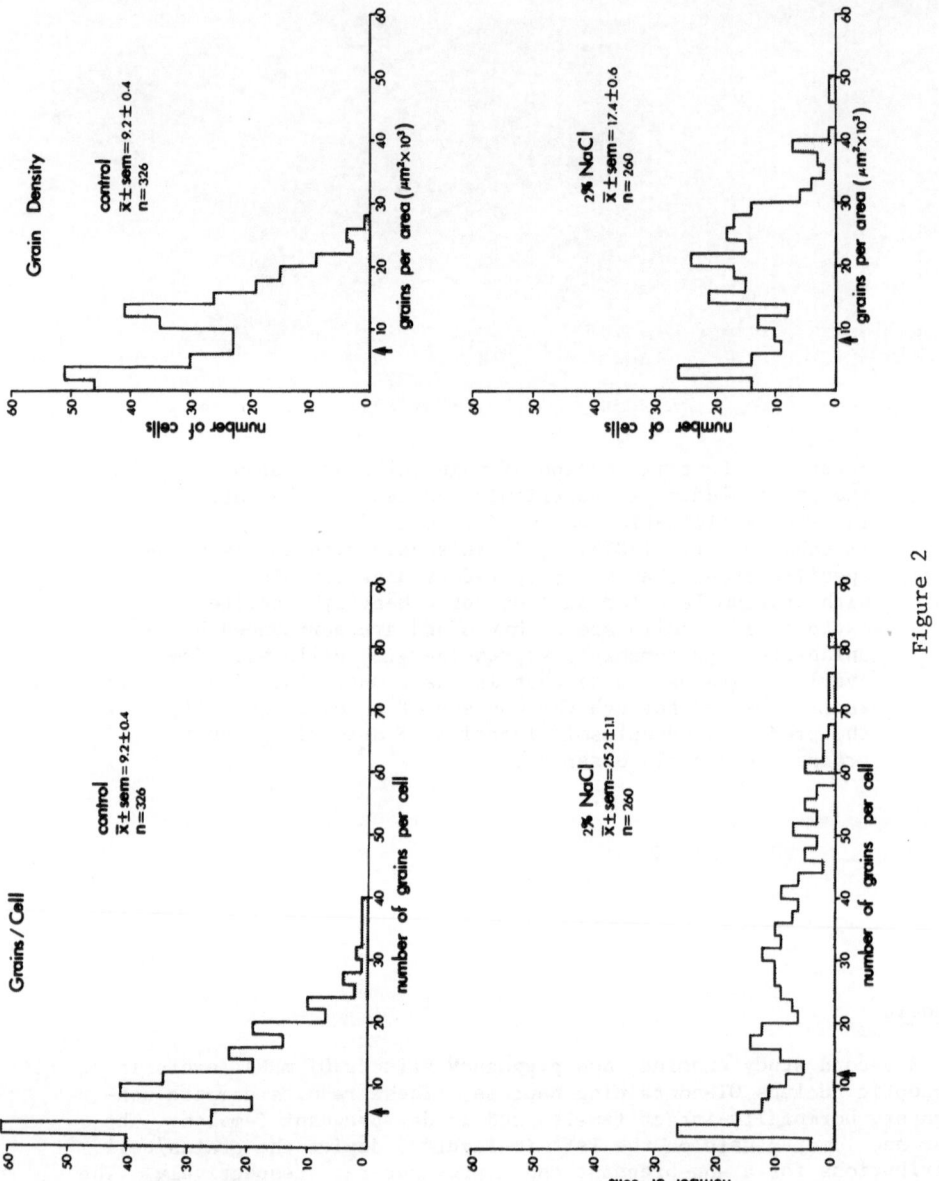

Figure 2

Figure 2. Histograms summarizing grain-counts over magnocellular neurons in the supraoptic nucleus of the hypothalamus from a control rat maintained on tap water (top panels) and a rat provided with a 2% sodium chloride solution to drink for 14 days (lower panels). In situ hybridization was conducted as described (McCabe, et al., 1985b,c) with vasopressin-specific probes which contained tritium-labelled nucleotides (Ivell & Richter, 1984; Schmale, et al., 1983). The left panels summarize grains counted in each cell, and on the right panels summarize "grain density" which was determined by ascertaining each cell's cross-sectional area, in square microns, and dividing the number of grains/cell by cellular area. For convenience, density was then arbitrarily multiplied by 1000. Kolmogorov-Smirnov tests (Siegel, 1956) were used to compare distributions from the control rat and the salt-loaded rat, and tests indicated a significant shift in the distributions of the salt-loaded rat whether measures are compared in terms of grains/cell or grains/area. The arrow below each abscissa indicates one criterion which has been used to estimate background labelling (see text).

OT probe (Figure 3) than with the VP probe (Figure 2). This may be the result of the longer strand-length of the OT probe compared with the VP probe (McCabe, et al., 1985b).

With respect to defining a labelled cell, our counts over adjacent neuropil provide a suitable criterion. The arrow just below the abscissa of each histogram indicates the point that is three times neuropil background. We find this criterion, which is a comparatively conservative criterion for identification of labelled cells, can be used to differentiate between labelled and unlabelled cells since it frequently falls in the middle of the region of overlap between two populations of oxytocin and vasopressin cells.

DISCUSSION

In both of our physiological studies, when control and experimentally-treated conditions are compared, we have noted an interesting phenomenon. It is evident in both Figures 1 and 2 that the second modal distribution in physiologically-challenged animals does not merely "shift" to the right, but that there is also an extended tail to the right indicating that some cells contain much higher levels of VP (Figure 2) or OT (Figure 3) mRNA. It appears from these observations that there may be subgroups of cells which are particularly responsive to physiological stimulation. This would imply that in different physiological states a group of cells categorized as a single peptide-containing cell-group do not necessarily respond homogeneously. Rather, particular subgroups of these cells , as a result of their synaptic and gap junction connections, their specific shape, size, or dendritic field, their individual inputs, their receptor functions, peptide-colocalization, or perhaps the number or arrangement of second messengers or organization of enhancer elements at the genomic level exhibit a particularly robust synthetic response. We propose these subsets of cells that most prominently respond may, in each case, constitute the subgroups which play a primary role in the regulation of that phase of hormonal action. We are currently studying this matter in more detail to determine whether this phenomenon is linked

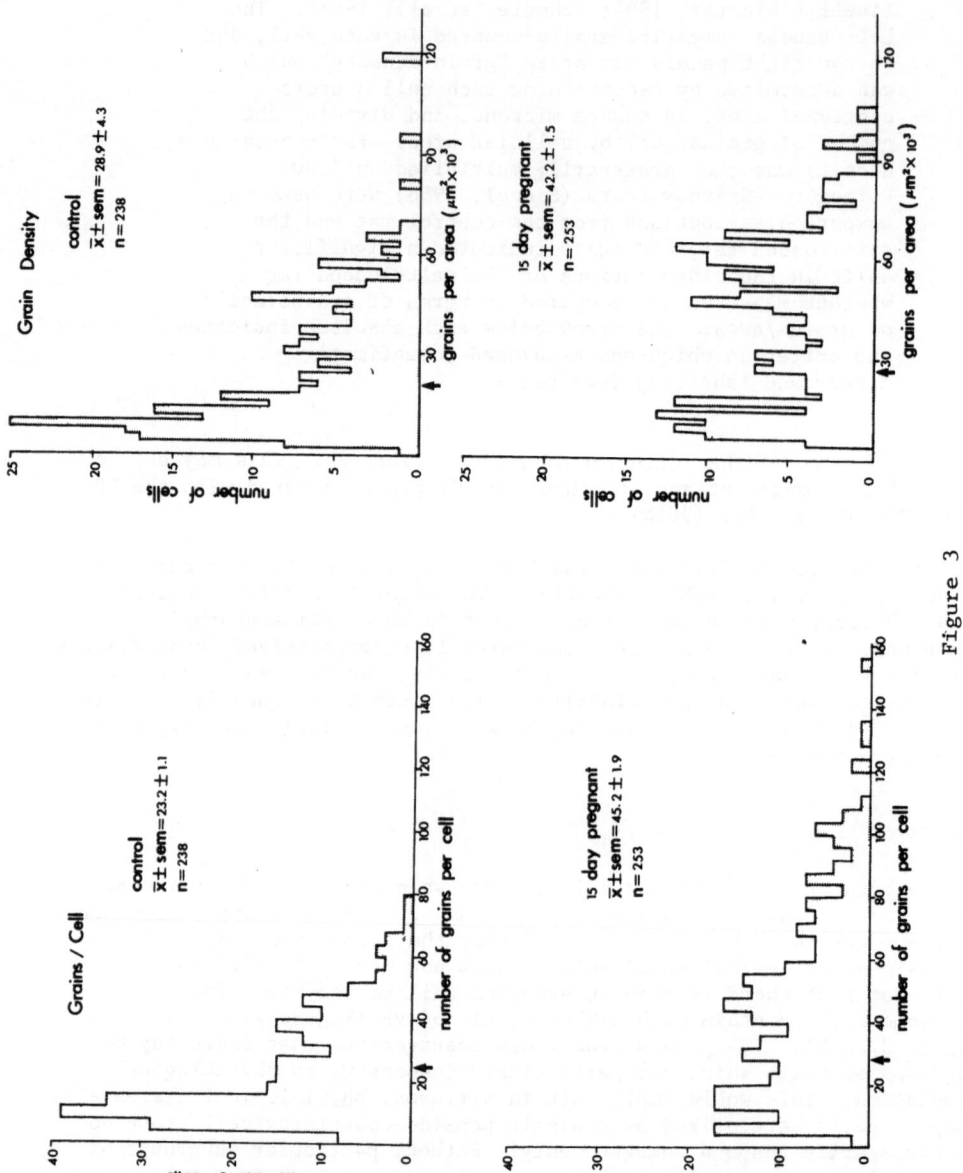

Figure 3

Figure 3. Histograms summarizing grain-counts over
magnocellular neurons in the supraoptic nucleus after
in situ hybridization with an oxytocin-specific probe
(Ivell & Richter, 1984). The histograms summarize
grains counted in each cell from a non-pregnant intact
female (upper panels) and a 15-day pregnant female rat
(lower panels). The panels on the left summarize
grain counts in terms of grains per cell, and on the
right these same data were summarized in terms of
grain density (see text and legend of Figure 1).
Kolmogorov-Smirnov tests indicated a significant
shift in the distribution of the pregnant rat compared
to the non-pregnant rat when data were analyzed in
terms of grains/cell or grain density. The arrow
below each abscissa indicates the criterion used to
estimate background labelling (see text).

to a cell's morphological condition or its anatomical position.

Detailed anatomical analyses with a technique such as in situ
hybridization will permit further study regarding this matter. Cellular
response to physiological challenges may be analogous to processes
described in the visual system, for example, where the feature-detection
capacities of single cells demonstrates the individuality of neuronal
function (Hubel & Wiesel, 1962). Therefore, in situ hybridization,
allowing cell-by-cell analyses, holds great promise for understanding
mechanistic details of neuroendocrine interactions.

ACKNOWLEDGEMENTS

Supported by National Institute of Health grant HD16327 to Joan I.
Morrell. The authors gratefully acknowledge the excellent technical
assistance of Florence Lowe and thank Adelaide Acquaviva for manuscript
preparation.

REFERENCES

Angerer, L. M., and Angerer, R. C., 1981, Detection of poly A+ RNA in sea
    urchin eggs and embryos by quantitative in situ hybridization,
    Nuc. Acids Res., 9:2819-2840.
Bauman, J. G. J., Van der Ploeg, M., and Van Duijn, P., 1984, Fluorescent
    hybridocytochemical procedures: DNA/RNA hybridization in situ, in:
    "Investigative Microtechniques in Medicine and Microbiology, vol. 1.,"
    J. Chayen and L. Bitensky, eds., pp. 41-88, Marcel Dekker, New York.
Brahic, M., and Haase, A. T., 1978, Detection of viral sequences of low
    reiteration frequency by in situ hybridization, Proc. Natl. Acad.
    Sci. USA, 75:6125-6129.
Brahic, M., Haase, A. T., and Cash, E., 1984, Simultaneous in situ
    detection of viral RNA and antigens, Proc. Natl. Acad. Sci. USA,
    81:5445-5448.
Burbach, J. P., De Hoop, M. J., Schmale, H., Richter, D., De Kloet, E. R.,
    and Ten Haaf, J. A., 1984, Differential responses to osmotic stress
    of vasopressin-neurophysin mRNA in hypothalamic nuclei, Neuroendo-
    crinology, 39:582-584.
Cox, K. H., DeLeon, D. V., Angerer, L. M., and Angerer, R. C., 1984,
    Detection of mRNAs in sea urchin embryos by in situ hybridization
    using asymmetric RNA probes, Dev. Biol., 101:485-502.

Gall, J. G., and Pardue, M. L., 1971, Formation and detection of RNA:DNA hybrid molecules in cytological preparations, Proc. Nat. Acad. Sci. USA, 63:378-383.

Gee, C. E., Chen, C.-L., Roberts, J. L., Thompson, R., and Watson, S. J., 1983, Identification of proopiomelanocortin neurones in rat hypothalamus by in situ cDNA-mRNA hybridization, Nature, 306:374-376.

Gee, C. E., and Roberts, J. L., 1983, In situ hybridization histochemistry: A technique for the study of gene expression in single cells, DNA, 2:157-163.

George, J. M., 1973, Localization in hypothalamus of increased incorporation of ^{3}H-cytidine into RNA in response to oral hypertonic saline, Endocrinology, 92:1550-1555.

Gilbert, C. D., 1983, Microcircuitry of the visual cortex, Ann. Rev. Neurosci., 6:217-247.

Godard, C., and Jones, K. W., 1979, Detection of AKR MULV-specific RNA in AKR mouse cells by in situ hybridization, Nuc. Acids Res., 6:2849-2861.

Griffin, W. S. T., Alejos, M., Nilaver, G., and Morrison, M. R., 1983, Brain protein and messenger RNA identification in the same cell, Brain Res. Bull., 10:507-601.

Haase, A. T., Brahic, M., Stowring, L., and Blum, H., 1984, Detection of viral nucleic acids by in situ hybridization, in: "Methods in Virology, vol. 7," K. Maramorosch and H. Koprowski, eds., pp. 189-226, Academic Press, Orlando, FL.

Henderson, A. S., 1982, Cytological hybridization to mammalian chromosomes, Intl. Rev. Cytol., 76:1-46.

Hubel, D. H., and Wiesel, T. N., 1962, Receptive fields, binocular interaction and functional architecture in the cat's visual cortex, J. Physiol., 160:106-154.

Ivell, R., and Richter, D., 1984, Structure and comparison of the oxytocin and vasopressin genes from rat, Proc. Natl. Acad. Sci. USA, 81:2006-2010.

Lawrence, J. B., and Singer, R. H., 1985, Quantitative analysis of in situ hybridization for the detection of actin gene expression, Nuc. Acids Res., 13:1777-1799.

Majzoub, J. A., Rich, A., van Boom, J., and Habener, J. F., 1983, Vasopressin and oxytocin mRNA regulation in the rat assessed by hybridization with synthetic oligonucleotides, J. Biol. Chem., 258:14061-14064.

McCabe, J. T., Morrell, J. I., and Pfaff, D. W., 1985a, Detection of rRNA and mRNA in rat hypothalamic neurons by in situ hybridization, Histochem. Soc. Abstr., in press.

McCabe, J. T., Morrell, J. I., Ivell, R., Schmale, H., Richter, D., and Pfaff, D. W., 1985b, In situ hybridization to localize rRNA and mRNA in mammalian neurons, J. Histochem. Cytochem., in press.

McCabe, J. T., Morrell, J. I., Richter, D., and Pfaff, D. W., 1985c, Localization of neuroendocrinologically-relevant RNA in brain by in situ hybridization, Frontiers of Neuroendocrinology, vol. 9, in press.

Morrell, J. I., Krieger, M. S., and Pfaff, D. W., 1985, Quantitative autoradiographic analysis of estradiol retention by cells in the preoptic area, hypothalamus and amygdala, Exper. Brain Res., submitted.

Nojiri, H., Sato, M., and Urano, A., 1985, In situ hybridization of the vasopressin mRNA in the rat hypothalamus by use of synthetic oligonucleotide probe, Neurosci. Lett., 58:101-105.

Pardue, M. L., and Gall, J. G., 1970, Chromosomal localization of mouse satellite DNA, Science, 168:1356-1358.

Pardue, M. L., and Gall, J. G., 1975, Nucleic acid and hybridization to the DNA of cytological preparations, in: "Meth. Cell Biol., vol. X," D. M. Prescott, ed., pp. 1-16.

Rakic, P. T., 1976, Local Circuit Neurons, MIT Press, Cambridge, MA.

Rogers, A. W., 1973, Techniques of Autoradiography, 2nd Ed., Elsevier, Amsterdam.

Schachter, B., Harlan, R., Pfaff, D., and Shivers, B., 1985, A practical guide to in situ hybridization, Histochem. Soc. Abstr., in press.

Schmale, H., Heinsohn, S., and Richter, D., 1983, Structural organization of the rat gene for the arginine vasopressin-neurophysin precursor, EMBO J., 2:763-767.

Sherman, T. G., and McKelvy, J. F., 1983, Cell-free biosynthesis of rat neurophysin polypeptides from poly(A)RNA isolated from individual hypothalamic nuclei, Neurosci. Abstr., 9:622.

Sherman, T. G., Watson, S. J., Herbert, E., and Akil, H., 1984, The co-expression of dynorphin and vasopressin: An in situ hybridization and dot-blot analysis of mRNAs during stimulation, Neurosci. Abstr., 10:358.

Siegel, S., 1956, Nonparametric Statistics, McGraw-Hill, New York.

Shivers, B. D., Schachter, B., and Pfaff, D. W., 1985, In situ hybridization for the study of gene expression in the brain, Meth. Enzymol., in press.

Uhl, G. R., Zingg, H. H., and Habener, J. F., 1985, Vasopressin mRNA in situ hybridization: Localization and regulation studied with oligonucleotide cDNA probes in normal and Brattleboro rat hypothalamus, Proc. Natl. Acad. Sci. USA., 82:5555-5559.

Wolfson, B., Manning, R. W., Davis, L. G., Arentzen, R., and Baldino, Jr., F., 1985, Co-localization of corticotropin releasing factor and vasopressin mRNA in neurones after adrenalectomy, Nature, 315:59-61.

, T., 1976, Local Circuit Neurons, MIT Press, Cambridge, Mass.

Rogers, L. J., 1971, Handbook of Motor Adaptation, Dekker, New York.

Schneider, D. Nielsen, M., Relill, D., and Salvato, D., 1985, A practical test of a statistical method for analysis in Soc. Neurosci., in press.

Sklar, L. A., Oainleve, F. J., and X. Miller, D., 1983, Structural organization of the protozoa gene during arginine repress in Drosophila in preparation. Exp. in, 31:15-34.

Smithwick T., Uh, and Salvato, D. M., 1984, DNA-type biosynthesis of rat carbohydrate polypeptide production uptake DNA isolated from individual fibroblastic cells, Immunol. Abstr., 19:24.

Swopes, D. G., Davion, S. J. Herbert, P., and Wolf, N., 1966, The arrangement of drosophila and eosin-phila DNA during hybridization and hybridization analysis of DNA's during accumulation, Nippon, Ntchry, 4:244.

Strand, P. E., 1964, Nonparametric Statistics, McGraw-Hill, New York.

Sukaevel, D., Gould, L., B., and Pratt, B. P., 1970, Limit behavior analysis for the study of gene expression in the field, Soc. Immunol., in press.

Vaal, H. S., Zipser, de R., and Hammersly, L. F., 1982, Geographical adaptation manipulation; localization and regulation studied with monoclonal antibodies to normal and Arabidopsis, int typescal disease, Proc. Natl. Acad. Sci. USA, World:9-115.

Watson, S. H., Robbing, H. W., Ro. J., G., Wengsham, G., and Bolduc, S., 1982 Multiple localization of serotonergic related factor and enkephalin and noradrenaline fiber distribution, Brain R., 45:79-81.

NEUROHYPOPHYSIAL PEPTIDES IN THE GONADS

B.T. Pickering, Sonia D. Birkett, H.M. Charlton*,
S.E.F. Guldenaar, Helen D. Nicholson, P.J. O'Shaughnessy,
R.W. Swann, D. Claire Wathes, and R.T.S. Worley

MRC Neuronal Peptides Research Group
Department of Anatomy, University of Bristol
Bristol BS8 1TD
*Department of Human Anatomy
University of Oxford, Oxford OX1 3QX

The first indication that the ovary is a source of oxytocin came from
the observations of Ott and Scott in 1910. However, the recent discovery
by Wathes and Swann (1982) that ovine corpus luteum contains significant
amounts of a substance with the biological, immunoreactive and chromatographic properties of oxytocin, and the observation by Flint and Sheldrick
(1982) that ovarian venous blood contained a higher concentration of the
hormone than the arterial supply to the organ, has focussed attention on
the presence of 'neurohypophysial' peptides in the gonads and their possible
roles in reproduction.

Oxytocin in the Corpus Luteum

Although oxytocin is present in greatest amounts in ruminant ovaries
(Wathes et al. 1983, 1986), it has been found, together with a vasopressin-
like peptide, in the ovaries of several other species including Man (Wathes
et al. 1982; Khan-Dawood & Dawood, 1983; Schaeffer et al. 1984; Khan-Dawood
et al. 1984; Pitzel et al. 1984). The ovarian concentration of oxytocin in
the cow shows great variation during the luteal cycle, reaching a peak in
mid-cycle (ca. 10 nmole/corpus luteum) which is 10^4 times the amount in the
preovulatory follicle, and decreasing to very low levels at the end of the
cycle even if the animal becomes pregnant and the progesterone output is,
thus, maintained (Wathes et al. 1984). Oxytocin has been localised immunocytochemically to the large luteal cells in sheep and cow (Watkins, 1983;
Guldenaar et al. 1984), and dispersed luteal cells *in vitro* incorporate
$[^{35}S]$ cysteine into oxytocin and a neurophysin by way of a precursor much
like that found in the hypothalamus (Swann et al. 1984). Indeed, Ivell and
Richter (1984) have shown that the oxytocin gene which is highly expressed
in bovine corpus luteum has a nucleotide sequence which is almost identical
to that in the hypothalamus.

Extracts from ovine luteal cells incubated with radioactive cysteine
for 7 hours showed two radioactive peptides which bound to neurophysin
affinity columns and could be separated by hplc (Fig.1). One of these
(Peak 2) has the chromatographic characteristics of authentic oxytocin and,
moreover, performic acid oxidation converted it into a product with the same

properties as performic acid - oxidised oxytocin (Swann *et al.* 1984). The other (Peak 1) emerged from the hplc column in the position of arginine vasopressin but, although these eluates did contain immunoreactive vaso-pressin, the radioactive peak does not represent this peptide since its oxidation products differed from oxidised AVP. Moreover, our oxytocin radioimmunoassays (which show no cross-reaction with vasopressin) also measured a component in Peak 1 but, while all our oxytocin antisera gave the same answer when used to assay Peak 2, they gave different answers for Peak 1, i.e. this peak contained an immunoreactive oxytocin-like molecule which was not oxytocin itself. We are not yet able to say whether this represents incompletely processed oxytocin precursor or a molecule contain-ing a substituted amino acid. In terms of oxytocic activity Peak 1 was a minor component and by far the greater part of the activity in the mid-cycle corpus luteum is due to oxytocin itself.

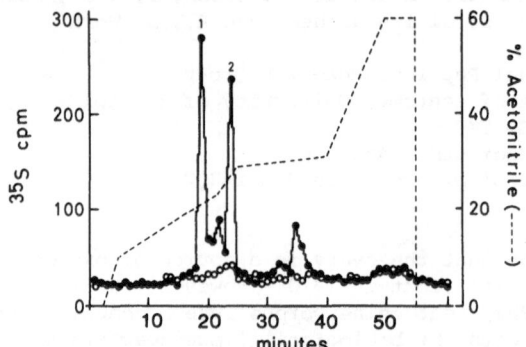

Fig.1. Hplc of an extract of ovine luteal
cells which had been incubated with
$|^{35}S|$cysteine for: O—O, 2h or
●—● 7h. Authentic oxytocin
co-eluted with Peak 2.

While we have found a vasopressin-like peptide in the ovaries of all the species we have studied, it has always been present at a much lower concentration than oxytocin (Wathes *et al.* 1986). Others, however, have reported larger amounts both of an AVP-like material and of its mRNA in rat ovaries (Lim *et al.* 1984; Fuller *et al.* 1985).

Oxytocin in the testis

The discovery of ovarian oxytocin and the reports that neurohypophysial peptides influence androgen production (Adashi & Hsueh, 1981) prompted an investigation of male reproductive tissues. Table 1 shows that oxytocin has been found in testes of all species which have been assayed, although at concentrations much lower than those found in ruminant corpora lutea. Much of the oxytocin immunoreactivity was associated with a component having the hplc characteristics of oxytocin itself (Nicholson *et al.* 1984) and immunocytochemistry showed it to be associated with interstitial tissue, very probably Leydig cells (Guldenaar & Pickering, 1985). Although we have not yet been able to demonstrate incorporation of radiotracer into oxytocin by testicular tissue, preliminary experiments have suggested that the hor-mone is synthesised by the organ. For example, when dispersed cells from human testis were cultured and sampled after 3, 5 and 11 days, the immuno-assayable oxytocin in the medium increased progressively while the content of the cells remained constant.

Table 1. *Oxytocin content of mammalian testes*

Species	n	Oxytocin I-R pg/g wet weight (mean ± S.E)
Human	31	402 ± 100
Wistar rat	23	820 ± 230
Brattleboro rat	3	1030 (70-4000)
Mouse	15	667 ± 300
Dog	4	212 ± 71
Cat	1	243
Rabbit	3	119 (78-141)
Sheep	3	111 (75-148)

Oxytocin-like peptides in prostate and semen

IR-oxytocin was found in prostate (human 1.8±0.4ng/g) and seminal fluid (human 1.3±0.4ng/ml) at levels which are too high to be accounted for by diffusion from the testis (Nicholson et al. 1985). Again, as in the corpus luteum, we found an immunoreactive component which is distinguished from authentic oxytocin both by its retention time in hplc and by its different-ial cross-reactivity with different antisera (Fig.2). Interestingly, Amico et al. (1984) have also drawn attention to an oxytocin-like peptide in human plasma which can be distinguished from authentic oxytocin by its differential cross-reaction and hplc characteristics, although there is no reason to suppose that this is related to the gonadal oxytocin.

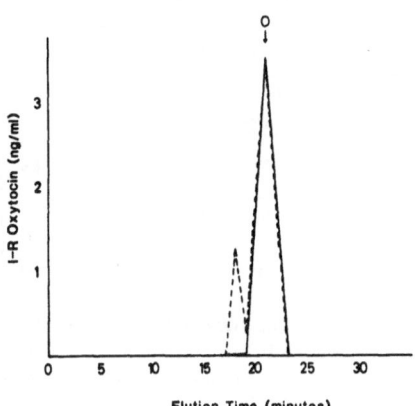

Fig.2. *Radioimmunoassay of the eluates from hplc of an extract of human prostate; ——, using antibody from I.C.A.F.Robinson (RIII/5); ----, using Bristol antibody (79/3). O, position of authentic oxytocin.*

A role for testicular oxytocin

As mentioned above, the initial rationale for looking for oxytocin in the testis was that, like the corpus luteum, it contained steroidogenic tissue and there was some suggestion of a relationship of neurohypophysial hormones and steroidogenesis. [Indeed, we now know that the adrenal, another steroidogenic organ, also contains oxytocin (Nicholson et al. 1984; Ang & Jenkins, 1984)] However, the most common action of oxytocin is to contract smooth muscle and the seminiferous tubules have a myoid layer which is responsible for their contractile activity (Roosen-Runge, 1951;

Niemi & Kormano, 1965). When rat seminiferous tubules were perifused and observed with videomicrography they could be seen to show two sorts of movement both of which were stimulated by small doses (ca. 1ng/ml) of oxytocin. Type A movements are localised dimpling of the tubule, and Type B movements more pronounced contractions of larger segments resulting in propulsion of the contents of the lumen along the tubule (Worley *et al.* 1985).

Testicular oxytocin levels and tubule contractility

Oxytocin is first detectable in the testes of the newborn rat on day 7 *post partum*, which also corresponds with a marked increase in the contractile activity of the seminiferous tubules *in vitro* (Worley *et al.* 1985). Is this a coincidental relationship, or is contractile activity related to the presence of oxytocin? The hypogonadal mouse (hpg/hpg) is a mutant described by Cattanach *et al.* (1977) in which the interstitial cells of the testis remain atrophic and spermatogenesis is arrested at the premeiotic stage because of the inability of the animal's hypothalamus to make GnRH and a consequent massive depletion of pituitary LH. Treatment of the mutant with gonadotrophin or testosterone results in testicular growth and development so that full spermatogenesis is stimulated (Lyon *et al.* 1981). Table 2 shows that the untreated hpg mouse has no detectable oxytocin in its testis and no observable contractile activity of its seminiferous tubules. Treatment with LH for 15 days, however, produced some testicular oxytocin and some tubular movement, while testosterone implants resulted in a normal testis with regard both to oxytocin content and tubule motility.

Table 2. Effects of LH and testosterone on
testicular oxytocin and tubular
movements in the hpg mouse

	Normal mouse	hpg mouse	LH			Testosterone implant		
			200ng/dy	2µg/dy	20µg/dy	2 weeks	6 weeks	12 weeks
1-R: oxytocin pg/g	945	<50	184	254	343	318	4200	1600
Tubule movements	++++	-	+	+	+	++	++++	++++

In this latter case the dose of testosterone used reduces pituitary LH content by over 50% in normal mice (Charlton *et al.* 1983). It is therefore unlikely that testosterone-treated hpg mice, already depleted in pituitary LH, will have significant levels of this hormone in the circulation. This argues that LH does not itself directly stimulate oxytocin production but that there is a relationship between oxytocin, androgen and seminiferous tubule contractility without indicating which of these two hormones has the primary action on motility.

Relationship of androgen, oxytocin and tubular motility

A single dose (75mg/kg i.p.) of ethane-1,2, dimethanesulphonate (EDS) leads to a temporary disappearance of Leydig cells from the testes of rats (Jackson & Jackson, 1984; Shanks *et al.* 1985). Three days after such treatment, the interstitial tissue of the testis was very sparse and no immunoreactive oxytocin could be detected in the organ, either immunocyto-chemically or by radioimmunoassay (Table 3). Moreover, there was also a dramatic reduction in the motility of the seminiferous tubules 3 days after EDS treatment. As had been described earlier (Jackson & Jackson, 1984; Shanks *et al.* 1985), the histological picture of the testis had returned towards normal 28 days after the injection and, as shown in Table 3, so

Table 3. *Testicular oxytocin (pg/g) and tubular movement at various times after a single injection of EDS in the rat*

	DAYS		
	3	10	28
Control	152 ± 10 ++++	198 ± 42 ++++	245 ± 46 ++++
EDS	<20 +	<20 +	180 ± 24 ++++
EDS + Testosterone	<20 +	<20 +	? ++++

had both the testicular oxytocin content and the seminiferous tubule motility. Depletion of the Leydig cells deprives the testis of androgen as well as oxytocin, but testosterone replacement (3mg/day T.P. s.c.) failed to restore the tubular motility of EDS-treated animals (Table 3).

Although the study is still in its infancy, we can conclude that tubular activity is related to oxytocin content and that this, in turn, may be related to androgen production. Preliminary studies with two other mutants help us to extend these conclusions into a hypothesis. Testicular feminisation (Tfm) mice are deficient in androgen receptors and secrete high levels of LH. The interstitial cells of their testes are hypertrophied and secrete androgens, but the seminiferous epithelium is disorganised and spermatogenesis is grossly abnormal. Sex reversed (Sxr,XX) mice lack germ cells in the seminiferous epithelium but secrete androgens from their Leydig cells (Lyon *et al.* 1981). The testes from both of these mutants were found to be devoid of immunoreactive oxytocin and to have seminiferous tubules which showed little or no movement. Thus it would seem that the testosterone produced in the Leydig cell is unable, by itself, to stimulate the production of oxytocin which, in turn, drives the tubule, but that some other factor, determined by the presence of normal germ cell development, is required for peptide synthesis in the testis.

How this relationship of steroid, peptide and tubular motility may be important for the conduction of sperm through the testis and, hence, for male fertility is still a matter for conjecture.

References

Adashi, E. Y. and Hsueh, A. J. W., 1981, Direct inhibition of testicular androgen biosynthesis revealing antigonadal activity of neurohypophysial hormones, *Nature*, 293:650.

Amico, J. A., Ervin, M. G., Leake, R. D., Fisher, D. A., Finn, F. M. and Robinson, A. G. 1984, A novel oxytocin-like and vasotocin-like peptide in human plasma after administration of estrogen, *J. Clin. Endocr. Metab.*, 60:5.

Ang, V. T. Y. and Jenkins, J. S., 1984, Neurohypophysial hormones in the adrenal medulla, *J. Clin. Endocr. Metab.*, 58:688.

Cattanach, B. M., Iddon, C. A., Charlton, H. M., Chiappa, S. A. and Fink, G., 1977, Gonadotrophin-releasing hormone deficiency in a mutant mouse with hypogonadism, *Nature*, 269:338.

Charlton, H. M., Halpin, D. M. G., Iddon, C., Rosie, R., Levy, G., McDowell, I. F. W., Megson, A., Morris, J. F., Bramwell, A., Speight, A., Ward, B. J., Broadhead, J., Davey-Smith, G. and Fink, G., 1983, The effects of daily administration of single and multiple injections of gonadotrophin-releasing hormone on pituitary and gonadal function in the hypogonadal mouse, *Endocrinology*, 113:535.

Flint, A. P. F. and Sheldrick, E. L., 1982, Ovarian secretion of oxytocin is stimulated by prostaglandin, *Nature*, 297:587.

Fuller, P. J., Clements, J. A., Tregear, G. W., Nikolaidis, I., Whitfeld, P. L. and Funder, J. W., 1985, *J.Endocr.*, 105:321.

Guldenaar, S. E. F. and Pickering, B. T., 1985, Immunocytochemical evidence for the presence of oxytocin in rat testis, *Cell Tissue Res.*, 240:396.

Guldenaar, S. E. F., Wathes, D. C. and Pickering, B. T., 1984, Immunocyto-chemical evidence for the presence of oxytocin and neurophysin in the large cells of the bovine corpus luteum, *Cell Tissue Res.*, 237:349.

Ivell, R. and Richter, D., 1984, The gene for the hypothalamic peptide hormone oxytocin is highly expressed in the bovine corpus luteum: biosynthesis, structure and sequence analysis, *EMBO J.*, 3:2351.

Jackson, C. M. and Jackson, H., 1984, Comparative protective actions of gonadotrophins and testosterone against the antispermatogenic action of ethane-dimethanesulphonate, *J. Reprod. Fert.*, 71:393.

Khan-Dawood, F. S. and Dawood, M. Y., 1983, Human ovaries contain immuno reactive oxytocin, *J. Clin. Endocr. Metab.*, 57:1129.

Khan-Dawood, F. S., Marut, E. L. and Dawood, M. Y., 1984, Oxoytocin in the corpus luteum of the cynomolgus monkey (*Macaca fascicularis*), *Endocrinology*, 115:570.

Lim, A. T. W., Lolait, S. J., Barlow, J. W., Autelitano, D. J., Toh, B. H., Boublik, J., Abrahams, J., Johnston, C. I. and Funder, J. W., 1984, Immunoreactive arginine-vasopressin in Brattleboro rat ovary, *Nature*, 310:61.

Lyon, M. F., Cattanach, B. M. and Charlton, H. M., 1981, Genes affecting sex differentiation in mammals, *in:* "Mechanisms of sex differentiation in animals and Man," C. R. Austin and R. G. Edwards, eds., Academic Press, London.

Nicholson, Helen D., Swann, R. W., Burford, G. D., Wathes, D. Claire, Porter, D. G. and Pickering, B. T., 1984, Identification of oxytocin and vasopressin in the testis and in adrenal tissue, *Reg. Pep.*, 8:141.

Nicholson, H. D., Peeling, W. B. & Pickering, B. T., 1985, Oxytocin in prostate and semen, *J. Endocr.*, 104 (Suppl.):127.

Niemi, M. and Kormano, M., 1965, Contractility of the seminiferous tubule of the postnatal rat testis and its response to oxytocin, *Ann. Med. Exp. Fenn.*, 43:40.

Ott, I. and Scott, J. C., 1910, The galactagogue action of the thymus and corpus luteum, *Proc. Soc. Exp. Biol.*, 8:49.

Pitzel, L., Welp, K., Holtz, W. and Konig, A., 1984, Neurohypophyseal hormones in the corpus luteum of the pig, *Neuroendocr. Lett.*, 6:1.

Roosen-Runge, E. C., 1951, Motions of the seminiferous tubules of rat and dog, *Anat. Rec.*, 153:109.

Schaeffer, J. M., Liu, J., Hsueh, A. J. W. and Yen, S. S. C., 1984, Presence of oxytocin and arginine vasopressin in human ovary, oviduct, and follicular fluid, *J. Clin. Endocr. Metab.*, 59:970.

Shanks, J. H., Dixon, J. S. and Lendon, R. G., 1985, Light and electron microscopic observations on rat Leydig cells following a single dose of ethylene-1,2-dimethanesulphonate (EDS), *J. Anat.*, 140:538.

Swann, R. W., O'Shaughnessy, P. J., Birkett, S. D., Wathes, D. C., Porter, D. G. and Pickering, B. T., 1984, Biosynthesis of oxytocin in the corpus luteum, *FEBS Letts.*, 174:262.

Wathes, D. Claire and Swann, R. W., 1982, Is oxytocin an ovarian hormone?, *Nature*, 297:225.

Wathes,D.C., Swann,R.W., Pickering,B.T., Porter,D.G., Hull,M.G.R. & Drife, J.O., Neurohypophysial hormones in the human ovary, *Lancet*(ii), 1982, 410.

Wathes, D. Claire, Swann, R. W., Birkett, S. D., Porter, D. G. and Pickering, B. T., 1983, Characterization of oxytocin, vasopressin and neurophysin from the bovine corpus luteum, *Endocrinology*, 113:693.

Wathes, D. Claire, Swann, R. W. and Pickering, B. T., 1984, Variations in oxytocin, vasopressin and neurophysin concentrations in the bovine ovary during the oestrous cycle and pregnancy, *J. Reprod. Fert.*, 71:551.

Wathes, D. Claire, Swann, R. W., Porter, D. G. and Pickering, B. T., 1986, Oxytocin as an ovarian hormone, *Current Topics Neuroendocr.*, 6 (In Press).

Watkins, W. B., 1983, Immunohistochemical localization of neurophysin and oxytocin in the sheep corpora lutea, *Neuropeptides,* 4:51.

Worley, R. T. S., Nicholson, Helen D. and Pickering, B. T., 1985, Testicular oxytocin: an initiator of seminiferous tubule movement?, in: "Recent progress in cellular endocrinology of the testis," (*INSERM* 123) J. M. Saez, M. G. Forest, A. Dazord and J. Bertrand, eds., INSERM, Paris.

MOLECULAR BIOLOGY OF OVARIAN OXYTOCIN

A.P.F. Flint, D.S.C. Jones, E.L. Sheldrick,
D.T. Theodosis*, and F.B.P. Wooding

A.F.R.C. Institute of Animal Physiology
Babraham
Cambridge CB2 4AT, U.K.
and *INSERM U.176, Domaine de Carreire
Rue Camille Saint-Saëns, 33077 Bordeaux, Cedex, France

INTRODUCTION

There is now a good deal of evidence to suggest that the ruminant corpus luteum synthesizes and secretes oxytocin. By radioimmunoassay and high performance liquid chromatography (HPLC) it has been shown that the corpus luteum contains high concentrations of oxytocin (Wathes and Swann, 1982; Flint and Sheldrick, 1982a), and that oxytocin is secreted into the ovarian vein (Flint and Sheldrick, 1982b). Peripheral circulating oxytocin concentrations rise and fall with the formation and lysis of the corpus luteum (Webb et al., 1981; Sheldrick and Flint, 1981; Schams et al., 1982), and large luteal cells incubated in vitro release oxytocin into the medium (Rodgers et al., 1983). The ovary also produces a neurophysin, which it secretes in a 1:1 molar stoichiometry with oxytocin (Watkins et al., 1984). These data on the synthesis of oxytocin have been confirmed by analysis of proteins labelled during incubation of dissociated ovine and bovine luteal cells with [^{35}S]cysteine, followed by HPLC (Swann et al., 1984).

IDENTITY OF LUTEAL OXYTOCIN

Even when carried out with the benefit of HPLC, radioimmunoassays are prone to misinterpretation as a result of their broad specificity; compounds related to oxytocin that cross react in radioimmunoassays for the peptide are known to circulate in peripheral plasma (see Amico et al., 1985), and these can give rise to anomalously high values for oxytocin in biological fluids. Therefore independent evidence must be sought for the identity of luteal oxytocin. The peptide has not been purified and sequenced by protein sequencing techniques; but there are now two pieces of evidence which, taken together, provide strong evidence that the luteal peptide is oxytocin. These are the sequence of the cloned oxytocin-neurophysin prohormone messenger RNA (mRNA), which is present in bovine (Ivell and Richter, 1984) and ovine (D.S.C. Jones, unpublished observations) corpora lutea in high concentrations, and the establishment of the molecular weight of luteal oxytocin by fast-atom bombardment mass spectrometry (Flint and Sheldrick, 1985). It is important to take these two pieces of evidence together because although the mRNA sequence indicates the prohormone in the corpus luteum has a structure identical to that in the hypothalamus (Ivell and Richter, 1984), this observation alone does not rule out the possibility that post-translational processing produces different secretory products in the two glands. The finding that the molecular weight of the peptide detected by radioimmunoassay in the corpus luteum is equal to that of oxytocin strongly indicates that post-translational processing occurs by identical mechanisms, and that the luteal peptide is authentic oxytocin. Restriction mapping suggests there is only one oxytocin-neurophysin gene in the ruminant genome (Ivell and Richter, 1984), so it appears the same gene is expressed in both tissues.

IDENTIFICATION OF OXYTOCIN IN SECRETORY GRANULES

It is well established that in the hypothalamus the oxytocin-neurophysin prohormone is packaged in secretory granules which undergo axonal transport to the neurohypophysis for storage and secretion. During the process of transport the prohormone is subjected to post-translational processing, which leads to the formation of the C-terminal amidated oxytocin molecule.

Although little is known about post-translational processing events in the corpus luteum, other than that they lead to the formation of a peptide identical to that in the neurohypophysis, there is evidence for the packaging of the hormone in secretory granules. When ultrathin sections of ovine corpora lutea were incubated with polyclonal antisera against either oxytocin or oxytocin-neurophysin, and bound antibody visualized using an immunogold complex, only secretory granules in the large luteal cells showed gold labelling (Theodosis et al., 1985). These are the cells which have been shown to contain oxytocin in cell dissociation (Rodgers et al., 1983) and immunohistochemical studies (Watkins, 1983). Labelling was blocked by pre-adsorption of the oxyto-

cin antibody with oxytocin, and was absent when sections were incubated
with antibody directed against vasopressin. The secretory granules were
slightly larger than those in the neurohypophysis (200-300 nm diameter)
and appeared to have a more complex structure.

In addition to identifying secretory granules inside cells,
Theodosis et al. (1985) also demonstrated the secretion of such granules
by exocytosis, and their presence in the extracellular space. Of particular significance was the observation that secretory granules which had
been exocytosed no longer bound oxytocin antibody, thus suggesting that,
following secretion, oxytocin is lost from the granules more rapidly
than their other constituents. This of course raises the question of
the nature of the granule matrix.

INTERACTION OF OXYTOCIN WITH THE OXYTOCIN RECEPTOR

Oxytocin secreted by the corpus luteum appears to be involved in
ensuring luteal regression, and it does this through an interaction with
the uterine oxytocin receptor. The evidence for these statements is
summarized below.

Five independent observations suggest that oxytocin plays a role in
controlling ovarian cyclicity in ruminants. 1) It causes premature
luteal regression on administration to heifers (Armstrong and Hansel,
1959), sheep (Milne, 1963; Dobrowolski, 1973; Hatjiminaoglou et al.,
1979) and goats (Cooke and Knifton, 1981). 2) It stimulates release of
prostaglandin $F_{2\alpha}$ ($PGF_{2\alpha}$) from the uterus (Sharma and Fitzpatrick, 1974;
Mitchell et al., 1975; Roberts et al., 1976). 3) It is released into
the circulation simultaneously with episodes of release of $PGF_{2\alpha}$ at
luteolysis (Fairclough et al., 1980; Flint and Sheldrick, 1983). 4)
Immunization against oxytocin, carried out either actively (Sheldrick et
al., 1980) or passively (Schams et al., 1983) delays luteal regression.
5) Secretion of oxytocin from the corpus luteum is stimulated by $PGF_{2\alpha}$
and the $PGF_{2\alpha}$-analogue, cloprostenol (Flint and Sheldrick, 1982b).

These observations suggest that oxytocin acts systemically, rather
than locally within the corpus luteum, and this is supported by the lack
of any consistent action of oxytocin on steroid secretion by luteal
tissues (see Flint and Sheldrick, 1985). A systemic action is further
confirmed by the absence of any alteration in circulating progesterone
concentrations in ewes following hysterectomy, when oxytocin is lost
from the corpus luteum (Sheldrick and Flint, 1983), and by the fact that
hysterectomy blocks the luteolytic action of oxytocin in heifers (Hansel

and Wagner, 1960). The stimulation, by oxytocin, of uterine production of $PGF_{2\alpha}$ is consistent with a mechanism of action involving the uterus, since this compound is luteolytic.

Based on these observations the suggestion has been made that oxytocin acts by ensuring the episodic nature of $PGF_{2\alpha}$ secretion, through a positive feedback loop (Flint and Sheldrick, 1983). The episodic nature of $PGF_{2\alpha}$ release is an important characteristic of its secretion, as judged by the increased luteolytic potency of $PGF_{2\alpha}$ administered at intervals, compared to that given continuously (Schramm et al., 1983), and its prevention, by such means as immunization against oxytocin, would be expected to result in delayed luteal regression (point 4, above).

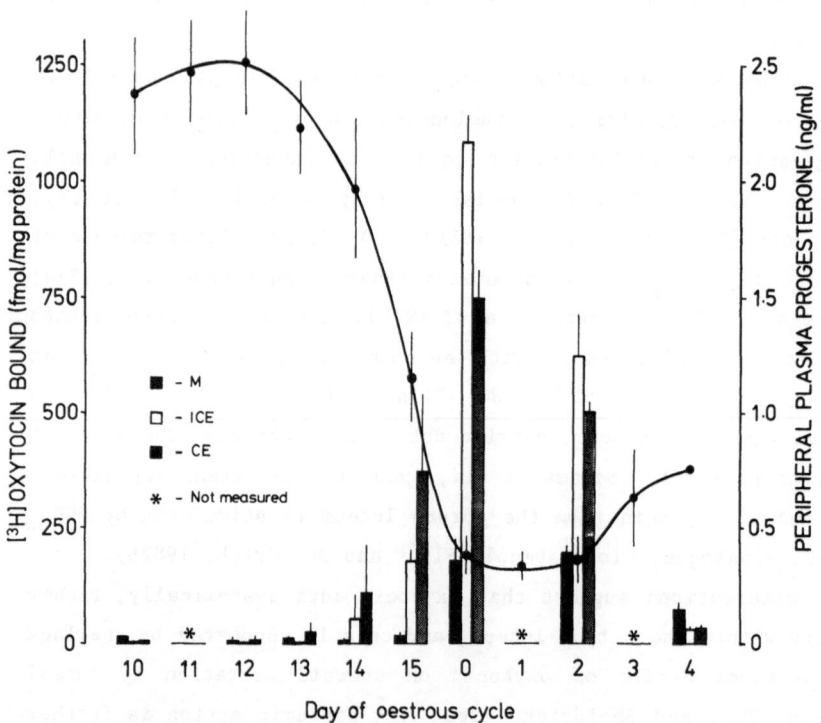

Figure 1. Concentrations of oxytocin receptor in myometrium (closed bars), intercaruncular endometrium (open bars) and caruncular endometrium (hatched bars) in sheep at varying stages of the oestrous cycle. Solid line indicates peripheral plasma progesterone concentration. Vertical bars indicate s.e.m. (From Sheldrick and Flint, 1985; with permission).

In order to characterize the response of the uterus to oxytocin, we have measured oxytocin receptor concentrations in caruncular and inter-caruncular endometrium and in myometrium, in cyclic ewes (Sheldrick and Flint, 1985). Concentrations in the endometrium exceeded those in the myometrium, at all stages tested (Figure 1), which is consistent with the high rate of $PGF_{2\alpha}$ secretion by endometrium (Findlay et al., 1981), and concentrations of receptor in all three tissues increased dramatically shortly after luteolysis. The increase in uterine response to oxytocin at luteolysis which these measurements indicate is clearly consistent with an action of the peptide at this stage of the cycle. These measurements of receptor concentration are in good agreement with those reported previously (Roberts et al., 1976).

MECHANISM OF ACTION OF LUTEAL OXYTOCIN

Although the oxytocin receptor has to some extent been charac-terized, there is little information available on the post-receptor events involved in the mechanism of action of oxytocin, and in particular on the mechanism by which oxytocin stimulates $PGF_{2\alpha}$ produc-tion in the endometrium. It is known on the other hand that arginine vasopressin acts through two receptors (V1 and V2, in liver and vascular smooth muscle, and in renal membranes respectively), and because of the high cross reaction of the uterine oxytocin receptor with vasopressin (Sheldrick and Flint, 1985) it appeared possible that the oxytocin receptor may resemble either the V1 or V2 receptor. Furthermore the V1 receptor is thought to act through stimulation of the hydrolysis of phosphatidylinositol (Michell et al., 1979), a process which can lead to release of arachidonic acid, the substrate for $PGF_{2\alpha}$ synthesis. In view of this background we have recently investigated whether oxytocin stimu-lates phosphatidylinositol breakdown.

Slices of caruncular endometrium from ewes treated sequentially with progestagen and oestrogen, a process which is known to stimulate oxytocin receptor synthesis (Sheldrick and Flint, 1985), accumulate [^{3}H]inositol in phosphatidylinositol on incubation with the labelled substrate in vitro. Following removal of unincorporated [^{3}H]inositol by chasing with unlabelled inositol, addition of oxytocin can be shown to cause accumulation of the water soluble inositol phosphates in the slices (Table 1). By reference to Figure 2 it can be seen that this indicates an increased breakdown of phosphatidylinositol, a byproduct of which is 1,2-diacylglycerol. Further hydrolysis of diacylglycerol releases fatty acids and since the fatty acid most frequently found in position-2 of phosphatidylinositol is arachidonate, this would be

Table 1. Accumulation of [³H]inositol in inositol phosphates following addition of oxytocin to slices of caruncular endometrium in which phosphatidylinositol had been labelled by pre-incubation with [³H]inositol.

Oxytocin concentration (M)	10^{-3} x d.p.m/g tissue		
	Inositol monophosphate (IP)	Inositol bis-phosphate (IP$_2$)	Inositol tris-phosphate (IP$_3$)
0	13.3 ± 1.7	3.0 ± 0.9	0.1 ± 0.1
10^{-11}	12.2 ± 0.5	2.2 ± 0.8	0.1 ± 0.1
10^{-9}	37.5 ± 6.4	13.8 ± 2.6	6.0 ± 3.7
10^{-7}	105.3 ± 9.3	49.2 ± 1.7	31.3 ± 3.5
10^{-6}	150.9 ± 23.1	72.8 ± 15.8	38.8 ± 8.6

Slices pre-incubated for 2 hr with [³H]inositol were rinsed and subsequently incubated with varying concentrations of oxytocin in the presence of LiCl (10 mM).

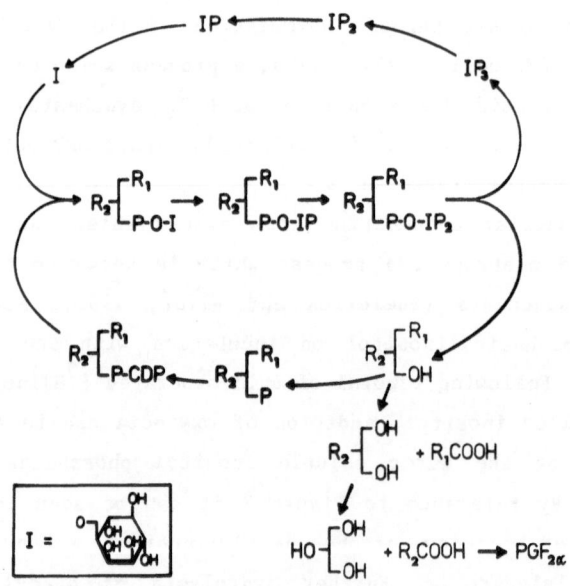

Figure 2.   The phosphatidylinositol cycle.

expected to lead to increased availability of arachidonic acid in the tissue. Arachidonic acid is rate-limiting in $PGF_{2\alpha}$ synthesis in ovine endometrium (Findlay et al., 1981), and its increased availability would therefore be expected to lead to increased $PGF_{2\alpha}$ production.

CONCLUSIONS

We have briefly discussed general aspects of the synthesis, secretion and mechanism of action of ovarian oxytocin from the point of view of the molecular interactions involved in its role in controlling oestrous cyclicity. It is clear that a good deal of information has accumulated in the short time since the discovery of luteal oxytocin on the way in which it contributes to ovarian function in ruminants. Important questions remain unanswered, however. For instance factors controlling the synthesis of oxytocin in the corpus luteum are largely unknown, and the clinical significance of the possible production of related peptides in the human ovary, and their roles in luteolysis and uterine dysfunction, remain largely unexplored. Let us hope that this symposium will stimulate work in these exciting areas.

ACKNOWLEDGEMENT

D.T.T. was supported by a generous grant from The Nuffield Foundation.

REFERENCES

Amico, J. A., Ervin, M. G., Leake, R. D., Fisher, D. A., Finn, F. M., and Robinson, A. G., 1985, A novel oxytocin-like and vasotocin-like peptide in human plasma after administration of estrogen, J. Clin. Endocr. Metab., 60:5.

Armstrong, D. T., Hansel, W., 1959, Alteration of the bovine estrous cycle with oxytocin, J. Dairy Sci., 42:533.

Cooke, R. G., and Knifton, A., 1981, Oxytocin-induced oestrus in the goat, Theriogenology, 16:95.

Dobrowolski, W., 1973, Wplyw oksytocyny na cykl plciowy owcy w sezonie rosplodu i ciszy seksualnej, Polski Arch Wet, 16:649.

Fairclough, R. J., Moore, L. G., McGowan, L. T., Peterson, A. J., Smith, J. F., Tervit, H. R. and Watkins, W. B., 1980, Temporal relationship between plasma concentrations of 13,14-dihydro-15-keto-prostaglandin F and neurophysin I/II around luteolysis in sheep, Prostaglandins, 20:199.

Findlay, J. K., Ackland, N., Burton, R. D., Davis, A. J., Maule Walker, F. M., Walters, D. E. and Heap, R. B., 1981, Protein, prostaglandin and steroid synthesis in caruncular and intercaruncular endometrium of sheep before implantation, J. Reprod. Fert., 62:361.

Flint, A. P. F., and Sheldrick, E. L., 1982a, Ovarian secretion of oxytocin in the sheep, J. Physiol. (Lond.), 330:61P.

Flint, A. P. F., and Sheldrick, E. L., 1982b, Ovarian secretion of oxytocin is stimulated by prostaglandin, Nature (Lond.), 297:587.

Flint, A. P. F., and Sheldrick, E. L., 1983, Evidence for systemic role for ovarian oxytocin in luteal regression in sheep, J. Reprod. Fert., 67:215.

Flint, A. P. F., and Sheldrick, E. L., 1985, Ovarian oxytocin, in: "Oxytocin: Clinical and Laboratory Studies" J. A. Amico and A. G. Robinson, eds, Elsevier Science Publishers B.V. (Biomedical Division).

Hansel, W., and Wagner, W. C., 1960, Luteal inhibition in the bovine as a result of oxytocin injections, uterine dilatation, and intra-uterine infusions of seminal and preputial fluids, J. Dairy Sci., 43:796.

Hatjiminaoglou, I., Alifakiotis, T., and Zervas, N., 1979, The effect of exogenous oxytocin on estrous cycle length and corpus luteum lysis in ewes, Ann. Biol. Anim. Biochem. Biophys., 19:355.

Ivell, R., and Richter, D., 1984, The gene for the hypothalamic peptide hormone oxytocin is highly expressed in the bovine corpus luteum: biosynthesis, structure and sequence analysis, The EMBO Journal, 3:2351.

Michell, R. H., Kirk, C. J., and Billah, M. M., 1979, Hormonal stimulation of phosphatidylinositol breakdown, with particular reference to the hepatic effects of vasopressin, Biochem. Soc. Trans., 7:861.

Milne, J. A., 1963, Effects of oxytocin on the oestrous cycle of the ewe, Aust. Vet. J., 39:51.

Mitchell, M. D., Flint, A. P. F., and Turnbull, A. C., 1975, Stimulation by oxytocin of prostaglandin F levels in uterine venous effluent in pregnant and puerperal sheep, Prostaglandins, 9:47.

Roberts, J. S., McCracken, J. A., Gavagan, J. E., and Soloff, M. S., 1976, Oxytocin-stimulated release of prostaglandin $F_{2\alpha}$ from ovine endometrium in vitro: correlation with estrous cycle and oxytocin-receptor binding, Endocrinology, 99:1107.

Rodgers, R. J., O'Shea, J. D., Findlay, J. K., Flint, A. P. F., and Sheldrick, E. L., 1983, Large luteal cells the source of luteal oxytocin in the sheep, Endocrinology, 113:2302.

Schams, D., Lahlou-Kassi, A., and Glatzel, P., 1982, Oxytocin concentrations in peripheral blood during the oestrous cycle and after ovariectomy in two breeds of sheep with low and high fecundity, J. Endocr., 92:9.

Schams, D., Prokopp, S., and Barth, D., 1983, The effect of active and passive immunization against oxytocin on ovarian cyclicity in ewes, Acta Endocr. Copenh., 103:337.

Schramm, W., Bovaird, L., Glew, M. E., Schramm, G., and McCracken, J. A., 1983, Corpus luteum regression induced by ultra-low pulses of prostaglandin $F_{2\alpha}$, Prostaglandins, 26:347.

Sharma, S. C., and Fitzpatrick, R. J., 1974, Effect of oestradiol-17β and oxytocin treatment of $PGF_{2\alpha}$ release in the anoestrus ewe, Prostaglandins, 6:97.

Sheldrick, E. L., and Flint, A. P. F., 1981, Circulating concentrations of oxytocin during the estrous cycle and early pregnancy in sheep, Prostaglandins, 22:631.

Sheldrick, E. L., and Flint, A. P. F., 1983, Regression of the corpora lutea in sheep in response to cloprostenol is not affected by loss of luteal oxytocin after hysterectomy, J. Reprod. Fert., 68:155.

Sheldrick, E. L., and Flint, A. P. F., 1985, Endocrine control of uterine oxytocin receptors in the ewe, J. Endocr., 106:249.

Sheldrick, E. L., Mitchell, M. D., and Flint, A. P. F., 1980, Delayed luteal regression in ewes immunized against oxytocin, J. Reprod. Fert., 59:37.

Swann, R. W., O'Shaughnessy, P. J., Birkett, S. D., Wathes, D. C., Porter, D. G., and Pickering, B. T., 1984, Biosynthesis of oxytocin in the corpus luteum, FEBS Letters, 174:262.

Theodosis, D. T., Wooding, F. B. P., Sheldrick, E. L., and Flint, A. P. F., 1985, Ultrastructural localisation of oxytocin and neurophysin in the ovine corpus luteum, Cell and Tissue Res. (In press).

Wathes, D. C., and Swann, R. W., 1982, Is oxytocin an ovarian hormone? Nature Lond., 297:225.

Watkins, W. B., 1983, Immunohistochemical localisation of neurophysin and oxytocin in the sheep corpora lutea, Neuropeptides, 4:51.

Watkins, W. B., Moore, L. G., Flint, A. P. F., and Sheldrick, E. L., 1984, Secretion of neurophysins by the ovary of sheep, Peptides, 5:61.

Webb, R., Mitchell, M. D., Falconer, J., and Robinson, J. S., 1981, Temporal relationships between peripheral plasma concentrations of oxytocin, progesterone and 13,14-dihydro-15-keto prostaglandin $F_{2\alpha}$ during the estrous cycle and early pregnancy in the ewe, Prostaglandins, 22:443.

CHARACTERIZATION, REGULATION AND FUNCTIONAL ACTIVITY OF SPECIFIC VASOPRES-

SIN RECEPTORS IN THE ANTERIOR PITUITARY GLAND

Bernard Koch and Bernadette Lutz-Bucher

Institut de Physiologie, UA CNRS 309, 21 rue René Descartes

F-67084 Strasbourg, France

INTRODUCTION

Vasopressin exerts multiple effects both at the periphery and at the central nervous system. Several lines of evidence show that, besides its well-defined action on the kidney, the liver and the cardiovascular system, the peptide also participates in behavioural processes (review by Forsling 1979), as well as in the control of pituitary corticotropic function (Yates et al.,1971; Gillies et al.,1982). Very recently, an adrenal site of action has even been described, as AVP was found to trigger steroid release from glomerulosa cells (Gallot-Payet et al.,1985).

Arginine-vasopressin (AVP) not only bears intrinsic corticotropin-re-leasing factor (CRF)-like activity, but also acts synergistically with syn-thetic CRF41 to augment pituitary ACTH secretion(Gillies et al.,1982). How-ever, in contrast to the effect of CRF41 that primarily involves cAMP as a messenger (Labrie et al.,1982; Bilezikjian and Vale,1983), the CRF-like pro-perties of AVP appear not to be directly linked to adenylate cyclase stimu-lation (Holmes et al.,1984); although it was able to cause a marked enhance-ment of CRF41-induced cAMP production (Giguère and Labrie,1982).

Recently, both pharmacological (Antoni,1984; Knepel et al.,1984; Baer-tschi and Friedli, 1985) and biochemical studies (Lutz-Bucher and Koch, 1983; Antoni,1984; Gaillard et al.,1984; Spinedi and Negro-Vilar,1984; Koch and Lutz-Bucher,1985) have revealed the presence in the anterior pituitary of highly specific AVP receptor sites, that are clearly of a novel type. The present report will summarize the current knowledge on the binding charac-teristics and the regulation of that binding system and, in addition, will also address the question of a possible relationship between AVP binding and biological response of pituitary cells.

CRF-LIKE PROPERTIES OF AVP AND ANALOGS

The ability of AVP to trigger ACTH release from anterior and intermedia pituitary tissues is clearly revealed by data displayed in Fig.1. When incu-bated in a perifusion system and pulsed with either AVP or CRF41, ACTH re-lease from both types of tissues was enhanced. However, whereas perifusion of anterior pituitary fragments in the presence of AVP along with CRF41 drama-tically elevated ACTH output, there was no such effect on hormonal release of the intermediate pituitary.

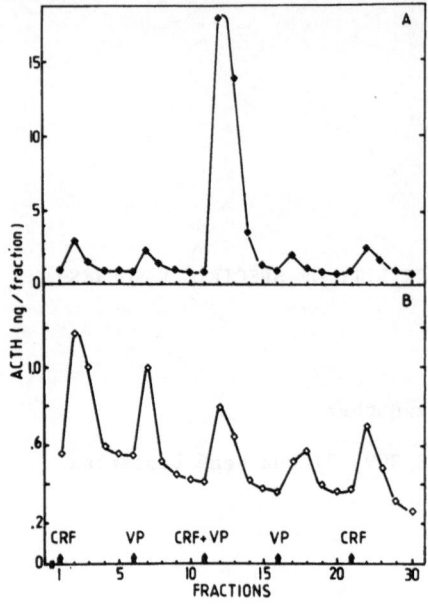

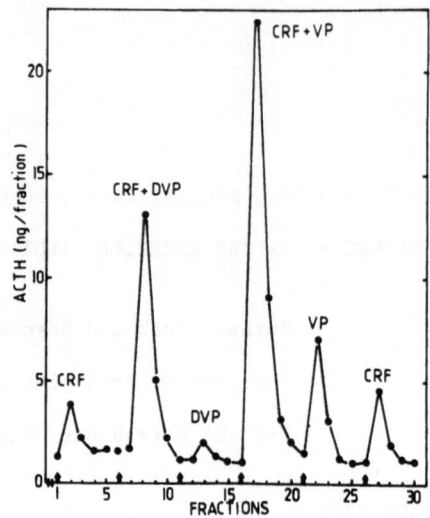

Fig.1. Pattern of ACTH secretion from perifused anterior (A) and intermediate (B) pituitaries of rats. Pulses of CRF41 (2nM), AVP (15nM) or both in combination(5min). Fractions were collected for 10min. (Lutz-Bucher & Koch,1983).

Fig.2. Profile of ACTH release from rat anterior pituitaries, as induced by pulses of 2nM CRF41, 1-deamino(8-D-AVP)=DVP (15nM) or both in combination. ACTH was bioassayed using isolated adrenal cells. (Lutz-Bucher & Koch,1983).

A striking observation was that the analog 1-deamino(8-D-AVP), that appeared to be much less potent a CRF-like agent than AVP and that was even reported by Anderson et al.(1972) to be devoid of activity, did nevertheless enhance CRF41-induced release of ACTH (Fig.2). Oxytocin was similarly reported to potentiate the effect of CRF41 (Antoni,1983). Also,conflicting results were obtained with antagonists against the V_1-type AVP receptor. The V_1-antagonist dPen Tyr(Me)AVP was found to antagonize ACTH secretion stimulated by AVP, while exhibiting a slight agonist effect as well (Antoni et al.,1984). However, the V_1-antagonist $d(CH_2)_5$Tyr(Me)AVP has been shown to be without influence on the CRF-like activity of AVP (Antoni et al.,1984), although other investigators reported an antagonistic effect (Knepel et al.,1984). On the basis of a study conducted with a number of AVP analogs, Aizawa et al. (1982) concluded that the CRF activity mainly resides in the vasopressor activity, independent of cAMP formation. More recent findings, together with binding studies,seem to support the issue that, in fact, the CRF-like effect of AVP involves neither classical antidiuretic nor pressor activities (Knepel et al.,1984).

CHARACTERIZATION OF PITUITARY AVP RECEPTORS

Interaction of ^{3}H-AVP with rat pituitary membranes reveals the presence of a single type of receptor sites, with K_d of nearly 1nM (Fig.3). This value is in good keeping with the AVP concentration found in hypophysial portal

Abbreviations: $d(CH_2)_5$Tyr(Me)AVP,1-Mercapto-β,β-cyclopentamethylene propionic acid(methyl-tyrosine)8-AVP; dPenTyr(Me)AVP, 1-deamino,pencillamine(methyl-tyrosine)8-AVP.

blood (Oliver et al.,1977, Gibbs,1985), as well as with the K_d values reported by others (Antoni,1984; Spinedo and Negro-Vilar,1984).

In Brattleboro homozygote rats, the pituitary AVP receptor system appears to be indistinguishible from that of heterozygote controls or normal Wistar rats (Fig.3B), in spite of a genetic lack of vasopressin. However, although the K_d values were identical, the number of sites found in homozygotes consistently exceeded by about 25% that seen in heterozygotes, reflecting the effect of circulating AVP on its own receptor.

The data obtained with pituitary membrane fractions were validated by using isolated cells in binding experiments. When isolated pituitary cells were reacted with ^{3}H-AVP, at either 15°C or 37°C, binding equilibrium was reached at 30-40min after incubation (inset to Fig.4). Interestingly, maximum binding appeared to be higher at 37°C than at 15°C, with Bmax values of 191 vs. 97 fmol/mg cell protein. That difference in binding most probably was due to internalization and accumulation of tracer within the cells, that are known to be favored at the higher temperature. The dissociation constants were calculated as 1.2 and 1.8nM at 37°C and 15°C respectively and were thus very close to those obtained with the particulate fraction.

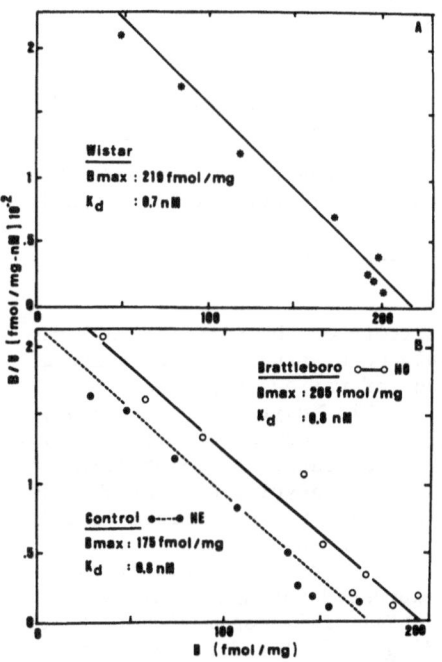 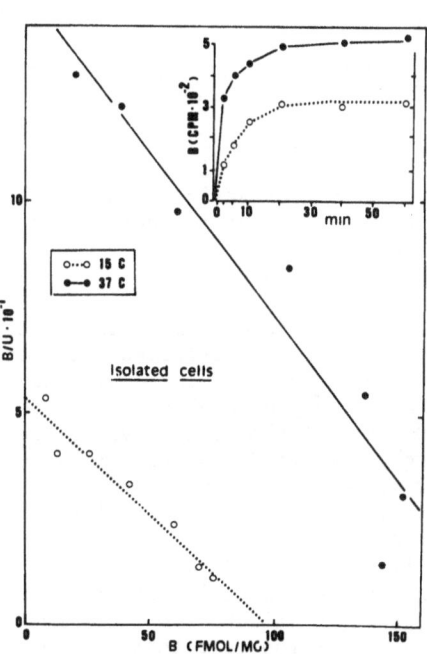

Fig.3. Scatchard plots of specific binding of ^{3}H-AVP to pituitary membranes of Wistar rats (A), as well as of Brattleboro homozygotes (HO) and heterozygotes (HE; B). Membrane suspensions in 50mM Tris-HCl buffer (pH 7.3) containing 5mM $MgCl_2$, 1mM EGTA and 0.1% BSA were incubated at 18°C for 40min. Bound and free moieties were separated by means of centrifugation (10,000xg/1min). Each point is derived from 4 pituitary equivalents of membranes.
(Lutz-Bucher & Koch,1983)

Fig.4. Scatchard plots of specific binding of ^{3}H-AVP to isolated pituitary cells, incubated at 37°C and 15°C. Cells were obtained by trypsinic digestion of tissues and incubated(1 pituitary equivalent or $1-1.5x10^6$ cells) in Krebs-Ringer bicarbonate buffer, containing 0.1% BSA. The reaction was stopped by diluting with buffer and cells were centrifuged at 10,000xg/1min. Inset: Time course of specific binding at 37°C and 15°C.
(Koch & Lutz-Bucher,1985).

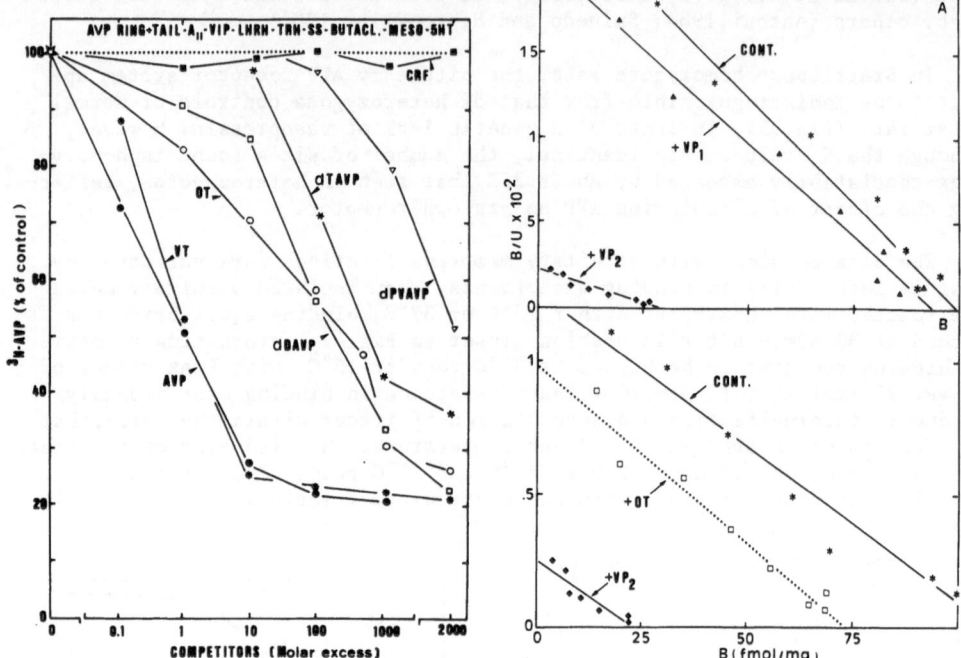

Fig.5. Competitive displacement of 2nM ^{3}H-AVP binding to pituitary membranes by various analogs and compounds. See text for definition of abbreviations.

Fig.6. Down-regulation of pituitary AVP receptors. Rats were treated chronically with LVP(3x10ug/day; VP_2) or with oxytocin (OT); or received a single injection of LVP (VP_1).

In order to determine the binding specificity of the pituitary AVP receptor system, we performed competitive displacement studies using a fixed amount of tracer and increasing concentrations of various peptides. The order of potency of analogs was AVP=LVP=vasotocin(VT) > oxytocin(OT) > 1-deamino-(8-D-AVP; dDAVP) > d(CH$_2$)$_5$Tyr(Me)-Val^4-AVP(dTAVP) > 1-deaminopencillamine-(Val^4 D-Arg^8)VP (dPVAVP). Neither ring or tail AVP fragments, nor mesotocin(MESO), that differ from VT by only one amino acid in the noncyclic part of the molecule, did compete for binding. Also, none of the various peptides tested, including CRF41 or other CRF-like compounds such as Angiotensin II (Capponi et al;,1982) or VIP (Westendorf et al.,1983)displaced the tracer. It is thus clear that CRF activity of AVP seems to be expressed through highly specific receptors, distinct from those that combine CRF41. This conclusion is further supported by the fact that, conversely, binding of ^{125}I-radiolabeled Tyr-CRF to pituitary membranes was unaffected by AVP (Wynn et al.,1983; Koch and Lutz-Bucher,1983).

Taken together with other recent findings (Antoni,1984; Gaillard et al., 1984), these data suggest that AVP pituitary receptors are distinct from those that have been previously described in the kidney, the liver and vascular smooth muscle(review by Jard,1983). They also seem to be different from OT receptors, since this peptide was found to interact with a high affinity-low capacity pituitary binder.

REGULATION OF PITUITARY AVP RECEPTORS

We examined the possibility of homologous down-regulation of pituitary AVP receptors and addressed this question by means of various experimental approaches.First, the effect of chronic treatment with either

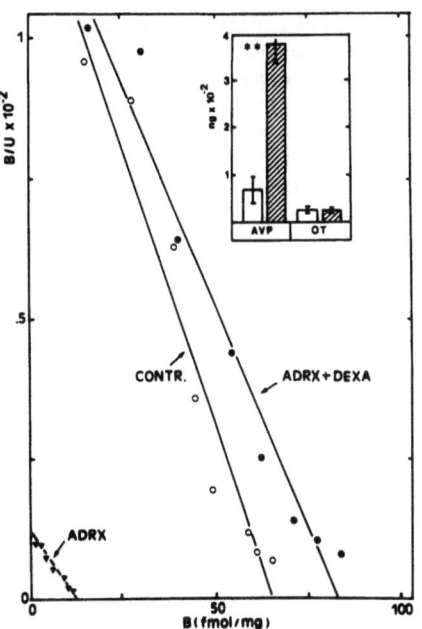

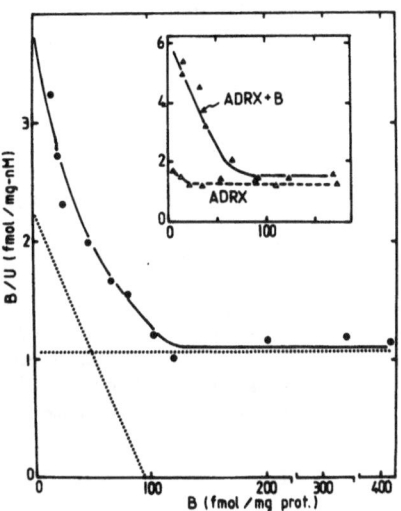

Fig.7. Effect of long-term adrenalec-
    tomy(ADRX),with or without
dexamethasone treatment(DEXA; 40ug/
day)on specific binding of 3H-AVP to
rat pituitary membranes.Inset: pituitary content of AVP and OT of control
(☐) or ADRX (▨; 1 week) rats.From
Koch & Lutz-Bucher (1985).

Fig.8. Scatchard plots of specific
    binding of 125I-Tyr-CRF to
pituitary membranes. Inset: Effect
of ADRX, with or without corticosterone therapy(ADRX+B)in drinking wa-
ter(10mg/100ml in 5% ethanol).From
Koch & Lutz-Bucher (1983).

vasopressin or OT was investigated. As depicted in Fig.6, long-term vasopressin administration (1 week) resulted in a dramatic fall in AVP receptor
number, that declined from a control level of 104 to 25 fmol/mg pituitary
membrane protein. There was no apparent change in the K_d values. Interestingly, when OT was administered under the same conditions as VP, the concentration of AVP receptors declined by only about 30%. That weaker effect
of OT is in accord with the demonstration of a much smaller affinity of the
peptide for AVP binding sites. As calculated from data in Fig.5, the Ki
values for AVP and OT were respectively 1 and 110 nM.

The loss of surface receptors may indicate internalization (Catt et al.,
1979) or merely reflect occupancy of sites by endogenous AVP. That possibility seems unlikely in view of the fact that a single and acute injection
of AVP decreased tracer binding by less than 10%, compared with the 80% fall
produced by chronic injections. This was further supported by measuring receptor capacity in pituitaries of Brattleboro homozygotes, 2 min after an
i.v.administration of a high dose of AVP (100ng/100g b.w.); a time lag at
which peptidic hormones such as CRF41(Leroux and Pelletier,1984) predomina-
te on the cell surface. Under these conditions, a difference in 3H-AVP binding of about 10% was observed between pituitary tissues of uninjected and
injected animals (Koch and Lutz-Bucher,1985).

In another set of experiments, we examined the effect of adrenalectomy on pituitary AVP receptor capacity, because lack of circulating glucocorticoids is known to cause accumulation of AVP in the median eminence (Stillman et al.,1977; Silverman et al.,1981), as well as in the anterior pituitary gland (Chateau et al.,1974; Lutz-Bucher et al.,1974).

253

As shown in Fig.7, long-term adrenalectomy (1 week) resulted in a dramatic decrease in pituitary AVP receptor density, which fell to about 20% of control values. That effect was completely reversed by dexamethasone treatment. Interestingly, receptor loss was associated with a nearly 4-fold increase in tissue AVP content, whereas the OT content was unaffected by adrenalectomy (inset). This finding agrees well with evidence that AVP immunoreactivity, in contrast to OT immunoreactivity, increases in the median eminence after adrenalectomy (Zimmerman etal.,1977).

It is of interest to note that adrenalectomy did likewise reduce high affinity pituitary binding sites for 125I-Tyr-CRF (Fig.8 and Wynn et al., 1983); an effect that was antagonized by corticosterone therapy.Moreover, recent findings point to a down-regulation of these sites due to enhanced release of CRF into hypophysial portal vessels(Wynn et al.,1985).

Since pituitary AVP receptors seem to undergo homologous down-regulation we reasoned that if AVP were the sole regulator of its own binder, then , in vasopressin-deficient Brattleboro homozygous rats, adrenalectomy should fail to produce as drastic an effect as in normal rats. The unexpected and striking observation was that, in homozygotes too, adrenalectomy decreased

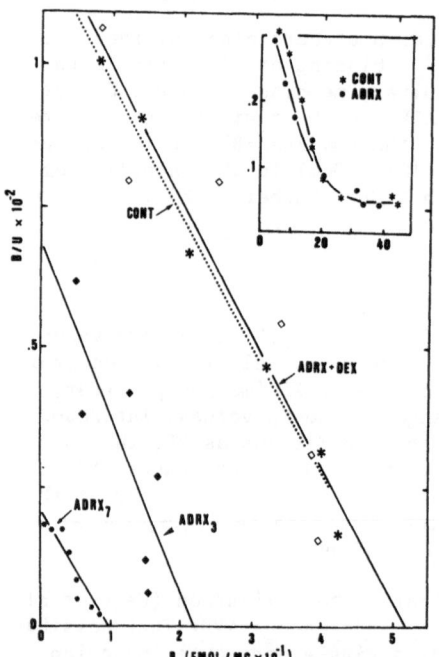

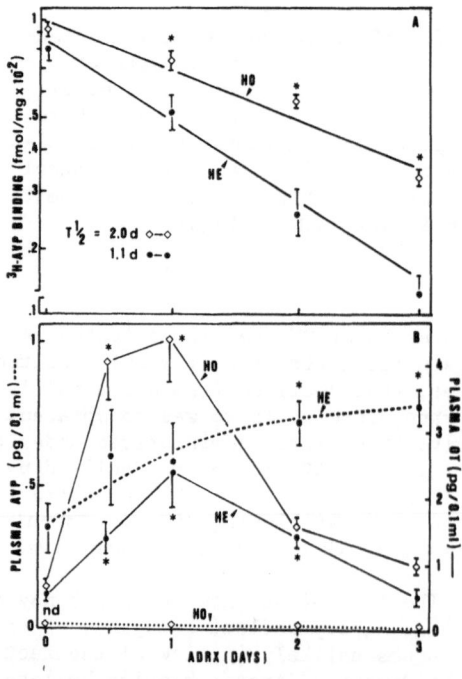

Fig.9. Effect of adrenalectomy,3days (ADRX3) and 7days (ADRX7) post-operation, on pituitary AVP receptor concentration in Brattleboro homozygous rats, injected or not with dexamethasone (40ug/day;DEX).Inset: Scatchard plots of specific binding of ^{3}H-AVP to membrane fractions of hippocampus of control and ADRX7 rats. No apparent effect of ADRX on that tissue was observed.

Fig.10. Kinetics of AVP receptor loss in pituitary of Brattleboro homozygotes (HO) and heterozygotes (HE) induced by ADRX. Binding was measured in the presence of a saturating amount (3nM) of tracer, together or not with 5uM unlabeled AVP(nonspecific binding):Part A. Effect of ADRX on plasma levels of AVP and OT in both groups of animals. Each point is the mean±SE of 3-6 animals:Part B.

^{3}H-AVP binding. The number of pituitary sites fell by about 60% and 80% at 3 and 7 days post-operation, compared with control values (Fig.9).Here again, glucocorticoid replacement reversed the depressing influence of adrenalectomy.

In an attempt to more precisely analyse the effect of adrenalectomy on AVP binding, we followed the kinetics of receptor disappearence both in homozygous and heterozygous rats and, at the same time, measured plasma levels of AVP and OT. As illustrated in Fig.10A, pituitary receptor density declined twice as rapidly in heterozygotes than in homozygotes, with half-life (T1/2) values of 1.1 and 2.0 days,respectively. As to plasma peptide concentrations, it was seen that,after adrenalectomy, plasma AVP rose progressively in heterozygotes, while it was undetectable in homozygotes. On the other hand, plasma OT content increased in both groups of animals, but peak secretions were transient and there was a return to control levels within 3 days post-operation (Fig.10B).

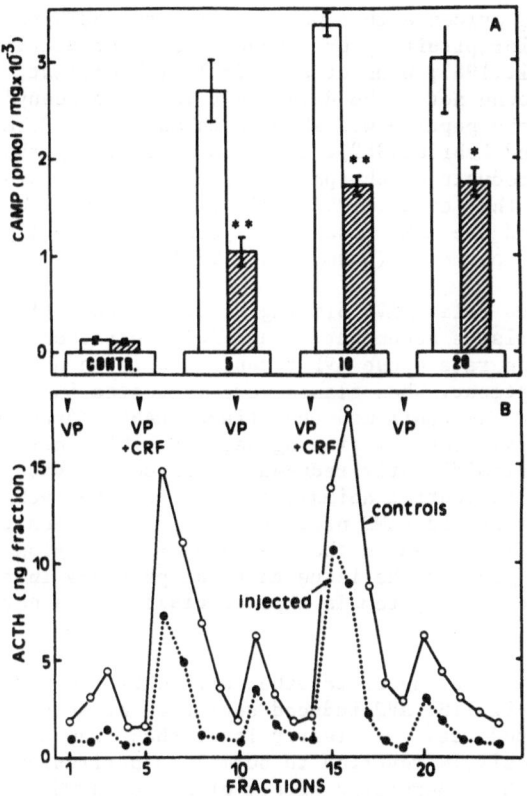

Fig.11. Concentration of cAMP in pituitary
        tissues of control (▢) and chro-
nically VP-treated rats (1 week).Glands
were incubated in the absence (CONTR) and
presence of 30nM AVP along with 4nM CRF41
for 5,10 and 20min.Means ± SE of 4 determinations: part A.
Profile of ACTH secretion of pituitary
fragments of control and VP-injected rats,
in response to 5min-pulses of 15nM AVP
alone or together with 2nM CRF41.Fractions
correspond to 10min collection periods;
Part B. (Koch & Lutz-Bucher, 1985).

These data thus suggest that, in addition to AVP, some other factors may intervene in the regulation of pituitary vasopressin receptors. OT may be such a regulator, but its weak affinity for the binder,together with the transient rise of plasma OT after adrenalectomy make it unlikely that it be the sole factor involved. Most unlikely, also, is the possibility that in homozygous Brattleboro rats the so-called "X" peptide may play a role.That peptide contains a precursor , whose tryptic digestion generates an AVP-like peptide (Russel et al.,1980). However, these investigators showed that the physico-chemical properties of the latter compound were completely different from those of "true" AVP and it is thus probably not recognized by the pituitary receptor. Although the identity of the putative additional regulatory factor(s) is (are) at present not known, it is clear nevertheless that it is (they are) under negative feedback control of glucocorticoids. Also, a direct effect of these steroids on the AVP binding system cannot be excluded.

FUNCTIONAL ACTIVITY OF THE PITUITARY AVP RECEPTOR SYSTEM

There is strong evidence that CRF41 stimulates ACTH release from cortitropes of the anterior pituitary by enhancing cAMP production (Giguère et al. 1982; Aguilera et al.,1983; Wynn et al.,1985). In contrast, the CRF-like activity of AVP was found not to be directly related to adenylate cyclase stimulation, although the peptide was able to enhance cAMP formation triggered by CRF41 (Giguère and Labrie,1982). It seems most likely that Ca ions, as well as metabolic products of phosphatidylinositol lipid breakdown may serve as messengers in that case (Kirk et al.,1982). It has however to be emphasized that the Ca and the cAMP messenger systems are closely interrelated, according to complex patterns (Rasmussen, 1984).

In an attempt to relate AVP binding to biological effect, we compared the extent of cAMP tissue accumulation and ACTH secretion of pituitaries from control rats and from rats whose AVP receptors were down-regulated by chronic vasopressin treatment. When glands were incubated in the presence of CRF41 along with AVP, we confirmed that tissue cAMP content was considerably enhanced (Fig.11A). The new finding was that this increment in nucleotide formation was significantly reduced in tissues of vasopressin-injected animals, compared with control animals treated with the vehicle. Moreover, a significant difference in cAMP production was still seen in both types of pituitaries when incubated in the presence of CRF41 alone (Table 1). These data strongly support the issue that the presence in the pituitary of a functional AVP receptor system is a prerequisite for a normal response to CRF.

The effect of AVP, alone or together with CRF41 on ACTH release, was also investigated (Fig.11B).AVP-induced secretion of ACTH from pituitaries of vasopressin-treated rats was clearly lower than from controls, although the response progressively reverted to normal as perifusion of glands proceeded. Similarly, the synergistic effect of AVP on ACTH peaks produced by CRF41 was also attenuated in down-regulated glands. CRF alone elicited a slightly but not significantly lower ACTH secretion in down-regulated compared to control pituitaries, in contrast to the significant effect on tissue cAMP content (Table 1).

These data strongly suggest the existence of a close correlation between loss of pituitary membrane receptors and attenuation of cAMP production and ACTH release elicited by secretagogues. In their recent study on pituitary CRF receptors, Wynn et al.(1985) reached a similar conclusion in that they showed that the fall in CRF receptor concentration induced by adrenalectomy was associated with a parallel decrease in cAMP formation. However, there was no reduction in ACTH release, that was even enhanced; an effect that was most probably due to the marked stimulation by adrenalectomy of the synthesis and release of ACTH.

Table 1. Effect of CRF on cAMP accumulation in pituitary tissues of control and VP-treated rats.

Incubation conditions	cAMP (pmol/mg protein)		P
	Controls	VP-treated	
Vehicle	152 ± 19	112 ± 10	NS
CRF			
5 min	657 ± 94	320 ± 17	< 0.01
10 min	687 ± 49	453 ± 39	< 0.01
20 min	359 ± 24	338 ± 4	NS

Pituitary fragments from control and chronically VP-treated rats were incubated in the presence or absence of 4nM CRF41. cAMP was radioimmunoassayed. Each value is the mean ± SE of 4 determinations measured in triplicate. (Koch & Lutz-Bucher, 1985).

However, since in our experiments receptor down-regulation was achieved by chronic vasopressin treatment, one may argue that , due to the CRF activity of vasopressin, the reduction in pituitary cAMP content and ACTH secretion may be the consequence of a feedback inhibition exerted by glucocorticoids. The possibility exists, indeed, that the secretion of these steroids may be finally enhanced. That pituitary desensitization is associated with AVP receptor internalization, rather than being indirectly due to glucocorticoids,is suggested by the following observations. We previously provided evidence that glucocorticoid inhibition of ACTH release from perifused pituitary glands occured, in fact, at a site distal to activation of the adenylate cyclase system (Koch et al.,1978). This was later confirmed by Giguère et al.(1982), using cultured pituitary cells incubated with CRF41. A recent report, however, shows that long-term treatment with dexamethasone significantly attenuated CRF-induced cAMP formation in cultured cells (Bilezikjian and Vale,1983). Finally and importantly, cAMP accumulation in pituitary cells was reported to be reduced not as a result of glucocorticoid administration, but after adrenalectomy (Wynn et al.,1985). Dexamethasone injections to adrenalectomized rats did restore a normal basal level of tissue cAMP, as well as the CRF41-induced accumulation of the nucleotide.

In target cells, desensitization is generally associated with internalization of specific receptors (Catt et al.,1979), although this may not always be the case (Peterson et al.,1978; Sheela Rani et al.,1983).In the pituitary gland, loss of CRF41 receptors was found to be related to reduction of CRF41-stimulated cAMP formation, but not to a fall in ACTH secretion (Wynn et al.,1985). We report here evidence for a close relationship between pituitary AVP receptor capacity and both the cAMP and ACTH responses to AVP and CRF41. In contrast, however, in the immature pituitary gland of neonatal rats, there was an apparent lack of correlation among these factors. As depicted in Fig.12, the concentration of high affinity AVP receptor sites was considerably lower in neonates than in adults: 11.7 ± 5.0 vs. 78.6 ± 6.9 fmol/mg (n=3), respectively.In spite of that difference, there was a striking similarity in the ability of pituitary immature and mature cells to respond to AVP, CRF41 and both peptides in combination, both in terms of cAMP production and ACTH release (Fig.13). This may be ascribed to the presence of "spare" receptors, which implies that occupancy of only a fraction of receptors triggers maximum response. Alternatively, the heterogeneous nature of the pituitary cell population may also be involved.

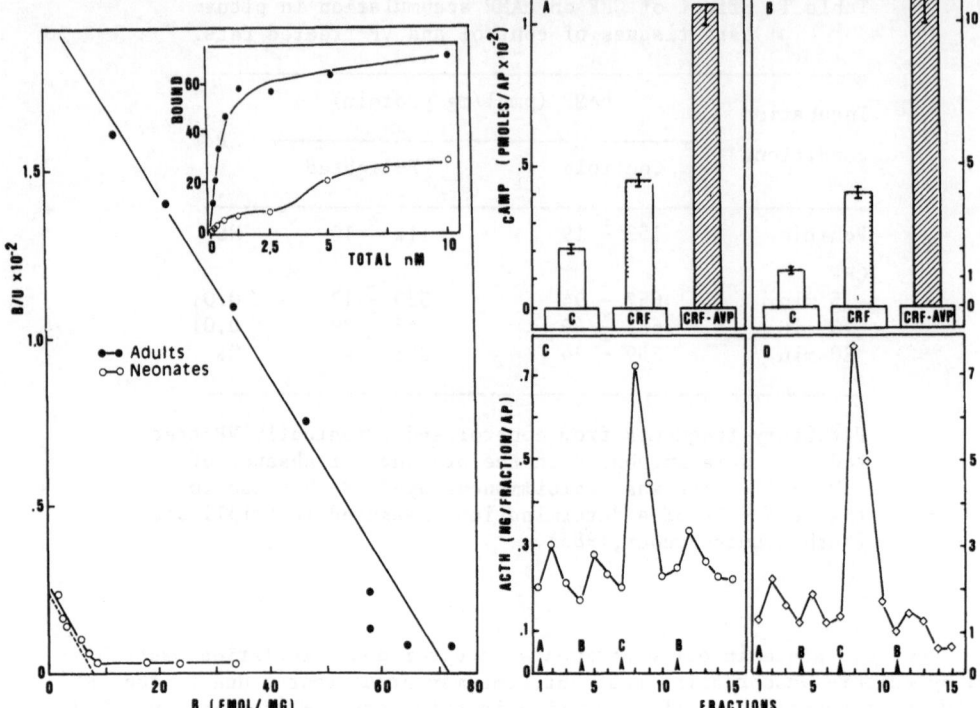

Fig.12. Scatchard plots of specific binding of ^{3}H-AVP to membranes of pituitaries of adult and 10-day-old rats. Inset: specific binding as a function of total ligand concentrations. Measurements were performed with 1 pituitary equivalent of membranes for adult glands, and 10 equivalents for immature ones (200-300 ug protein).(Lutz-Bucher & Koch,1985).

Fig.13. Effect of CRF41, alone or together with AVP on cAMP content of neonatal (part A) and adult pituitaries (part B). ACTH release from perifused neonatal (part C)and adult glands (part D)in response to CRF(A),AVP(B) or both peptides(C). Note the difference of one order of magnitude between cAMP and ACTH values of mature and immature glands. Peptide concentrations were 2nM for CRF41 and 10nM for AVP. AP= anterior pituitary.(Lutz-Bucher & Koch,1985).

CONCLUSION

Among the various factors that, besides CRF, act in concert to affect ACTH secretion, AVP appears to play a major role. That peptide was shown to interact with highly specific receptors in the pituitary gland, that are clearly distinct from those that combine CRF41 and other CRF-like peptides. This AVP receptor system seems to be of a novel type, because its peculiar binding specificity makes it different from classical V1- or V2-types.The concentration of these binding sites appeared to be down-regulated not only by AVP and OT, but also by other factors that are under negative control of glucocorticoids; although a direct effect of these steroids on the receptor cannot be excluded. Evidence was found that suggested that loss of pituitary AVP binder was associated with cell desensitization, assessed both in terms of cAMP production and ACTH secretion in response to AVP,CFR41 and both peptides in concert. The conclusion emerged that full expression of pituitary CRF41-binding sites required the presence and functional activity of AVP receptors as well.

REFERENCES

Aguilera,G.,Harwood,J.P.,Wilson,J.X.,Morell,J.,Brown,J.H.and Catt,K.J.,1983
Mechanisms of action of CRF and other regulators of ACTH release in the
rat pituitary cells,J.of Biol.Chemistry,258:8039.

Aizawa,T.,Yasuda,N.,Greer,M.A.and Sawyer,W.H.,1982, In vivo adrenocorticotropin-releasing activity of neurohypophyseal hormones and their analogs,
Endocrinology, 110:98.

Andersson,K.E.,Arner,B.,Hedner,P.and Mulder,J.L.,Effects of 8-lysine-vasopres-
sin and synthetic analogs on release of ACTH, Acta Endocrinol.69:640.

Antoni,F.E.,Holmes,M.C. and Jones,M.T.,1983,Oxytocin as well as vasopressin
potentiate ovine CRF in vitro, Peptides 4:411.

Antoni, F.E.,1984,Novel ligand specificity of pituitary vasopressin receptors
in the rat,Neuroendocrinology 39:186.

Antoni,F.E.,Holmes,M.C.,Makara,G.B.,Karteszi,M.and Maszlo,F.A.,1984,Evidence
that the effects of arginine-8-vasopressin on pituitary ACTH release
are mediated by a novel type of receptor, Peptides 5:519.

Baertschi,A.J.and Friedli,M.,1985, A novel type of vasopressin receptor in
anterior pituitary corticotrophs? Endocrinology 116:499.

Bilezikjian,L.M.and Vale,W.,1983,Glucocorticoids inhibit CRF-induced production of adenosine 3',5'-monophosphate in cultured anterior pituitary
cells,Endocrinology113:657.

Capponi,A.Favrod-Loune,L.A.,Gaillerd,R.C.and Muller,A.F.,1982,Binding and
activation properties of angiotensin II in dispersed rat anterior pituitary cells. Endocrinology 110:1043.

Catt,K.J.,Harwood,J.P.,Aguilera,G.and Dufau,M.L.,1979, Hormonal regulation
of peptide receptors and target cell responses,Nature 280:109.

Chateau,M.A.,Burlet,A.and Marchetti,J.,1974, La vasopressin-like du lobe
antérieur de l'hypophyse :isolement et identification par son activité
biologique et immunologique,J.Physiol.(Paris)68:10.

Forsling,M.L.,1979,Antidiuretic hormone,Eden Press Inc.,Edinburgh.

Gaillard, R.C.,Schoenenberg,P.,Favrod-Coune,C.A.,Muller,A.F,Marie,J.,Bockaert,
J.and Jard,S.,1984,Properties of rat anterior pituitary vasopressin re-
ceptors: relation to adenylate cyclase and the effect of CRF. Proc.Natl.
Acad.Sci.USA 81:2907.

Gallo-Payet, N.,Escher,E.,Guillon,G.and Jard,S.,1985,Mechanism of vasopressin
stimulated adrenocortical function. Evidence for binding sites. The Endocrine Society, abstract Nr.55.

Gibbs, D.M.,1985,Measurement of hypothalamic corticotropin-releasing factors
in hypophyseal portal blood,Federation Proc.44:203.

Giguère,V.and Labrie,F.,1982,Vasopressin potentiates cyclic AMP and ACTH release induced by CRF in rat anterior pituitary cells,Endocrinology
111:1752.

Giguère,V.,Labrie,F.,Cote,J.,Coy,D.H.Sueiras-Diaz,J.and Schally,A.V.,1982,
Stimulation of cAMP accumulation and corticotropin release by synthetic ovine CRF in rat anterior pituitary cells: site of glucocorticoid
action, Proc.Natl.Acad.Sci. USA79:3466.

Gillies,G.E.,Linton,E.A.and Lowry,P.J.,1982, Vasopressin and the corticoliberin complex,in: Vasopressin, Corticoliberin and Opiomelanocortins,A.J.
Baertschi and J.J.Dreifuss,ed.,Academic Press,New York,p.239.

Holmes,M.C.,Antoni,F.A.and Szentendrei,T.,1984, Pituitary receptors for
CRF: no effect of vasopressin on binding or activation of adenylate
cyclase, Neuroendocrinology 39:162.

Jard,S.,1983, Vasopressin: mechanisms of receptor activation,in: The neurohypophysis: structure, function and control,B.A.Cross and G.Leng,ed.,
Elsevier, Amsterdam,p.383.

Kirk,C.J.,Creba,J.A.,Hawkins,P.T.and Michell,R.H.,Is vasopressin-stimulated
inositol lipid breakdown intrinsic to the mechanism of Ca^{2+}-mobilization
at Vl vasopressin receptors ?,in: The neurohypophysis:structure, function and control, B.A.Cross and G.Leng,ed.Elsevier, Amsterdam,p.405.

Knepel,W.,Homolka,L.,Vlaskovska,M.and Nutto,D.,1984,Stimulation of adreno-
corticotropin/beta-endorphin release by synthetic ovine CRF in vitro.

Enhancement by various vasopressin analogs. Neuroendocrinology 38:344.

Koch,B.,Lutz-Bucher,B.,Briaud,B.and Mialhe,C.,1978,Relationshipe between ACTH secretion and corticoid binding to specific receptors in perifused adenohypophyses,Neuroendocrinology 28:169.

Koch,B.and Lutz-Bucher,B.,1983, Characterization and modulation of high affinity receptors for CRF in the pituitary gland,Neuroendocrinol.Lett.5:227

Koch,B.and Lutz-Bucher,B.,1985, Specific receptors for vasopressin in the pituitary gland: evidence for down-regulation and desensitization to adrenocorticotropin-releasing factors,Endocrinology 116:671.

Labrie,F.,Veilleux,R.,Lefevre,G.,Coy,D.H.,Sueiras-Diaz,J.and Schally,A.V., 1982,CRF stimulates accumulation of adenosine 3',5'-monophosphate in rat pituitary corticotrophs,Science 216:1007.

Leroux,P.and Pelletier,G.,1984,Radioautographic study of binding and internalization of CRF by rat anterior pituitary corticotrophs, Endocrinology,114:14.

Lutz-Bucher,B.,Koch,B. and Mialhe,C.,1974, Présence et mode d'action de l'hormone antidiurétique au niveau de l'antéhypophyse du rat, C.R.Acad.Sci. (Paris),279:1903.

Lutz-Bucher,B. and Koch,B.,1983, Characterization of specific receptors for vasopressin in the pituitary gland, Biochem.Biophys.Res.Comm.115:492.

Lutz-Bucher,B. and Koch,B.,1983,Failure of vasopressin to potentiate the effect of synthetic CRF on ACTH output from intermediate pituitary, Neuroendocrinol.Lett. 5:111.

Lutz-Bucher,B. and Koch,B.,1985, Reduced number of specific vasopressin receptors in the neonatal pituitary gland fails to be associated with parallel changes in cell activity,Neuroendocrinol.Lett.7:67.

Oliver,C.,Mical,R.S.and Porter,J.C.,1977,Hypothalamic-pituitary vasculature evidence for retrograde blood flow in the pituitary stalk,Endocrinology, 101:598.

Petersen,B.,Beckner,S.and Blecher,M.,1978,Hormone receptors.VII.Characteristics of insulin receptors in a new line of cloned neonatal rat hepatocytes,Biochim.Biophys.Acta,542:470.

Rasmussen,H.and Barret,P.Q.,1984,Calcium messenger system:an integrated view,Physiol.Rev. 64:938.

Russel,J.T.,Brownstein,M.J.and Gainer,H.,1980,Biosynthesis of vasopressin, oxytocin and neurophysins: isolation and characterization of two common precursors,Endocrinology 107:1880.

Sheela Rani,C.S.,Keri,G.and Ramachandran,J.,1983,Studies on corticotropin-induced desensitization of normal rat adrenocortical cells,Endocrinology 112:315.

Silvermann,A.J.,Hoffman,D.,Gadde,C.A.,Krey,L.C. and Zimmerman, E.A., 1981, Adrenal steroid inhibition of the vasopressin-neurophysin neurosecretory system to the median eminence of the rat.Neuroendocrinology 32:129.

Spinedi,E. and Negro-Vilar,A.,1984, Arginine vasopressin and adrenocorticotropin release:correlation between binding characteristics and biological activity in anterior pituitary dispersed cells,Endocrinology 114: 2247.

Stillman,M.A.,Recht,L.D.,Rosario,S.L.,Seif,S.M.,Robinson,A.G. and Zimmerman,E.A.,1977,The effect of adrenalectomy and glucocorticoid replacement on vasopressin and vasopressin-neurophysin in the zona externa of the rat,Endocrinology 101:42.

Westendorf,J.M.,Phillips,M.A.and Schonbrunn,A.,1983,Vasoactive intestinal peptide stimulates hormone release from corticotropic cells in culture, Endocrinology 112:550.

Wynn,P.C.,Aguilera,G.,Morell,J.and Catt,K.J.,1983, Properties and regulation of high affinity pituitary receptors for CRF, Biochem.Biophys.Res.Commun.110:602.

Wynn, P.C.,Harwood,J.P.,Catt,K.J. and Aguilera,G.,1985, Regulation of CRF receptors in the rat pituitary gland: effects of adrenalectomy on CRF receptors and corticotroph responses, Endocrinology 116:1653.

Yates,F.E.,Russel,S.M.,Dallman,M.F.,Hedge,G.A.,McCann,S.M.and Dhariwal,A.P., 1971,Potentiation by AVP of ACTH release induced by CRF.Endocrinol.88:3.

THE REGULATION OF PROOPIOMELANOCORTIN GENE EXPRESSION BY

ESTROGEN IN THE RAT HYPOTHALAMUS

James L. Roberts, Josiah N. Wilcox, and Mariann Blum

Center for Reproductive Sciences
Columbia University
630 West 168th Street
New York, NY   10032 USA

INTRODUCTION

Many, if not all, neuropeptides are derived biosynthetically from larger precursor proteins. Proopiomelanocortin (POMC) is an example of such a precursor protein from which beta LPH, ACTH, the MSH's, and beta-endorphin are derived (reviewed in Eipper and Mains, 1980). Immunohistochemical staining using ACTH or endorphin antibodies indicates that these peptides are present in cell bodies in the periarcuate region of the rat hypothalamus extending from the retrochiasmatic area to the premammilary nucleus (Block et al., 1979; Bloom et al., 1978; Finley et al., 1981; Joseph, 1980; Pelletier and Leclerc, 1979; Watson et al., 1978; Watkins, 1980; Zimmerman et al., 1978). It was initially thought that the POMC peptides in the brain were derived by transport and uptake of these proteins from the pituitary, a tissue rich in ACTH and beta-endorphin. However, it has recently been shown that the hypothalamus actually synthesizes POMC (Liotta et al., 1979), contains POMC mRNA (Civelli et al., 1982; Gee et al., 1983), and releases beta-endorhin into the portal blood (Wardlaw et al., 1980; Sarkar and Yen, 1984), suggesting that the hypothalamic POMC cells synthesize POMC independent of the pituitary.

An accumulating body of evidence suggests that the POMC synthesized in the hypothalamus is related to the reproductive function of the animals. Beta-endorphin administration has profound influences on pituitary secretion, resulting in decreases in plasma luteinizing hormone and follicle stimulating hormone, with an increase in plasma growth hormone and prolactin levels (reviewed in Meites et al., 1979). Estrogen administration has been shown to decrease radioimmunoassayable endorphin levels in the basal hypothalamus of rats (Wardlaw et al., 1982), but whether this represents a direct effect on endorphin synthesis is not clear. Beta-endorphin levels decrease in the periarcuate nucleus and increase in the median eminence on the afternoon of proestrus in rats (Barden et al., 1981), suggesting that the ovarian hormones may alter release and/or transport of this neuropeptide from the cell bodies in the arcuate.

In order to resolve this controversey, we set out to study the effects of ovariectomy and estrogen replacement on hypothalamic POMC mRNA levels. While, in general, mRNA levels closely parallel the synthesis of neuropeptides, there is no evidence that neuropeptide mRNAs are axonally transported in the brain, so are not subject to the same problems of interpretation as the neuropeptide levels. Thus, any changes observed in POMC mRNA levels after estrogen replacement may be interpreted as reflecting alterations in POMC biosynthesis. In addition, we used an in vitro nuclear runon transcription assay in order to measure the number of RNA polymerase II complexes on the POMc gene in the arcuate POMC neurons as a function of estrogen treatment. These studies have allowed us to begin to interpret the molecular mechanism by which estrogen acts to alter POMC mRNA levels.

The POMC neurons of the periarcuate hypothalamus have an extensive set of projections to a variety of regions of the brain. For example, they project forward into the preoptic area, laterally to the amygdaloid area, and posteriorly down the brain stem. Recent studies coupling POMC peptide immunohistochemistry with retrograde tracing of axonal projections indicate that the periarcuate POMC neurons may be divided into discrete subgroups based on brain region to which they project. For example, the POMC neurons from the middle region of the periarcuate nucleus are, in general, the ones that project to the preoptic area of the rat (N. McCluskey, Yale Univ., personal communication). If one was able to functionally subdivide the POMC neurons prior to analysis, it may be easier to interpret changes in POMC gene expression. Unfortunately, this has not been possible based upon such techniques such as Palkovits' punch dissection of the periarcuate region. Alternatively we have used a single cell mRNA assay which we have developed from the in situ cDNA:mRNA hybridization procedure. By coupling our semi-quantitative in situ hybridization procedure with fluorescent dye retrograde transport techniques, we have been able to identify and quantitate the amount of POMC mRNA in neurons which project to the preoptic area of the rat hypothalamus. It is using this type of technique that we feel will allow neurobiologists to begin to measure the effects of various types of environmental, behavioral, and hormonal influence on expression of specific gene in discrete populations of neurons in the brain.

MATERIALS AND METHODS

Animals

Female Sprague-Dawley rats, 150-200g body weight were housed with a 14/10 light/dark cycle (lights on at 0600), and given food and water ad libitum. Animals were bilaterally ovariectomized under ether anesthesia 2 or 3 wk prior to experimentation. Animals receiving estrogen were implanted with silastic capsules containing 262ug estradiol 17-beta/ml in sesame oil (10mm/100g b.w.) according to the method or Moreines (1980). The capsules were implanted s.c. in the lower abdomen under light ether anesthesia. In our laboratory, these implants produce mean serum estrogen levels of 100pg/ml measured 1 or 4 d after implantation. Control animals received similar silastic implants containing sesame oil alone.

Brain Dissection

At selected times after hormonal treatment, animals were sacrificed by decapitation, the brain removed and dissected on a cold glass plate as follows: a coronal slice limited anteriorally by the posterior border of the optic chiasm and posteriorally by the anterior border of the mammilary bodies was removed. The arcuate median eminence region (Arc-ME) was dissected from this coronal section by making two diagonal cuts extending from the hypothalamic fisures to the third ventricle, about 3mm from the ventral surface of the brain. The Arc-ME weighed approximately 5mg and represented the smallest dissection possible that contains all the hypothalamic POMC neurons as determined by immunocytochemistry.

Isolation of RNA

Total nucleic acid was isolated from the Arc-ME sample by proteinase-K/SDS/phenol extraction (Chen et al., 1982) as follows: after dissection the tissue was immediately placed in SET buffer (0.5% SDS, 10mM EDTA, 20mM Tris, pH 8.0) containing 65ug/ml proteinase-K, and homogenized by vigorous pipetting with a large bore micropipet. The homogenates were then incubated for 2hr at 42C, 3ul of 3% PMSF and alcohol added, the samples frozen in liquid nitrogen, and stored at -70C. After all the samples had been collected, the homogenates were thawed and extracted twice with phenol/chloroform, followed by a single ethanol/sodium chloride precipitation overnight at -20C. Nucleic acids were collected by centrifugation. The pellet was redissolved in 300ul of sterile 1XTE (10mM Tris, pH 8.0, 1mM EDTA) and aliquots taken for the estimation of RNA concentration by ethidium bromide/agarose assay described previously (Chen et al., 1982). Individual Arc-ME dissections yielded approximately 15-20ug of total RNA by this procedure.

Quantitation of mRNA

POMC mRNA was determined in individuals or pools Arc-ME samples by filter hybridization/dot blot (Wilcox and Roberts, 1985). Aliquots of each total nucleic acid sample cntaining approximately 2-4ug total RNA were diluted in sterile 1xTE to 100ul total volume. The samples were heated at 65C for 10min to denature the RNA and diluted with 100ul of cold 20xSSC (1xSSC = 150mM sodium chloride, 15mM sodium citrate, pH 7.2). Samples were then bound to nitrocellulose filters using a manifold dot blot apparatus (Schleicher and Schuell) using a gentle vacuum and rinsed with 100ul of 10xSSC. The nitrocellulose filters were then air-dried and baked under vacuum at 80C for 3 hr.

Prehybridization and hybridization was carried out according to the method of Wahl et al. (1979). Hybridization was performed using a ^{32}P labeled cDNA specific for the POMC gene (Chen et al., 1982). After hybridization, the unhybridized probe was rinsed off of the blot, the nitrocellulose dried and exposed against x-ray film. The intensities of the different dots were measured by quantitative autoradiography. Normalization of all dots for total amount of mRNA spotted was made using a 32P labeled oligo dT probe, as previously described (Murphy et al., 1983). Such normalization greatly enhanced our ability to meausre small changes in POMC mRNA levels (Wilcox and Roberts, 1985).

Nuclear Transcription Runon Assays

The transcription assay was performed using a previously published procedure (Evans et al., 1981). The only significant modification was the inclusion of an ATP regenerating system using creatin phosphokinase and creatin phosphate. Nuclei were isolated from the Arc-ME of 3 week ovariectomized animals treated with similar silastic tubing implants as described above. POMC mRNA transcripts from the in vitro transcription reaction were quantitated by hybridization for 48hr to a POMC DNA filter containing the entire protein coding region of the exon region 3 of the POMC rat gene (Eberwine and Roberts, 1984). Background hybridization was measured using a pBR322 "nonspecific" DNA control filter. Data is expressed relative to the total amount of incorportation in the in vitro transcription assay.

In Situ cDNA:mRNA Hybridization

The in situ cDNA:mRNA hybridization procedure has been described previously in detail (Wilcox et al., 1985). Briefly, this procedure involved perfusing the animal with 4% paraformaldehyde for 20 min, removing the brain, immersing it in a sucrose/PBS solution for 1 hr, freezing the tissue in OTC embedding compound, and taking 10 micron frozen sections on a cryostat. The sections were then thaw-mounted onto subbed slides and immediately stored at -70C with dessicant. To prepare for hybridization, the tissue was thawed in a 10ug/ml proteinase-K solution to inhibit ribonucleases as well as permeablilize the tissue. The proteinase-K solution was rinsed off and replaced 2 hr later with a hybridization solution containing the same components plus 10-20,000 CPM of tritiated POMC cDNA probe. After 24 hr of hybridization, the unhybridized probe is rinsed off by overnight washing with 0.5xSSC, and the tissue dried and coated with a Kodak NTB2 emulsion. After exposure for 2-6 wk, the sections are developed and counterstained with eosin and hemotoxalin.

In experiments where retrograde axonal tracing was performed, the procedure was essentially the same with the exception that the animals were injected stereotaxically with 100nl over 5 minutes of Fast Blue dye in the preoptic area according to stereotaxic coordinates 3 days prior to perfusion with paraformaldehyde. The location of the Fast Blue dye was then identified by fluorescence microscopy.

RESULTS

We examined the effect of estrogen or oil implants left in place for 1 or 3 days on Arc-ME POMC mRNA levels in 2 week ovariectomized rats (Wilcox and Roberts, 1985). The nucleic acid extract originating from the Arc-ME of each group of animals was combined into a single sample and spotted five times on a nitrocellulose filter. The filter was then hybridized to the ^{32}P probe and normalized to the total poly-A RNA, as described in the Methods section. There is no observable effect of estrogen treatment in the amount of total RNA isolated or the amount of poly-A+ RNA detected in the dot blot assay, relative to the non-estrogen treated castrated controls. One day of estrogen treatment did not significantly affect the Arc-ME POMC mRNA levels relative to oil controls (Table 1). However, 3 days of estrogen treatment had significantly reduced POMC levels in the Arc-ME to 61% of oil treated controls (T=5.21, df8, p<0.001). This observation was verified in two similar experiments.

The effect of estrogen on POMC gene transcription in castrated animals was also analyzed in a similar type of experiment (Table 2). Female rats were ovariectomized for 3 wk and then given silastic implants as described for the mRNA studies. The animals were subsequently sacrificed and nuclei isolated from the periarcuate region of the hypothalamus. These nuclei were shown to specifically transcribe the POMC gene at a level of 0.012% of total

Table 1. Relative POMC mRNA Levels in Arc-ME

	1 Day	3 Day
Oil	1.00+0.11	0.82+0.03
Estrogen	0.93+0.06	0.50+0.06

Relative amounts of POMC mRNA in Arc-ME after treatment with estrogen or oil for 1 or 3 days. Nucleic acid samples were pooled from each group of rats (N=5), and 2ug total RNA spotted 5 times on nitrocellulose filters. Relative amounts of POMC mRNA are expressed as mean ug equivalents of POMC mRNA per ug equivalents of poly-A RNA as described in Methods (+/- standard deviation).

Table 2. POMC Transcription in Arc-ME Nuclei

Treatment	POMC Transcription
OVX-control	115 PPM
OVX-20'E	98 PPM
OVX-60'E	68 PPM
OVX-4d E	59 PPM

Levels of POMC transcription were analyzed as described in the Methods section. Data is expressed in parts per million (PPM) as the ratio of CPM bound specifically to the POMC DNA filter relative to the total number of CPM put into the hybridization reaction.

transcription. Animals treated with estrogen were shown to have a decreased level of POMC gene transcription, detectable after only 1 hr of treatment.

In situ hybridization analysis of POMC neurons shows that one can identify by this method those neurons which contain POMC mRNA (Fig. 1). By coupling this procedure with retrograde axonal tracing, three classes of neurons can be identified (data not shown). There are cells in which there is blue dye present, but no POMC in situ autoradiographic grains; cells in which there are autoradiographic grains only showing the presence of POMC mRNA; and a group of cells which show both blue dye and POMC in situ autoradiographic grains, implying that these cells both express the POMC gene and project to the preoptic area of the rat brain. Only in the medial areas of the periarcuate regions of the hypothalamus showed substantial numbers of POMC neurons which colocalized with the retrogradely transported dye, in agreement with the immunohistochemical studies discussed above.

DISCUSSION

We have shown that 3 days of estrogen treatment to a 2 week ovariectomized rat reduces hypothalamic mRNA lvels by 40%. These results are consistent with those of Wardlaw and collegues (1982), who observed reduction in beta-endorphin in ovariectomized rat hypothalamus after estrogen treatment. It was unclear from their studies whether estrogen was acting on the synthesis or release of POMC peptides from the hypothalamus. Since neuropeptides are secreted and transported from their site of synthesis, the observed decrease in neuropeptide at the cell body may not reflect decreased synthesis, but may instead reflect an increased rate of synthesis coupled with an enhanced transport of the neuropeptide to other regions of the brain. That the decline in POMC peptide has been associated with the decline in POMC mRNA levels strongly supports the interpretation that estrogen acts to reduce

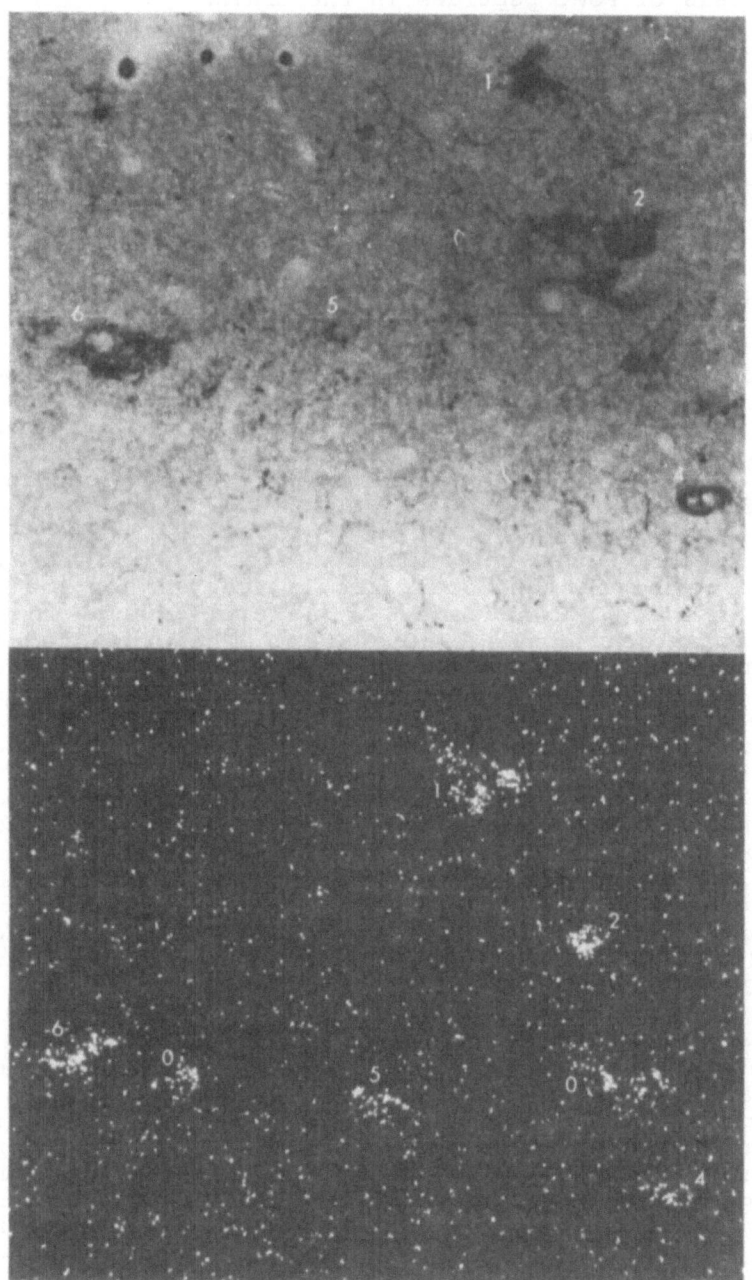

Figure 1. POMC _in situ_ cDNA:mRNA hybridization and immunocytochemistry (ICC) on serial section in the periarcuate region of the rat hypothalamus. 10 micron coronal sections from rat hypothalamus were (top) processed for beta-endorphin ICC or (bottom) processed for _in situ_ hybridization. Note the correspondance of autoradiographic grains and ICC (numbered cells) over a discrete subset of the neurons in this figure.

the synthesis of POMC peptides in the brain.

The changes in POMC mRNA levels elicited by estrogen treatment appear to be mediated, at least in part, by similar effects of estrogen treatment on transcription of the POMC gene in the periarcuate region of the hypothalamus. Estrogen clearly inhibited POMC gene transcription within 60 min after hormone administration. We hypothesize that the change in transcription of the POMC gene after estrogen treatment eventually causes a reduction in the level of POMC mRNA, which would reduce the synthesis of POMC peptides. The fact that estrogen induced changes in POMC gene expression were detected the thousands of copies of mRNA in the cell, changes in gene transcription often require many hours to days to be manifested as alterations in the level of the large pool first at the level of transcription is not suprising when compared with other steroid responsive gene expression systems. Since a single gene is responsible for producing of specific mRNA.

The small changes in POMC mRNA levels in the Arc-ME after estrogen treatment could result from either small changes in all POMC neurons, or a large change in a particular subset of the POMC system. The presence of large unresponsive populations of cells would dilute the observed effect of estrogen on POMC mRNA levels. This would be analogous to the effects of gluccorticoids on POMC gene expression see in the anterior and intermediate lobes of the rat pituitary gland. It had originally been reported that glucocorticoid treatment of adrenalectomized rats caused a 3-4 fold increase in whole pituitary POMC mRNA levels (Nakanishi et al., 1977). However, when anterior and intermediate lobes are separated and assayed independently, a 30-50 fold differences are observed in the anterior lobe, with no change detected in the intermediate lobe (Schachter et al., 1982; Birnberg et al., 1983). The presence of a large population of non-glucocorticoid responsive POMC cells in the original pituitary assay had diluted out the observed effect. Since less than 10% of the POMC cells in the arcuate nucleus accumulate radioactive estrogen (Morrell et al., 1983), the possibility that a large population of nonresponsive cells reduce the overall effect of estrogen in the brain cannot be discounted. Complete delineation of this problem awaits the study of POMC gene expression in the arcuate neurons by in situ mRNA hybridization coupuled with techniques to functionally subdivide the POMC neurons.

Preliminary studies have shown the feasibility of coupling the POMC mRNA single cell quantitation assay with retrograde axonal tracing techniques. We were able to identify those POMC neurons which projected to the preoptic area by the retrograde tracing methodology coupled with the in situ hybridization methodology on the same section. We are curently using this technique to examine the effect of estrogen on only those POMC cells projecting to the preoptic and/or other areas of the rat brain.

AKNOWLEDGEMENTS

This research was supported by grant AM-27484 to JLR and grant HD-07093 to JNW. We would like to thank Carleen Ippolitto for technical assistance, and Edith Kupsaw for secretarial assistance. Preliminary retrograde dye tracing experiments were performed in collaboration with T. O'Donohue and B. Chronwall of NICDS.

REFERENCES

Barden, N., Merand, Y., Rouleau, D., Garon, M., and Dupont, A., 1981, Brain Res., 204:441.

Birnberg, N.C., Lissitzky, J.D., Hinman, M., and Herbert, E., 1983, Proc. Natl. Acad. Sci. USA, 80:6982.

Block, B., Bugnon, C., Fellmann, D., Lenys, D., and Gouget, A., 1979, Cell. + Tiss. Res., 204:1.

Bloom, F., Battenberg, E., Rossier, J., Ling, N., and Guillemin, R., 1978, Proc. Natl. Acad. Sci. USA, 75:1591.

Chen, C.L.C., Dionne, F.T., and Roberts, J.L., 1982, Proc. Natl. Acad., Sci. USA, 80:2211.

Civelli, O., Birnberg, N., and Herbert, E., 1982, J. Biol. Chem., 257:6783.

Eberwine, J.H. and Roberts, J.L., 1984, J. Biol. Chem., 259:2166.

Eipper, B.A., and Mains, R.E., 1980, Endo. Rev., 1:1.

Evans, M.I., Hager, L.J., and McKnight, G.S., 1981, Cell, 25:187.

Finley, J.C.W., Lindstrom, P., and Petrusz, P., 1981, Neuroendo., 33:28.

Gee, C.E., Chen, C.L.C., Roberts, J.L., Thompson, R., and Watson, S.J., 1983, Nature, 306:374.

Joseph, S.A., 1980, Am. J. Anat., 158:553.

Liotta, A.S., Gildersleeve, D., Brownstein, M.J., and Kreiger, D.T., 1979, Proc. Natl. Acad. Sci. USA, 76:1448.

Meites, J., Bruini, J.F., Van Vugt, D.A., and Smith, A.E., 1979, Life Sci., 24:1325.

Moreines, J.D.K., 1980, Ph.D. thesis, Univ. of Michigan.

Morrell, J.I., McGinty, J., and Pfaff, D.W., 1983, Soc. Neurosci. 13th Ann. Mtg., Abs. #27.1.

Murphy, D., Brickell, P.M., Latchman, D.S., Willison, K., and Rigby, P.W.J., 1983, Cell, 35:865.

Nakanishi, S., Kita, T., Taii, S., Imura, H., and Numa, S., 1977, Proc. Natl. Acad. Sci. USA, 74:3283.

Pelletier, G. and Leclerc, R., 1979, Endo., 104:1426.

Sarkar, D.K. and Yen, S.S.C., 1984, Seventh Int. Cong. Endo., Abs. #2062.

Schachter, B.S., Johnson, L.K., Baxter, J.D., and Roberts, J.L., 1982, Endo., 110:1442.

Wahl, G.M., Stern, M., and Stark, G.R., 1979, Proc. Natl. Acad. Sci. USA, 76:3683.

Wardlaw, S.L., Thoron, L., and Frantz, A.G., 1982, Brain Res., 245:327.

Wardlaw, S.L., Wehrenberg, W.B., Ferin, M., Carmel, P.W., and Frantz, A.G., 1980, Endo., 106:1323.

Watkins, W.B., 1980, Cell Tiss. Res., 207:65.

Watson, S.J., Richard, C.W., and Barchas, J.D., 1978, <u>Science</u>, 200:1180.

Wilcox, J.N., Gee, C.E., and Roberts, J.L., 1985, <u>Methods in Enzymology</u>, M. Conn, ed., Academic Press, New York.

Wilcox, J.N. and Roberts, J.L., 1985, <u>Endo.</u>, in press.

Zimmerman, E.A., Liotta, A., and Kreiger, D.T., 1978, <u>Cell. Tiss. Res</u>, 186:393.

DIFFERENTIAL REGULATION OF ACTH AND β-ENDORPHIN BY CONTROLLED PROCESSING

Jack Ham and Derek G. Smyth

National Institute for Medical Research
The Ridgeway
Mill Hill, London NW7 1AA

INTRODUCTION

It is well known that the biologically active peptides ACTH and β-endorphin are formed from a single prohormone, pro-opiomelanocortin (POMC), and there have been several reports that the two peptides are released concomitantly (Allen et al., 1982; Mains and Eipper, 1981). However it appears that these precursor related peptides are not always released in equivalent amounts. It has been shown, for example, that stimulation of AtT-20 tumour cells with CRF leads to a non-stoichiometric release (Hook et al., 1982) and in vivo evidence has been obtained that application of certain forms of stress to rats leads to preferential secretion of ACTH (DeSouza and Van Loon, 1985). If it can be shown that mechanisms exist for the excision of one region of a prohormone rather than another, it would be reasonable to expect that the processing of multifunctional prohormones may be flexible and that various combinations of activities may be generated to fulfil complex physiological functions.

We present here the results of experiments employing monolayer cultures of anterior pituitary cells which show that the pattern of the various forms of ACTH and β-endorphin that occur intracellularly can differ strikingly from the forms that are released during the course of stimulated secretion. The data point to the existence of different intracellular pools of peptides which can be secreted independently.

PROCESSING OF ACTH AND β-ENDORPHIN IN CULTURED ANTERIOR PITUITARY CELLS

The anterior region of the pituitary gland contains two overlapping peptides derived from the C-terminal region of the POMC prohormone. They are the 31 residue peptide β-endorphin which is a potent analgesic agent (Feldberg and Smyth, 1977; Bradbury et al., 1977; Loh et al., 1976) and its precursor polypeptide lipotropin which is devoid of opiate activity. Although the lipotropin (LPH) and β-endorphin in the anterior pituitary generally occur in similar concentrations (Smyth and Zakarian, 1980; Zakarian and Smyth, 1982a), recent investigations have shown that the proportions of the two peptides can vary. In a series of experiments on the intracellular levels of LPH and β-endorphin in rat anterior pituitary, it has been found that the relative concentrations of the two peptides exhibit circadian variation, with values ranging from

approximately 4:1 to 1:1.   These findings indicated that proteolysis of
the β-endorphin prohormone is a dynamic process and that physiological
factors are probably involved in its regulation.

The initial gene product from which β-endorphin is derived is a
complex prohormone, pro-opiomelanocortin, which undergoes post-translational
processing to give a series of bioactive peptides.   A surprising finding,
made in early studies of this prohormone (Bradbury et al., 1975), is that
the processing reactions do not go to completion.   For example, in the
anterior pituitary β-endorphin occurs in company with a larger polypeptide
(lipotropin) which is formed as an intermediate; in the pars intermedia it
is found in company with a number of more heavily processed derivatives
comprising N-acetylated peptides and C-terminally shortened forms
(Zakarian and Smyth, 1982b).   There is thus a substantial difference in
the degree of proteolytic processing exhibited by pro-opiomelanocortin in
these two regions of pituitary.   In cultures of anterior pituitary cells,
on the other hand, in addition to lipotropin and β-endorphin there is a
significant amount of the intact prohormone and this in vitro system
offers an attractive model for the study of factors that affect prohormone
processing.

Suspensions of pituitary cells were obtained by digestion of 1mm^3
pieces of anterior pituitary tissue using trypsin and collagenase and
they were plated out in tissue culture flasks in the presence of medium
containing 10% foetal calf serum.   The cells were allowed to divide until
75% confluency was attained and prior to experimental investigation the
media were replenished with 2% foetal calf serum (Ham and Smyth, 1984).
Incubations were carried out in the presence and absence of corticotropin
releasing factor (CRF), epinephrine and the inhibitory factor dexamethasone.
The immunoreactive (ir) ACTH and β-endorphin were determined in the cell
extracts and in the culture media.

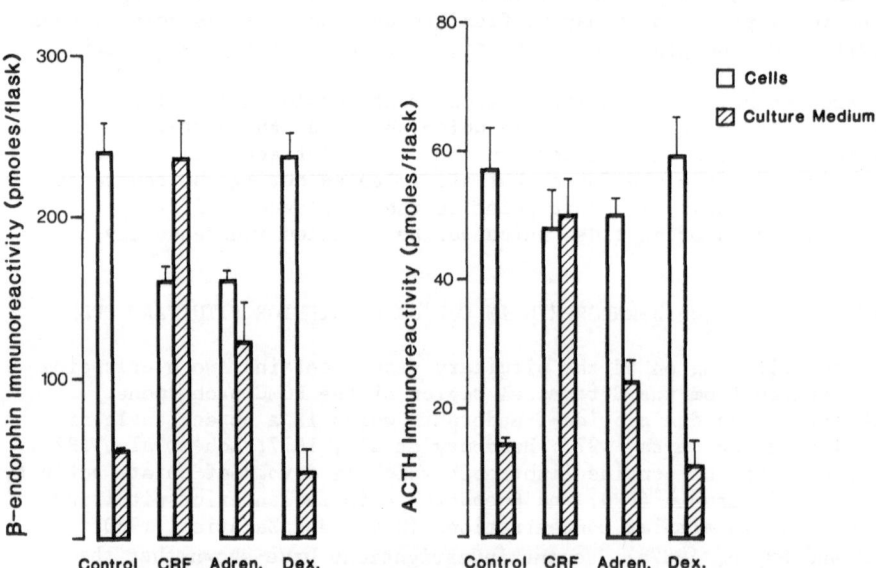

Fig. 1   Influence of corticotropin-releasing factor, epinephrine and
          dexamethasone on intra- and extra-cellular β-endorphin and
          ACTH in primary cultures of rat anterior pituitary cells.

It was shown that 10^{-8}M CRF stimulated the release of both ACTH and
β-endorphin, as was expected, and this was associated with a decrease in
the intracellular levels of the peptides.   Stimulation was also observed
with 10^{-7}M epinephrine, though the effects were less marked.   Dexamethasone,
in contrast, had little influence on basal release of ACTH and β-endorphin
(Fig. 1).   A rather surprising finding was that in these relatively short
incubations there appeared to be an imbalance in the concentrations of the
ACTH and β-endorphin related peptides that were released.   Inclusion of
the ^{125}I-labelled peptide markers in the culture media demonstrated that
no significant degradation took place during a 2 h incubation and the
apparent imbalance in the amounts of the peptides released cannot be due
to differential degradation.   However, ACTH and β-endorphin are undoubtedly present in stoichiometric proportions in their common prohormone so
the differing amounts of the two peptides secreted can be attributed to
differential packaging in the secretory granules or to their exposure to
intracellular enzymes that affect one peptide more than the other.

To determine the molecular forms of ACTH and β-endorphin produced in
the pituitary cells, the cell extracts and supernatants were fractionated
by size exclusion chromatography in a dissociating solvent which prevented
possible aggregation phenomena.   In the culture media obtained from cells

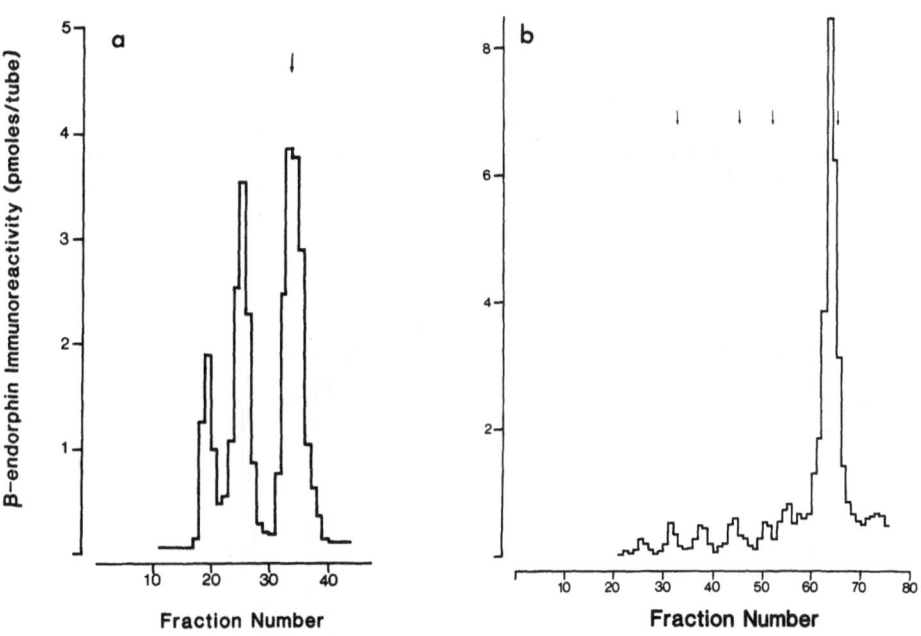

Fig. 2a   Pattern of β-endorphin immunoreactive peptides released from rat
            anterior pituitary cells during basal secretion.   The peptides
            were resolved by gel filtration on Sephadex G75 in 20% acetic
            acid; the arrow indicates the elution position of ^{125}I β-endorphin.
            The two peaks emerging ahead of β-endorphin correspond to pro-
            opiomelanocortin and lipotropin.

Fig. 2b   Identification of β-endorphin 1-31 as the principal form of
            β-endorphin produced in rat anterior pituitary cells.   The
            peptides were resolved by ion exchange chromatography on
            SP Sephadex C25.   The arrows indicate the elution positions
            (L to R) of α,N-acetyl β-endorphin 1-27, β-endorphin 1-27,
            α,N-acetyl β-endorphin 1-31 and β-endorphin 1-31.

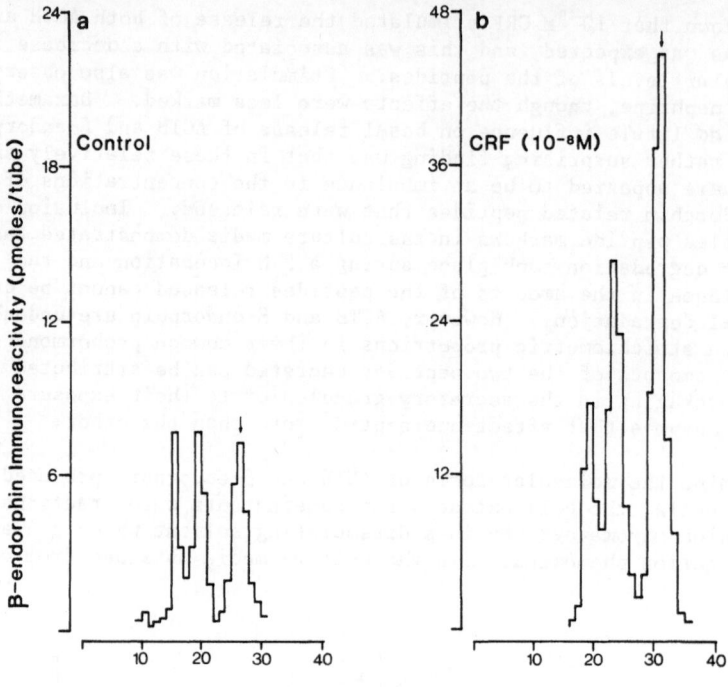

Fig. 3   Comparison of the molecular forms of immunoreactive β-endorphin released in a 2 h period from rat anterior pituitary cells during (a) basal and (b) CRF stimulated secretion.   The arrow indicates the elution position of $|^{125}I|$-β-endorphin.

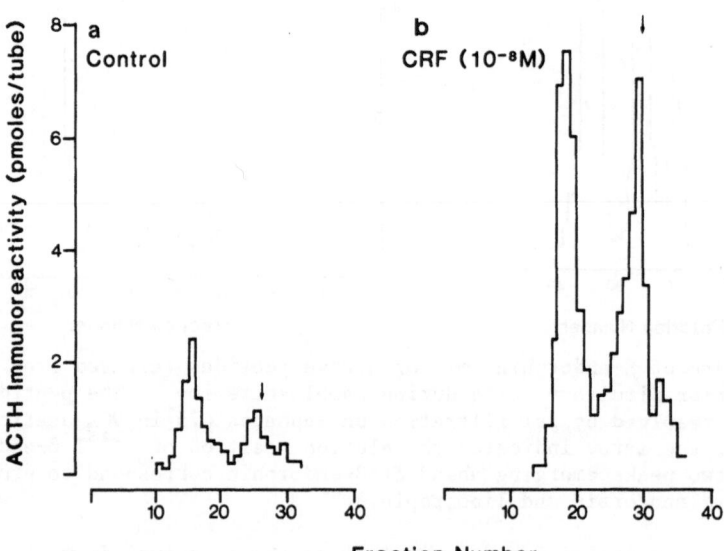

Fig. 4   Comparison of the molecular forms of immunoreactive ACTH released over 2 h from rat anterior pituitary cells during (a) basal and (b) CRF stimulated secretion.

undergoing basal secretion, three forms of β-endorphin immunoreactivity were identified: the two major forms corresponded to lipotropin and β-endorphin and the minor form to the intact prohormone (Fig. 2a). Further resolution of the peptides with the molecular size of β-endorphin was carried out by ion exchange chromatography, which showed that β-endorphin 1-31 was produced with a high degree of specificity (Fig. 2b); notably a negligible amount of its shortened or acetylated forms was present.

Analysis of the peptides released from the anterior pituitary cells stimulated by 10^{-8}M CRF showed that the same three peptides were elaborated (Fig. 3); however they were present in different proportions from the basal pattern, with β-endorphin clearly the major component (Ham and Smyth, 1985).

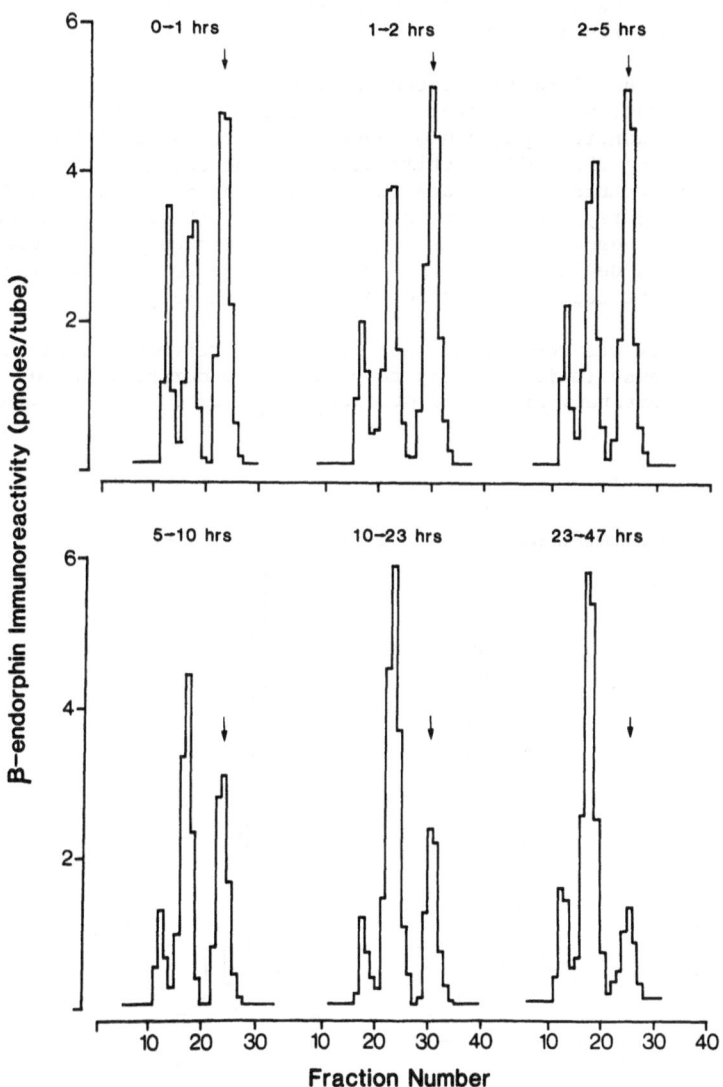

Fig. 5   Time dependent changes in β-endorphin immunoreactive peptides released during stimulation of rat anterior pituitary cells by CRF.

Similar analysis of the molecular forms of ACTH in the control cultures
showed that the major peptide released was a high molecular weight
precursor and it was accompanied by only small amounts of 4.5K bioactive
ACTH. However on CRF stimulation the peptides released were found to
include a significantly higher proportion of the bioactive form (Fig. 4),
in line with the results obtained with the different forms of β-endorphin.
The total immunoreactive β-endorphin released again proved to exceed the
total immunoreactive ACTH in agreement with other reports (De Souza and
Van Loon, 1985; Allen et al., 1978). Both for the various forms of
β-endorphin and of ACTH, the experiments show that the pattern of peptides
released under basal conditions can differ from the peptides released
during stimulated secretion. The data obtained are consistent with the
view that a variety of peptide pools exist which are secreted under
different conditions. This is in accord with evidence that there are
two routes of POMC processing in AtT-20 cells, a constitutive pathway
which is involved in basal secretion and a regulated pathway which is
influenced by the action of secretagogues (Gumbiner and Kelly, 1982).

    To obtain more detailed information on a possible connection between
processing and secretion, experiments were carried out to investigate
whether or not an identical pattern of peptides is released at various
time intervals during CRF (10^{-7}M) stimulation. The results revealed
that a progressive change takes place in the patterns of the released
peptides. In the early stages, as shown previously (Fig. 2a), the main
component was 3.5K β-endorphin. After 10 h of continuous stimulation,
however, there was a decline in the proportion of β-endorphin to lipotropin
and by 22 h the major peptide released was lipotropin (Fig. 5).

    The possibility was considered that prolonged treatment of the cells
with CRF might exhaust their capacity to produce β-endorphin, in which
case release of immaturely processed products might take place. Examination

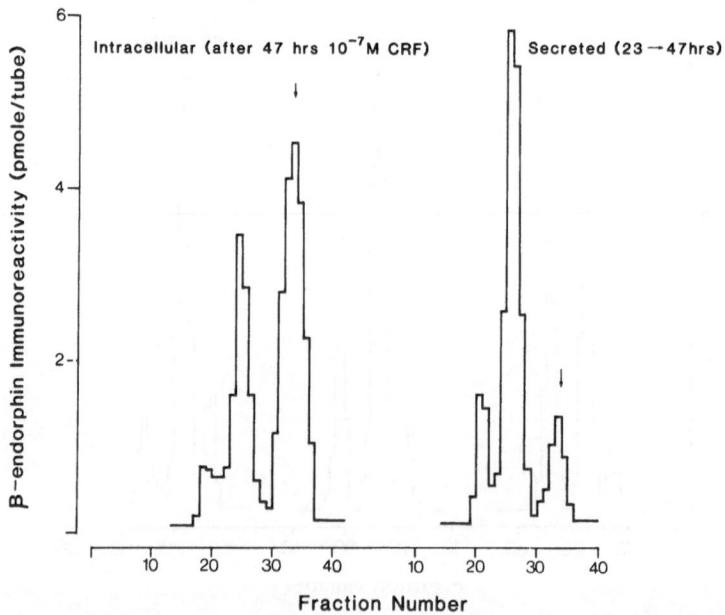

Fig. 6   Comparison between intracellular and released β-endorphin
related peptides after chronic stimulation of rat anterior
pituitary cells with CRF.

of the peptides remaining in the cell after chronic stimulation, however, in fact showed that the intracellular peptides contained a high proportion of fully processed bioactive β-endorphin (Fig. 6).   It is again clear that the pattern of peptides secreted can be quite different from the pattern of peptides remaining in the cell and the results point to the existence of secretory granules that contain different segments of the POMC seque..ce.

With the demonstration that CRF can promote the release of a specific group of peptides derived from POMC, it became of interest to learn whether a factor that has an inhibitory influence on secretion might lead not only to the release of a decreased amount but of an altered pattern of peptides. Acute treatment of the anterior pituitary cells with 10^{-6}M dexamethasone was found to produce no visible change in the amounts of immunoreactive ACTH and β-endorphin released but prolonged treatment led to a reduced rate of peptide release and a consequent accumulation within the cell. Examination of the different forms of ACTH and β-endorphin secreted revealed a significant decrease in the extent of processing, the peptides containing an unusually high concentration of intact prohormone (Fig. 7). The pattern of the peptides remaining in the cells differed from the secreted peptides in that β-endorphin was the major component.   Thus again the nature of the peptides released was not representative of the average intracellular content of the cells.   In addition it is of particular interest that the processing patterns exhibited by the peptides released in the presence of dexamethasone were the converse of the patterns released from cells that were actively secreting.   It is clear that during active secretion, and during inhibited secretion, the processing patterns are modified.   The results support a view that processing and secretion are inter-linked.

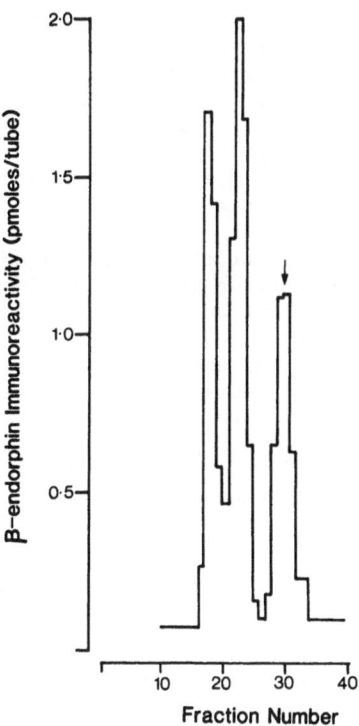

Fig. 7   Effect of dexamethasone on the forms of ir-β-endorphin released from rat anterior pituitary cells.   Note the prelavence of the higher molecular weight forms.

The principal results of the experiments on the molecular forms of ACTH and β-endorphin in cultured cells was that the pattern of released peptides can be quite different from that of the intracellular peptides. The simplest explanation is that secretory vesicles may be heterogeneous in respect of the peptides they contain; furthermore the different granules may be called independently into the secretory pathway, leading to exocytosis in response to specific environmental stimuli. It is always possible, though less likely, that heterogeneity occurs at the cellular level, with the existence of a variety of cell types containing different processing enzymes and generating different peptide products (Deftos and Catherwood, 1980).

Further study will be required to elucidate the mechanisms that are involved in what appears to be intragranular selection. One possibility is that a variety of receptors occur on the cell surface which are linked to different secretory pools. The receptors may respond to individual secretagogues, such as CRF, vasopressin and angiotensin, or to combinations of these releasing factors (Rivier and Vale, 1983; Gillies et al., 1982). It should also be borne in mind that sub-populations of receptors may exist which have a range of affinities for a single ligand. This would allow differential peptide secretion to take place during mild or prolonged exposure to a releasing factor.

The way in which a particular receptor might be linked to a given set of granules could involve coupling through the action of a specific secondary messenger. It is already known, for example, that the effects of CRF on hormone secretion are mediated through adenyl cyclase (Giguere and Labrie, 1983) whereas vasopressin acts via the phospho-inositol pathway (Raymond et al., 1985). It would therefore be possible to envisage that the release of different packages of peptides might be triggered by CRF and by vasopressin and a similar possibility could be envisaged in respect of the release of ACTH- and β-endorphin-related peptides by other secretagogues. An alternative possibility is that a single intracellular messenger might be channelled along different routes in a cell making connections between an initial receptor on the cell surface and its complementary granule.

In the longer term, replacement of secreted peptides involves increased biosynthesis of the peptide prohormone, which in the case of POMC has been demonstrated at the transcriptional level. Elevated levels of POMC mRNA have been observed during chronic treatment of AtT-20 cells with CRF (Bruhn et al., 1984) and it is to be expected that this would be accompanied by a corresponding increase in messenger RNA for the specific processing enzymes, since it appears that the processing enzymes are secreted along with the peptide products (Mains and Eipper, 1985). Homeostasis would require, of course, that both the products of processing and the enzymes that generate them should be replenished.

A further level at which differential regulation may take place is through the selective proteolysis of 'stored' intracellular peptides. Thus the presence of unprocessed lipotropin in a cell that generates β-endorphin, or the conversion of β-endorphin 1-31 to the essentially inactive β-endorphin 1-27 or 1-26 in the intermediate lobe of the pituitary, undoubtedly diminishes the availability of opiate activity relative to the activity of ACTH, with which it is co-produced. Similarly α,N-acetylation of β-endorphin eliminates its opiate activity (Smyth et al., 1979) whereas the acetylation of ACTH 1-13 (α-MSH) greatly increases its potency (Eberle, 1980). In this way the acetylation reaction can alter the balance of the activities of α-MSH and β-endorphin.

Another interesting possibility is that the peptides in certain types of granule may be more susceptible to intracellular degradation and this could lead to an imbalance in the products available for secretion, even though the peptides are initially released stoichiometrically from the same prohormone. Such an explanation may account for the 'non equivalent' amounts of ir-ACTH and ir-β-endorphin that are released from cultured anterior pituitary cells during the early stages of CRF stimulated secretion.

The general mechanisms of differential peptide release discussed here require that the form of a peptide entering the secretory pathway will be determined by the recruitment of pre-existing granules and not by a direct effect of a secretagogue on the processing enzymes. Since the granules lost by secretion are likely to be replaced, the activity of the appropriate processing enzymes may be expected to increase concomitantly but it is important to note that this would occur as a *secondary* event. Thus we propose that releasing factors may influence regulation by a process of granule selection.

These concepts have emerged from our finding that different packages of peptides can be released from anterior pituitary cells and that the secreted peptides may differ from the peptides that are retained intracellularly. The way is now open to examine further whether different releasing factors, or combinations of releasing factors, promote the secretion of different biological activities and to investigate the mechanisms that are involved at this new level of hormone regulation.

REFERENCES

Allen, R.G., Herbert, E., Hinman, M., Shibuya, H. and Pert, C.B. (1978), Coordinate control of ACTH, β-LPH and β-endorphin release in mouse pituitary cell cultures, Proc. Natl. Acad. Sci. U.S.A., 75:4972-4976.

Allen, R.G., Hinman, M. and Herbert, E. (1982), Forms of corticotropin and β-endorphin released by partially purified CRF in mouse anterior pituitary cultures, Neuropeptides, 2:175-184.

Bradbury, A.F., Smyth, D.G. and Snell, C.R. (1975), Biosynthesis of α-MSH and ACTH, in "Peptides, chemistry, structure and biology", R. Walter and J. Meienhofer, eds., Ann Arbor Sci. Publishers Inc., Michigan.

Bradbury, A.F., Smyth, D.G., Deakin, J.F.W. and Wendlandt, S. (1977), Comparison of the analgesic properties of lipotropin C-fragment and stabilized enkephalins in the rat, Biochem. Biophys. Res. Commun., 64:748-753.

Bruhn, T.O., Sutton, R.E., Rivier, C.L. and Vale, W.W. (1984), Corticotropin releasing factor regulates pro-opiomelanocortin messenger ribonucleic acid levels in vivo, Neuroendocrinol., 39:170-175.

Deftos, L.J. and Catherwood, B.D. (1980), Dissociation between ACTH and β-endorphin immunoreactivity in cells of the rat pituitary gland, Life Sci., 27:223-228.

De Souza, E.B. and Van Loon, G.R. (1985), Differential plasma β-endorphin β-lipotropin and adrenocorticotropin responses to stress in rats, Endocrinol., 116:1577-1586.

Eberle, A. (1980), Structure and chemistry of the peptide hormones of the intermediate lobe, in "Peptides of the pars intermedia", G. Lawrence and D.C. Evered, eds., Pitman Medical London (Ciba Found. Symp. 81).

Feldberg, W.S. and Smyth, D.G. (1977), C-Fragment of lipotropin: an endogenous potent analgesic peptide, Brit. J. Pharmacol., 60:445-454.

Giguere, V. and Labrie, F. (1983), Additive effects of epinephrine and corticotropin releasing factor (CRF) on adrenocorticotropin release in rat anterior pituitary cells, Biochem. Biophys. Res. Commun., 110:456.462.

Gillies, G.E., Linton, E.A. and Lowry, P.J. (1982), Corticotropin releasing activity of the new CRF is potentiated several times by vasopressin, Nature, 299:355-357.

Gumbiner, B. and Kelly, R.B. (1982), Two distinct intracellular pathways transport secretory and membrane glycoproteins to the surface of pituitary tumour cells, Cell, 28:51-59.

Ham, J. and Smyth, D.G. (1984), Regulation of bioactive β-endorphin processing in rat pars intermedia, Febs Letts., 175:407-411.

Ham, J. and Smyth, D.G. (1985), β-Endorphin and ACTH-related peptides in primary cultures of rat anterior pituitary cells: evidence for different intracellular pools, Febs Letts., in the press.

Hook, V.J.H., Heisler, S., Sabol, S.L. and Axelrod, J. (1982), Corticotropin releasing factor stimulates adrenocorticotropin and β-endorphin release from AtT-20 mouse pituitary tumour cells, Biochem. Biophys. Res. Commun., 106:1364-1371.

Loh, H.H., Tseng, L.F., Wei, E. and Li, C.H. (1976), β-Endorphin is a potent analgesic agent, Proc. Natl. Acad. Sci. U.S.A., 73:2895-2898.

Mains, R.E. and Eipper, B.A. (1981), Co-ordinate, equimolar secretion of small peptide products derived from Pro-ACTH/endorphin by mouse pituitary cells, J. Cell Biol., 89:21-28.

Mains, R.E. and Eipper, B.A. (1985), Peptidyl-glycine α-amidating activity in tissues and serum of the adult rat, Endocrinol., 116:2497-2504.

Raymond, V., Leung, P.C.K., Veilleux, R. and Labrie, F. (1985), Vasopressin rapidly stimulates phosphatidic acid-phosphatidylinositol turnover in rat anterior pituitary cells, Febs Letts., 182:196-200.

Rivier, C.L. and Vale, W.W. (1985), Interaction of corticotropin releasing factor and arginine vasopressin on adrenocorticotropin secretion in vivo, Endocrinol., 113:939-942.

Smyth, D.G. (1983), β-Endorphin and related peptides in pituitary, brain, pancreas and antrum, Brit. Med. Bull., 39:25-30.

Smyth, D.G., Massey, D.E., Zakarian, S. and Finnie, M.D.A. (1979) Endorphins are stored in biologically active and inactive forms; isolation of α,N-acetyl peptides, Nature, 279:252-254.

Smyth, D.G. and Zakarian, S. (1980), Selective processing of β-endorphin in regions of porcine pituitary, Nature, 288:613-615.

Zakarian, S. and Smyth, D.G. (1982a), Distribution of β-endorphin related peptides in rat pituitary and brain, Biochem. J., 202:561-571.

Zakarian, S. and Smyth, D.G. (1982b), β-Endorphin is processed differently in specific regions of rat pituitary and brain, Nature, 296:250-253.

STRATEGIES IN THE REGULATION OF SECRETORY SIGNALS FROM

PROOPIOMELANOCORTIN-PRODUCING CELLS

Bruce G. Jenks[*], Hans J. Leenders[*], B.M. Lidy Verburg-van Kemenade[*],
Marie-Christine Tonon[◊] and Hubert Vaudry[◊]

[*] Department of Zoology, Catholic University, Toernooiveld, 6525 ED
    Nijmegen, The Netherlands and
[◊] Laboratory of Molecular Endocrinology, UA CNRS 650, Faculty of
    Sciences, University of Rouen, 76130 Mont-Saint-Aignan, France

SUMMARY

    Proteolytic cleavage of the precursor protein proopiomelanocortin
(POMC) has the potential to yield a number of different bioactive peptides. The current status of the possible control points in regulating
the bioactive output of POMC-producing cells is considered in this review.
Attention is payed to a possible role of glycosylation of POMC in directing the cleavage process but we conclude there is insufficient evidence
to support this hypothesis. Detailed attention is also given to the regulatory role of the N-terminal acetylation of two of the POMC-derived peptides, α-melanotropin and β-endorphin. In particular, the possibility is
considered that the acetylation process within a POMC-cell could be
regulated and thus used to modulate the secretory signal of the cell.

INTRODUCTION

    Proopiomelanocortin (POMC) is the precursor protein for a number of
peptide hormones and neuropeptides. POMC-derived peptides are found in the
pituitary gland, both in corticotroph cells of the pars distalis and in
melanotroph cells of the pars intermedia, as well as in diverse neurons
of the central nervous system (reviews : Eipper and Mains, 1980 ; Chrétien
and Seidah, 1981 ; Herbert, 1981 ; O'Donohue and Dorsa, 1982). They have
also been found to be present in secretory-cells (Larsson 1981) and neuronal cells (Wolter, 1985) of the gastro-intestinal tract, in the adrenal
medulla (Evans et al., 1983), the placenta (Liotta and Krieger, 1980) the
testis (Margioris et al., 1983 ; Chen et al., 1984) and in the pancreatic
islets (Grube et al., 1978 ; Watkins et al., 1980). The discovery of low
levels of POMC-like immunoreactivity in diverse tissues has lead Saito and
Odell (1983) to even suggest that POMC-related peptides may also be involved in cell to cell communication at the tissue level in mammals.
Proteolytic cleavage of POMC (fig. 1) has the potential of producing peptides with diverse bioactivities such as the opiate activity possessed by
β-endorphin, the melanotropic activity possessed by melanocyte-stimulating
hormone (MSH) and the adrenocorticotropic-activity possessed by adreno-
corticotropic hormone (ACTH). In view of this potential to deliver a wide
range of bioactive peptides, there obviously must be stringent control
systems within a POMC-producing cell to regulate which of the potential
spectrum of bioactive peptides are in fact synthesized and released. In

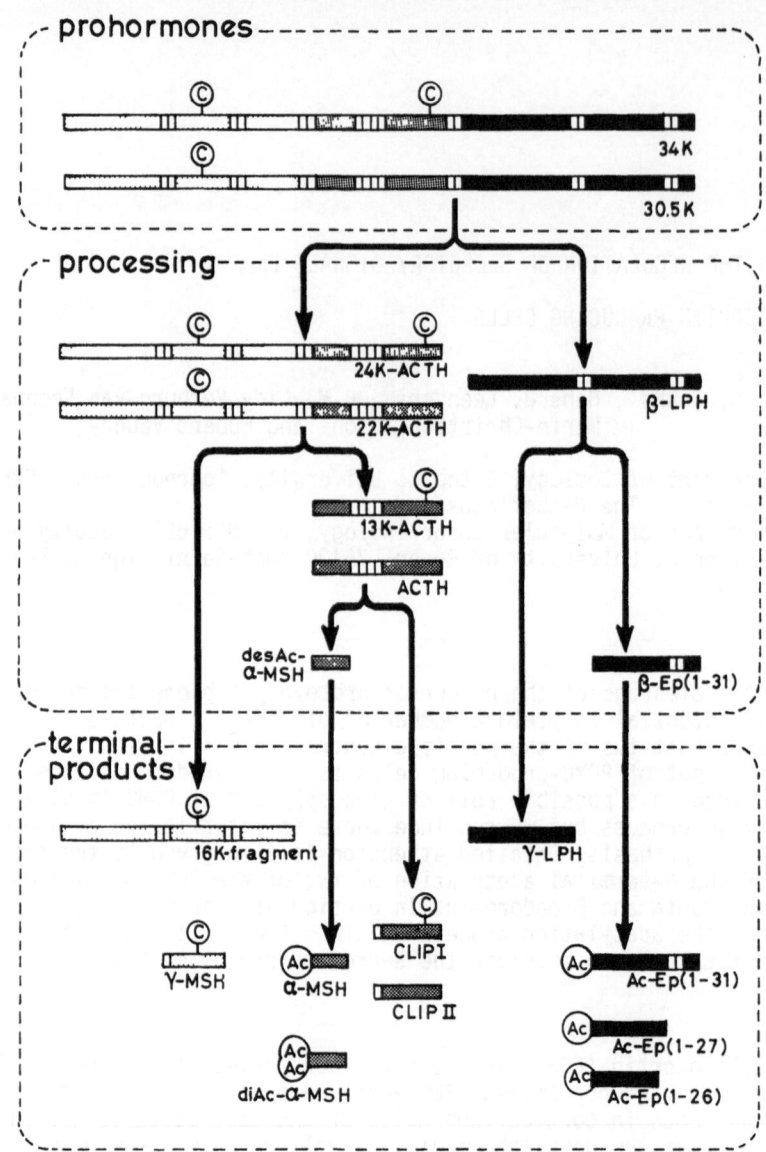

Fig. 1 : General scheme for processing of POMC in melanotrophs of the pars
intermedia of the rat or mouse. Scheme is based largely on studies referenced in text. The overall structure of the prohormone
has been confirmed using nucleotide sequencing techniques for
both the rat (Drouin and Goodman, 1980) and mouse (Uhler and
Herbert, 1983). A white box within the peptide structures indicates the position of a lysine or arginine ; those locations
where two or more of these basic amino acids occur sequentially
are indicated. Trypsin-like cleavages at these sites are involved
in generating the POMC-derived peptides. Both species produce two
forms of the prohormone which differ in their degree of glycosylation. Molecular weights given are for mouse pars intermedia
POMC (Jenks et al 1983a, b; 1986)
Abbreviations : C, carbohydrate group ; Ac, N-terminal acetylation ; K, kilodalton ; numbers after endorphin (Ep) indicate the
number of amino acids in the peptide.

this review we examine which of the sequence of steps involved in production and release of secretory signals could be considered regulatory points within POMC-producing cells.

GENERAL REGULATORY STRATEGIES : AN OVERVIEW

Regulation of POMC-cell output could come, in theory at least, at any one of a number of steps involved in the synthesis and release of the secretory signals (fig 2). Regulation at the level of POMC-gene transcription is undoubtly important in determining the availability of secretory peptides. From reports that glucocorticoids inhibit POMC-gene transcription in corticotrophs (Roberts et al., 1983 ; Schachter et al., 1982 ; Birnberg et al., 1983), and that dopamine receptor agonists exert a similar effect on melanotrophs (Chen et al., 1983, Holt et al., 1982), it is apparent that factors which are known to regulate release of POMC-related peptides also influence, directly or indirectly, the rate of biosynthesis of these peptides. Regulation of POMC-gene transcription could be considered a control point in influencing the "signal-strength" of POMC-cell output (i.e. a quantitative role in that a depleated cell would have a low signal strength). It would seem unlikely, however, that regulation of POMC-gene itself would have a qualitative function in determining which bioactive peptides are ultimately formed by the cell. Regulation of transcription of genes coding for processing-enzymes, however, would be expected to play a crucial qualitative role in establishing POMC-cell out put (see discussion of POMC cleavage).

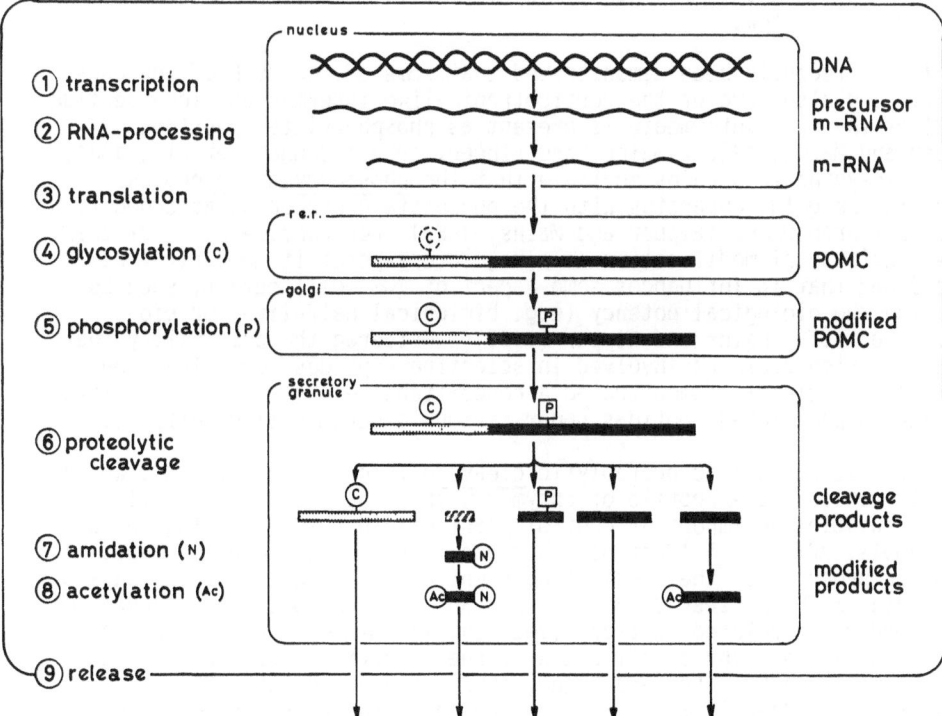

Fig. 2 : Schematic diagram giving the potential sites at which the ultimate secretory signal of POMC-cells could be regulated. Each of these potential sites, listed on the left in the figure, is briefly discussed in the text. Note : the potential glycosylation in the CLIP-site is not shown in this figure.

Concerning regulation at the level of RNA processing, it should be mentioned that there already is an example of tissue-specific processing in the hormone/neuropeptide field ; calcitonin-gene precursor mRNA is processed in the thyroidal "C" cells to mRNA coding for calcitonin while alternative processing of the RNA in neural tissue gives rise to a mRNA coding for calcitonin gene-related peptide rather than calcitonin (Rosenfeld et al., 1983). Whether a similar potential exists for expression of the POMC-gene remains to be determined. Interestingly, there have been reports that some POMC-containing neuronal networks in the hypothalamus possess MSH-like immunoreactivity but are devoid of endorphin-like immunoreactivity (Guy et al., 1980, 1981, 1982 ; Jégou et al., 1983 ; Watson et al., 1978, Watson and Akil, 1979, 1980). While a number of explanations for this phenomenon are conceivable, alternative processing of POMC mRNA is an interesting possibility.

Another potential site of regulation which warrents attention is the process of translation. With recently developed recombinant techniques to measure mRNA levels and turn-over, it is to be expected that regulatory mechanisms functioning at this level will eventually be elucidated.

The glycosylation of POMC begins in the rough endoplasmic reticulum with attachment of core sugars to the prohormone, and is completed in the golgi apparatus with a number of modifications to the carbohydrate side chain (Phillips et al., 1981), including sulfation (Bourbonnais and Crine, 1985, Hoshina et al., 1982). It has been suggested that oligosaccharides of POMC could be involved in determining the direction of proteolytic cleavage of the prohormone. Because there has been some controversy concerning this function, this subject will receive more detailed attention later in this review.

It has recently been demonstrated that some of the ACTH of the pars distalis and also some of the corticotropin-like intermediate lobe peptide (CLIP) of the pars intermedia is present as phosphorylated peptides (Eipper and Mains, 1982 ; Mains, and Eipper, 1983 ; Bennett et al., 1981, 1982). Biosynthetic studies indicate that the phosphorylation process occurs in the golgi apparatus with the phosphate group being attached to the intact prohormone (Eipper and Mains, 1982). For consideration of this post-translational modification as a regulatory-point it must first be established that it influences some aspect of POMC-cell output, such as affecting the biological potency (e.g. biological half-life) or bio-activity of ACTH. Mains and Eipper (1983) considered the possibility that phosphorylation could be involved in selecting peptides for release but their results clearly demonstrated a concomitant release of phosphorylated and non-phosphorylated peptides under various experimental conditions.

The tissue specific proteolytic cleavage (processing) of POMC, which probably reflects the profile of enzymes within the secretory granules, is without question an important control point within POMC-producing cells. For example, ACTH and β-lipotropic hormone (β-LPH) are the primary biosynthetic products of the corticotrophs while, in the cells of the pars intermedia, ACTH is cleaved to yield MSH and CLIP and β-LPH is cleaved to give β-endorphin-related peptides (see fig. 1). Attempts are now being made to isolate and characterize the tissue specific processing enzymes (Seidah et al., 1983 ; Loh and Chang, 1982 ; Chang and Loh, 1984 ; Pelaprat et al., 1984). Still to be answered is the question of whether fully differentiated POMC-cells have the potential to alter their enzyme profiles or the activity of these enzymes and thus alter the spectrum of peptides available for secretion.

While α-amidation of many regulatory peptides is an essential step, necessary for these peptides to achieve full biological activity (Eipper et al., 1983a,b ; Emeson 1984) it would seem doubtful whether amidation of α-MSH constitutes a control point, in that, with the exception of some fish species (Kawauchi et al., 1980 ; Bennett et al., 1974),no naturally occuring non-amidated α-MSH has been reported. Rather, the amidation of α-MSH seems to be a rapid and obligatory step in the production of α-MSH. The reported loss of α-MSH amidating activity in primary cultures of rat intermediate lobe cells (Eipper et al., 1983a), while in itself is an interesting observation, could very well reflect a change in the cells induced by the in vitro situation.

The process of acetylation, which occurs on the N-terminal amino acid of both α-MSH and β-endorphin, meets a number of the necessary criteria to be considered an important regulatory point in determining POMC-cell output. These criteria are : (1) naturally occuring acetylated and non-acetylated forms of these peptides have been found as terminal products of POMC-processing (acetylation tends to be tissue specific, with the cells of the pars intermedia producing acetylated peptides while POMC-containing neuronal systems of the hypothalamus produce non-acetylated peptides) ; (2) the bioactivities of these peptides are altered when they are acetylated (acetylation potentiates the melanotropic activity of MSH and eliminates the opiate activity of β-endorphin) ; (3) recent studies have indicated that the acetylation system within POMC-cells may be subject to regulation and thus, under different experimental conditions, there is a preponderance of either the acetylated or non-acetylated form of these peptides as secretory signals. The first two points have been extensively reviewed by O'Donohue and Dorsa (1982) and will not be considered further. More details concerning the last point will be given later in this review.

Regulation of the secretory signal of the POMC-cell could, in theory, come at the level of release. Evidence for this, however, is very limited. Differential release of POMC-related peptides (i.e. selective release of one group of peptides over another) has been reported for secretion of intermediate lobe peptides of both the rat (Randle et al., 1983) and the amphibian, Xenopus laevis (Loh and Jenks, 1981). In the case of the rat it was suggested that the altered secretory signal was achieved by intervention of a secretagogue with the direction of processing of POMC. The rat lobe is know to contain two POMC-cell types (Stoeckel et al., 1981). Therefore, a perhaps simpler explanation for the differential release is differences in the response of these two cell-types to the secretagogue. For Xenopus, it was suggested that selective degradation of peptides was occuring in intermediate lobes treated with dopamine, thus altering the profile of peptides available for secretion. In another amphibian species, Rana ridibunda, we find no evidence for such degradation (Jenks et al., 1985). Thus, it would seem doubtful whether differential-release of peptides within a POMC-producing cell, is of general importance as a regulatory mechanism. Far more numerous are reports of concomitant release of POMC-related peptides and their coordinate regulation during the secretory process (see Jenks et al., 1985, for references).

FUNCTIONAL SIGNIFICANCE OF GLYCOSYLATION OF POMC

A valuable tool in the study of the functional significance of glycosylation has been the use of the antibiotic tunicamycin, which blocks asparagine-linked N-glycosylation of proteins. Loh and Gainer (1978, 1979) were the first to apply this antibiotic to studies concerning POMC-producing cells. They reported that, in tunicamycin treated pars intermedia of the amphibian Xenopus laevis, non-glycosylated POMC undergoes random proteolysis to form "atypical" peptides which suggested that the carbohydrate of the prohormone may be involved in determining the direction of processing. The biosynthetic products in these studies were characterized primarily on the basis of electrophoretic mobility and, from the results, it is difficult to determine to what degree peptides synthesized in tunicamycin-treated tissue represent "incorrect" cleavage products of POMC. In a similar investigation, using a corticotrophin-producing tumor cell line, Budarf and Herbert (1982) report that processing of POMC is not grossly altered when its glycosylation is prevented. In this study some effort was made to characterize further the non-glycosylated prohormone and intermediate products of processing, primarily through tryptic mapping. More recently biosynthetic studies with the mouse pars intermedia showed that each peptide synthesized by tunicamycin-treated tissue had identical chromatographic characteristics as those of a peptide synthesized by control tissue (Jenks et al., 1983a, 1986). That correct cleavage peptides were being produced from the non-glycosylated prohormone was confirmed by tryptic mapping of the various peptides cleaved from the prohormone.

We have recently extented these studies to the pars intermedia of the frog, Rana ridibunda (Vaudry et al., 1986). Again, no evidence was found for a random proteolysis of the non-glycosylated prohormone. Each peptide produced by tunicamycin-treated tissue corresponded to a peptide synthesized in the control tissue. We therefore conclude that there is little evidence to indicate that glycosylation is important in determining the direction of processing of POMC.

Within a given species there have been no reports, to our knowledge, of any major differences occuring in the nature of the glycosylation of POMC from intermediate lobe cells versus distal lobe corticotroph cells. The very different cleavage patterns achieved by these two POMC-cell types from similar if not identically glycosylated prohormones reinforces the argument, discussed earlier, that the proteolytic enzyme profile is a crucial factor in determining the direction of processing. It should also be pointed out that there is considerable differences among species in the number of carbohydrate side-chains attached to POMC and in the position of their attachment within the prohormone structure (reviewed by Vaudry et al., 1986). One might have expected a higher degree of conservation in the nature of the glycosylations if they were to perform vital intracellular functions.

One generalization possible concerning glycosylation is that POMC of most species is glycosylated in the γ-MSH region and that a glycosylated N-terminal fragment of POMC (the so-called 16 K-fragment) and glycosylated γ-MSH are established as secretory products of POMC-cells. While γ-MSH structures are reported to be very weak melanotropins (Ling et al., 1979), it has been suggested that the 16 K-fragment or γ-MSH itself could be involved in regulation of the adrenal gland in mammals (Pedersen and Brownie, 1980 ; Estivariz et al., 1982 ; Farese et al., 1983) and in sub-mammalian vertebrates (Takahashi et al., 1985 ; Leboulenger et al., 1986). Possibily, the carbohydrate moiety of these peptides could have an extracellular function, such as influencing their biological half-life or play a role in the interaction of these peptides with receptors at the level of the target tissue.

ACETYLATION MAY BE A REGULATED PROCESS

Beyond the question of tissue-specific acetylations, which is undoubtly an important regulatory point in distinguishing the secretory signal of different POMC-producing systems (see earlier discussion), an important consideration is whether regulation of the acetylation process within an individual POMC-cell is possible such that (1) a POMC-cell can acetylate its MSH and endorphin under one set of circumstances but produce non-acetylated peptides under different circumstances, or (2) a POMC-cell can regulate independently the acetylation of its melanotropin and opiate. While it is too early to give a definite answer to these question, recent

studies, reviewed below, indicate that mechanisms to regulate the
acetylation process may indeed be functioning within POMC-producing
cells.

Characterization of the acetyltransferase activity responsible for the
acetylation of MSH and endorphin in the rat intermediate lobe has led
to the suggestion that the same enzyme is responsible for the acetylation
of both peptides (Chappell et al., 1982 ; Glembotski, 1982 ; Barnea and
Cho, 1983). While this suggestion seems to preclude independent acetylation for these peptides, it has recently been shown that the major
biosynthetic- and immunoreactive-form of MSH in the fetal intermediate lobe
of the mouse is the non-acetylated peptide (des-Nα-acetyl α-MSH) while, in
this same lobe, the major biosynthetic and immunoreative form of
β-endorphin is the N-terminal acetylated peptide (Leenders et al., 1986).
How these observations can relate to the concept of a single acetylation
enzyme is unclear. However this study shows firstly that it is possible for
a POMC-producing cell to switch from synthesis and release of nonacetylated
MSH to acetylated forms of this peptide (although admitedly this may simply
reflect the ontogenesis of the MSH-acetylation system) and secondly, that
acetylation of MSH and endorphin are not necessarily concomitant events.
Unfortunately at the moment it is not possible to attach any physiological
significant to these findings as the function of the pars intermedia in
mammals is unknown.

A second indication for regulation of the acetylation process comes
from studies with the amphibian intermediate lobe. It has been shown
in two species of amphibians that the acetylation of MSH is associated
with the release process (Martens et al., 1981 ; Vaudry et al., 1983,
1984). These studies reveal that the major tissue-form of MSH is the non-
acetylated peptide while, following in vitro incubations, α-MSH is found
to be a major secretory product of the pars intermedia. We have suggested
that electrochemical changes in vicinity of the secretory granule during
the process of exocytosis may activate an acetyltransferase, thus "linking"
acetylation to release. More recent studies indicate that, through storage
of MSH in the non-acetylated form, the POMC-cell of the pars intermedia may
have the potential to regulate the acetylation status of the MSH released
(Jenks et al., 1985). These studies, conducted with the intermediate lobe
of the frog Rana ridibunda, showed that treatment with dopamine, which is
thought to be the physiological MSH release inhibiting factor in amphibians,
inhibited not only the release of MSH but also its acetylation. As a
consequence of this dual effect of dopamine treatment there is, under
partial dopamine inhibition, a shift to a secretory signal more predominant
in non-acetylated peptide (fig. 3). This switch in the nature of the signal
may be of physiological importance. MSH is released from the pars intermedia of amphibians on a black background and, consequently, there is an
expansion of black pigment in dermal melanophores. The palor observed in
the skin of white-adapted animals may not only be due to dopamine-induced
lowering of circulating levels of MSH but also due to the signal-switch to
the less melanotropic non-acetylated form of MSH.

There has recently been several additional reports of possible
dopaminergic regulation of acetylation. Millington et al. (1985) have
found that chronic treatment of rats with the dopamine receptor antagonist
haloperidol leads to an elevation in intermediate lobe acetyltransferase
activity. Their data indicate that interruption of the dopaminergic
 inhibition of the melanotrophs produces an increase in the rate of biosynthesis of the acetylation enzyme. In studies also concerning the rat
intermediate lobe, Ham et al. (1984 a, b) report that non-acetylated forms
of endorphins are released from cells that have been in culture for several
days and that dopamine-treatment restored the acetylation of these
peptides. There is an accelerated release of endorphins and other POMC-

related peptides in these cultured cells and it is therefore quite possible that dopamine, by slowing down this accelerated release, is simply allowing sufficient time for the normal intracellular acetylation of these peptides to occur. Nonetheless, the results of Ham et al. (1984 a, b) could be indicative of a potential physiological mechanism. Biosynthetic studies with mouse and rat intermediate lobe cells show that cleavage of POMC to produce non-acetylated MSH and endorphin occurs much more rapidly than the subsequent acetylation of these peptides. If mechanisms can be found which selectively stimulate the release of newly synthesized POMC-related peptides, relative to peptides in the more mature secretory granules, then one might expect the peptide population released to be enriched in the nonacetylated form of MSH and β-endorphin. In such a case the secretory signal would switch from a potent melanotropic-signal to the opiate-active signal.

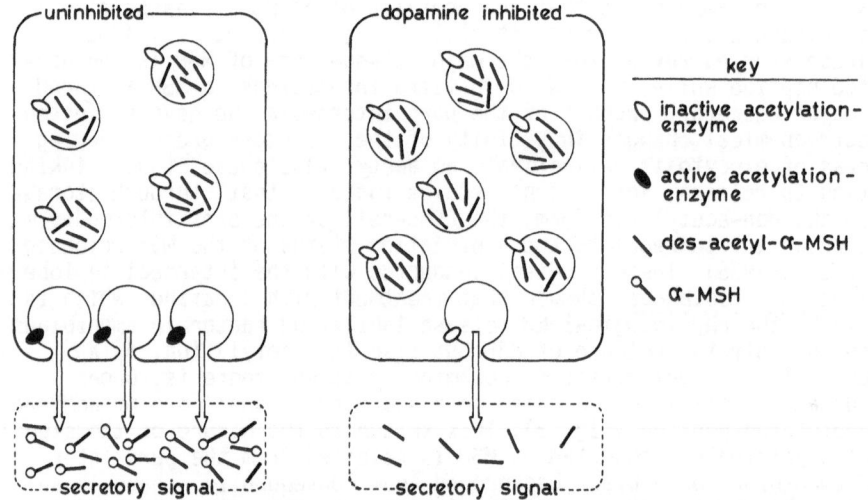

Fig. 3 : The effect of dopamine on the release of MSH from the malanotroph cell of the pars intermedia of the frog, Rana ridibunda. In the frog melanotrophs MSH is present entirely in the non-acetylated form and acetylation to produce α-MSH is associated with release (Vaudry et al., 1983). In vitro, these cells release a mixture of acetylated and non-acetylated peptides (left figure). Dopamine has been shown not only to inhibit release but also to inhibit the acetylation process (Jenks et al., 1985) ; consequently, treatment with a high concentration of dopamine leads to a low level of release and the MSH released in this case is almost entirely in the non-acetylated form (right figure). Several possible mechanisms for this dopaminergic regulation of the acetylation process have been presented (Jenks et al., 1985). Although the acetylation enzyme is shown in the figure to be associated with the membrane, the location of this enzyme in amphibians is unknown.

In conclusion, the concept that N-terminal acetylation of peptides constitutes an important regulation point in distinguishing the secretory signals of different POMC-producing systems, originally proposed by the groups of O'Donohue (O'Donohue and Dorsa, 1982) and Smyth (Zakarian and Smyth, 1979 ; 1982), is well established. There is now experimental evidence to suggest that the acetylation system may even be used to modulate the secretory signal from a single POMC-cell type.

ACKNOWLEDGEMENTS

H.V. was a Visiting Professor in the Faculty of Sciences in Nijmegen. This work was supported by research grants from the Institut National de la Santé et de la Recherche Médicale (82-4019), the Centre National de la Recherche Scientifique (UA 650), the European Economic Community (STI-084-JC-CD). We thank Mr. P.M.J.M. Cruijsen for technical assistance and Mrs I. Pastor for typing the manuscript.

REFERENCES

Barnea, A., and Cho, G., 1983, Acetylation of adrenocorticotropin and β-endorphin by hypothalamic and pituitary acetyltransferases, Neuroendocrinology, 37:434.

Benett, H. P. J., Lowry, P. J., Mc Marten, C., and Scott, A. P., 1974, Structural studies of alpha-melanocyte stimulating hormone and a novel beta-melanocyte stimulating hormone from the neurointermediate lobe of the pituitary of the dogfish, Squalus acanthias, Biochem. J., 141:439.

Bennett, H. P. J., Browne, C. A., and Solomon S., 1981, Biosynthesis of phosphorylated forms of corticotropin-related peptides, Proc. Natl. Acad. Sci. USA, 78:4713.

Bennett, H. P. J., Browne, C. A., and Solomon S., 1982, Characterization of eight forms of corticotropin-like intermediate lobe peptide from the rat intermediate pituitary, J. Biol. Chem., 257:10096.

Birnberg, N. C., Lissitzky, J. C., Hinman, M., and Herbert. E., 1983, Glucocorticoids regulate proopiomelanocortin gene expression in vivo at the level of transcription and secretion, Proc. Natl. Acad. Sci. USA, 80:6982.

Bourbonnais, Y., and Crine, P., 1985, Post-translational incorporation of [35S] sulfate into oligosaccharide side chains of pro-opiomelanocortin in rat intermediate lobe cells. J. Biol. Chem., 260:5832.

Burdarf, M. L., and Herbert, H., 1982, Effect of tunicamycin on the synthesis, processing and secretion of pro-opiomelanocortin peptides in mouse pituitary cells, J. Biol. Chem., 257:10135.

Chang, T. L., and Loh, Y. P., 1984, In vitro processing of proopiocortin by membrane-associated and soluble converting enzyme activities from rat intermediate lobe secretory granules, Endocrinology, 114:2092.

Chappell, M. C., Loh, Y. P., and O'Donohue, T. L., 1982, Evidence for an opiomelanotropin acetyltransferase in the rat pituitary intermediate lobe, Peptides, 3:405.

Chen, C. L. C., Dionne, F. T., and Roberts, J. L., 1983, Regulation of pro-opiomelanocortin mRNA levels in rat pituitary by dopaminergic compounds, Proc. Natl. Acad. Sci. USA, 80:2211.

Chen, C. L. C., Mather, J. P., Morris, P. L., and Bardin, C. W., 1984, Expression of pro-opiomelanocortin-like gene in the testis and epididymis, Proc. Natl. Acad. Sci. USA, 81:5672.

Chrétien, M., and Seidah, N. G., 1981, Chemistry and biosynthesis of pro-opiomelanocortin, Mol. Cell. Biochem., 34:101.

Drouin, J., and Goodman, H. M., 1980, Most of the coding region of rat ACTH/β-LPH precursor gene lacks intervening sequences, Nature, 288:610.

Eipper, B. A., and Mains, R. E., 1980, Structure and biosynthesis of pro-adrenocoticotropin/endorphin and related peptides, Endocrine Reviews, 1:1.

Eipper, B. A., and Mains, R. E., 1982, Phosphorylation of pro-adreno-corticotropin-endorphin-derived peptides. J. Biol. Chem., 257:4907.

Eipper, B. A., Glembotski, C. C., and Mains, R. E., 1983a, Selective loss of α-melanotropin-amidating activity in primary cultures of rat intermediate pituitary cells, J. Biol. Chem., 258:7292.

Eipper, B. A., Glembotski, C. C., and Mains, R. E., 1983b, Bovine inter-mediate pituitary α-amidation enzyme : preliminary characteriza-tion, Peptides, 4:921.

Emeson, R. B., 1984, Hypothalamic peptidylglycine α-amidation monooxygen-ase : preliminary characterization, J. Neurosci., 4:2604.

Estivariz, F. E., Iturriza, F., Mc Lean, C., Hope, J., and Louwry, P. J., 1982, Stimulation of adrenal mitogenesis by N-terminal proopio-cortin peptides, Nature, 297:419.

Evans, C. J., Erdelyi, E., Weber, E., and Barchas, J. D., 1983, Identifi-cation of pro-opiomelanocortin-derived peptides in the human adrenal medulla, Science, 221:957.

Farese, R. V., Ling, N. C., Sabir, M. A., Larson, R. E., and Trudeau, W. L., 1983, Comparison of effects of adrenocorticotropin and Lys-γ3-melanocyte stimulating hormone on steroidogenesis, adenosine 3',5'-monophosphate production, and phospholipid metabolism in rat adrenal fasciculata reticularis cells in vitro, Endocrinology, 112:129.

Glembotski, C. C., 1982, Characterization of the peptide acetyltransferase activity in bovine and rat intermediate pituitaries responsible for the acetylation of β-endorphin and α-melanotropin. J. Biol. Chem., 257:10501.

Grube, D., Voigt, K. A., Weber, E., 1978, Pancreatic glucagon cells con-tain endorphin-like immunoreactivity, Histochemistry, 59:75.

Guy, J. Leclerc, R., Vaudry, H., and Pelletier, G., 1980, Identification of a second category of α-melanocyte-stimulating hormone (α-MSH) neurons in the rat hypothalamus, Brain Res., 199:135.

Guy, J., Vaudry, H., and Pelletier, G., 1981, Differential projections of two immunoreactive α-melanocyte stimulating hormone (α-MSH) neu-ronal systems in the rat brain, Brain Res., 220:199.

Guy, J., Vaudry, H., and Pelletier, G., 1982, Further studies on the identification of neurons containing immunoreactive alpha-melanocyte stimulating hormone (α-MSH) in the rat brain. Brain Res., 239: 265.

Ham, J., Mc Farthing, K. G., Toogood, C. I. A., and Smyth, D. G., 1984a, Influence of dopaminergic agents on β-endorphin processing in rat pars intermedia, Biochem. Soc. Transact., 12:927.

Ham, J., and Smyth, D. G., 1984b, Regulation of bioactive β-endorphin processing in rat pars intermedia, Febs Lett., 75:407.

Holt, V., Haarmann, I., Slizinger, B. R., and Herz, A., 1982, Chronic haloperidol treatment increases the level of in vitro translatable m-RNA coding for β-endorphin/ACTH precursor POMC in the pars intermedia of the rat, Endocrinology, 110:1885.

Hoshina, H., Hortin, G., and Boime, I., 1982, Rat pro-opiomelanocortin contains sulphate, Science, 217:63.

Herbert, E., 1981, Discovery of pro-opiomelanocortin, a cellular polyprotein, Trends Biochem. Sci., 6:184.

Jégou, S., Tonon, M.C., Guy, J., Vaudry, H., and Pelletier, G., 1983, Biological and immunological characterization of α-melanocyte-stimulating hormone (α-MSH) in two neuronal systems of the rat brain, Brain Res., 260:91.

Jenks, B. G., Cruijsen, P. M. J. M., Feyen, J. H. M., Martens, G. J. M., and van Overbeeke, A. P., 1983a, Effects of tumicamycin on biosynthesis of pars intermedia peptides in the mouse, in : "Integrative Neurohumoral Mechanisms : Physiological and Clinical Aspect", Hungaria Academic press, Budapest pp 281-287.

Jenks, B. G., van Daal, J. H. H. N., Scharenberg, J. G. M., Martens, G. J. M., and van Overbeeke, A. P., 1983b, Biosynthesis of pro opiomelanocortin-derived peptides in the mouse neurointermediate lobe, J. Endocrinol., 98:19.

Jenks, B. G., Verburg van Kemenade, B. M. L., Tonon, M. C., and Vaudry, H., 1985, Regulation of biosynthesis and release of pars intermedia peptides in Rana ridibunda : dopamine affects both acetylation and release of α-MSH. Peptides, in press.

Jenks, B. G., Ederveen, A. G. H., Freyen, J. H. M., van Overbeeke, A. P., 1986, The functional significance of glycosylation of proopiomelanocortin in melanotrophs of the mouse pituitary gland, J. Endocrinol., in press.

Kawauchi, H., Adachi, Y., and Tsubokawa, M., 1980, Occurence of a new melanocyte stimulating hormone in the salmon pituitary gland, Biochem. Biophys. Res. Commun., 96:1508.

Larsson, L. I., 1981, Adrenocorticotropin-like and α-melanotropin-like peptides in a subpopulation of human gastrin cell granules : Bioassay, immunoassay and immunocytochemical evidence, Proc. Natl. Acad. Sci. USA, 78:2990.

Leboulenger, F., Lihrmann, I., Netchitailo, P., Delarue, C., Perroteau, I., Ling, N., and Vaudry, H., 1986, In vitro study of frog (Rana ridibunda Pallas) interrenal function by use of a simplified perifusion system. VIII. Structure-activity relashionship of synthetic ACTH fragments and γ-MSH, Gen. Comp. Endocrinol., in press.

Leenders, H. J., Janssens, J. J. W., Theunissen, H. J. M., Jenks, B.G., and van Overbeeke, A. P., 1986, Acetylation of melanocyte stimulating hormone and β-endorphin in the pars intermedia of the perinatal pituitary gland in the mouse. Neuroendocrinology, in press.

Ling, N., Ying, S., Minick, S., and Guillemin, R., 1979, Synthesis and biological activity of four γ-melanotropin peptides derived from the cryptic region of the adrenocorticotropin β-lipotropin precursor, Life Sci., 25:1773.

Liotta, A. S., and Krieger, D. T., 1980, In vitro biosynthesis and comparative posttranslational processing of immunoreactive precursor corticotropin/β-endorphin by human placentral and pituitary cells, Endocrinology, 106:1504.

Loh, Y. P., and Gainer, H., 1978, The role of glycosylation on the biosynthesis, degradation and secretion of ACTH-β-lipotropin common precursor and its peptide products, Febs Lett., 96:269.

Loh, Y. P., and Gainer, H., 1979, The role of the carbohydrate in the stabilization, processing, and packaging of the glycosylated adrenocorticotropin-endorphin common precursor in toad pituitaries. Endoccrinology, 105:474.

Loh, Y. P., and Jenks, B. G., 1981, Evidence for two different pools of adrenocorticotropin, α-melanocyte stimulating hormone, and endorphin-related peptides released by the frog pituitary neurointermediate lobe. Endocrinology, 109:54.

Loh, Y. P., and Chang, T. L., 1982, Pro-opiocortin converting activity in rat intermediate and neural lobe secretory granules, Febs Lett., 137:57.

Mains, R. E., and Eipper, B. A., 1983, Phosphorylation of rat and human adrenocorticotropin-related peptides : physiological regulation and studies of secretion, Endocrinology, 112:1986.

Margioris, A., Liotta, A. S., Vaudry, H., Bardin, C. W., and Krieger, D. T., 1983, Characterization of immunoreactive proopiomelanocortin related peptides in rat testis, Endocrinology, 113:663.

Martens, G. J. M., Jenks, B. G., and van Overbeeke, A. P., 1981, Nα-acetylation is linked to α-MSH release from pars intermedia of the amphibian pituitary gland, Nature, 294:558.

Millington, W. R., Chappell, M. C., and O'Donohue, T. L., 1985, Induction of opiomelanocortin acetyltransferase activity in the pars intermedia of rat pituitary, 67th Ann. Meet. Endocrine Society, Abst. 651.

O'Donohue, T. L., and Dorsa, D. M., 1982, The opiomelanotropinergic neuronal and endocrine system, Peptides, 3:353.

Pedersen, R. C., and Brownie, A. C., 1980, Adrenocortical response to corticotropin is potentiated by part of the amino-terminal region of pro-corticotropin/endorphin, Proc. Natl. Acad. Sci. USA, 77:2239.

Pelaprat, D., Seidah, N. G., Sikstrom, R.A., Lambelin, P., Hamelin, J., Lazure, C., Cromlish, J. A., and Chrétien, M., 1984, Subcellular fractionation of pituitary neurointermediate lobes : revelation of various basic proteases, Endocrinology, 115:581.

Phillips, M. A., Budarf, M. L., and Herbert, E., 1981, Glycosylation events in the processing and secretion of pro-ACTH-endorphin in mouse pituitary tumor cells, Biochemistry, 20:1666.

Randle, J. C. R., Moor, B. C., and Kraicer, J., 1983, Differential control of the release of proopiocortin-derived peptides from the pars intermedia of the rat pituitary, Neuroendocrinology, 37:131.

Roberts, J. L., Eberwine, H., and Gee, C. E., 1983, Analysis of POMC gene expression by transcription assay and in situ hybridization histochemistry, in : "Cold Spring Harbor Symposia on Quantitative Biology" Vol. 28, pp 385-391.

Rosenfeld, M. G., Mermod, J. J., Amara, S. G., Swanson, L. W., Sawchenko, P. E., Rivier, J., Vale, W., and Evans, R. M., 1983, Production of a novel neuropeptide encoded by the calcitonin gene via tissue-specific RNA processing, Nature, 304:129.

Saito, E., and Odell, W. D., 1983, Corticotropin/lipotropin common precursor-like material in normal rat extrapituitary tissues. Proc. Natl. Acad. Sci. USA, 80:3792.

Schachter,B. S., Johnson, L. K., Baxter, J. D., and Roberts, J. L., 1982, Differential regulation by glucocorticoids of proopiomelanocortin mRNA levels in the anterior and intermediate lobes of the rat pituitary, Endocrinology, 106:1442.

Seidah, N. G., Pélaprat, D., Rochemont, J., Lambelin, P., Dennis, M., Chan, J. S. D., Hamelin, J., Lazure, C., and Chrétien, M., 1983, Enzymatic maturation of pro-opiomelanocortin by anterior pituitary granules, J. Chromatogr., 266:213.

Stoeckel, M. E., Schmitt, G., and Porte, A., 1981, Fine structure and cytochemistry of the mammalian pars intermedia, in "Peptides of the Pars Intermedia", Ciba Foundation Symposium, Pitman Medical, pp 101-127.

Takahaski, A., Kubota, J., Kawanchi, H., and Hirano, T., 1985, Effects of N-terminal peptide of salmon proopiocortin on interrenal function of the rainbow trout, Gen. Comp. Endocrinol., 58:328.

Uhler, M., and Herbert, E., 1983, Complete amino acid sequence of mouse pro-opiomelanocortin derived from the nucleotide sequence of pro-opiomelanocortin cDNA, J. Biol. Chem., 258:257.

Vaudry, H., Jenks, B. G., and van Overbeeke, A. P., 1983, The frog pars intermedia contains only the non-acetylated form of α-MSH : acetylation to generate α-MSH occurs during the release process, Life Sci., 33:97.

Vaudry, H., Jenks, B. G., and Overbeeke, A. P., 1984, Biosynthesis, processing and release of pro-opiomelanocortin related peptides in the intermediate lobe of the pituitary gland of the frog (Rana ridibunda), Peptides, 5:905.

Vaudry, H., Jenks, B. G., Verburg-van Kemenade, B. M. L., and Tonon, M. C.

Vaudry, H., Jenks, B. G., Verburg-van Kemenade, B. M. L., and Tonon, M.C., 1986, Effect of tunicamycin on biosynthesis, processing and release of proopiomelanocortin-derived peptides in the intermediate lobe of the frog Rana ridibunda, submitted.

Watkins, W. B., Bruni, J. F., and Yen, S. S. C., 1980, β-endorphin and somatostatin in the pancreatic D-cell ; colocalization by immuno-cytochemistry, J. Histochem. Cytochem., 28:1170.

Watson, S. J., Akil, H., Richard, C. W., and Barchas, J. D., 1978, Evidence for two separate opiate peptide neuronal systems and the co-existence of β-lipotropin, β-endorphins and ACTH immunoreactivities in the same hypothalamic neurons, Nature, 275:226.

Watson, S. J., and Akil, H., 1979, The presence of two α-MSH positive cell groups in the rat hypothalamus, Europ. J. Pharmacol., 48:101.

Watson, S. J., and Akil, H., 1980, α-MSH in rat brain : occurence within and outside of β-endorphin neurons, Brain Res., 182:217.

Wolter, H. J., 1985, α-melanotropin, β-endorphin and adrenocorticotropin-like immunoreactivities are colocalized within duodenal myenteric plexus perikarya, Brain Res., 325:290.

Zakarian, S., and Smyth, D. G., 1979, Distribution of active and inactive forms of endorphins in rat pituitary and brain, Proc. Natl. Acad. Sci. USA, 76:5972.

Zakarian, S., and Smyth, D. G., 1982, β-endorphin is processed differently in specific regions of rat pituitary and brain, Nature, 296:250.

EXPRESSION OF PREPROSOMATOSTATIN GENES IN HETEROLOGOUS CELLS

Dennis Shields and Reza F. Green

Departments of Anatomy & Structural Biology and
Developmental Biology and Cancer
Albert Einstein College of Medicine, Bronx, NY 10461

INTRODUCTION

The biosynthesis, intracellular transport and posttranslational processing of peptide hormone precursors is a very useful model for investigating cellular and molecular aspects of the secretory pathway. Since the discovery of proinsulin, it is evident that most polypeptide hormones are synthesized as larger precursors. DNA sequencing has established the sequence of numerous hormone and neuropeptide precursors (Douglass et al., 1984), many of which contain the sequences of several bioactive peptides. To generate a biologically active peptide from its precursor, the prohormone must be proteolytically cleaved and in addition, may undergo several post-translational modifications including glycosylation, sulfation, phosphorylation, amidation and acetylation (Mains, et al., 1983). These reactions occur sequentially as the precursor is transported from the ER to the Golgi apparatus and secretory granules prior to secretion. A major goal of our laboratory is to identify sorting and processing sequences in the precursor to the peptide hormone somatostatin.

Sorting and proteolytic processing domains in preprosomatostatin

Somatostatin (SRIF) is a 14 amino acid polypeptide hormone that inhibits the secretion of other polypeptide hormones including growth hormone, insulin and glucagon. A second form of somatostatin, SRIF-28 which is the SRIF tetradecapeptide containing 14 additional amino acids at its NH_2-terminus, has also been isolated from several tissues (Pradayrol, et al., 1980; Noe and Spiess, 1983); both forms are synthesized as larger precursors, preproSRIF (Shields, 1980). The primary amino acid sequence of several preproSRIFs molecules has been deduced from cloned cDNAs generated from a number of species (Reichlin, 1983). The rat and human preproSRIFs are virtually identical whereas, surprisingly, two forms of angler fish precursor share little homology to each other or to the mammalian precursors in either their signal peptides or prohormone regions. In all species however, the organization of the precursor is identical in that it consists of an N-terminal signal peptide followed by a proregion of 80-90 residues. The mature hormone is located at the carboxyl terminus of the propeptide and is preceded by a single pair of basic

amino acids; ArgLys. These basic residues are part of a subset of six
amino acids (AlaProArgGluArgLys) which are conserved in all SRIF
precursors, suggesting they may have functional significance.
Furthermore, despite differences in the primary amino acid sequence of
preproSRIFs, secondary structure predictions (Argos et al., 1983)
suggest that some domains are highly conserved. These could function
in (a) providing the correct protein conformation to facilitate
proteolytic processing of proSRIF to SRIF-14 or SRIF-28; or (b)
intracellular transport of proSRIF from its site of synthesis on the
rough ER through organelles of the secretory apparatus where
prohormone processing occurs. To test these hypotheses directly we
used DNA transfection (Warren and Shields, 1984b) and demonstrated,
that cells which do not normally synthesize prohormones (COS-monkey
kidney cells) have the proteolytic enzymes and secretory apparatus to
cleave proSRIF and secrete mature SRIF. We have now extended these
studies by introducing mutations into defined regions of preproSRIF
and expressing these molecules in a variety of cells including yeast.

RESULTS

Expression of preproSRIF cDNA

    To obtain expression of angler fish preprosomatostatin DNA in
cells that do not normally synthesize the hormone, an SV-40-pBR322
expression vector pJC119 was used, this vector has previously been
used in studies on the expression of several membrane proteins
(Sprague et al., 1983). During construction of pJC119 most of the
coding region of the SV-40 late gene, VP-1, including its initiation
codon was deleted. The remainder of the late region is intact
including the sites for transcription initiation and polyadenylation.
pJC119 contains several convenient cloning sites near the VP-1
promoter, in particular a Bam HI site. Also during construction of
pJC119 the large T antigen of the SV-40 early region was inactivated,
this is normally required for viral gene expression. We therefore
used COS cells, which are constitutive for this early gene function,
to obtain expression of the SRIF cDNA which was cloned into pJC119.
The angler fish cDNA clone used in these experiments was pLaS1
(Hobart et al., 1980) which contains the complete preproSRIF I cDNA
cloned into the PstI site of pBR322, and encodes a preproSRIF that
migrates with an apparent M_r 18,000 upon SDS-PAGE (Warren and Shields,
1984a). Cleavage of pLaS1 with Bam HI results in a DNA fragment which
encompasses about 40 nucleotides upstream from the initiator Met to
the beginning of the poly A addition site i.e., the complete coding
region of preproSRIF I (Fig. 1). The Bam HI fragment was ligated into
a synthetic Bam HI site in the promoter region of the VP-1 late gene
(Fig. 1), the resulting plasmid was designated pSVppS18.

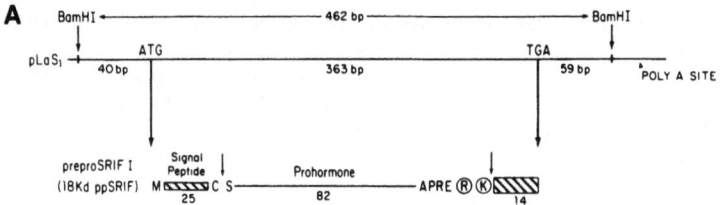

A

B   pSVppS18

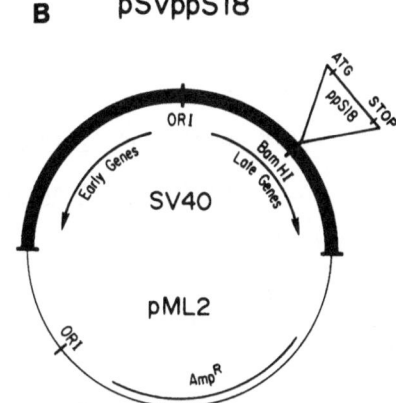

Fig. 1.   Upper: A. Structure of angler fish preprosomatostatin. The
upper line represents the Bam HI fragment of plasmid pLaS₁
(Hobart et al., 1980) encoding the entire sequence of
preproSRIF; the initiator methionine (ATG) and the stop codon
(TGA) are indicated. The organization of the preproSRIF
polypeptide and the site of the signal peptide cleavage
between Cys and Ser residues (arrow) and the sequence of the
conserved hexapeptide APRERK are also shown. The second
arrow indicates the start of mature SRIF-14 which is located
at the carboxyl terminus of the molecule (shaded area). The
one letter amino acid code is: C=Cys; Ser=Ser; A=Ala; P=Pro;
R=Arg; E=Glu and K=Lys. Based on data from Hobart et al.
(1980) and Warren and Shields (1984a,b).

Lower: B. Construction of pSVppS18: The Bam HI fragment of
pLaS₁ (above) was ligated into the promoter region of the
SV-40 VP-1 gene contained in the expression pasmid pJC119
(Sprague et al., 1983). Plasmid pJC119 consists of pML2
which is pBR322 deleted for poison sequences between Ori and
the Tet^R gene that inhibit replication in eukaryotic cells.
The SV-40 VP-1 late gene is deleted for about 1000 bp
(deletion from ~10 bp 5' of the VP-1 start codon to near the
stop codon of VP-1); insertion of the preproSRIF cDNA was via
a synthetic Bam HI site at the point of deletion; during
construction of pJC119 the large T antigen is also
inactivated (Sprague et al., 1983).

COS cells were transfected with pSVppS18 (Rose and Bergmann, 1982) and at various times following transfection aliquots of the culture medium were assayed for somatostatin immunoreactivity using a highly sensitive radioimmunoassay (Green and Shields, 1984). Two days following transfection immunoreactive SRIF was present in the culture medium and significant levels were found three days after transfection (Table 1). In contrast, SRIF immunoreactivity was not detected in medium obtained from cells transfected with the vector pJC119, or with a vector designated pSVI-40, in which the preproSRIF cDNA was inserted into the VP-1 gene in the incorrect orientation with respect to the SV40 late promoter.

Table 1

SRIF Immunoreactivity pg/10^6 cells

		Days after Transfection		
Experiment	Plasmid	1	2	3
A	pJC119	0	ND	0
	pSVI-40	0	ND	0
	pSVppS18	0	ND	1488
B	pSVppS18	0	279	809
C	pSVppS18	0	272	955

COS-7 cells, were transfected with the indicated plasmid and at 1, 2 and 3 days after transfection, aliquots of the culture medium were removed and the amount of SRIF immunoreactive material determined by radioimmunoassay (Green and Shields, 1984) using rabbit antiSRIF serum, RSS1 (Warren and Shields, 1984a, 1984b).

Translation of angler fish preproSRIF mRNA in COS cells

Having shown that cells transfected with plasmid pSVppS18 synthesized and secreted SRIF-immunoreactive material into the tissue culture medium, we wanted to determine the molecular basis of this immunoreactivity. It is possible that it resulted from synthesis and secretion of either proSRIF or the mature hormone. We therefore analyzed both the intracellular and extracellular SRIF-immunoreactive material using several different techniques. Cells transfected with plasmids pSVppS18, pSVI-40 or pJCI19 were labelled with [35S] Met and following incubation were lysed and treated with either immune or rabbit antiSRIF antiserum; the resulting immunoprecipitates were resolved by sodium dodecyl sulfate-polyacrylamide gel electrophoresis (SDS-PAGE) (Fig. 2). Only cells transfected with pSVppS18 had detectable immunoreactive SRIF and this polypeptide exactly comigrated with the M_r 17,000 proSRIF synthesized in the in vitro system supplemented with microsomal membranes (compare lanes 1 and 5); a result suggesting that the COS cell signal peptidase cleaved the nascent signal peptide correctly. COS cells transfected with either pJC119 or pSVI-40 nor untreated cells had no detectable proSRIF polypeptides (lanes 3,7 and 9).

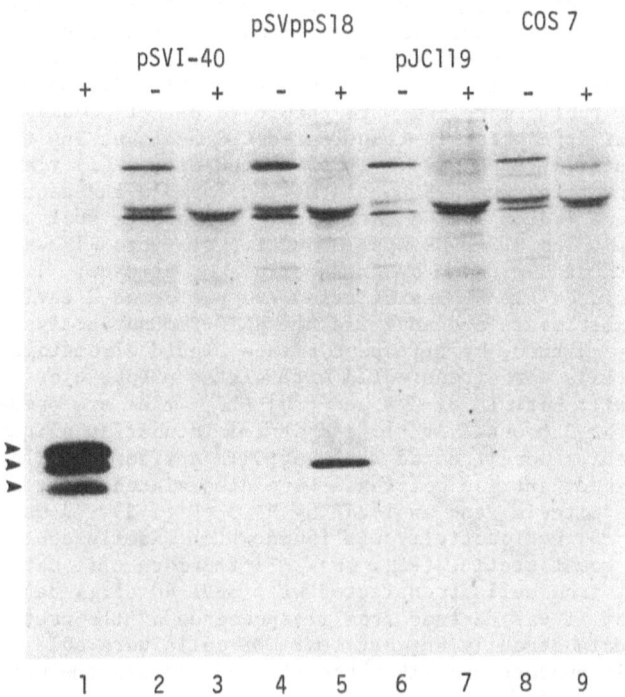

Fig. 2. Expression of prosomatostatin in COS-7 cells. Semiconfluent
cultures of COS-7 cells were transfected with the appropriate
plasmid DNA using a chloroquine/DEAE Dextran protocol (Warren
and Shields, 1984b). Three days after transfection the cells
were pulse-labelled for 3 hours with [^{35}S] Met. Following
labelling the medium was removed and the cells washed
extensively, harvested and lysed in PBS containing detergent
and a cocktail of protease inhibitors. The cell lysate was
divided into two equal aliquots which were treated with
rabbit antiSRIF serum RSS-1 (+), or with pre-immune serum
(-), and the immunoprecipitates resolved by SDS-PAGE. Lane
1: SRIF-immunoreactive translation products of angler fish
preproSRIF mRNA synthesized in vitro in the presence of
microsomal membranes (Warren and Shields, 1984a). Arrows
indicate the migration of M_r 18,000 preproSRIF I, M_r 17,000
proSRIF I and preproSRIF II, respectively. (The
proteolytically processed form of preproSRIF II was not
resolved). Lanes 2-9: Intracellular SRIF-immunoreactive
translation products synthesized in vivo by COS 7 cells
following transfection with the indicated plasmid DNA.

Identification of SRIF immunoreactive polypeptides

Since appropriately transfected COS cells synthesize proSRIF we wanted to determine the nature of the immunoreactive SRIF present in the tissue culture medium (Table 1). This material could be due to secretion of proSRIF or might result from proteolytic processing of the propeptide to the mature hormone followed by its secretion. During the course of our experiments, we observed that with increasing periods of chase of up to 3 hours, labelled proSRIF disappeared from cell lysates. Furthermore, we were unable to detect proSRIF in the chase medium of appropriately transfected COS cells at any time following a pulse of [^{35}S]-Cys. This implied either (i) that the COS cells were converting proSRIF to the mature hormone and secreting it into the culture medium and we were unable to resolve such a small polypeptide upon our SDS-PAGE system or (ii) that proSRIF was secreted into the medium and degraded by extracellular proteases. To distinguish between these possibilities, we performed a series of pulse-chase experiments and analyzed the SRIF-immunoreactive material, present in the culture, by high performance liquid chromatography (HPLC). COS cells were transfected with either pSVppS18 or pSVI-40 and labelled with both [^{35}S]-Cys and [^{3}H]-Phe, which are present in mature SRIF, for 3 hours. At the end of the incubation aliquots of the culture medium were treated with antiSRIF antiserum. The resulting antibody antigen complexes were dissociated and the SRIF immunoreactive material and analyzed by HPLC (Fig. 3). A major peak of [^{3}H[and [^{35}S] radioactivity was found which exactly co-eluted with native mature somatostatin, (Fig. 3b). Furthermore this material was totally absent from cells transfected with pSVI-40 (Fig. 3a) suggesting that it was derived from the presence of the preproSRIF cDNA. These data strongly suggest that COS cells were able to proteolytically process proSRIF correctly and secrete the mature hormone into the tissue culture medium.

Identification of processing domains

As a first step towards identifying putative sorting domains in the proregion of the SRIF precursor, we constructed a deletion mutation in which 63 nucleotides (171 to 234 inclusive) were excised from preproSRIF cDNA by digestion with PstI. Following ligation, the truncated cDNA molecule was sequenced and was found to contain a deletion of 21 amino acids in the middle of the proregion, (residues 45 to 65 inclusive were deleted). To determine if this truncated preproSRIF molecule could be translocated into the lumen of the endoplasmic reticulum, the cDNA was inserted into a transcription-translation expression vector, pDS5 (Steuber et al., 1984). This vector which is analogous to the commercially available SP-6 system, contains the phage T5 promoter and a polylinker sequence containing a Bam HI restriction site downstream from the promoter. Capped mRNA was synthesized by transcribing the resulting plasmid with E. coli RNA polymerase in the presence of 7mGpppA and four nucleotide triphosphates. Aliquots of the resulting mRNA were then translated directly in the wheat germ system containing either [^{35}S]Cys or [^{35}S]Met in the absence and presence of microsomal membranes (Shields

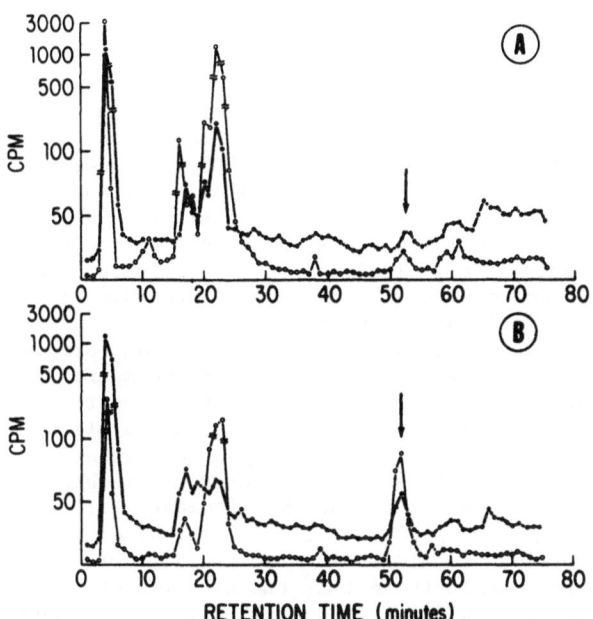

Fig. 3. Characterization of SRIF-immunoreactive material secreted by
COS cells transfected with pSVppS18 or pSVI-40: Analysis by
HPLC. COS 7 cells transfected with pSVI-40 (Panel A) or
pSVppS18 (Panel B) were incubated with [35S] Cys and [3H]
Phe. The medium was collected and treated with RSS 1
antiSRIF serum. The immunoprecipitates were washed
extensively, dissociated by incubation in 2N acetic acid and
applied directly to a Waters μ Bondapak C_{18} reverse phase
column. The column was eluted with a gradient of
acetonitrile in 0.1% trifluoracetic acid (Warren and Shields,
1984a,b). Gradient fractions, corresponding to 30 sec, were
collected, dried and counted directly. The arrow indicates
the elution of native SRIF.
[35S]Cys: ●———●; [3H]Phe: o———o.

and Blobel, 1977). The goal of these experiments was to determine if truncated preproSRIF could be translocated into the lumen of the endoplasmic reticulum membrane vesicles in vitro, (Fig. 4).

The in vitro synthesized products obtained by translating the native and truncated preproSRIF mRNA's were then treated with antiSRIF serum RSS-1 and the SRIF-immunoprecipitable products analysed by SDS-polyacrylamide gel electrophoresis (Fig. 4). As expected the translation products obtained via transcription of the native full-length preproSRIF and those obtained from angler fish islet mRNA had identical mobilities. (Compare "met" lanes, S18 and AF mRNA.) In contrast translation of S18 Δ-1 mRNA resulted in the synthesis of a molecule somewhat smaller (Mr $\approx$ 17,000) than the native precursor which has an apparent molecular weight of 18,000 (Shields, 1980; Warren and Shields, 1984a; compare lanes "cys" S18Δ-1 and met S18, downward pointing arrows). Similarly when translation was performed in the presence of microsomal membranes, which cleave the preproSRIF signal peptide (Goodman et al., 1983; Warren and Shields, 1984a) the native precursor is proteolytically processed to proSRIF (lane "met" +mb S18). Similarly the truncated preproSRIF is also co-translationally processed to a "proSRIF" which is of correspondingly lower molecular weight than the native prohormone (compare lanes S18Δ-1 and S18, upward pointing arrows). It is noteworthy that S18Δ-1 preproSRIF has methionine residues only in the signal peptide (Hobart et al., 1980; Warren and Shields, 1984a) consequently synthesis in the presence of [^{35}S]Met and microsomal membranes resulted in no [^{35}S]Met labelled pro S18Δ-1 (compare left two lanes). In contrast synthesis in the presence of [^{35}S]Cys, which is present in the mature hormone yielded [^{35}S]Cys-labelled pro S18Δ-1. Taken together these results, demonstrate that the deletion of 21 amino acids from the middle of proSRIF (residues 45-61) are not required for translocation of the molecule into the E.R. Currently, the Δ-1 mutant cDNA is being transfected into COS cells to determine if the truncated proSRIF can be processed and secreted in similar fashion the native molecule.

DISCUSSION

Most small polypeptide hormones are synthesized as part of larger precursors, many of which are polyproteins containing the sequence of several different hormones usually flanked by pairs of basic amino

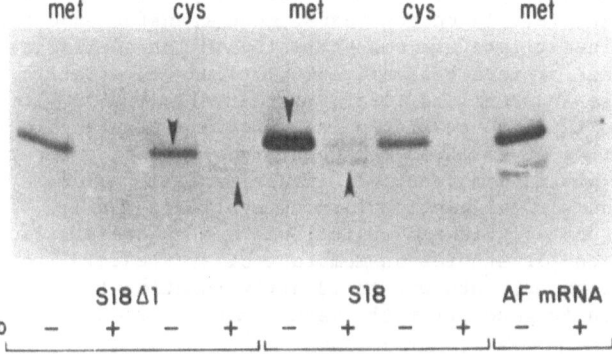

Fig. 4.   Cell-free translation of native and truncated preproSRIF
mRNAs. cDNA's encoding full-length native preproSRIF (S18)
or the Δ -1 deletion (S18 Δ1, see text) were cloned into the
Bam HI site of pDS5 (Steuber et al., 1984). Following in
vitro transcription by E. coli RNA polymerase, the resulting
mRNA was translated in the wheat germ cell-free protein
synthesizing system containing either [^{35}S]-Cys or [^{35}S]-Met
in the absence (-) or presence (+) of canine pancreatic
microsomal membranes (Warren and Shields, 1984a,b). Total
angler fish islet mRNA was similarly translated in vitro,
(AF mRNA). The translation products were treated with
antiSRIF serum and the immunoprecipitable products were
analysed by SDS-PAGE; shown is an autoradiograph of the dried
gel. Lane: met, downward pointing arrow: translation
products of full length preproSRIF mRNA apparent (Mr 18,000).
Lane: cys, downward pointing arrow: products of the Δ -1
preproSRIF mRNA (M_r 17,000). Lane: met, upward pointing
arrow: translation products of full length preprosRIF mRNA
synthesized in the presence of microsomal membranes, arrow
indicates mobility of proSRIF. Lane: Cys, upward pointing
arrow: products of truncated Δ 1 preproSRIF mRNA synthesized
in the presence of microsomal membranes, arrow indicates
migration of truncated proSRIF.

acids (Douglas et al., 1984). These prohormones may have several functions including correct folding of the precursor molecule to facilitate proteolytic cleavage and/or post-translational modification and in intracellular transport to ensure that the precursor is sorted to the correct organelle. In recent years, gene transfer experiments have been used to investigate the biosynthesis and intracellular transport of a number of integral membrane proteins (e.g. Gething and Sambrook, 1981; Rose and Bergmann, 1982; Yost et al., 1983). These experiments have defined important domains which may play a role in regulating the sorting of integral membrane glycoproteins prior to insertion into the plasma membrane. To investigate the role of propeptides in precursor polypeptide hormone synthesis and post-translational proteolytic processing, we have used a similar approach to that used for studies on membrane biosynthesis. The rationale for these experiments was to identify functional domains that may be involved in generating the mature hormone from its precursor.

We have chosen the somatostatin precursor as a model for these studies because: (i) the complete amino acid sequence of preprosomatostatin has been deduced in several species including angler fish, catfish, rats and humans; (ii) the primary events in co-translational processing of the nascent preproSRIF have been elucidated (Shields, 1980; Goodman et al., 1983; Warren and Shields, 1982; 1984a) as well as distal processing steps (Fletcher et al., 1980; Noe and Spiess, 1983), and (iii) the organization of the precursor is identical in all species i.e., the mature hormone resides at the carboxyl terminus of preproSRIF and is preceded by a pair of basic amino acids: ArgLys (Fig. 1). These residues are a subset of six amino acids, AlaProArgGluArgLys, that are highly conserved in most SRIF precursors. The conservation of this hexapeptide suggests that it may have functional significance, and since it contains a pair of basic amino acids, it is most likely the site of proteolytic cleavage of the mature hormone. Furthermore in all preproSRIFs examined to date, there is considerable conservation of hydrophobic and hydrophilic domains as well as the potential for formation of helical domains (Argos et al., 1983). The conservation of these structural domains is surprising since the fish and mammalian precursors, for example, share only about 40% amino acid sequence homology in their signal peptides and proregions. Finally, preproSRIF may represent the simplest of the known peptide hormone precursors because in most species they have only a single set of basic amino acids and hence most likely encode only a single bioactive peptide. The conservation of domains as well as the proteolytic processing site suggests that structural features of the precursor may be important for generating the mature hormone. It is possible that the conserved structural domains may also function in the intracellular targeting of the nascent prohormone through the organelles of the secretory apparatus in order to facilitate post-translational modification.

As a first step towards decoding the structural information in proSRIF, we introduced angler fish preproSRIF I cDNA into mammalian cells to examine the synthesis, proteolytic processing and secretion of mature somatostatin. The data demonstrate that the SRIF cDNA was transcribed from the SV-40 late gene VP-1 promoter into a functionally active mRNA that was significantly larger than the "wild type" mRNA (Warren and Shields, 1984b). Translation of this mRNA in a cell-free system yielded products that were indistinguishable from those obtained with natural mRNA (Warren and Shields, 1984b). Translation of this angler fish preproSRIF mRNA in vivo in COS cells resulted in synthesis of proSRIF since the nascent signal peptide had been cotranslationally cleaved correctly yielding the authentic prohormone. However, the finding that COS cells could process proSRIF to the mature hormone and secrete it was rather unexpected and implies that these cells have the enzymes and secretory apparatus necessary to fulfill this function. In agreement with these observations Low et al., (1985) recently demonstrated that when preproSRIF cDNA was expressed in transgenic mice, SRIF-14 and SRIF-28 peptides were expressed in several tissues including the anterior pituitary which does not normally synthesize this hormone.

Similarly Hellerman et al., (1984) demonstrated that when preproparathyroid hormone cDNA was introduced into growth hormone secreting cells (GH4C1), using a retrovirus expression vector, mature parathyroid hormone was cleaved from its precursor and secreted into the culture medium. These cells do not normally synthesize parathyroid hormone or peptide hormones that undergo proteolytic processing at paired basic amino acids. Thus several recent observations from gene transfer experiments have demonstrated that foreign prohormones can be accurately cleaved and secreted from heterologous cells, even those exhibiting tissue specific processing (Thomas et al., 1985). The data appear particularly intriguing since (a) they indicate that the protease(s) which normally cleave peptide hormone precursors are not specific to tissues that express the hormone and (b) correct processing occurred even with proSRIF, which has the least favored paired basic residues (ArgLys) at its processing site. Obviously, our understanding of the factors regulating processing specificity would be greatly enhanced if all the processing enzymes could be isolated.

Taken together these observations suggest that cells such as COS cells, which do not normally secrete polypeptide hormones contain the necessary proteolytic processing enzymes to convert precursor peptide hormones to the mature hormone and the cellular apparatus to facilitate their secretion. These observations suggest that propeptide processing might be a property of many diverse cells and raise the question as to what factors determine tissue-specific processing of peptide hormone precursors. The use of gene transfer techniques in combination with the construction and availability of hitherto unknown "mutant" precursors should greatly facilitate elucidation of this problem.

ACKNOWLEDGEMENTS

We thank Mr. M. Brenner for technical assistance, Ms. E. Horowitz and Ms. C. Hubertus for typing the manuscript. This work was supported by NIH grant AM21860. DS is the recipient of a Research Career Development Award AM01208 and an Irma T. Hirschl Salary Award.

REFERENCES

Argos, P., Taylor, W. L., Minth, C. D., and Dixon, J. E., 1983,
    Nucleotide and amino acid sequence comparisons of
    preprosomatostatins, J. Biol. Chem., 258:8788.
Douglass, J., Civelli, O., and Herbert E., 1984, Polyprotein gene
    expression: Generation of diversity of neuroendocrine peptides,
    Ann. Rev. Biochem. 53:665.
Fletcher, D. J., Noe, B. D., Bauer G. E., and Quigley, J. P., 1980,
    Characterization of the conversion of a somatostatin precursor
    to somatostatin by islet secretory granules, Diabetes, 29:593.
Gething, M-J. and Sambrook, J., 1981, Cell-surface expression of
    influenza haemagglutinin from a cloned DNA copy of the RNA gene,
    Nature, 293:620.
Goodman, R. H., Aron, D. C., and Roos, B. A., 1983, Rat
    pre-prosomatostatin, J. Biol. Chem., 258:5570.
Green, R. and Shields, D., 1984, Sodium butyrate stimulates
    somatostatin production by cultured cells, Endocrinology,
    114:1990.
Hellerman, J. G., Cone, R. C., Potts, J. T., Rich, A., Mulligan, R. C.
    and Kronenberg, H.M., 1984, Secretion of human parathyroid
    hormone from rat pituitary cells infected with a recombinant
    retrovirus encoding preproparathyroid hormone, Proc. Natl. Acad.
    Sci. USA 81:5340.
Hobart, P., Crawford, R., LuPing, S., Raymond, P., and Rutter, W. J.,
    1980, Cloning and sequence analysis of cDNAs encoding two
    distinct somatostatin precursors found in the endocrine pancreas
    of anglerfish, Nature, 288:137.
Low, M. J., Hammer, R. E., Goodman, R. H., Habener, J. F., Palmiter,
    R. D., and Brinster, R. L., 1985, Tissue-specific
    posttranslational processing of preprosomatostatin encoded by a
    metallothionein-somatostatin fusion gene in transgenic mice,
    Cell, 41:211.
Noe, B. D. and Spiess, J., 1983, Evidence for biosynthesis and
    differential post-translational proteolytic processing of
    different (pre)prosomatostatins in pancreatic islets, J. Biol.
    Chem., 258:1121.
Pradayrol, L., Jornvall, H., Mutt, V., and Ribet, A., 1980,
    N-terminally extended somatostatin: the primary structure of
    somatostatin-28, FEBS Letters, 109:55.
Reichlin, S., 1983, Somatostatin, New Engl. J. Med., 309:1495.
Rose, J. K. and Bergmann, J. E., 1982, Expression from cloned cDNA of
    cell-surface and secreted forms of glycoprotein of vesicular
    stomatitis virus in eucaryotic cells, Cell, 30:753.
Shields, D., 1980, In vitro biosynthesis of fish islet
    preprosomatostatin: Evidence of processing and segregation of a
    high molecular weight precursor, Proc. Natl. Acad. Sci. USA,
    77:4074.
Shields, D. and Blobel, G., 1977, Cell-free synthesis of fish
    preproinsulin, and processing by heterologous mammalian
    microsomal membranes, Proc. Natl. Acad. Sci. USA, 74:2059.
Sprague, J., Condra, J. H., Arnhetter, H. and Lazzarini, R. A., 1983,
    Expression of a recombinant DNA gene coding for the vesicular
    stomaitis virus nucleocapsid protein, J. Virol., 45:773.
Steuber, D., Ibrahimi, I., Cutler, D., Dobberstein, B., and Bujard, H.
    1984, A novel in vitro transcription-translation system:
    accurate and efficient synthesis of single proteins from cloned
    DNA sequences, EMBO J. 3:3143.

Thomas, G., Hodges, W., Hruby, D., and Herbert, E., 1984, This
    volume.
Warren, T. G. and Shields, D., 1984a, Cell-Free Biosynthesis of
    Multiple Preprosomatostatins: Characterization by hybrid
    selection and amino-terminal sequencing, Biochem., 23:2684.
Warren, T. G. and Shields, D., 1984b, Expression of preprosomatostatin
    in heterologous cells: Biosynthesis, posttranslational
    processing, and secretion of mature somatostatin, Cell, 39:547.
Yost, C. S., Hedgpeth, J., and Lingappa, V. R., 1983, A Stop Transfer
    sequence confers predictable transmembrane orientation to a
    previously secreted protein in cell-free systems, Cell, 34:759.

DEGRADATION OF LUTEINIZING HORMONE-RELEASING HORMONE (LHRH)
BY PITUITARY PLASMA MEMBRANE AND BY PITUITARY CELLS IN
CULTURE

Y. Koch, S. Elkabes and M. Fridkin*

Departments of Hormone Research and *Organic
Chemistry
The Weizmann Institute of Science
Rehovot, 76100, Israel

INTRODUCTION

Luteinizing hormone releasing hormone (LHRH), a decapeptide pGlu-His-Trp-Ser-Tyr-Gly-Leu-Arg-Pro-Gly NH_2 is synthesized and released by the hypothalamus. It is carried
via the portal blood vessels to the pituitary where it induces a cascade of reactions within the cell that results in
the release of LH and FSH. The first step in the action of
LHRH at the pituitary gland is its binding to high affinity
cell surface receptors (Clayton and Catt, 1981; Meidan and
Koch, 1981). Therefore the presence of pituitary plasma
membrane-bound LHRH-degrading enzymes may be of potential
physiological significance. These peptidases may be involved in the mechanism of action of LHRH, by regulating the
amount and the duration of action of the decapeptide at its
receptor site.

Evidence that the anterior pituitary contain peptidases, capable of degrading LHRH, were first reported by Kochman et al. 1975. Studies on the localization of LHRH degrading enzymes have indicated their existance in various
subcellular fractions such as the cytosol (Fridkin et al.,
1977; McDermott et al., 1982), mitochondria (McDermott et
al., 1982; Bauer and Horsthemke, 1984) and plasma membranes
(Clayton et al., 1979; Elkabes et al., 1981).

In addition to the anterior pituitary, LHRH-degrading
enzymes have been reported also in the hypothalamus (Fridkin
et al., 1977), in the kidney (Stetler-Stevenson et al.,
1981), liver (Towatari and Katunuma, 1983), serum (McDermott
et al., 1981), and brain (Wilk et al., 1979). Evidence suggesting the possible involvement of the kidney in the metabolism and clearance of LHRH was provided by the observation
that in patients with renal failure the metabolic clearance
rate of LHRH was decreased (Pimstone et al., 1977). It was
later demonstrated that rat renal homogenates hydrolyse LHRH
into pyroglutamic acid, (1-2)LHRH, (1-3)LHRH and (1-4)LHRH
(Stetler-Stevenson, 1981). Lysosomal thiol proteases, Ca-

thepsin L and B, purified from rat liver, have been shown to degrade LHRH in various sites (Towatari and Katunuma, 1983). The major sites of degradation by Cathepsin B were the Gly^6-Leu^7, Ser^4-Tyr^5 and His^2-Trp^3 bonds of LHRH whereas the major site of Cathepsin L degradation was the Gly^6-Leu^7 bond.

In this study we describe the pattern of LHRH degradation by pituitary plasma membrane associated peptidases and by peptidases confined to the mitochondrial fraction. We demonstrate that degradation of LHRH occurs when the neurohormone is added to pituitary cells in culture and looked for fluctuations in the enzymic activity due to steroid hormone treatment and for a possible physiological role for one of the degradation products of LHRH.

METHODS

Animals

3 month-old Wistar derived rats of the departmental colony were used. They were housed in air-conditioned quarters, illuminated between 05:00 and 19:00 h. Pelleted food (Ralson Purina Co.) and water were offered ad libitum. In experiments which involved the use of cycling rats, only females which exhibited at least two normal 4-day cycles, as determined by daily vaginal smears, were used. Female rats were bilaterally ovariectomized at random stages of the estrous cycle and used 3-4 weeks after castration.

Chemicals

LHRH was kindly donated by Hoffman-La Roche (Basel, Switzerland). [Pyroglutamyl 3,4-^{3}H]-LHRH (42.3 Ci/mmole), [L-Proline-2,3,4,5 ^{3}H(N)]-TRH (90.0 Ci/mmole) and (3-Me-His^2)- [^{3}H]TRH (70.4 Ci/mmole), were purchased from New England Nuclear (Massachusetts, U.S.A.). Estradiol benzoate (EB, B-estradiol-3-benzoate) was purchased from Sigma, U.S.A. and progesterone from Ikapharm, Israel.

Purification of plasma membranes and mitochondria

Plasma membranes and mitochondria of rat anterior pituitaries were purified by a modification (Koch et al., 1986) of the procedure employed by Fleischer, and Kervina 1974) for the fractionation of rat liver.

Enzyme Marker Assays

5'-nucleotidase activity was used as a plasma membrane enzyme marker and was determined by a modification of the procedure given by Aronson, and Touster (1974). Succinate-cytochrome C reductase, an enzyme marker for the mitochondrial fraction, was determined according to Tisdale (1967).

LHRH receptor binding assay was performed according to a procedure described by Liscovitch et al. (1984).

LHRH degrading activity in plasma membrane and mitochondrial preparations

Aliquots of the preparations were incubated in a shaking bath with synthetic LHRH (10 ug) and ^{3}H–LHRH (250,000 cpm) in a final volume of 0.3 ml of 10 mM Trisma buffer (pH 7.2) for 20 minutes at 37°C. The degradation reaction was stopped by heating the samples in a boiling bath for 10 minutes. After centrifugation the supernatant was applied onto a reversed phase high pressure liquid chromatography (HPLC) column. Samples that were later subjected to amino acid analysis for the identification of degradation products were prepared by incubating the preparations (750 ug/ml) with LHRH (0.5 mg/ml) in a final volume of 0.3 ml Tris buffer. The incubation time and temperature were the same as stated above. The degradation reaction was stopped by cooling the tubes to 4°C and immediate centrifugation for 15 min at 100,000 x g. The supernatant obtained was applied on the high pressure liquid chromatography column after millipore filtration. Control tubes containing the same amount of plasma membranes, with or without LHRH were treated and analyzed under identical conditions.

The LHRH degradation mixture was injected to a reversed-phase Lichrosorb RP-18 column (Merck, Darmstadt, Germany, 0.4x2.5 cm, particle size 10 u). All solvents were filtered through 0.22 u Millipore filters before use. The column was eluted with a linear gradient of isopropanol in 0.05M ammonium acetate pH 5.5 starting with 5% isopropanol and ending with 20% isopropanol. The flow rate was 1 ml/min and the gradient lasted for 50 min. Elution was followed continuously by monitoring UV absorbance at 230 nm. When the samples contained labeled LHRH the radioactivity of the eluent was determined in a Flow One Model HS Radioactive Flow Detector (Radiomatic Instruments and Chemical Co. Inc.). In studies aiming the identification of the degradation products by amino acid analysis, the peaks eluted were collected, lyophilized and repurified on HPLC under isocratic conditions.

Anterior pituitary cells dispersion and culture procedure

Pituitary tissue was dispersed essentially according to the method of Brazeau et al. (1981). After dispersion cells were washed twice by centrifugation and plated in Falcon multiwell plates (5-7x10^5 cells/well) using F-10 nutrient mixture and Dulbecco-modified minimum essential media (DMEM) in a ratio of 2:1, and supplemented with 8 mM HEPES, BSA (2 mg/ml), 5% fetal calf serum (FCS), and gentamycin (50 g/ml). The cultures were placed in a water jacketed incubator at 37°C under a water saturated atmosphere of 5% CO_2, 95% air. On the second day after plating, the medium was changed, using medium-199 containing 10 mM HEPES and 25 mM $NaHCO_3$ but without FCS. On the fourth day the plates were washed three times with 1 ml medium-199. The cells were then incubated for different periods of time. At the end of the incubation period 0.5 ml of the medium was transferred to separate wells (conditioned medium). ^{3}H–LHRH (10 nM) was introduced both to wells containing conditioned medium and those plated with cells. Incubation was continued for various time intervals as indicated in the Results section and reaction was terminated by removing the medium and heating it in a boiling bath for 5 min.

Bioactivity of (1-3)LHRH

The N-terminal tripeptide (pGlu-His-Trp) in saline solution was administered i.v. under light ether anaesthesia at 10 a.m. to proestrous females or to male rats that were treated with estradiol benzoate (10 ug in 0.2 ml corn oil) subcutaneously for four days. Controls were injected only with saline. Blood samples were taken by heart puncture under light ether anesthesia 15 minutes after injection. Serum levels of prolactin and TSH were determined by radioimmunoassay using kits kindly supplied by NIMDD, Rat Pituitary Hormone Program. Results are expressed in terms of the RP-1 reference preparation.

Brain plasma membranes were prepared by mincing and homogenizing two rats brains in 30 ml of 0.32 M sucrose in a Dounce homogenizer, type B. The homogenate was centrifuged at 1200 x g for 10 min, the pellet was discarded and the supernatant centrifuged for 20 min at 12000 x g. The pellet was then washed three times with the assay buffer. The radioreceptor assay for TRH was performed as follows: Brain membranes (1 mg protein) were incubated with increasing concentrations of unlabeled TRH or (1-3)LHRH in the presence of ^{3}H-TRH (80000 cpm) in a final volume of 0.5 ml (25 mM phosphate buffer pH 7.5/7.5 mM KCl/1 mM $MgCl_2$) for 75 min at $4°C$. Reaction was terminated by filtering the samples through Whatman GF/B filters. The filters were washed with cold PBS (3x3 ml), placed in vials with 10 ml scintillation fluid and counted in a liquid scintillation spectrometer. Assays, using (3-Me-His^2)-[^{3}H]TRH instead of ^{3}H-TRH were performed identically.

RESULTS

The plasma membrane and the mitochondrial fractions of the pituitary glands were characterized by using assays of marker enzymes and by a radioreceptors assay for LHRH. The fraction containing the membranes is several fold enriched (4-8 folds in different experiments) in 5'-nucleotidase activity as compared to the homogenate, while the activity of the mitochondrial marker enzyme Succinate-Cytochrome C reductase was low, indicating that there is not a significant contamination of this fraction by mitochondria. Moreover, the specific binding of the LHRH analog is 30 fold higher in the plasma membrane enriched fraction, than in the starting homogenate, and 54 fold higher than in the mitochondrial fraction (data not shown).

Identification of the degradation products

Incubation of aliquots of plasma membranes containing 100 ug protein with synthetic LHRH (10 nmoles) at neutral pH (10 mM Tris buffer pH 7.2) for 20 min at $37°C$, resulted in the partial degradation of LHRH. The main products observed were the N-terminal tripeptide (pGlu-His-Trp) and N-terminal hexapeptide (pGlu-His-Trp-Ser-Tyr-Gly). However, the N-terminal tetrapeptide (pGlu-His-Trp-Ser), the N-terminal penta-

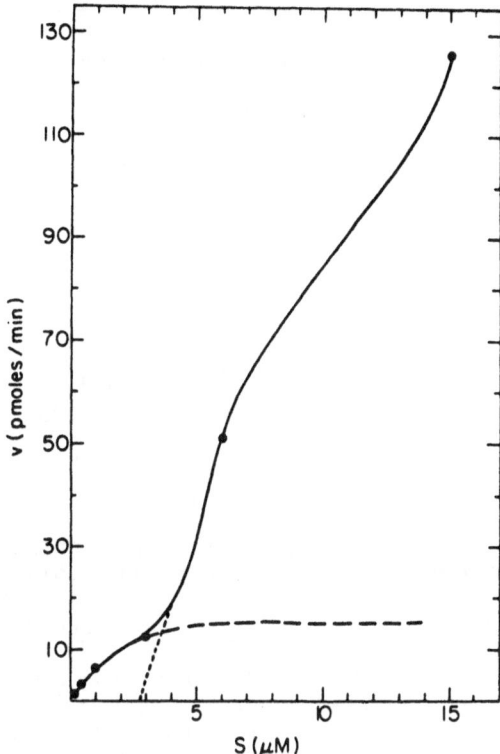

Fig. 1: The log v vs log s plot of the enzymatic degradation of LHRH.

peptide (pGlu-His-Trp-Ser-Tyr) and the (6-10) C-terminal
pentapeptide (Gly-Leu-Arg-Pro-Gly) were also present, but in
smaller ratios. In addition, pGlu was eluted with the solvent front at 3.2 min.

        The K_m and V_{max} were determined by carrying out the
degradation reaction over a wide range of substrate doses
(0.15-450 uM). The log v vs log s plots (Fig. 1) indicated
a sigmoidal curve suggesting the presence of an allosteric
enzyme exhibiting positive cooperativity or the existance of
two enzymes. Further experiments and calculations (1/v vs
1/s plots over different substrate ranges) have indicated
the presence of two enzymes. The values thus calculated
were: K_{m1} = 0.9×10^{-6} M, V_{max1} = 10 pmoles/min; K_{m2} =
6.7×10^{-4} M, and V_{max2} = 6.6 nmoles/min.

        The presence of LHRH degrading enzymes in the mitochon-
drial fraction was also investigated. Incubation of LHRH
with aliquots of the mitochondrial fraction, followed by the
isolation of the degradation products indicated that the
pattern of degradation of LHRH by this fraction is similar
to that exhibited by purified plasma membranes. However,
the activity of the different enzymes in the mitochondrial
fraction is not identical to the plasma membrane enzymes.
Fig. 2 demonstrates that the activity of the enzyme cleaving
the Trp^3-Ser^4 bond is lower in the mitochondrial fraction
than in the plasma membranes, whereas the specific activity
of the enzyme cleaving the Tyr^5-Gly^6 bond is higher in the
mitochondrial as compared to the plasma membrane fractions.

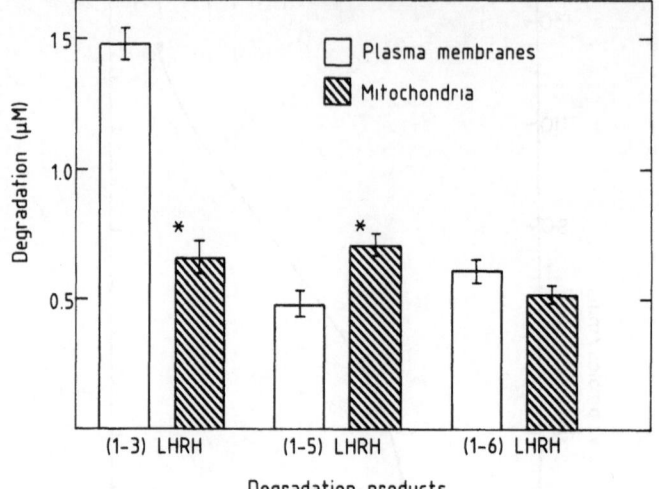

Fig. 2: Degradation of LHRH by enzymes present in subcellular fractions of rat anterior pituitary.

Enzymatic activity of pituitary cells in culture

Incubation of ^{3}H-LHRH (10 nM) with pituitary cells in
culture, leads to its degradation and the pattern obtained
is similar to that exhibited by plasma membranes. In addition, ^{3}H-Glu is also eluted from the column as one of the
main degradation products. To investigate whether the enzymatic activity is due to membrane bound enzymes, we preincubated cells with medium M-199 in the absence of ^{3}H-LHRH.
Aliquots of the medium (conditioned medium) were then removed and ^{3}H-LHRH added to both cells and conditioned medium. Incubation was continued for different periods of time
as indicated in Table 1.

The results obtained suggested that the enzymatic activity also resides in the conditioned medium. However, the
concentrations of the N-terminal tripeptide is much higher
in the cell culture rather than in the conditioned medium.
The amount of (1-3)LHRH and (1-6)LHRH increases with time
both in cells and conditioned medium while the quantity of
(1-5)LHRH and glutamic acid remains relatively constant. In
addition the ratio of the (1-3)LHRH fragment produced by
cells to that generated by conditioned medium during 30, 60,
120 and 240 minutes increases respectively from 1.8, 2.4,
2.7, to 3.1, while the contrary is true for the (1-6)LHRH
hexapeptide which decreases from 0.97, 0.79, 0.68 to 0.47.
The release of the enzyme into the incubation medium may be
due to various factors. Although this may be part of the
mechanism of enzyme action, it may also result from damage
of the cells during the preparation, culture and incubation
period.

Effects of gonadectomy and gonadal steroid replacement on the activity of the plasma membrane bound enzymes

To elucidate the effects of gonadal steroids on the enzymatic activity, we studied the degradation of LHRH by pi-

Table 1: Degradation of ^{3}H-LHRH by pituitary cells or conditioned medium.

| | Time of Incubation (min) | Degradation products (pM) | | | |
		Glu	(1-3)LHRH	(1-5)LHRH	(1-6)LHRH
Conditioned	30	235	50	60	200
medium	60	230	110	65	335
	120	230	210	65	510
	240	270	390	65	715
Cells					
	30	235	90	60	195
	60	225	265	65	265
	120	260	575	80	350
	240	345	1210	95	340

Cells were preincubated with 1 ml medium M-199 for 4 hours at 37°C. 0.5 ml medium was then transferred to other wells and ^{3}H-LHRH (10 nM) was added to both cells and medium. Incubation was continued for different periods of time as indicated below.

tuitary plasma membranes derived from ovariectomized rats treated with steroid hormones. Female rats were ovariectomized at random stages of the estrous cycle and used 3 weeks postoperation. They were primed with estradiol benzoate in peanut oil (20 g/rat sc) 2 days prior to sacrifice. On the morning of the experiment (10:00 h), some of the animals were injected with progesterone 2 mg/rat, sc). A group of ovariectomized rats that were injected only with peanut oil served as control. The rats were sacrificed at 13:00 h and pituitaries were excised immediately. Purification of plasma membranes and degradation of LHRH were carried out as described earlier. Fig. 3 demonstrates that administration of estradiol benzoate alone to ovariectomized rats or of progesterone to ovariectomized EB treated animals did not significantly affect cleavage at the Trp^3-Ser^4 and Tyr^5-Gly^6 bonds. The concentration of (1-6)LHRH in EB treated animals is however higher than in ovariectomized rats. Administration of progesterone to the ovariectomized EB treated rats further increased the cleavage at the Gly^6-Leu^7 bond.

Studies on the bioactivity of a metabolite of LHRH, the pGlu-His-TRP fragment

The (1-3)LHRH fragment (pGlu-His-Trp) has some structural similarities with TRH (pGlu-His-Pro NH_2). We therefore investigated a possible action of this tripeptide in the brain or on the pituitary gland, where TRH acts to release prolactin and TSH. Intravenous administration of (1-3)LHRH (12.5 ug/kg) to proestrous rats inhibited the basal release of prolactin by 40% (Fig. 4). When male rats that had been treated with B-estradiol-3-benzoate for four days were injected with (1-3)LHRH (25 ug/kg) a similar ef-

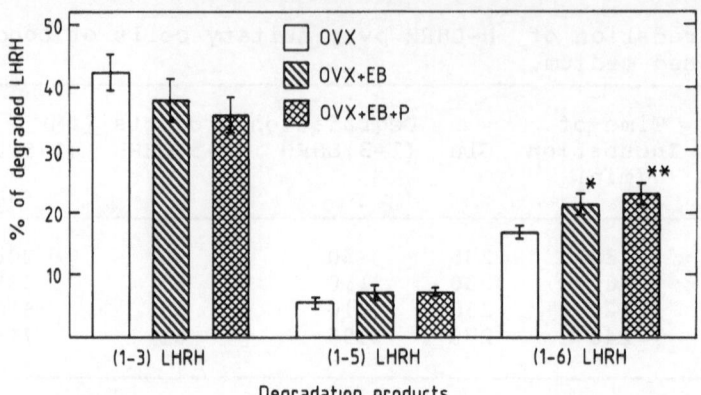

Fig. 3: Degradation of LHRH by pituitary plasma membranes derived from ovariectomized estradiol benzoate primed (OVX+EB) and ovariectomized, estradiol benzoate and progesterone treated (OVX+EB+P) rats. Total degradation is expressed as the ratio of LHRH degraded to total LHRH introduced to the incubation mixture (multiplied by 100). The degradation is defined as 100% and the degradation products are expressed as percent of total LHRH degraded. Values are means ± SEM of results of 9 different experiments each performed in duplicate. * Significantly different from OVX (p<0.05), + Significantly different from OVX (p<0.01), by t-test.

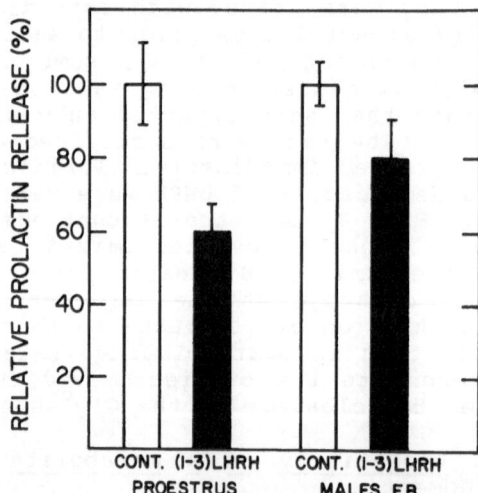

Fig. 4: Inhibition of basal prolactin release by (1-3)LHRH in proestrous females or estradiol benzoate (EB) treated male rats. Prolactin levels of saline treated animals in proestrous (235 ng/ml) and in male EB rats (425 ng/ml) were taken as 100%. Values are mean ± SEM of three independent experiments with 5-6 rats in each group. * Significantly different from control (p<0.01). + Significantly different from control (p<0.05).

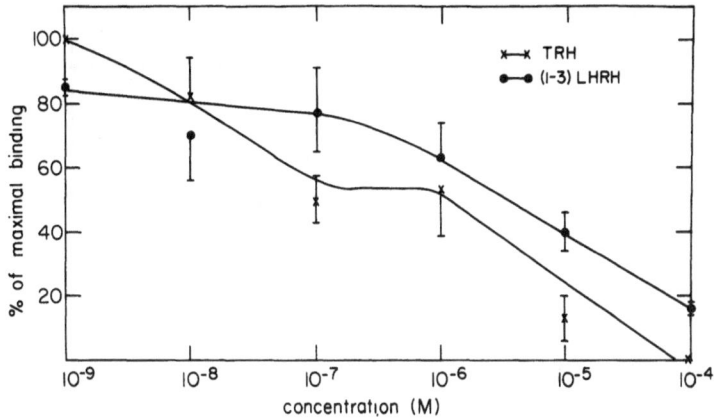

Fig. 5: Displacement of [^{3}H]TRH by TRH or by (1-3)LHRH.

fect was observed. However, the basal release of prolactin was decreased only by 20% (Fig. 4). TSH levels were not affected (data not shown) by the administration of the N-terminal tripeptide.

Binding studies

The presence of low and high affinity binding sites for TRH was reported by Burt, and Snyder (1975). Binding of [^{3}H]TRH to brain membranes in the presence of increasing concentrations of non-labeled TRH indicated, as already reported, the presence of low and high affinity binding sites (Fig. 5). Incubation of increasing doses of (1-3)LHRH with brain membranes in the presence of [^{3}H]TRH indicated that the tripeptide competed only with the low affinity binding site. In addition, its affinity for the receptor was lower than that of TRH. Since the presence of low affinity high capacity binding sites may interfere with the accurate measurement of the high affinity low capacity sites, we made use of an analog of TRH, [3-Me-His^2]TRH, that binds only to the high affinity site. Competition studies with TRH or (1-3)LHRH confirmed the results obtained with [^{3}H]TRH. TRH competed for the high affinity binding site while (1-3)LHRH could not displace the labeled TRH analog (Fig. 6).

DISCUSSION

The neuropeptide LHRH initiates its action at the target organ, the pituitary, by binding to specific receptors located on the cell surface (Clayton and Catt, 1981). This leads, inside the cell, to a series of reactions which are not yet elucidated, eventually resulting in the secretion of LH. The bioactivity of LHRH can be terminated by different mechanisms such as: a) dissociation of the hormone-receptor complex followed by escape, dilution and metabolism of the neurohormone in the general circulation, liver and kidney; b) degradation of the neurohormone at its binding site by plasma membrane associated enzymes (Clayton et al., 1979; Elkabes et al., 1981; Koch et al., 1984). The presence of

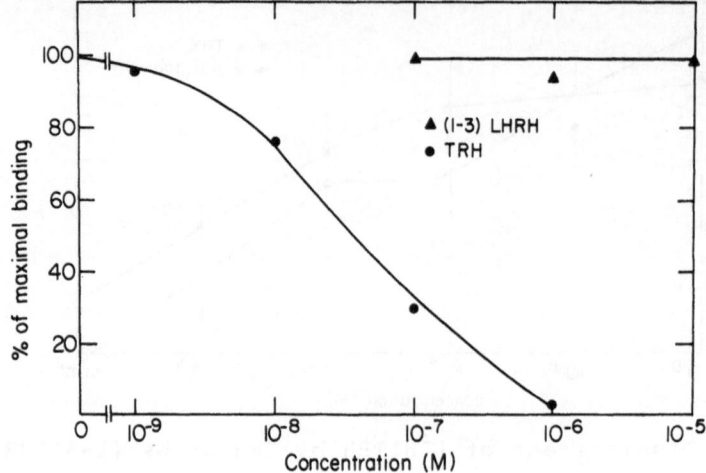

Fig. 6: Displacement of ^{3}H(3-Me-His^2)TRH by TRH or by (1-3)LHRH.

LHRH degrading enzymes on plasma membranes, where the LHRH binding sites are also located, may suggest that these enzymes may be involved in the mechanism of action of the neurohormone by regulating its concentration, and duration of action at the receptor site. c) internalization of the hormone-receptor complex (Hazum, 1982), followed by degradation of the hormone by lysosomal and cytosolic enzymes. It is possible, however, that plasma membrane associated enzymes are important also for the degradation of the internalized hormone since it has been demonstrated by autoradiographic techniques that only part of the internalized complexes are found in lysosomes whereas a significant portion of the radioactive material is localized on secretory vesicles (Table 2). No significant accumulation of radioactivity was found to be associated with mitochondria (Hazum et al., 1985). This may indicate a mechanism for rapid recycling of the receptor which may occur upon fusion of the secretory granules with the plasma membrane. It is possible that degrading enzymes in the internalized plasma membrane vesicle are involved in the process of the dissociation of the hormone-receptor complex.

Therefore, in addition to studies aiming at the characterization of LHRH binding sites and the elucidation of the cascade of reactions that start upon binding of the neurohormone to its receptor, it is necessary to clarify the role of the membrane-bound peptidases inactivating the decapeptide. The first step in the characterization of the LHRH degrading enzymes is the elucidation of its degradation products. The main fragments generated by the cleavage of LHRH are the N-terminal tripeptide (pGlu-His-Trp), pentapeptide (pGlu-His-Trp-Ser-Tyr) and hexapeptide (pGlu-His-Trp-Ser-Tyr-Gly).

The pattern of degradation of LHRH by pituitary mitochondrial fraction is similar to that exhibited by purified pituitary plasma membranes, although the ratios of the various N-terminal fragments generated by the two fractions are

Table 2: Subcellular distribution of ^{125}I-(azidobenzoyl-D-Lys^6)LHRH in rat pituitary gonadotropes.

Subcellular	Incubation		
components	90 min, 4°C	15 min, 37°C	45 min, 37°C
Plasma membrane	93.1+4.8	69.5+4.0	51.4+3.7
Endocytic vesicles	0.0	3.2+0.8	2.3+0.8
Golgi	0.0	1.9+0.6	1.8+0.7
Lysosomes	0.7+0.4	11.2+1.6	20.0+1.9
Rough endoplasmic reticulum	2.8+0.8	1.0+0.4	1.8+0.7
Mitochondria	1.0+0.5	0.0	0.9+0.5
Secretory granules	0.3+0.2	12.1+1.7	19.1+2.2
Nuclei	2.1+0.7	1.1+0.5	2.7+0.8

Silver grains were scored over profiles of 40 cells for each treatment. Only cells that were cut at the level of the nucleus and showed all cell components were examined. The total number of grains was 1223. The data + SE are expressed as % of total silver grains scored at the indicated time (modified from Hazum et al., 1985).

quite different (Table 2). It seems that the LHRH degrading enzymes present in these two subcellular fractions are not due to contamination of either of the fractions by the other since the specific activity of the enzymes is similar in the mitochondrial and the plasma membrane fractions. This notion is further supported by assay of marker enzymes in the purified subcellular fractions as well as by demonstration that the LHRH-receptor binding activity was enriched only in the pituitary plasma membrane fraction. Therefore, in addition to the plasma membrane fraction, where LHRH receptors are located, other subcellular fractions also contain enzymes that are capable of metabolizing the decapeptide. The functional significance of such peptidases in the mitochondria and their association with the low-affinity binding sites for LHRH (Liscovitch and Koch, 1982) has still to be elucidated. Nevertheless, the absence of any significant accumulation of LHRH on mitochondria, as measured by radioautography (Hazum et al., 1985), casts some doubt on the physiological relevance of these mitochondrial enzymes.

Incubation of LHRH with pituitary cells in culture resulted in the breakdown of the decapeptide and in the generation of degradation products that were similar to those obtained with the plasma membranes. The reasons for the failure of Nikolics et al. (1983) to detect degradation of LHRH by pituitary cells in culture are not known. In order to investigate whether the cleavage of the neurohormone is only due to enzymes located on the cell surface, LHRH was incubated with medium that was derived from the preincubation of the cells (conditioned medium). The results indicated that the activity resides also in the conditioned medium although the production of (1-3)LHRH is higher in cells (Table 1). The enzymatic activity found in the medium may be due to damage of the cells followed by secretion of their content, e.g. the enzyme containing organelles into the incubation mixture. This damage could be the consequence of the various manipulations to which the cells were subjected during dispersion and culture. However, this does not seem to be likely, since similar results were obtained with cells that were prepared by another procedure (data not shown). In addition, if the secretion of enzymes was due to damage of cells, the cytosolic peptidase, which produces different degradation fragments, would also be released into the conditioned medium. As shown in Table 1, the concentration of the (1-5)LHRH fragment was much lower than that of the (1-3)LHRH or (1-6)LHRH and remained low even when the preincubation time was extended from four to eight hours (data not shown). This was not due to its further breakdown since the production of the (1-6)LHRH fragment which also undergoes redegradation (Koch et al., 1984), was increased over the same period of time. Therefore, the enzymatic activities recovered in the conditioned medium may be the consequence of the gradual selective release of some pituitary peptidases and may reflect part of their mechanism of action.

The effects of steroid hormones was studied in an experimental animal model. The enzymatic activity in pituitaries of ovariectomized rats was compared to those of estradiol benzoate or estradiol benzoate and progesterone administered rats (Fig. 3). No significant difference in the enzymatic activities cleaving the Trp^3-Ser^4 or the Tyr^5-Gly^6 bonds of OVX vs. OVX+EB or OVX vs. OVX+EB+P rats was observed. The activity of the enzyme cleaving the Gly^6-Leu^7 bond was slightly higher in OVX + EB rats as compared with OVX rats ($p < 0.05$ by t-test), and administration of progesterone to estradiol primed ovariectomized rats, further increased this activity ($p < 0.01$ by t-test).

The N-terminal tripeptide is the main degradation product of LHRH by the pituitary plasma membrane preparation. This tripeptide (pGlu-His Trp) has a primary structure which is similar to that of TRH (pGlu-His-Pro NH_2). This similarity has prompted us to look whether this peptide may have TRH-like bioactivities. Administration of (1-3)LHRH caused a decrease of about 40% in serum prolactin levels of proestrus rats and of about 20% in the serum of male rats pretreated with estradiol benzoate (Fig. 4). These findings may suggest a possible role for (1-3)LHRH as one of the factors that are involved in the regulation of prolactin release. Since Burt, and Taylor (1980) have suggested that brain, and pituitary TRH receptors are identical, we have studied the competition of (1-3)LHRH for TRH binding sites in the rat brain (Figs. 5 and 6). The results indicated that (1-3)LHRH competes only for the low-affinity binding sites of TRH. However, it is possible that the (1-3)LHRH may exert its effects through its own receptors.

SUMMARY

Incubation of synthetic luteinizing hormone-releasing hormone (LHRH) with pituitary plasma membranes resulted in the partial degradation of the neurohormone. Isolation of the degradation products by HPLC followed by amino acid analysis, indicated that the N-terminal tripeptide (pGlu-His-Trp), N-terminal hexapeptide (pGlu-His-Trp-Ser-Tyr-Gly) and N-terminal pentapeptide (pGlu-His-Trp-Ser-Tyr) sequences of LHRH are the main metabolites generated by the breakdown of the neurohormone. In addition pGlu, the N-terminal tetrapeptide (pGlu-His-Trp-Ser) and the C-terminal pentapeptide (Gly-Leu-Arg-Pro-Gly NH_2) were also present, but in smaller quantities. The (1-3) LHRH fragment is mainly generated by the cleavage of the intact neurohormone at the Trp^3-Ser^4 bond, although it can also be produced by further breakdown of (1-5) LHRH and (1-6) LHRH fragments. Pituitary cells in culture are capable of degrading exogenous LHRH in a similar pattern to pituitary plasma membrane preparations. Incubation of LHRH with the mitochondrial fraction of the pituitary gland results in the appearance of degradation products that are identical to those yielded by purified pituitary plasma membranes, but in different ratios.

The involvement of steroid hormones in the regulation of enzymatic activity was further investigated in an experimental model. The peptidase levels of plasma membranes derived from pituitaries of gonadectomized rats was compared to that of gonadectomized, estradiol benzoate primed animals. These studies indicated that steroid replacement slightly enhanced the production of the (1-6) LHRH fragment while the production of (1-3) LHRH and (1-5) LHRH fragments were not altered. Administration of progesterone to gonadectomized-estradiol benzoate treated rats further increases the levels of the N-terminal hexapeptide. This suggests a possible involvement of steroid hormones in the regulation of the enzyme cleaving the Gly^6-Leu^7 bond.

The N-terminal tripeptide, which is the main degradation product of LHRH, has some structural similarities to TRH and is capable to inhibit partially prolactin secretion when administered to proestrous or to estradiol-primed male rats. This tripeptide competes with TRH for its low-affinity, but not for its high-affinity, binding sites.

ACKNOWLEDGEMENTS

We thank Mrs. R. Levin for excellent text-processing of the manuscript. This study was supported by a grant from the Rockefeller Foundation New York. Y.K. is the incumbent of the Adlai E. Stevenson III Professor of Endocrinology and Reproductive Biology.

REFERENCES

Aronson, N.N., and Touster, O., 1974, Isolation of rat liver plasma membrane fragments in isotonic sucrose, in: "Methods in Enzymology", 31, S. Fleischer and L. Packer, eds., 90, Academic Press, New York.

Bauer, K., and Horsthemke, B., 1984, Degradation of LHRH. in: "Hormonal Control of the Hypothalamo-Pituitary-Gonadal Axis", K.W. McKerns and Z. Naor, eds., 101, Plenum Press, New York.

Brazeau, P., Ling, H., Esch, F., Bohlen, P., Benoit, R., and Guillemin, R. 1981, High biological activity of the synthetic replicates of somatostatin-28 and somatostatin-25. Regul. Peptides, 1:255.

Burt, R.B., and Snyder, S.H., 1975, Thyrotropin releasing hormone (TRH): Apparent receptor binding in rat brain membranes. Brain Res., 93:309.

Burt, D.R., and Taylor, R.L., 1980, Binding sites for thyrotropin-releasing hormone in sheep nucleus accumbens resemble pituitary receptors. Endocrinology, 106:1416.

Clayton, R.N., Shakespear, R.A., Duncan, J.A., and Marshall, J.C., 1979, Luteinizing hormone-releasing hormone inactivation by purified pituitary plasma membranes: Effects on receptor binding studies. Endocrinology, 104:1484.

Clayton, R.N., and Catt, K.J., 1981, Gonadotropin-releasing hormone receptors: Characterization, physiological regulation and relationship to reproductive function. Endocr. Rev., 2:186.

Elkabes, S., Fridkin, M., and Koch, Y., 1981, Studies on the enzymic degradation of luteinizing hormone releasing hormone by rat pituitary plasma membranes. Biochem. Biophys. Res. Commun., 103:240.

Fleischer, S., and Kervina, M., 1974, Subcellular fractionation of rat liver. in: "Methods in Enzymology", 31, S. Fleischer and L. Packer, eds., 6, Academic Press, New York.

Fridkin, M., Hazum, E., Baram, T., Lindner, H.R., and Koch, Y., 1977, Hypothalamic and pituitary LRF-degrading enzymes: Characterization, purification and physiological role. in: "Peptides", M. Goodman and J. Meinhofer, eds., 193, J. Wiley and Sons, Inc., New York.

Hazum, E., 1982, Receptor regulation by hormones: Relevance to secretion and other biological functions. in: "Cellular Regulation of Secretion and Release", P.M. Conn, ed., 3, Academic Press, London.

Hazum, E., Koch, Y., Liscovitch, M., and Amsterdam, A., 1985, Intracellular pathways of receptor-bound GnRH agonist in pituitary gonadotropes. Cell Tissue Res., 239:3.

Koch, Y., Elkabes, S., and Fridkin, M., 1984, Degradation of GnRH by enzymes associated with rat pituitary plasma membranes. in: "Hormonal Control of the Hypothalamo-Pituitary-Gonadal Axis", K.W. McKerns and Z. Naor, eds., 115, Plenum Press, New York.

Koch, Y., Elkabes, S., and Fridkin, M., 1986, Biodegradation of luteinizing hormone-releasing hormone by pituitary plasma membrane and by pituitary cells in culture. in: "Proceedings of the 4th International Symposium on Psychoneuroendocrinology in Reproduction", P. Pancheri and L. Zichella, eds., in press, Hamisphere Publishing Co.

Kochman, K., Kerdelhue, B., Zor, U., and Jutisz, M., 1975, Studies of enzymatic degradation of luteinizing hormone-releasing hormone by different tissues. FEBS Lett., 50:190.

Liscovitch, M., and Koch, Y., 1982, Characterization and subcellular localization of GnRH analog binding in rat brain. Peptides, 3:55.

Liscovitch, M., Ben-Aroya, N., Meidan, R., and Koch, Y., 1984, A differential effect of trypsin on pituitary gonadotropin-releasing hormone receptors from intact and ovariectomized rats: Evidence for the existence of two distinct receptor populations. Eur. J. Biochem., 140:191.

McDermott, J.R., Smith, A.I., Biggins, J.A., Hardy, J.A., Dodd, P.R. and Edwardson, J.A., 1981, Degradation of luteinizing hormone-releasing hormone by serum and plasma in vitro. Regul. Peptides, 2:69.

McDermott, J.R., Smith, A.I., Biggins, J.A., Edwardson, J.A. and Griffiths, E.C., 1982, Mechanism of luteinizing hormone-releasing hormone degradation by subcellular fractions of rat hypothalamus and pituitary. Regul. Peptides, 3:257.

Meidan, R., and Koch, Y., 1981, Binding of luteinizing hormone-releasing hormone to dispersed rat pituitary cells. Life Sci., 28:1961.

Nikolics, K., Szoke, B., Keri, G., and Teplan, I., 1983, Gonadotropin releasing hormone (GnRH) is not degraded by intact pituitary tissue in vitro. Biochem. Biophys. Res. Commun., 114:1028.

Pimstone, B., Epstein, S., Hamilton, S.M., LeRoith, D. and Hendricks, S. 1977, Metabolic clearance and plasma half disappearance time of exogenous gonadotropin releasing hormone in normal subjects and in patients with liver disease and chronic renal failure. J. Clin. Endocrinol. Metab., 44:356.

Stetler-Stevenson, M.A., Yang, D.C., Lipkowski, A., McCartney, L., Peterson, D.R., and Fluoret, G., 1981, An approach to the elucidation of metabolic breakdown products of the luteinizing hormone-releasing hormone. J. Med. Chem., 24:688.

Tisdale, H.D., 1967, Preparation and properties of succinic-cytochrome C reductase (complex II-III). in: "Methods in Enzymology", 10, R.W. Estabrook and M.E. Pullman, eds., 213, Academic Press, New York.

Towatari, T., and Katunuma, N., 1983, Selective cleavage of peptide bonds by Cathepsin L and B from rat liver. J. Biochem., 93:1119.

Wilk, S., Benuck, M., Orlowski, M., and Marks, N., 1979, Degradation of luteinizing hormone-releasing hormone (LHRH) by brain prolyl endopeptidase with release of des-glycinamide LHRH and glycinamide. Neurosci. Lett., 14:275.

ANALYSIS OF HORMONE SECRETION FROM INDIVIDUAL PITUITARY CELLS

J.D. Neill, P.F. Smith, E.H. Luque, M. Munoz de Toro,
G. Nagy, and J.J. Mulchahey

Department of Physiology and Biophysics, University of
Alabama at Birmingham, Birmingham, AL 35294 USA

INTRODUCTION

Analysis of the function of individual neurons and immune cells (B lymphocytes) has been enormously profitable for the understanding of the nervous and immune systems. Indeed, specialization of individual cells is central to theories underlying the function of those systems. In contrast, homogeneity among cells subserving various functions in endocrine and exocrine secretory systems is the common assumption. These differences in viewpoint may arise simply from the availability of methods for the study of individual neurons (single unit electrical recordings) and immune cells (hemolytic plaque assay for detection of antibody secretion) on the one hand, and their absence for individual endocrine and exocrine cells, on the other. Thus, the extent to which heterogeneity in cell function contributes to the function of endocrine and exocrine systems is unknown.

With these considerations in mind, we developed a reverse hemolytic plaque assay which permits the detection and measurement of hormone secretion from individual pituitary cells in culture (Neill and Frawley, 1983). Derivative of the hemolytic plaque assay of Niels Jerne (Jerne et al., 1974) for detection of antibody secretion from individual B lymphocytes in culture, the reverse hemolytic plaque assay has been used for detection of antigen secretion (Molinaro and Dray, 1974) and is based on complement mediated lysis of antibody-coated erythrocytes co-incubated with antigen-secreting cells. Antigen secretion results in hemolysis of erythrocytes surrounding the secretory cells so that clear areas of lysis (plaques) identify cells secreting antigen recognized by the antibody used to coat the erythrocytes (see Fig. 1). We have used the reverse hemolytic plaque assay to detect and measure prolactin (Neill and Frawley, 1983), LH (Smith et al., 1984), growth hormone (Frawley and Neill, 1984), ACTH (Smith and Neill, 1984), and TSH (unpublished) secretion from individual adenohypophysial cells. In the report that follows, we present the results of studies on heterogeneity of secretory activity among pituitary cells belonging to the same class (primarily gonadotropes) and attempt to relate its contribution to the secretory patterns of LH secretion observed during the rat estrous cycle.

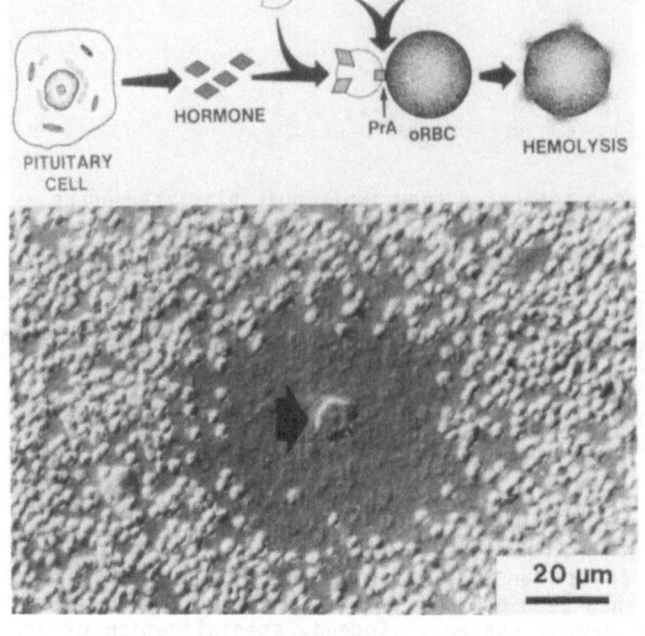

Fig. 1. Upper panel: Schematic diagram of the mechanism of hormone-directed hemolysis in the reverse hemolytic plaque assay. Lower panel: Photomicrograph of a plaque (zone of hemolysis containing erythrocyte ghosts) surrounding a gonadotrope (arrow). Small, round cells are unlysed erythrocytes.

REVERSE HEMOLYTIC PLAQUE ASSAY

This assay was performed as described in detail previously (Neill and Frawley, 1983; Smith et al., 1985). Anterior pituitary glands from female rats at various stages of the estrous cycle are dispersed into single cells using trypsin (Hymer and Hatfield, 1983). The cells are mixed with ovine erythrocytes (oRBC) previously coupled to protein A using chromium chloride and infused into a Cunningham chamber coated with poly-L-lysine to promote adherence of the two cell types to the floor of the chamber as a homogeneous monolayer (see Figs. 1 and 2). Cunningham chambers have a volume of about 30 μl, a depth of about 90 μm, and are constructed with a cover slip spanning 2 pieces of double-stick tape across the surface of a glass microscope slide. This configuration maximizes the lateral diffusion of secreted hormone so that high concentrations of it are achieved in the secretory cell surround containing oRBC. After adherence of the cells, hormone antibody and secretagogues as appropriate are infused into the chamber; incubation occurs for a period sufficient for all secretory cells to form plaques (usually 1-4 hours). Then, complement is added, and after 30 minutes the cells are fixed for microscopic examination and measurement of plaque area.

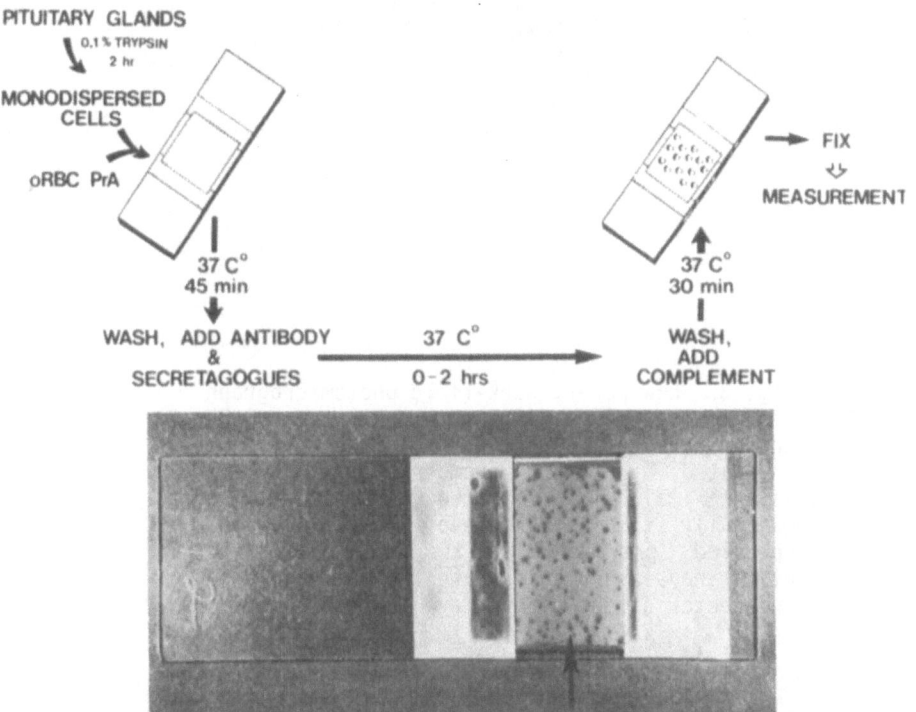

Fig. 2. <u>Upper panel</u>: Flow diagram illustrating the methods used to perform the reverse hemolytic plaque assay. oRBC-PrA=protein A coupled ovine erythrocytes which are mixed with pituitary cells and infused under the cover-slip into the Cunningham chamber. <u>Lower panel</u>: Microscope slide with a Cunningham chamber constructed at one end. Large LH plaques (black spots in chamber indicated by an arrow) are evident without magnification.

Gonadotropes are responsive to gonadotropin releasing hormone (GnRH) when used immediately after dispersion: maximal secretory responses (GnRH >10^{-8}M) of 15-20 fold increases over baseline secretion (no GnRH) are observed (Fig. 3). Plaques form around approximately 5% of the total pituitary cell population after 2 hours when an ovine LH antiserum is used, and a similar fraction of the same cells stain immunocytochemically using the NIADDK rLH β-subunit antiserum (Smith et al. 1984). Immunocytochemistry can be performed directly on plaque forming cells if the LH antibodies remaining on the oRBC from the plaque reaction are eluted with 0.2M sodium acetate/0.5M sodium chloride (10 min at 4°C) before fixation and processing for avidin/biotin peroxidase immunocytochemistry (Fig. 4). When incubated in the presence of a maximal dose of GnRH (10^{-7}M), gonadotropes form plaques within 15 min after incubation is initiated; thereafter, plaques have formed around all gonadotropes from proestrus rats at 2 hours and no additional ones form with longer incubation times (Fig. 5). The sizes of plaques measured as areas of hemolysis in μm² continue to enlarge until the practical limit of diffusion is reached but 2-4 hr incubations are optimal. The specificity of the plaque reaction for LH has been detailed previously (Smith et al., 1984; Smith et al., 1985).

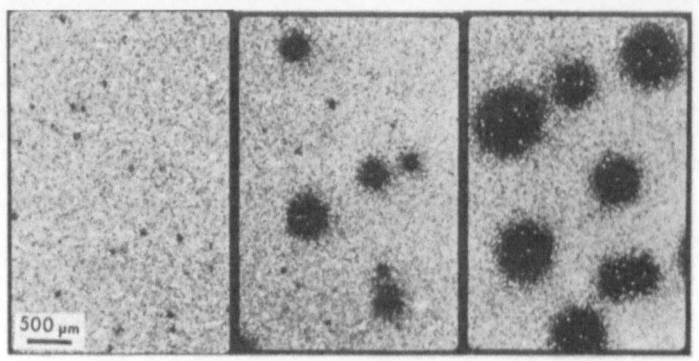

Fig. 3. Low power, dark-field photomicrograph of LH plaques formed in response to a GnRH analog (Buserilin). Left chamber remained untreated, center chamber received a half-maximal dose, and the right chamber a maximally stimulatory dose of the analog. The speckled, gray area is the oRBC lawn, the black areas are plaques, and the white dots in the center of the plaques are gonadotropes.

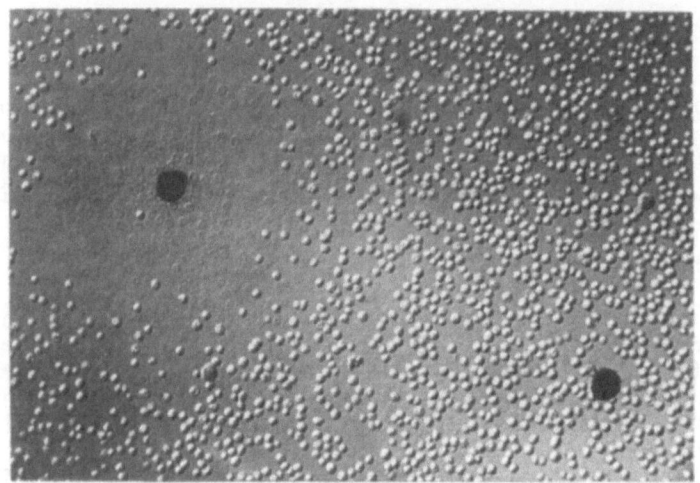

Fig. 4. A photomicrograph illustrating the plaque assay combined with immunocyto-chemistry. Both cells shown contain immuncyto-chemically detectable LH but only one secreted LH (formed a plaque) after 2 hours of incubation in the presence of a maximally stimulatory dose of GnRH (10^{-7}M).

QUANTIFICATION OF LH SECRETION FROM INDIVIDUAL GONADOTROPES

Several initial observations and considerations indicated that the plaque assay might be used to quantify LH secretion: 1) increases in plaque area measured either as total area of all plaques or as mean plaque area, matched increases in dose of GnRH (Fig. 3), 2) plaque area increased as a function of length of incubation (Fig. 5), and 3) the mechanism of the plaque reaction is hormone-directed hemolysis so

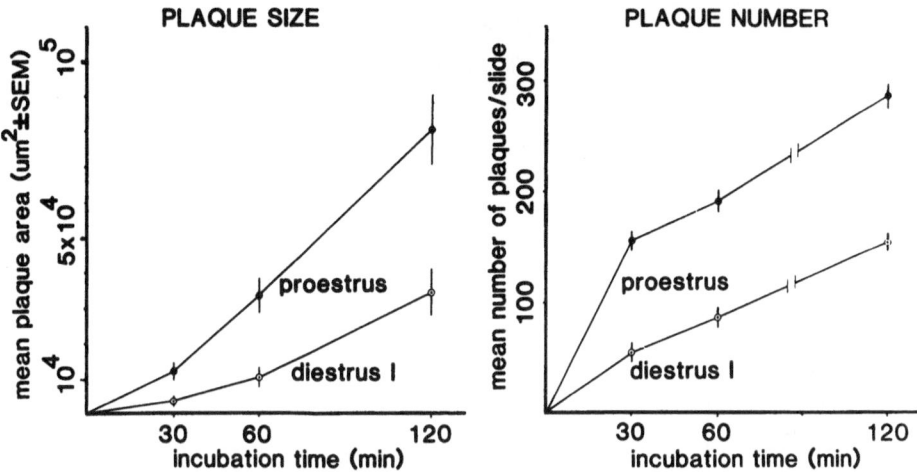

Fig. 5. Rate of plaque formation by LH gonadotropes. <u>Left panel</u>:
Rate of increase in plaque area over time in proestrous and diestrous
rat pituitary cells. <u>Right panel</u>: Number of plaques formed by cells
from proestrous and diestrous rat pituitary cells over time. Longer
incubation times do not increase the numbers of plaques formed.

that more or less hormone should result in more or less hemolysis;
indeed, plaque area could be expressed as the number of oRBC lysed
since they are uniformly distributed as a monolayer on the surface of
the Cunningham chamber.

To confirm that LH secretion could be quantified, we incubated 14
cohort batches of pituitary cells under similar conditions either in
Petri dishes for measurement of LH secretion using radioimmunoassay or
in Cunningham chambers for measurement of plaque area (Smith and Neill,
1983). The LH secretory activity was manipulated over a 50-fold range
using various lengths of incubations combined with various doses of
GnRH. Total radioimmunoassayable LH release in Petri dishes was
corrected so that equal numbers of cells were being considered in the
RIA and the plaque assay. This value was then divided by number of
gonadotropes present (determined by immunocytochemistry) to derive LH
secretion per gonadotrope. LH secretion/gonadotrope was then plotted
against the mean plaque area/gonadotrope for each of the 14 treatment
groups (Fig. 6). In 4 such assays, significant statistical regressions
were observed with correlation coefficients ranging from 0.93-0.98.
The sensitivity of the plaque assay (smallest detectable plaque) is
thus about 2-5 pg of NIADDK rLH-RP1. Since the abundance of LH in this
preparation is about $1\frac{1}{2}$-3% of preparations containing only rLH, the
sensitivity of the plaque assay (smallest detectable plaque) is about
$1-5 \times 10^{-18}$ moles (attamoles).

These findings demonstrate that plaque area is linearly related to
amount of LH secreted, thus permitting the use of plaque area as a
relative measure of LH secretion without the necessity of measuring LH
release by RIA simultaneously on cohort cells. Also, they permit the
plaque assay to be used as a convenient bioassay for GnRH and its
congeners; in this case, total plaque area (which is comparable to
determinations of medium LH content by RIA in Petri dish incubations of
cells) is measured using an automated video image analysis system
(Image Technology Corporation Model 3000; distributed by Nikon) (see
Fig. 7). This approach requires 2 orders of magnitude fewer cells per

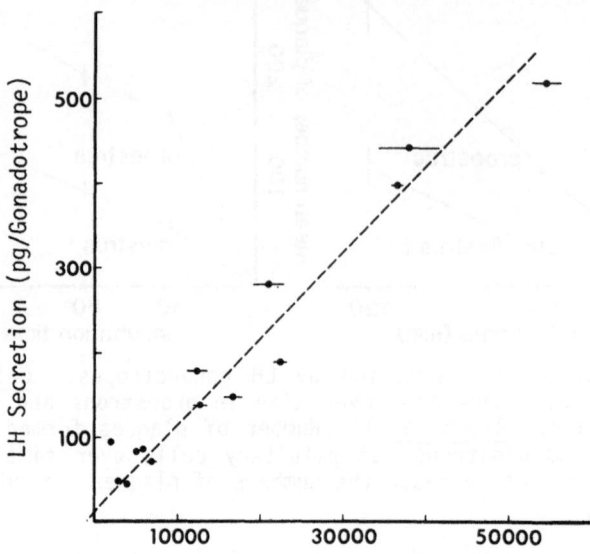

LH PLAQUE ASSAY
STANDARD CURVE

Mean Plaque Area (μm^2 $\pm$ SEM)/Gonadotrope

Fig. 6. Comparison of LH secretion from cells
incubated in Petri dishes and measured by
radioimmunoassay (ordinate) with plaque areas
(abscissa) of cells incubated in Cunningham
chambers. Cohort cells (14 batches) were
treated similarly with GnRH to generate
secretory responses covering an approximate
50-fold range. Correlation coefficients in 4
such assays ranged from 0.93-0.98.

determination and assay results are available on the same day that cell
dispersion is performed.

QUANTITATIVE HETEROGENEITY OF GONADOTROPE SECRETION

Not all gonadotropes respond equally to GnRH as assessed by
several functional parameters 1) lag-time to onset of secretion 2)
thresh-holds for stimulation 3) responsiveness, and 4) maximum
secretory output. The first parameter (lag-time) is illustrated in
Fig. 5 (right panel) where additional plaques appear during the 30-60
min interval and again during the 60-120 min interval after incubation
was initiated with maximal concentrations of GnRH (10^{-7}M). Clearly,
some cells secrete LH more rapidly than others and probably are
identical with the cells forming the largest plaques. When cells are
treated with intermediate or low doses of GnRH, not all of the
gonadotropes form plaques (data not shown), suggesting that varying
thresh-holds exist among gonadotropes for activation of LH secretion.
In females at stages of the estrous cycle other than proestrus, and in
males, some gonadotropes (40-50% of the total) are totally unresponsive

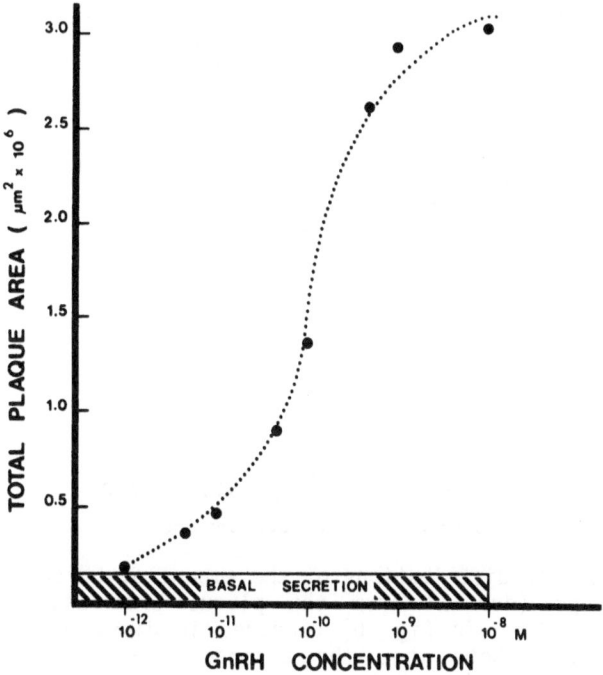

Fig. 7. GnRH dose-response curve in the reverse hemolytic plaque assay illustrating its utility as a bioassay for agents that alter LH secretion. Total plaque area was measured with an automatic electronic image analysis system.

to maximally stimulatory concentrations of GnRH (10^{-7}M) even when incubated for several hours longer than required for all gonadotropes from proestrus rats to become secretory (see Fig. 5, right panel; Smith et al., 1984). This unresponsiveness occurs in gonadotropes which can be shown to have ample stores of LH using immunocytochemistry (Fig. 4). The maximum LH secretory output is heterogeneous among gonadotropes treated with 10^{-7}M GnRH as illustrated in Fig. 8; the difference between the cell secreting the smallest amount of hormone and the one secreting the largest amount ranges from 500-1000 fold (note that in this comparison all cells are derived from the pituitary of a single rat).

A final heterogeneity arising among gonadotropes is the apparent existence of sub-populations based on frequency distributions of the amount of LH secreted (Figs. 8 and 9). Two such subpopulations are observed in gonadotropes derived from proestrus rat pituitaries whereas a single sub-population in diestrous rats coincides with the smaller proestrous sub-population (Fig. 8). The two sub-populations of gonadotropes in proestrous rat pituitaries can be observed only after 2 hours of incubation (Fig. 9); before then, single sub-populations only are exhibited.

FUNCTIONAL ROLE OF GONADOTROPE SUBPOPULATIONS IN THE OVULATORY LH SURGE

The pattern of LH secretion during the 4 day estrous cycle of the rat is composed of low, relatively unvarying levels throughout except

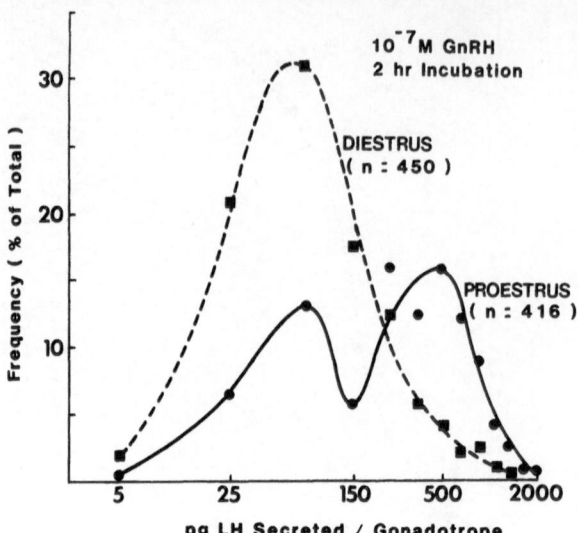

Fig. 8. Frequency distributions of the amount of LH secreted per gonadotrope (derived from the plaque area using the standard curve shown in Fig. 6) compared in diestrous-1 and proestrous rats. Note that the unimodal distribution at diestrus shifts to a bimodal distribution at proestrus.

for an 8-10 hour period on the afternoon of proestrus when there is an approximate 50 fold increase (the preovulatory LH surge) (M.S. Smith et al. 1975). These two components of the pattern are thought to reflect two phases of LH secretion, the tonic pattern of LH secretion resulting from the inhibitory effects of low estradiol levels on GnRH and LH secretion, and the phasic pattern (preovulatory surge) resulting from the stimulatory effects of the rising estradiol levels to increase GnRH secretion and to increase adenohypophysial responsiveness to GnRH (Fink, 1979). This increased responsiveness is characterized by an elevation in the fraction of LH content that maximal doses of GnRH can release (the "readily releasable pool") (Fink, 1979; Pickering and Fink, 1979).

Do the findings of functional heterogeneity among gonadotropes (see above) have explanatory value for the physiologic regulation of LH secretion? We have previously demonstrated (Smith et al, 1984) that estradiol stimulates LH secretion through increasing the amount of hormone released from individual gonadotropes (see Figs. 5, and 8, and 9) as well as by increasing the fraction of gonadotropes that are secretory. Furthermore, we suggested (Smith et al., 1984) that recruitment by estradiol of gonadotropes into the secretory pool might explain the difference between releasable and non-releasable forms of LH; non-releasable stores of LH might simply be contained in the non-secretory gonadotropes found at diestrus. Since there is a several-fold increase in the "readily releasable pool" of LH at proestrus without a concomitant change in total LH stores (Pickering

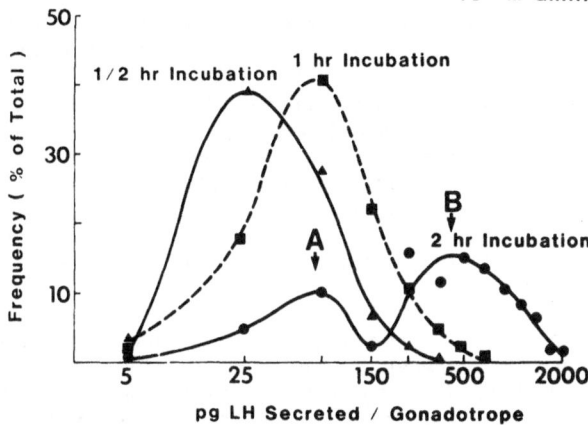

Fig. 9. Development over time of incubation of the bimodal
frequency distribution of amount of LH secreted per
gonadotrope at proestrus.   Frequency distributions were
unimodal at ½ and 1 hour of incubation but became bimodal (A
and B) at 2 hours. LH secreted/gonadotrope was derived from
plaque areas using the standard curve shown in Fig. 6.

and Fink, 1979) non-secretory gonadotropes at diestrus would have to
contain more LH than their secretory cohorts to be able to fully
account for the phenomenon.   Although LH stores are ample in
non-secretory   gonadotropes   (see   Fig.   4),   quantitative
immunocytochemical studies of LH levels in secretory and non-secretory
gonadotropes are required to resolve this issue.

Cellular   mechanisms   underlying   the   property   of   estradiol,
depending on whether its concentrations are low or high, to inhibit or
stimulate LH secretion, have defied elucidation. We have considered
the possible existence of two subpopulations of gonadotropes with
opposite secretory polarities, one responding to estradiol with an
inhibition of LH release and the other responding with an increase in
LH release. For example, the non-secretory gonadotrope at diestrus
might be a cell that is stimulated by estradiol since treatment with
that steroid simultaneously induces the LH surge and recruits such
cells into the secretory pool (Smith et al., 1984). On the other hand,
the secretory gonadotrope at diestrus might be a cell that is inhibited
by estradiol since it seems to be active when the tonic pattern of
secretion is observed. This hypothesis remains untested because we
have been unable thus far to measure hormone secretion from the same
gonadotropes repeatedly by application of sequential plaque assays
(Smith et al., 1985); affirmation of the hypothesis would require that
non-secretory gonadotropes at diestrus become the sub-population of
gonadotropes secreting large amounts of LH at proestrus (see Fig. 8)
and that the secretory sub-population of gonadotropes at diestrus
continue secreting small amounts of LH at proestrus (see Fig. 8).

The induction of the preovulatory surge of LH secretion at proestrus by the rising plasma levels of estradiol is associated with a doubling in the number of GnRH receptors (Clayton et al. 1980). The observation reported earlier (see above) that only 50-60% of the gonadotropes are secretory at diestrus, and that estradiol recruits these cells into the secretory pool at proestrus raised the question of whether non-secretory gonadotropes lacked GnRH receptors which were subsequently induced by estradiol. Also, we wondered whether the marked heterogeneity in secretory response observed among individual gonadotropes at proestrus (Figs. 8 and 9) was due simply to differences in GnRH receptor number.

To address these and other pertinent issues, we combined the plaque assay with GnRH receptor autoradiography to examine the relationship between GnRH binding and LH release at the single cell level. A GnRH agonist, Buserelin (Hoechst) (called GnRH-A), was radioiodinated to a specific activity of about 1500 μCi/μg using the methods described by Clayton et al. (1980). Monodispersed pituitary cells were used immediately after trypsin dispersion and were mixed with protein-A coated oRBC for infusion into a Cunningham chamber as described earlier (Fig. 2). The plaque assay was then performed using a 2 hour incubation of the LH antiserum and ^{125}I-GnRH-A as the secretagogue for LH plaque formation. After complement treatment (Fig. 2), cold glutaraldehyde (2% in Sorensen's phosphate buffer) was infused into the Cunningham chamber for a fixation time of 30 min. The cover-slips were then removed from the Cunningham chamber and the cells on the slide were dehydrated with alcohol, air-dried, and dipped in Kodak NTB-2 emulsion. After an exposure time of 14 days at - 20°C, autoradiograms were developed in Kodak D-19, fixed, and then mounted. Areas of individual LH plaques were measured using a video-based image processing system (Bioquant, Leitz); with this system the plaque is outlined using a digitizing pad. ^{125}I-GnRH-A binding to the same cell was measured using a Leitz MPV-Compact Microphotometer; developed silver grains over the cell were measured with a 63x oil objective and reflectance illumination. Paired plaque area and reflectance measurements were made on 50 plaque-forming cells per slide with 3 slides per treatment group (total=150 paired measurements).

Shown in Fig. 10 are three plaque-forming gonadotropes derived from groups of cells treated with ^{125}I-GnRH-A at varying doses illustrating the finding that mean plaque size and mean number of autoradiographic grains increased with increasing doses of ^{125}I-GnRH-A. Figure 11 illustrates this finding more explicitly, showing that mean LH secretion per gonadotrope was highly correlated (r=0.95) with mean radioiodinated GnRH-A binding per cell. The binding of radioiodinated GnRH-A to gonadotropes was specific because it could be completely inhibited with excess unlabelled GnRH-A (see Fig. 12) or GnRH.

These results demonstrate that specific, high affinity binding sites for an iodinated analog of GnRH can be measured on individual gonadotropes derived from recently trypsin-dispersed rat pituitary cells. Since GnRH inhibits the binding of the radioiodinated analog, and the mean amount of LH released/gonadotrope is highly correlated with mean GnRH-A binding/gonadotrope, we may suggest that the level of ^{125}I-GnRH-A binding is a relative measure of the number of GnRH receptors/gonadotrope. Exact quantification of GnRH receptor number and affinity constants are not possible because the studies required to convert autoradiographic grain number (reflectance) to dpm are not yet completed.

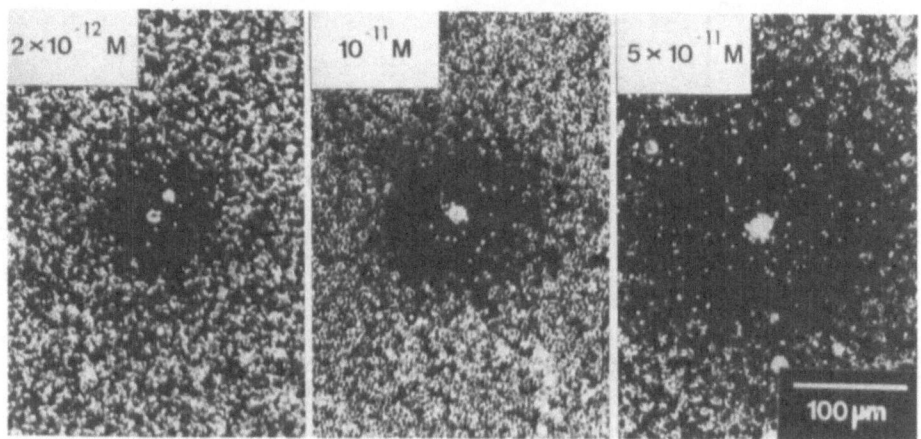

Fig. 10. Photomicrographs illustrating the relationship between mean plaque size and mean autoradiographic grain number per gonadotrope. Silver grains (white dots in this dark-field photomicrograph) were concentrated over the cell in the center of the plaque. Varying concentrations of radioiodinated GnRH-A were incubated with the gonadotropes from a proestrus rat pituitary gland for 2 hours to evoke LH secretion and to permit measurement of ^{125}I-GnRH-A binding.

RELATIONSHIP OF GnRH-A BINDING AND LH RELEASE

MEANS OF 150 CELLS PER TREATMENT GROUP

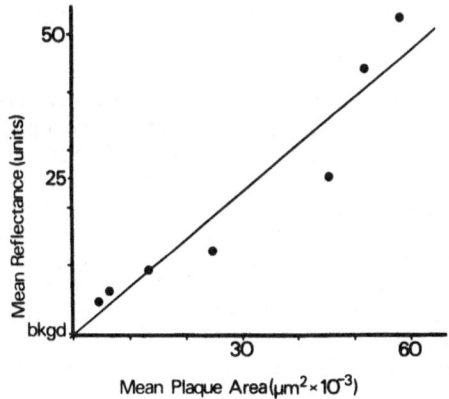

Fig. 11. This graph illustrates the relationship between mean ^{125}I-GnRH-A binding/gonadotrope (reflectance) and mean LH release/gonadotrope (plaque area) (paired measurements on 150 cells/data point). Each data point represents one concentration of ^{125}I-GnRH-A from a dose curve (from lowest to highest of 10^{-12}, 2 x 10^{-12}, 5 x 10^{-12}, 10^{-11}, 2 x 10^{-11}, 5 x 10^{-11}, and 10^{-10}M). Proestrous rat pituitary cells were used. Correlation coefficient=0.95.

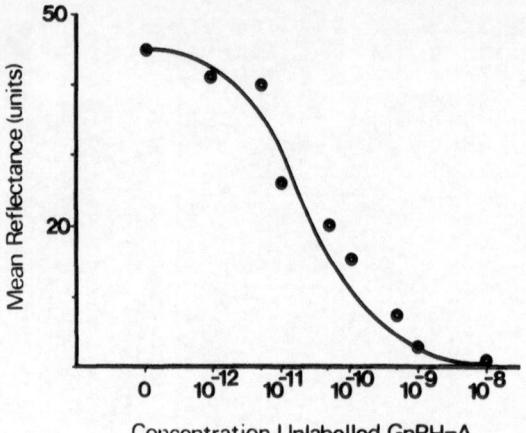

Fig. 12. Inhibition of ^{125}I-GnRH binding to
gonadotropes by unlabeled GnRH-A.
Radioiodinated GnRH-A (50 pM) was incubated
together with unlabeled GnRH-A for 2 hours
(proestrus rat pituitary cells) before
autoradiography was performed. Reflectance
measurements (autoradiographic grain number)
were made on 150 cells per data point;
reflectance was used as a measure of GnRH
binding.

DISTINCTION BETWEEN FUNCTIONAL AND NON-FUNCTIONAL BINDING SITES FOR
GnRH

Previous reports have indicated that trypsin-dispersed rat
pituitary cells have no detectable GnRH receptors whereas
collagenase-dispersed cells do, using standard liquid radioreceptor
assays (Naor et al., 1980). We have confirmed that cells prepared with
trypsin as described earlier (see above) do not bind detectable amounts
of ^{125}I-GnRH-A when they are homogenized and incubated with
radiolabeled GnRH-A using the methods described by Naor et al. (1980);
however, whole pituitaries analyzed similarly provide binding data
similar with those reported by Naor et al. (1980). Thus, cells
dispersed with trypsin secrete LH copiously in response to GnRH (see
Fig. 7 for example) but do not exhibit binding sites in standard liquid
radio-receptor assays.

Another unreconcilable observation is the report of Naor et al.
(1980) that the dose of GnRH-A required for full receptor occupancy
exceeded the dose required for maximal LH secretion by about 5-fold;
i.e., only about 20% of the receptors need be occupied to obtain a full
biological response, indicating the presence of large numbers of
non-functional or "spare" receptors. In contrast, using trypsin-

dispersed cells for the study of binding by autoradiographic detection of ^{125}I-GnRH and of LH secretion by the plaque assay, we find essentially superimposable secretory and binding curves (Fig. 13). These results suggest that non-functional ("spare") receptors are preferentially inactivated by trypsin and that functional receptors are resistant to this enzyme. The explanation for this difference is unknown but may have a morphologic origin rather than relecting physiochemical differences in the receptors; for example, since functional receptors must somehow be coupled to the effector mechanisms for secretion they could be cryptic.

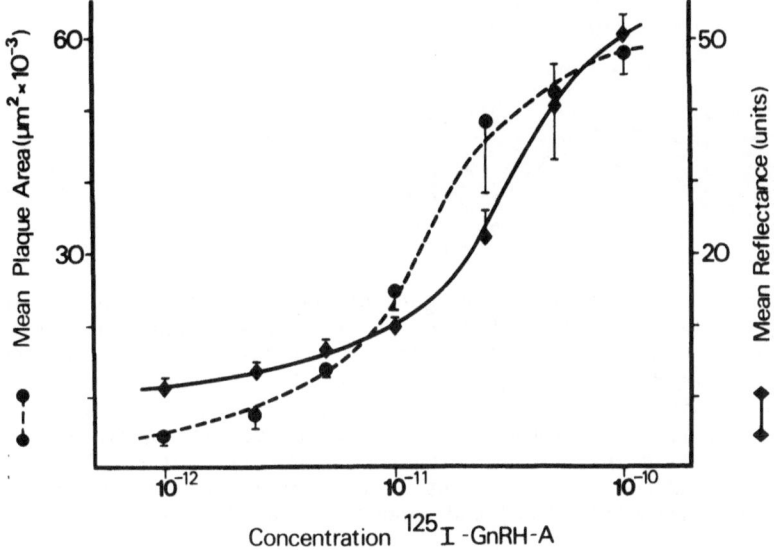

Fig. 13 This graph illustrates the dose-response and dose-binding relationship using ^{125}I-GnRH-A as both the secretagogue and binding agent for individual proestrus cells. Reflectance (silver grains) is a measure of GnRH-A binding and plaque area is a measure of LH secretion. A total of 3 similar experiments were performed.

RELATIONSHIP OF GnRH BINDING SITES TO LH SECRETION BY INDIVIDUAL GONADOTROPES

Earlier we demonstrated a high correlation between mean GnRH-A receptor binding and mean LH secretion measured in groups of gonadotropes (see Figs. 10, 11, and 13). This high correlation coefficient (r=0.95) is observed only among group means; when individual gonadotropes within those groups are studied, the correlation between LH secretion (plaque area) and radioiodinated GnRH binding (reflectance of silver grains) by the same cells falls to low levels (overall correlation coefficient = 0.33; range of correlations within individual groups = 0.02 to 0.38). This low correlation is illustrated in Fig. 14 where two large plaques are seen to be associated with either high or low grain numbers, and two small plaques are also associated with high or low grain numbers. The interpretation of this finding is that GnRH receptor number for any individual

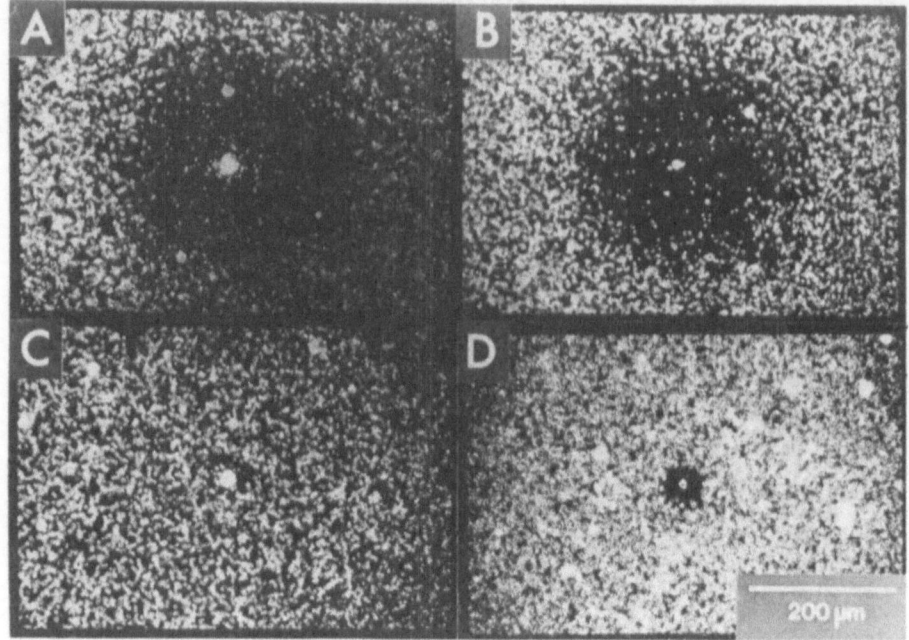

Fig. 14 Darkfield photographs of gonadotropes demonstrating the low correlation between GnRH-A binding (grain number) and LH release (plaque area) by individual cells within a treatment group. All four cells were treated with 100 pM ^{125}I-GnRH-A for 2 hours. A: Large plaque-high grain number. B: Large plaque-low grain number. C: Small plaque-high grain number. D: Small plaque-low grain number. Correlation coefficients were calculated for each of 7 different radioiodinated GnRH-A treated groups (150 pairs of data were included in each GnRH-A treatment group) and ranged from 0.02 to 0.38.

gonadotrope is a poor predictor of the amount of LH it can secrete; nevertheless, full occupancy of all its GnRH receptors (whether high or low in number) is required for a gonadotrope to reach its full LH secretroy capacity (which may be high or low in amount). Thus, the somewhat surprising conclusion is reached that for any individual gonadotrope the level of other factors in the chain of events subserving the activiation of secretion are the primary determinants of the amount of hormone secreted; GnRH receptors are essential for secretion but their absolute number accounts for only a small fraction of the variance associated with level of secretion.

RELATIONSHIP OF GnRH RECEPTOR BINDING WITH PHYSIOLOGIC EVENTS

Measurement of receptor binding and LH secretion by individual cells derived from proestrus and diestrus-1 rat pituitary glands revealed silver grains located only over secretory gonadotropes (plaque forming cells). In these experiments, as in those reported earlier (see Fig. 5), all of the gonadotropes were secretory at proestrus but only 50-60% were at diestrus-1. Thus, non-secretory gonadotropes at diestrus lack GnRH binding sites. Since estradiol recruits such diestrous non-secretory gonadotropes into the secretory pool at proestrus (Smith et al. 1984), it is apparent that this recruitment

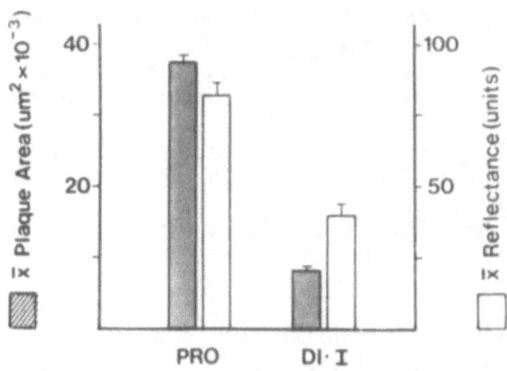

Fig. 15 Measurement of GnRH receptor binding and LH secretion by individual cells derived from proestrus and diestrus-1 rat pituitary glands. Incubation was for 2 hours in the presence of maximally stimulatory concentrations of ^{125}I-GnRH-A (100pM). Only one experiment has been performed.

occurs in part by the stimulation of GnRH receptor formation in non-secretory gonadotropes. Apparently, this appearance of new receptors accounts partly for the increased number of receptors reported to be associated with the preovulatory LH surge (Clayton et al., 1980). We also have found an increase in the number of GnRH binding sites at proestrus (see Fig. 15); the number is doubled from diestrus-1 to proestrus when the data are presented as mean reflectance per gonadotrope (Fig. 15) but approximately quadrupled when total binding sites are calculated; i.e., when correction is made for the fact that only 50-60% of the gonadotropes at diestrus exhibit radioiodinated GnRH-A binding. These results (Fig. 15) represent only a single experiment and hence must be considered preliminary; if repeatable, they suggest that factors in addition to increases in GnRH receptor number are required to account for the approximate 50-fold increase in LH secretion rate during the preovulatory LH surge.

REFERENCES

Clayton, R.N., Solano, A.R., Garcia-Vela, A., Dufau, M., and Catt, K.J., 1980, Regulation of pituitary receptors for gonadotropin-releasing hormone during the rat estrous cycle, Endocrinology 107:699.

Fink, G., 1979, Feedback actions of target hormones on hypothalamus and pituitary with special reference to gonadal steroids, Annu Rev Physiol 41:571.

Frawley, L.S., and Neill, J.D., 1984, A reverse hemolytic plaque assay for microscopic visualization of growth hormone release from individual cells: evidence for somatotrope heterogeneity, Neuroendocrinology 39:484.

Hymer, W.C., and Hatfield, J.M., 1983, Separation of cells from the rat anterior pituitary gland, in "Methods in Enzymology", P.M. Conn, ed., Academic Press, New York, 103:257.

Jerne, N.K., Henry, C., Nordin, A.A., Fuji, H., Koros, A.M.C., and Lefkovits, I., 1974, Plaque forming cells: methodology and theory, Transplant. Rev. 18:130.

Molinaro, G.A., and Dray, S., 1974, Antibody coated erythrocytes as a manifold probe for antigens, Nature 248:515.

Naor, Z., Clayton, R.N., and Catt, K.J., 1980, Characterization of gonadotropin-releasing hormone receptors in cultured rat pituitary cells, Endocrinology 107:1144.

Neill, J.D., and Frawley, L.S., 1983, Detection of hormone release from individual cells in mixed populations using a reverse hemolytic plaque assay, Endocrinology 112:1135.

Pickering, A.J.M.C., and Fink, G., 1979, Variation in the size of the 'readily releasable pool' of luteinizing hormone during the oestrous cycle of the rat, J. Endocrinol 83:53.

Smith, P.F., Frawley, L.S., and Neill, J.D., 1984, Detection of LH release from individual pituitary cells by the reverse hemolytic plaque assay: estrogen increases the fraction of gonadotropes responding to GnRH, Endocrinology 115:2484.

Smith, M.S., Freeman, M.E., and Neill, J.D., 1975, The control of progesterone secretion during the estrous cycle and early pseudopregnancy in the rat: prolactin, gonadotropin, and steroid levels associated with rescue of the corpus luteum of pseudopregnancy, Endocrinology 96:219.

Smith, P.F., Luque, E.H., and Neill, J.D., 1985, Detection and measurement of secretion from individual neuroendocrine cells using a reverse hemolytic plaque assay, in "Methods in Enzymology", P.M. Conn, ed., Academic Press, New York (in press).

Smith, P.F., and Neill, J.D., 1983, Heterogeneity in secretory response to GnRH within the rat pituitary gonadotrope population: quantitative analysis at the single cell level, Program of the 13th Annual Meeting of the Society for Neuroscience, Boston, MA, p. 17.

Smith, P.F., and Neill, J.D., 1984, The reverse hemolytic plaque assay for detection of hormone secretion by individual cells, in "Excerpta Medica International Congress Series (7th International Congress of Endocrinology)" F. Labrie and L. Proulx, eds., Elsevier Science Publishers, Amsterdam, pp. 1113-1116.

LHRH PRIMING IN GONADOTROPHS : A MODEL SYSTEM FOR THE ANALYSIS OF NEUROENDOCRINE MECHANISMS AT THE CELLULAR LEVEL

John F. Morris, Claire E. Lewis and George Fink*

Department of Human Anatomy, Oxford OX13QX, and
*M.R.C. Brain Metabolism, Edinburgh EH89J2

INTRODUCTION

The priming effect of Luteinizing Hormone Releasing Hormone (LHRH), ie. the capacity of LHRH to increase the responsiveness of pituitary gonadotrophs to itself, is an important component of the mechanism by which changes in output of gonadotrophins during the oestrous cycle, in particular the preovulatory luteinizing hormone (LH) surge are controlled (Fink, 1979). LHRH priming can be elicited both in vivo (rat: Aiyer et al., 1974a; Fink et al.,1976; mouse: Fink et al., 1982) and in vitro (Pickering and Fink, 1976a; Waring and Turgeon, 1980). It is specific to LHRH and not elicited by secretagogues such as 40 mM K^+ (Pickering and Fink, 1976b) and is not dependent on the initial release of LH (which normally occurs during the same 1h period as LHRH priming), since it occurs when LH release is blocked by removing calcium from the incubation medium (Pickering and Fink 1979). The priming effect of LHRH requires the synthesis of new protein (Pickering and Fink, 1979; Curtis et al., 1985) which is not LH (de Koning et al., 1976; Pickering and Fink, 1976a,1979), and is dependent on the functional integrity of microfilaments (Pickering and Fink, 1979; Lewis et al., 1985). However, the subcellular mechanisms whereby the readily releasable pool of gonadotrophin is increased are still poorly understood. Previous ultrastructural analyses of LHRH stimulated gonadotrophs in vivo (Rommler et al., 1978) and in vitro (Tixier-Vidal et al., 1975; Childs, 1985) have reported changes in the position of granules but this has not been systematically or quantitatively analysed with respect to the priming effect of LHRH. The purpose of this report is to describe a series of quantitative ultrastructural analyses of gonadotrophs in different experimental and physiological states related to the augmented responsiveness of gonadotrophs produced by LHRH priming. The experiments have used normal mice, hypogonadal (hpg) mutants which do not produce detectable hypothalamic LHRH (Cattanach et al., 1977) and testicular feminised (tfm) mutants which do not respond to androgenic negative feedback (Naik et al., 1984). In addition, rats at different stages of the oestrous cycle have been studied to correlate changes in the morphology of gonadotrophs with changes in pituitary responsivness to LHRH. Our results indicate that augmentation of gonadotroph responsiveness to LHRH, such as that which occurs during the priming effect of LHRH involves a reversible migration of secretory granules to the surface of the gonadotrophs and a change in granule dimension indicative of increased intra-granular processing, associated with changes in the length and orientation of cytoskeletal microfilaments.

MATERIALS AND METHODS

Animals

Mice have been used for most experiments in order to facilitate
comparison of results obtained with normal adult, hypogonadal (hpg) and
testicular feminised (tfm) mice. All mice were aged 60-100 days. For the
study of the oestrous cycle in vivo, adult female Wistar rats were used to
take advantage of the greater ease with which the stage of the oestrous
cycle can be determined in regularly cycling rats, compared with mice.

Incubation of Pituitaries and Assay of LH Released

The methods of pituitary incubation and assay of LH were similar to
those of Pickering and Fink (1976a, 1979; see Lewis et al., 1985). Bi-
sected pituitaries were preincubated for 2h, the period during which a
high initial washout of LH into the medium occurs, then incubated with
8.5nmol LHRH/L or with control media for one or two successive periods of
one hour each. The hemipituitaries were either assayed for LH or fixed
for electron microscopic analysis at the end of both the first and second
hour of incubation, and 200μl samples of the incubation medium were assayed
for LH using the ovine-ovine double antibody radioimmunoassay of Niswender
et al. (1968). The increment of LH released into the medium during each
hour of incubation was then calculated (see Lewis et al., 1985).

Identification and Morphometric Analysis of Gonadotrophs

Gonadotrophs containing LH have been identified by immunocytochemical
staining using antisera to LHβ of 0.5μm sections of epoxyresin embedded
pituitary glands, conventionally prepared for electron microscopy by fixation with a buffered mixture of 0.5% glutaraldehyde and 2.5% paraformaldehyde, osmication, dehydration and embedding. Gonadotrophs thus identified were located on the serial ultrathin sections (the serial thin-
semithin technique of Polak et al., 1976), and complete transnuclear
profiles of gonadotrophs were sampled by a systematic random method
(Weibel, 1969). Standard morphometric techniques were used for the analysis
on micrographs (x 10,000) of the sizes of cells and their granule content.
The distribution of granules was analysed by defining an arbitrary marginal
zone of cytoplasm extending 500nm inward from the plasmalemma and determining the number and size of granules in both this marginal zone, and the
more central cytoplasm. The size of granules, and the number, length and
orientation of the microfilaments (identified by their 5-9nm diameter)
were analysed on tracing of highly magnified (x52,000) cell profiles (see
Lewis et al., 1985).

Pituitaries from groups of 5-6 animals were exposed to each experimental
or control treatment. For morphometric analyses, data from individual
cell profiles derived from any one animal were pooled to give an animal
mean which then formed the basis for subsequent statistical analysis of
differences between experimental treatments. Results are expressed as
mean + SEM (n = 5-6) unless otherwise stated.

RESULTS AND DISCUSSION

Peripheral Migration of Gonadotroph Granules and a Reduction of Granule
Size Accompany LHRH-Priming in vitro

Pituitaries from proestrous normal mice, like those from rats, can be
primed in vitro by exposure to LHRH. The first hour of incubation with
8.5nmol LHRH/L caused the release of 10±2 ng LH, whereas 31±4 ng LH was

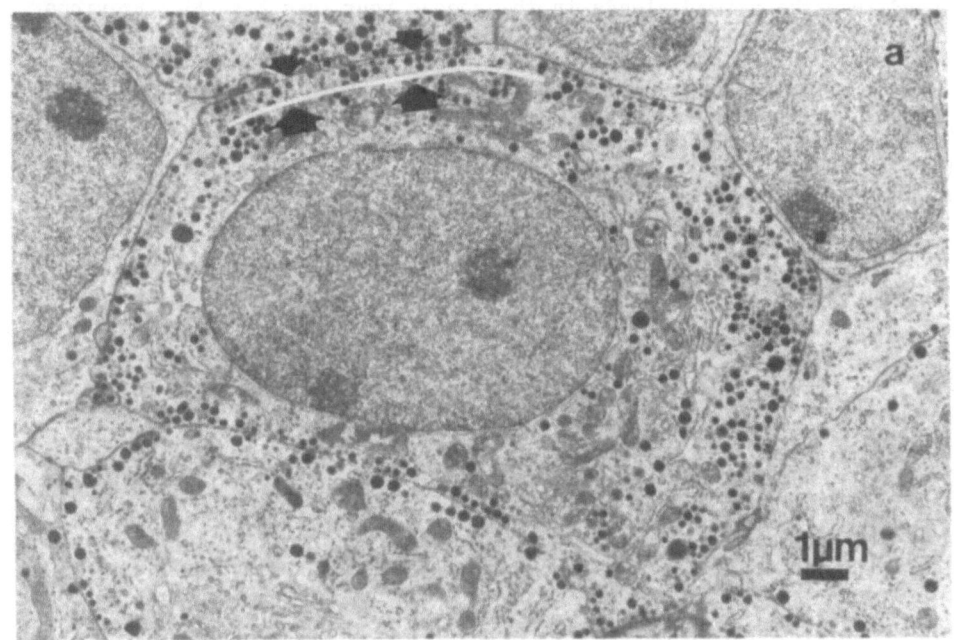

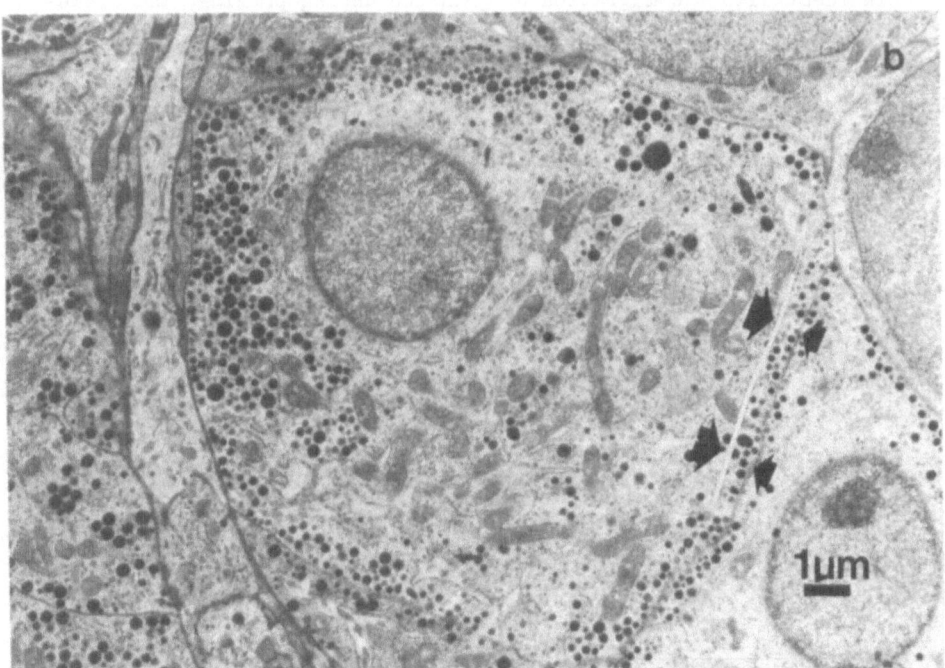

Fig. 1. Electron micrographs (x 7,000) of LHβ-immunoidentified gonadotrophs from proestrous female mice preincubated for 2h and then incubated for 1h in the presence of either a) control medium or b) medium containing 8.5nmol LHRH/L. The 500nm wide marginal zone is indicated by the white line and arrowheads. After the priming effect of an initial exposure to LHRH increased numbers of granules which are predominantly small in size can be seen in the marginal zone.

released from the primed glands in the second hour (see Fig. 3). Previous
exposure to LHRH and ovarian steroids is not a prerequisite for the priming
effect, because priming of pituitaries from hpg mice, which lack hypothalamic LHRH and have inactive ovaries) is also demonstrable: 6±1 ng LH,
and 26±4 ng LH were released in the first and second hours respectively of
incubation with LHRH. Gonadotrophs were unaltered in size as a result of
either the initial or the primed release of LH. In neither normal nor hpg
animals was there a demonstrable increase or decrease of pituitary LH content
or the average number of granules per cell as a result of the release of LH
during the first hour, in which priming is occurring. However, in normal
animals, the pituitary content of LH decreased significantly as a result of
LHRH-primed release, from 287±36 ng LH/gland (control) to 202±25 ng LH/gland
(after primed release), and this was associated with a decrease in the amount
of secretory granules in gonadotrophs, from 7.4±0.5 μm^2/cell (control) to
5.1±0.3 um²/cell (after primed release). The most striking change was in
the distribution of the secretory granules within the gonadotrophs (Fig. 1).
At the end of the first hour of incubation granules in gonadotrophs were
dispersed rather evenly throughout the cell cytoplasm, whereas those in
pituitaries primed by incubation with LHRH were preferentially located at the
periphery. Morphometric analysis revealed that 33±2% of granules were located in the 500nm wide marginal zone of gonadotrophs incubated in control
media, and 59±3% and 53±3% at the end of the first and second hours of
incubation with LHRH. Thus, migration of granules to the periphery of the
cell occurs during the priming period and continues during the period of
primed release.

    Coincident with the migration of granules during priming, the diame-
ter of granules - especially those in the marginal zone - was decreased.
Fig. 2 illustrates the number of granule profiles of different sizes in
the marginal zone of glands incubated for 1h or 2h with either control or
LHRH-containing medium. The increased numbers of smaller granules in the
marginal zone of primed (1h incubation with LHRH) gonadotrophs is seen.
The granule population of the central zone shows a decrease in the larger
(200-300nm diameter) granules (not illustrated; see Lewis et al., 1986)
indicating that there is an overall decrease in the size of granules associated with priming. During the second hour of incubation with LHRH, when
primed release of LH occurs, it was predominantly the small granules that
were lost from the marginal zone (Fig. 2).

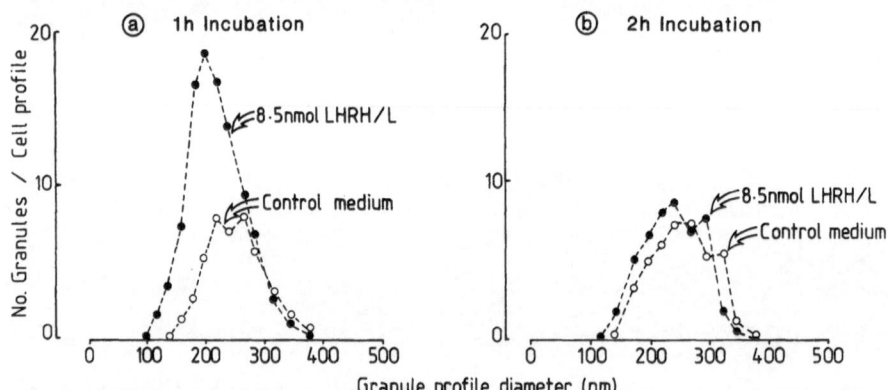

Fig. 2. Diameters of granule profiles in the marginal zone of gonadotrophs
        from proestrus mice preincubated for 2h and then incubated for
        either a) 1h or b) 2h in control medium or medium containing
        8.5nmol LHRH/L. Note the increase in the number of small granules
        in the marginal zone after the priming effect of an initial expo-
        sure to LHRH (a), and the loss of these small granules after the
        primed release of LH (b).

Granules in the marginal zone of gonadotrophs of both normal and hpg mice were smaller than those in the central zone (data not shown, see Lewis et al.,1986). This indicates that as granules migrate to the periphery of the cells they decrease in size, probably as a result of intragranular processing, and suggests that when the migration of granules is increased during priming, intragranular processing is also increased. An entirely similar pattern of subcellular events was seen in gonadotrophs from pituitaries of mice primed by injection of LHRH in vivo (Lewis, Morris and Fink, unpublished data).

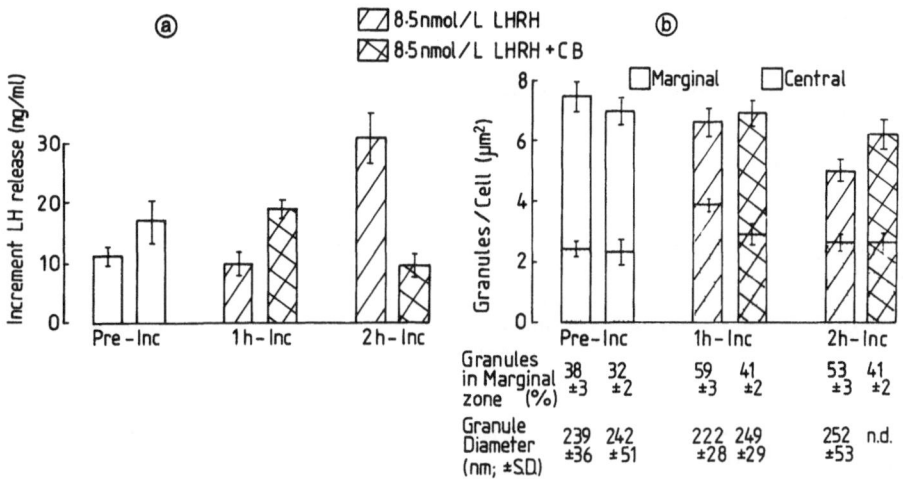

Fig. 3. Effects of cytochalasin B (CB) on (a) LH release, or (b) granule content, distribution and size in gonadotrophs after preincubation for 2h (Pre-Inc) and then incubation in the presence of 8.5nmol LHRH/L for 1 (1h-Inc) or 2h (2h-Inc).

The microfilament poison cytochalasin B is known to block priming by LHRH of gonadotrophs from rats (Pickering and Fink, 1979). Fig. 3 shows that cytochalasin B not only blocks priming in mice, but also blocks the peripheral migration of granules, the reduction of granule diameter which accompanies margination, and the reduction in the number of granules which results from primed release. An analysis of the microfilaments (Lewis et al., 1985) reveals that priming by LHRH results in an increase in the length of individual microfilaments in gonadotrophs, but no change in their number. The microfilament profiles also become orientated more parallel to the plasmalemma as a result of stimulation by LHRH. All these changes are also blocked by cytochalasin B. These data indicate that both the margination of granules and the reduction in their size are important components of the mechanism of the priming effect of LHRH.

Peripheral Migration of Granules is a Necessary Component of Priming by LHRH but Not a Sufficient Explanation of the Phenomenon

In another series of experiments we have examined the effects on the margination and size of gonadotroph granules of depolarisation by 40mM K^+, exposure to LHRH under conditions in which calcium cannot enter the cell, and exposure to LHRH in the presence of the protein synthesis inhibitor cycloheximide. The results are summarised in Table 1, which shows that incubation with cytochalasin B and prevention of calcium entry by addition of cobalt ions, both of which block margination of granules also block priming. This indicates that the greater availability of granules at the margin of the cell is an important component of the priming process. Increased availability of granules cannot, however, provide a sufficient explanation for the increase in the readily-releasable pool of LH produced

Table 1. Effects of various conditions known to influence the LH-
releasing and priming actions of LHRH in vitro on the
margination, and the reduction in size of granules in
gonadotrophs of pituitaries from proestrous female mice

Addition to incubation medium in first hour	Release of LH	Priming by LHRH	Margination of granules	Reduction in size of granules
8.5nmol LHRH/L	occurs	occurs	occurs	occurs
8.5nmol LHRH/L + 14.3μmol/L CB	occurs	blocked	blocked	blocked
8.5nmol LHRH/L + 10mmol/L CoCl$_2$	blocked	occurs	occurs	occurs
8.5nmol LHRH/L + 7.1μmol/L CH	occurs	blocked	blocked	occurs
40mmol/L KCl	occurs	does not occur	occurs	occurs

(CB - Cytochalasin B; CoCl$_2$ - Cobalt chloride; CH - Cycloheximide; KCl - Potassium chloride)

by LHRH priming, since stimulation by 40mmol/L potassium ions, which does not cause priming, nevertheless causes both margination of granules and a reduction in their size. The dissociation between the margination of granules and the reduction in their size seen in gonadotrophs incubated with LHRH in the presence of cycloheximide indicates that margination and increased intragranular processing are separately controlled phenomena which normally occur together in response to different second messenger systems stimulated by LHRH.

The Peripheral Migration of Granules Occurs Rapidly and is Reversible in the Absence of LHRH

Since pituitaries are normally exposed to pulses of hypothalamic LHRH which reach them at intervals of one hour or less (Fink, 1979), the preincubation period used by many workers to 'stabilise' loss of gonadotrophin from incubated pituitaries could result in important changes in the granule population of gonadotrophs. We therefore analysed the number and position of granules in gonadotophs from proestrous mice during the preincubation period, and after 5 min, 1h and 2h exposure to 8.5nmol/L LHRH. Fig. 4 shows the result of this experiment. The unchanged total granule content of gonadotrophs, coupled with the reduced proportion of granules in their marginal zones, indicates that the peripheral migration of granules induced by LHRH is a reversible process in the absence of LHRH. Peripheral migration of granules in response to LHRH must be a very rapid process since after only 5 min exposure to LHRH 54% of granules are found in the marginal 500nm zone, and this proportion does not alter over the first hour of incubation with LHRH. In fact, priming can be demonstrated as early as 20-30 min after the first exposure to LHRH (Aiyer et al., 1974a; Waring and Turgeon, 1983).

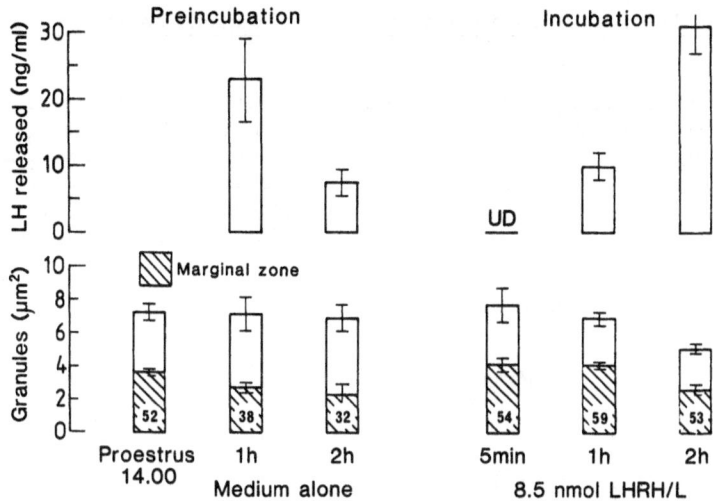

Fig. 4. Comparison of the amount of LH released (top panel; only
data for incubated tissues shown) and the cellular content
and distribution of granules in gonadotrophs (bottom panel;
proportion of granules in the marginal zone indicated at
the base of each column) of glands removed from mice at
1400h on the day of proestrus and either processed for
electron microscopy immediately (Proestrus 14.00) or preincubated for 1 or 2h, and then incubated for 5 min, 1h or
2h in the presence of 8.5nmol LHRH/L.

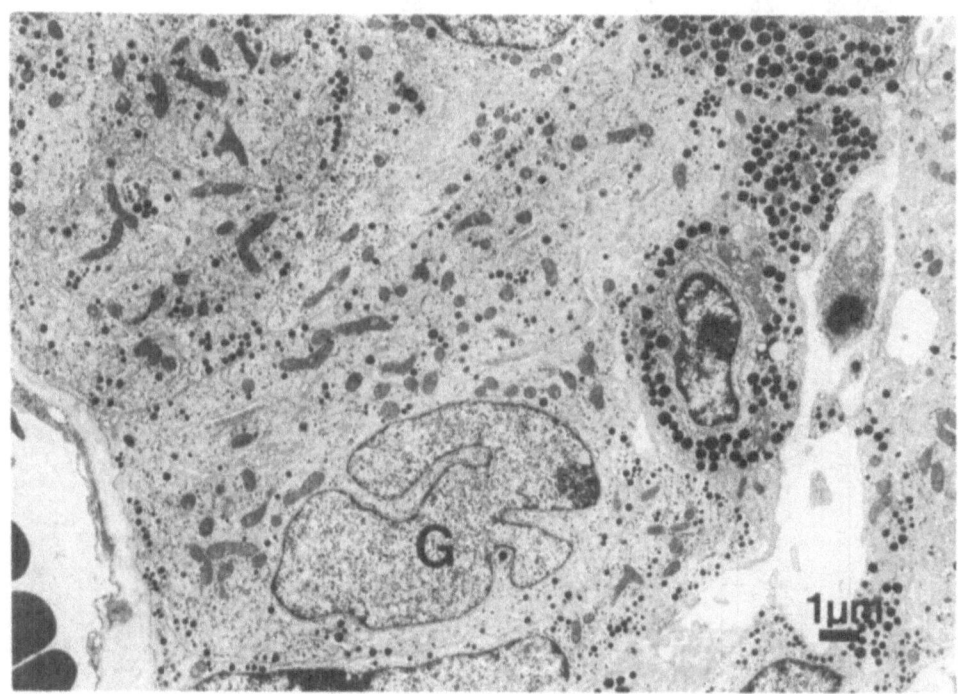

Fig. 5. Electron micrograph (x 6,200) of an immunoidentified gonadotroph
(G) from a tfm mouse. Note the presence of large quantities of
dilated rough endoplasmic reticulum, and of only a few, marginated
granules in the gonadotrophs of such animals.

Gonadotrophs of Testicular Feminised (tfm) Mice Respond as if Primed by LHRH

The failure of androgenic negative feedback in mutant tfm results in high plasma levels and reduced pituitary stores of gonadotrophins (Naik et al., 1984). Electron microscopic examination of pituitaries from tfm mice revealed gonadotrophs with dilated rough endoplasmic reticulum and a population of secretory granules that was markedly more marginated and the granules significantly smaller in size than that those of normal male mice (Figs. 5, 6).

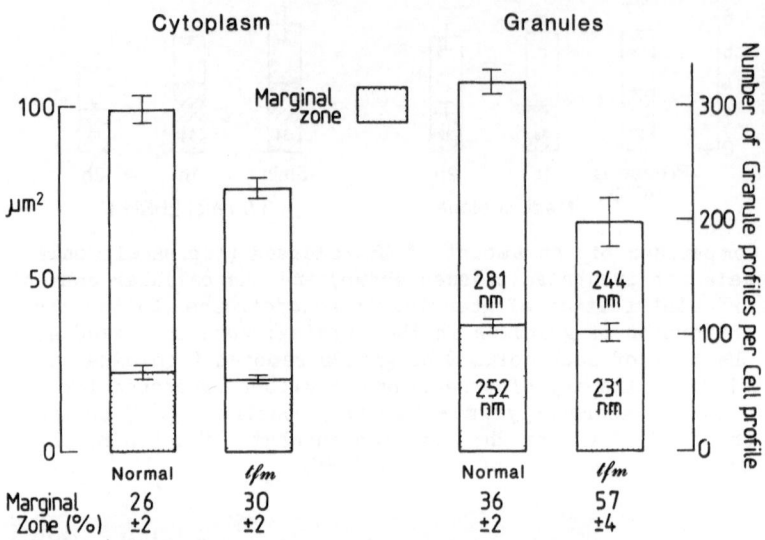

Fig. 6. Comparison of the cytoplasmic size, and the content, distribution and size of granules in gonadotrophs from normal male and tfm mice.

Since this appearance was reminiscent of that seen in LHRH-primed gonadotrophs of female mice, pituitaries from tfm and normal male mice were challenged in vitro with 8.5nmol LHRH/L. Fig. 7 shows that, although glands from tfm and normal mice lost equal amounts of LH into the medium during a 1h preincubation, glands from tfm, but not normal male, mice

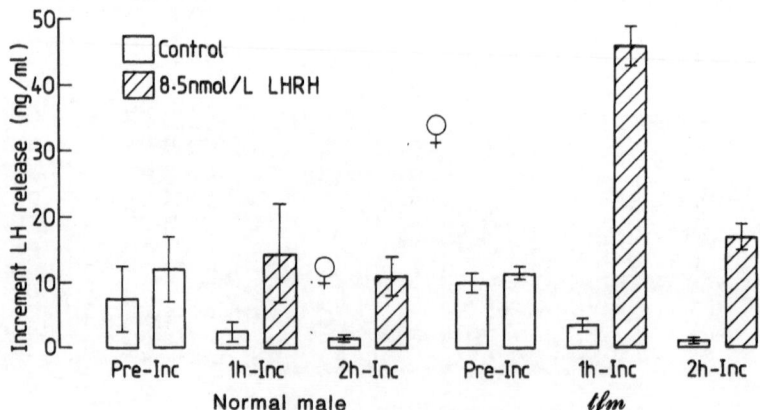

Fig. 7. Comparison of the amount of LH released by pituitary glands from normal and tfm male, and from proestrous female mice after pre-incubation for 2h (Pre-Inc) and then incubation for 1h (1h-Inc) or 2h (2h-Inc) in the presence of 8.5nmol LHRH/L.

exposed to LHRH for 1h released an amount of LH equal to that released by LHRH-primed pituitaries of proestrous female mice.

It will be seen that priming of gonadotrophs from pituitary glands of normal male mice did not occur in this experiment and that, during the second hour of incubation with LHRH, only a small amount of LH was released from pituitaries of tfm mutants. The reasons for these two findings are, at present, unclear.

These data indicate that lack of androgenic negative feedback in tfm mice results in both the morphological and functional characteristics of the LHRH-primed condition. This would be consistent with the claim that the major site of action of testosterone is the hypothalamus (Charlton et al., 1982), because it would be predicted that pituitaries of tfm mice would receive greater than normal LHRH pulses from the hypothalamus. These pulses would have to be of such a frequency and magnitude as to cause increased responsiveness by a priming-like action, rather than the desensitisation which can occur in some species when the pituitary is exposed to excessive amounts of LHRH (Knobil, 1980). Whether or not there are additional, direct effects on the pituitary of the lack of androgenic feedback has still to be determined.

Changes in Gonadotroph Granule Populations during the Rat Oestrus Cycle

The responsiveness of the pituitary gland of the rat to LHRH varies markedly during the oestrous cycle (Aiyer et al., 1974a; Fink, 1979). We have therefore examined the granule populations of gonadotrophs at different stages of the oestrous cycle using tissues from animals killed at 16.00-17.00 on the days of oestrus, metoestrus and dioestrus, at 16.00-17.00 on proestrus (i.e. at the start of the preovulatory LH surge) and at 22.00-23.00 on proestrus (i.e. at the end of the LH surge). Assays of pituitary gonadotrophin content revealed a significant decrease from 6.0 ± 0.8 ug LH/pituitary at 16.00-17.00 on proestrus to 2.1 ± 0.5 ug LH/pituitary at the end of the LH surge at 22.00-23.00, with an increase in LH content through metoestrus and dioestrus to peak content at 16.00-1700 on proestrus. The FSH content of the pituitary followed a similar pattern but with less pronounced changes. The total granule content of gonadotrophs reflected the gonadotrophin content of the pituitary, with a significant decrease in the amount of granules per cell as a consequence of the preovulatory LH surge. Analysis of the size of granules in gonadotrophs revealed a) that granules were smallest at 16.00-17.00 on proestrus, suggesting increased intragranular processing at this time, b) that the proestrous LH surge resulted in the loss of the smaller rather than the larger granules from the cells, and c) that an increase in the larger granules, which should indicate increased granule production, was found at metoestrus, the day on which the assays showed significant replenishment of pituitary gonadotrophin stores. Gonadotrophs varied in size through the oestrous cycle, being largest in dioestrus animals. This complicated the assessment of the margination of granules by analysis of the granule content of the 500nm peripheral zone of cytoplasm. However, when differences in the size of the marginal zone (as defined) were taken into account, a greater than random degree of granule margination was demonstrated in gonadotrophs from proestrus, oestrus and metoestrus, but not dioestrus animals. This pattern of margination correlates well with the pattern of responsiveness to LHRH during the cycle (Aiyer et al., 1974b). Granule margination was not shown to be clearly greater on proestrus than on any other day of the cycle, but responsiveness alters very rapidly during proestrus (Aiyer et al., 1974b) and, by sampling at the start and at the end of the LH surge, we do not yet have data for the time of maximal responsiveness at 18.00 on proestrus. Therefore, although the analysis of the proportion of granules in the marginal zone can give only a partial insight into flux of granules through the marginal zone for release, and although our other experiments indicate that margination per se is not a complete explanation for gonadotroph responsiveness, this analysis of gonadotrophs during the oestrous cycle has provided clear quantitative evidence of changes in the granule population that can be linked directly to physiological changes in the hormone content, granule production and responsiveness of the cells to LHRH.

Conclusions

Analysis of gonadotrophs exposed to the priming action of LHRH shows that an essential part of the mechanism of priming is a rapid peripheral migration of the secretory granules. Since it occurs in hpg mice, this migration cannot depend on previous exposure either to LHRH or to ovarian factors. The increase in peripheral granules which results from the migration would provide a physical correlate for the increased readily-releasable pool of gonadotrophin. The migration appears to be reversible in culture in the absence of LHRH. If the same obtains in vivo then the granule populations of gonadotrophs must exist in a dynamic state, their exact distribution being dependent on the time that has elapsed since the previous pulse of hypothalamic LHRH reaching the pituitary gland via the portal circulation.

The migration of gonadotroph granules is normally associated with a decrease in size of the granules that signifies some sort of intragranular processing, as yet undefined. This 'maturation' would appear to result from the action of a different intracellular messenger system to that which produces migration, since it can be dissociated from the migration of the granules. Analysis of gonadotrophs from tfm mice and rats at different stages of the oestrous cycle shows that margination of granules and reduced granule size correlate well with physiological states of increased responsiveness to LHRH.

The analysis presented above takes no account of heterogeneity among gonadotrophs, but treats them as a single population of cells, which is appropriate when considering the overall response of the pituitary gland. Heterogeneity among gonadotrophs undoubtedly exists, and it is hoped that similar morphological analyses will provide insights into the mechanisms whereby gonadotrophs differ in their functional capacities and the ways in which these differences are controlled.

ACKNOWLEDGEMENTS

We thank Dr. G.D. Niwender and Prof. L.E. Reichert Jr., and the National Pituitary Hormone Programme of the NIADDK (Baltimore, Md, U.S.A) for radioimmunoassay materials used in the measurement of LH, and J. Bennie, M. Arkley and M. Surtees for their expert technical assistance. J. E. Pearson and Alison Gray assisted with the studies on tfm mice and the rat oestrous cycle, respectively. C.E.L. was supported by a MRC Partnership Training Award, and the study was financed in part by AFRC grant AG43/94 and MRC grant G608/263 to J.F.M.

REFERENCES

Aiyer, M. S., Chiappa, S. A., and Fink, G., 1974a, A priming effect of luteinizing hormone releasing factor on the anterior pituitary gland of the female rat, J. Endocrinol., 62: 573.

Aiyer, M. S., Fink, G., and Greig, F., 1974b, Changes in the sensitivity of the pituitary gland to luteinizing hormone releasing factor (LRF) during the oestrous cycle of the rat. J. Endocrinol., 60:47

Cattanach, B. M., Iddon, C. A., Charlton, H. M., Chiappa, S. A., and Fink, G., 1977, Gonadotrophin releasing hormone deficiency in a mutant mouse with hypogonadism, Nature, 260:338.

Charlton, H. M., Halpin, D. M. G., Iddon, C., Rosie, R., Levy, G., McDowell, I. F. W., Megson, A., Morris, J. F., Bramwell, A., Speight, A., Ward, B. J., Broadhead, J., Davey-Smith, G., and Fink, G., 1983, The effects of daily administration of single or multiple injections of gonadotropin-releasing hormone on pituitary and gonadal function in the hypogonadal (hpg) mouse, Endocrinology, 113:535.

Childs, G. V., 1985, Shifts in gonadotropin storage following GnRH stimulation in vitro, Peptides, 6:103.

Curtis, A., Lyons, V., and Fink, G., 1985, The priming effect of LH-releasing hormone; effects of cold and involvement of new protein synthesis, J. Endocrinol., 105:163.

Fink, G., 1979, Neuroendocrine regulation of gonadotrophin secretion, Brit. Med. Bull., 35:155.

Fink, G., Chiappa, S. A., and Aiyer, M. S., 1976, Priming effect of luteinizing hormone releasing factor elicited by preoptic stimulation and by intravenous infusion and multiple injections of the synthetic decapeptide, J. Endocrinol., 69:359.

Fink, G., Sheward, W. J., and Charlton, H. M., 1982, Priming effect of luteinizing hormone releasing hormone in the hypogonadal mouse, J. Endocrinol., 94: 283.

Knobil, E., 1980, The neuroendocrine control of the menstrual cycle, Rec. Prog. Horm. Res., 36:53.

deKoning, J., van Dietan, O. A. M. J., and van Rees, G. P., 1976, LHRH dependent release of protein synthesis necessary for LH release from rat pituitary glands in vitro, Mol. Cell Endocrinol., 5:151.

Lewis, C. E., Morris, J. F., and Fink, G., 1985, The role of microfilaments in the priming effect of LH-releasing hormone: an ultrastructural study using cytochalasin B, J. Endocrinol., 106:211.

Lewis, C. E., Morris, J. F., Fink, G., and Johnson, M., 1986, Changes in the granule population of gonadotrophs of hypogonadal (hpg) and normal female mice associated with the priming effect of LH-releasing hormone in vitro. J. Endocrinol., 109: In Press.

Naik, S. I., Young, L. S., Charlton, H. M., and Clayton, R. N., 1984, Pituitary gonadotrophin-releasing hormone receptor regulation in mice II: Females, Endocrinology, 115:114.

Niswender, G. D., Midgley, A. R. Jr., Monroe, S. E., and Reichert, L. E. Jr., 1968, Radioimmunoassay for rat luteinizing hormone with antiovine LH serum and ovine LH-^{131}I, Proc. Soc. Exp. Biol. Med., 128:807.

Pickering, A. J. M. C., and Fink, G., 1976a, Priming effect of luteinizing hormone releasing factor: in vitro and in vivo evidence consistent with its dependence upon protein and RNA synthesis, J. Endocrinol., 69:373.

Pickering, A. J. M. C., and Fink, G., 1976b, Priming effect of luteinizing hormone releasing factor: in vitro studies with raised potassium concentrations, J. Endocrinol., 69:453.

Pickering, A. J. M. C., and Fink, G., 1979, Priming effect of luteinizing hormone releasing factor in vitro: role of protein synthesis, contractile elements, Ca^{2+} and cyclic AMP, J. Endocrinol., 81:223.

Polak, J. M., Bloom, S. R., McCrossan, M., Timson, C. M., Arimura, A., and Pearse, A. G. E., 1976, Studies on G cell pathology, Gut 17:400.

Rommler, A., Seinsch, W., Hasan, A. S., and Haase, F., 1978, Ultrastructure of rat pituitary gonadotrophs in relation to serum and pituitary LH levels following repeated LHRH stimulation, Cell Tiss. Res., 190:135.

Tixier-Vidal, A., Gourdji, A. D., and Tougard, C., 1975, A cell culture approach to the study of anterior pituitary cells, Int. Rev. Cytol., 41:173.

Waring, D. W., and Turgeon, J. L., 1980, Luteinizing hormone releasing hormone secretion in vitro: cyclic changes in the responsiveness and self-priming, Endocrinology, 106:1430.

Waring, D. W., and Turgeon, J. L., 1983, Self-priming effect of luteinizing hormone releasing hormone on gonadotrophin secretion, Am. J. Physiol., 244:C410.

Weibel, E. R., 1973, Stereological techniques for electron microscopic morphometry, In: Principles and techniques of electron microscopy. Ed. M. A. Hayat. New York. Van Nostrand Reinhold.

MEMBRANE TRAFFIC IN RELATION WITH RELEASE MECHANISMS IN

NEUROENDOCRINE CELLS IN CULTURE

A. Tixier-Vidal, C. Tougard and A. Faivre Bauman

Groupe de Neuroendocrinologie Cellulaire et Moléculaire
C.N.R.S. U.A. 04 1115, Collège de France
11 Place Marcelin Berthelot 75231 PARIS CEDEX 05

INTRODUCTION

In eucaryotic cells the release of secretory products involves at least two membrane pathways : a biosynthetic route along which membrane proteins and secretory proteins travel from their site of synthesis to the plasma membrane and an endocytic route which serves for the recycling and/ or degradation of membrane components. All of these pathways converge to the Golgi zone where membrane components and secretory material are sorted and directed to their final sites.

The molecular mechanisms underlying these phenomena are still poorly understood. However important progresses have been recently made in the analysis of structures and pathways involved in intracellular transport. This was made possible with the development of antisera directed against membrane components and secretory products, respectively, in conjonction with technical advances that permit access of antibody to membrane bound intracellular compartments. Moreover this was also favored by the use of cultured cells which are easily accessible to reagents and to ligands which modulate or perturb secretory activity.

Prolactin (PRL) cells in culture were among the very first ones to which such methods were applied. The secretory proteins, prolactin (Tougard et al, 1982), on the one hand, and, on the other hand, some membrane components involved either in the biosynthetic pathway (Tougard et al, 1983a) or in the endocytic pathway (Tougard et al, 1985) could be localized at the subcellular levels in relation with variations of the release of prolactin (see rev. Tixier-Vidal and Tougard, 1984). In contrast, neuro-secretory neurons have been, so far, the object of very few investigations in that respect. Indeed, although they offer an original model of secretory cells, their anatomical location and structural organization have made them less accessible to manipulations.

In this chapter we first review our recent studies on prolactin compartments and membrane compartments in cultured PRL cells. Then we present, as a promising system for further analysis and reflection, hypothalamic neurons cultured in a chemically defined medium. Indeed we have recently shown that when cultured in appropriate conditions, these neurons acquire the capacity to release thyroliberin (TRH) in a calcium dependent manner in response to potassium evoked depolarization (Loudes et al, 1983a). This was correlated with the development of axon terminals and synapses which exhibited a vesicular depletion under exposure to high potassium concentration and a massive and rapid restoration of vesicles following return to normal potassium chloride concentration (Tixier-Vidal et al, in press).

I. MEMBRANE TRAFFIC IN RELATION WITH PRL RELEASE

Our studies were performed on two types of PRL cells in culture : clonal, tumor derived GH3 cells and normal rat PRL cells in primary cultures. Both systems release large amount of PRL into the medium. However the first one differs from the second one by a low intracellular PRL store, consistent with the small number and size of secretory granules (see rev. Gourdji et al,1982). In both systems, thyroliberin (TRH) acutely stimulates PRL release (see rev. Gourdji et al, 1982) whereas monensin inhibits basal PRL release without affecting the response to TRH (Tougard et al, 1983b).

Using mostly an immunocytochemical approach at the electron microscope level and various immunological probes, we have obtained evidence for a compartmentalization of the components of the biosynthetic pathway, that is the secretory protein, PRL, and the membranes of the endoplasmic reticulum.

PRL compartments

The distribution of PRL into several, membrane bound, compartments which are interconnected and can be differentially regulated, has been revealed by two complementary approaches : immunoelectron microscope visualization of PRL and pulse chase experiments followed by immuno-precipitation of labeled PRL.

Using a preembedding method associated to saponine permeabilization, PRL could be visualized not only in secretory granules but also along the successive compartments of the endoplasmic reticulum, that is the rough endoplasmic reticulum (RER) cisternae and the Golgi cisternae (Tougard et al, 1980, 1982). Such localizations are consistent with the present concept on the transit of secretory proteins in eucaryotic cells (rev. Farquhar and Palade, 1981) as well as with the results of electron microscope auto-radiographic analysis performed on normal PRL cells (Farquhar et al, 1978). PRL was also found inside small vesicles associated to the Golgi zone or dispersed in the cytoplasm (Tougard et al, 1982). This pointed out for the first time the possible role of small vesicles as carriers for PRL release, a role particularly prominent in GH3 cells which possess very few secretory granules.

Treatment of GH3 cells with TRH triggered the intraluminal transit of PRL as revealed by the transient destaining of the endoplasmic reticulum observed within the first hour of exposure to TRH. At the same time a spreading of the Golgi zone occurred together with de multiplication of PRL loaded vesicles (Tougard et al, 1982). In contrast, monensin, a carboxylic monovalent ionophore, which is known to interrupt the transit of secretory proteins at the level of the Golgi zone, induced an accumulation of PRL on the luminal face of large Golgi derived vacuoles, consistent with a strong inhibition of basal PRL release. However monensin did not inhibit, neither the TRH induced stimulation of PRL release, nor the TRH induced appearance of PRL loaded small vesicles. These effects of monensin suggested the possibility of several intracellular routes for the basal or stimulated release of PRL (Tougard et al, 1983b).

Pulse chase experiments performed on same cell types also demonstrated that PRL is intracellularly distributed into several compartments and can be differentially regulated. At least two intracellular PRL pools could be distinguished by their different half-lives : 15 min and 3 hrs respectively in GH3 cells instead of 2.5 hrs and 22 hrs in normal PRL cells. Moreover in basal conditions newly synthesized PRL was rapidly released whereas in the presence of TRH an old PRL store was preferentially released (Morin et al, 1984 a, b). This may reflect a functional heterogeneity of the cells, as previously suggested by Walker and Farquhar (1980), or the existence of several intracellular routes, or both. Strong evidences in favor of the latter hypothesis were provided by the study of monensin treated GH3 cells. In basal conditions monensin prevented the release of stored PRL but not that of newly synthesized PRL, thus suggesting the existence of a direct route for newly formed molecules, originated from an early Golgi step, insensitive to monensin. In contrast, in the presence of TRH, monensin did not inhibit the TRH induced release of stored PRL, thus suggesting the existence of another route for the regulated release of a PRL store located downstream the monensin blockade (Morin et al, 1984a). Such a distinction between a constitutive and a regulated pathway of hormone release was previously found using other approaches in the AtT20 mouse pituitary cell line which secretes ACTH and its unprocessed precursor (Gumbiner and Kelly, 1982).

Membrane compartments involved in the biosynthetic pathway

Electron microscope observations have long ago postulated the involvement of several membrane compartments in the secretory pathway. However whether this structural compartmentalization corresponded or not to biochemically different domains was for a long time debated (see rev. Farquhar and Palade, 1981). Direct evidences for the existence of an immunological specificity of membrane domains along the secretory pathway were recently obtained in PRL cells (Tougard et al, 1983a). This was made possible with the development of immunological probes specific on the one hand, to several polypeptides of the dog pancreas RER (A-RER), and, on the other hand, to a 135 Kd polypeptide of rat liver light Golgi fraction (A-Go) (Louvard et al, 1982). These tools were applied to intact PRL cells in culture, both GH3 cells and normal rat pituitary cells, using same procedures as for PRL subcellular localization.

The A-RER labeled, in both cell systems, exclusively the membrane of the RER cisternae, including the perinuclear cisternae. The distribution of A-RER labeled membrane was not affected by acute modifications of PRL release in response to TRH, or monensin, or both.

The A-Go gave a very different picture. In cells cultured in basal conditions it labeled some Golgi saccules, in a medial position and, with a decreasing intensity, the inner saccules of the trans-face. In addition, it labeled the membrane of small vesicles, mainly located in the Golgi zone, as well as that of lysosome-like structures. In some cells, a few labeled vesicles were found beneath the plasma membrane and a few positive patches at the cell surface. In contrast the membrane of secretory granules was unstained, except in the core of the Golgi zone of normal pituitary cells, at the level of segregating secretory granules and in a very few pictures.

In contrast to that of the A-RER, the distribution of the A-Go antigen underwent modifications following alterations of PRL release in GH3 cells. During acute stimulation of PRL release by TRH an extension of Golgi labeled saccules occurred as well as an increase in number of labeled vesicles in the cytoplasm. Profiles of vesicles fusing with the plasma membrane became frequent resulting in a conspicuous labeling of the cell surface. In monensin treated cells, the A-Go stained the membrane of most of the large vacuoles induced by the drug, indicating that they are indeed derived from a Golgi subcompartment.

In conclusion, the subcellular localization of membrane antigens specific for RER membranes and Golgi membranes, respectively, has provided the first direct evidences for a rather narrow immunological specificity of membrane domains along the biosynthetic pathway in PRL cells. Indeed there is no overlap between the distribution of RER antigens and Golgi antigens, on the one hand, as well as between the Golgi antigen and some components of the secretory granule membrane, on the other hand. The latter finding suggests a change in membrane composition, or conformation, at an early step of the packaging of secretory granules. Although the role of the 135 Kd Golgi antigen in the transit of PRL is still unknown, the fact that its distribution is modified concomitantly with PRL release in response to TRH treatment suggests that it might be involved in the sorting and directing of PRL. A direct demonstration of a role of the 135 Kd Golgi antigen in the sorting of membrane components was elegantly performed in BHK cells microinjected with mRNA coding for an anti-Golgi antibody. This resulted in the accumulation of antibodies inside the endoplasmic reticulum and an inhibition of the cell surface expression of a plasma membrane protein (G protein of VSV)(Burke and Warren, 1984). However, in GH3 cells one cannot exclude, so far, that the A-Go labeled vesicles represent another centrifuge membrane flow, not involved in the release of PRL. Such a distinction between the intracellular pathways of secretory and membrane proteins could be recently made in GH secreting GH3 cells infected with VSV (Green and Shields, 1984).

Membrane components involved in the endocytic pathway

The initial evidence for an endocytic pathway, that is a centripete route, in PRL cells was obtained using electron dense tracers of adsorptive endocytosis such as cationized ferritin (Farquhar, 1978) and concavalin A (Tougard et al, 1980) or of fluid phase endocytosis such as horse radish peroxidase (Farquhar et al, 1975, Tixier-Vidal et al, 1976, Vila Porcile and Olivier, 1980). These studies were mostly directed towards the demonstration of the current concept (see rev. Steinman et al, 1983) that following exocytosis, membrane added in excess to the plasmalemma were retrieved by endocytosis and reached the Golgi stacks to be then reutilized in the packaging of secretory granules (see rev. Farquhar, 1983). The internalization of these tracers followed the same inital sequential steps of endocytosis : plasma membrane pits and small vesicles, coated or not, large irregularly shaped vacuoles, multivesicular bodies. The final step of this pathway, however, varied depending on the tracer and the studies : it was located either on lysosomes only or both in lysosomes and Golgi cisternae of the trans-face, thus suggesting in a very indirect manner a possible recycling of membrane components towards the biosynthetic pathway. However such studies were limited by the fact that tracers of adsorptive endocytosis may interfere on membrane traffic (see rev. Steinman et al, 1983) and cannot be considered as true membrane markers.

A direct approach to the analysis of membrane components involved in the endocytic pathway was recently provided by the preparation of anti-bodies directed against a purified lysosomal membrane fraction from rat liver. These polyclonal antibodies (A-Ly) recognize a 100 Kd antigen and cross react with a purified H^+, K^+-ATPase from gastric mucosa (Reggio et al, 1984).

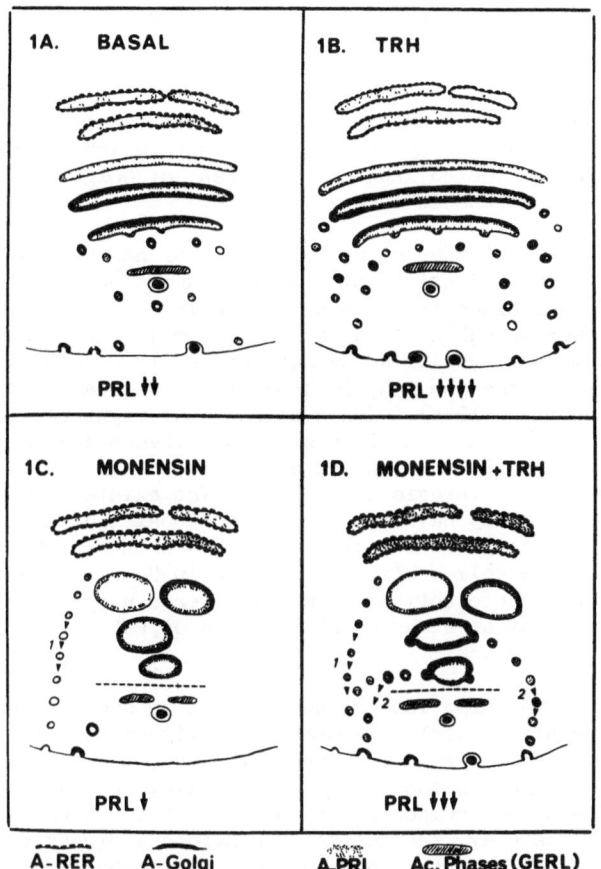

Fig. 1 - Schematic representation of the subcellular distribution of PRL, of RER antigens and of the 135 Kd Golgi antigen in relation with modifications of the secretory activity in GH3 cells. The two routes for PRL release which are represented in fig. 1C and 1D are postulated from the results of both immunocytochemical localization and pulse chase experiments Route 1 which would serve for the release of newly synthesized PRL and would originate from an early Golgi step is not blocked by monensin. Route 2 which serves for the stimulated release of PRL in response to TRH would originate from a late Golgi step, downstream the monensin blockade.

When applied to both GH3 cells and normal rat pituitary cells using same procedure as for A-RER and A-Go antibodies, the A-Ly labeled not only the membrane of lysosome-like structures, but also that of large, electron luscent vacuoles and of multivesicular bodies. In the Golgi zone it clearly stained 2-3 saccules, in a rather medial location, and slightly stained some vesicles in the core of the Golgi zone. In contrast the secretory granule membranes were unstained as well as the plasma membrane. However, a conspicuous staining was observed when living cells were exposed to the A-Ly before fixation, thus indicating the presence of the antigen at the cell surface (Tougard et al, 1985).

As compared to the subcellular distribution of the 135 Kd Golgi antigen, there is a large overlap between the distribution of these two antigens, since both antibodies label the membrane of Golgi medial saccules, lysosomes and endosome-like structures.

From a functional point of view, this new tool has been applied, so far, to the immunological identification of membrane components involved in the endocytosis of cationized ferritin by PRL cells. This study has revealed the existence in PRL cells of a prelysosomal compartment which possesses in its membrane the 100 Kd Ly antigen but is devoid of acid phosphatase activity. Since the A-Ly is known to cross react with a gastric proton pump (Reggio et al, 1984), such a compartment might represent an acidified prelysosomal or endosome-like structure (Tycko and Maxfield,1982, see rev. Steinman et al, 1983) which was also detected in other cell types using same A-Ly antibodies (Reggio et al,1984). The existence of such a compartment in PRL cells may have important functional implications in particular for ligand-receptor dissociation. Moreover the presence of the 100 Kd Ly antigen at the cell surface as well as in Golgi cisternae and vesicles suggests that these antibodies might offer a useful tool to study membrane traffic in relation with the stimulation of PRL release. This is presently under investigation.

In conclusion, the comparison of the respective subcellular distribution of several intracellular membrane antigens reveals either no overlap – for the RER antigens – or a large overlap – for the Golgi and lysosomal membrane antigens respectively. The coexistence in some Golgi subcompartments of membrane components involved in the biosynthetic pathway and in the endocytic pathway, respectively, provides a direct evidence for the role of this organelle in sorting and directing membrane components. The development in the future of new immunological tools specific to other membrane compartments involved in the secretory process appears as an important step towards a better understanding of membrane traffic in relation with PRL release.

II. MEMBRANE TRAFFIC IN HYPOTHALAMIC NEURONS IN CULTURE

Present Concept on vesicular traffic in neurons

Studies on secretory processes in gland cells have offered a back-ground for the understanding of membrane and vesicular traffic in neurons (see rev. Holtzman, 1977). However because of several limits inherent to neuronal systems most of the informations are concerned with vesicular traffic in axon terminals, a structure which, at a first sight, has no equivalent in glandular cells. Moreover the pilote model for such studies was the neuromuscular junction. This led to the concept that transmitter release involves the exocytosis of transmitter filled vesicles which fuse with the presynatpic membrane at specialized regions, are then retrieved by endocytosis and recycled. This concept was mostly based on morphological studies using either freeze fracture techniques or tracers of endocytosis such as horse radish peroxidase (see rev. Holtzman, 1977, Steinman et al, 1983). Formerly established for the neuromuscular junction, it was then found valid for a few other systems : photoreceptors of the retina (Schaeffer and Raviola, 1978), sympathetic nerve terminals in culture (Buckley and Landis, 1983) and, at least under certain conditions, neuro-secretory PC 12 cells (Pozzan et al, 1984, Saito et al, 1985).

As concerns hypothalamic neurosecretory neurons, studies on membrane traffic in relation with hormone release were limited, so far, to two systems of short-term in vitro incubations : neural lobes of the rat pituitary gland (Nordmann and Morris, 1976, Morris and Nordmann, 1980, Lescure and Nordmann, 1980) and superfused adult rat median eminences (Zamora and Ramirez, 1983, Zamora et al, 1984). Both systems take advantage of a naturally occuring concentration of axon terminals in these parts of the hypothalamus but they may present some drawbacks since the neuronal perikarya are absent and the axons have been freshly sectioned. In both systems exposure to high potassium ion concentration gives rise to hormone release which was correlated with exocytosis and depletion of neuro-secretory granules followed by membrane endocytosis. However the latter event does not involve the same structures in both systems which may be surprising in view of their relative similarities of origin and structure.

Hypothalamic cell cultures initiated with cells taken at late fetal stage in the rat or mouse have been recently shown to offer fruitful models to study the release of several neuropeptides such as somatostatin, opiates, TRH (see rev. Loudes et al, 1983b, Shoemaker et al,1983). Such systems present the advantage over the two previous ones, to respect the integrity of the neuron organization from cell bodies to synapses. We have recently investigated the ultrastructural correlates of the response of hypothalamic cell cultures to high potassium depolarization using serum free medium cultures of mouse fetal hypothalamic cells taken on the 16th day of gestation. The functional validity of this system has been previously established by showing that when cultured in an appropriate medium for 10-12 days these cells acquire the capacity to release TRH in a calcium dependent manner (Loudes et al, 1983 a,b).

Ultrastructural correlates of potassium evoked depolarization in cultured hypothalamic neurons
(Tixier-Vidal et al, in press)

After being cultured for 10-12 days in an appropriate serum free medium, hypothalamic neurons have developed fully differentiated synapses as attested to by the differentiation of the presynaptic and the post-synaptic membranes, by the presence of many synaptic vesicles of regular diameter (40-50 nm) and by the diversity of synaptic configurations. As compared to adult neurosecretory axon terminals, these axon endings contain however only few small neurosecretory granules. After incubation for 3 min in the release medium containing 3 mM KCl, the synapse ultrastructure remains unchanged.

Exposure of the cells for 3 min to a release medium containing 60 mM KCl induced vesicular depletion and axon swelling in any axon ending. The remaining synaptic vesicles displayed a decrease in diameter and were either packed into small clusters closely associated to the presynaptic membrane, or dispersed in the axoplasma. A few pictures of synaptic vesicles fusing with the presynaptic membrane could be seen between dense projections . However we did not see any fusion of dense core vesicles with the plasmalemma. In addition other membrane bound structures appeared in the depolarized axon terminals : a few irregular cisternae or vacuoles and coated vesicles of variable diameter. Moreover the axoplasm exhibited a conspicuous and complex filamentous network, sometimes interconnected with the remaining small vesicles. In contrast to this striking reorganization of axon terminals, the axon itself whenever it could be followed in the same section, was not clearly modified. The neuron perikarya were particularly healthy in such cultures with a large Golgi zone and conspicuous polysomes.

Following return to 3 mM KCl for only 3 min a striking restoration of the vesicle population occurred in all axon terminals and synaptic boutons. The restored vesicles either occupied the entire surface of the section of a synaptic bouton or were polarized onto the presynaptic differentiation. Such "restored" vesicles moreover displayed features different from those of classical synaptic vesicles in control cultures. Their diameter was less homogeneous (40 to 60 nm), their shape was slightly irregular and they often displayed irregular pits at their cytoplasmic face. In addition the cytoplasm seemed to contain an electron dense granular material. In case of limited restoration of the vesicle population, the rest of the axoplasm was occupied by large vacuoles or large dense core vesicles. However, no particular modification of the smooth reticulum of the corresponding axon could be seen, whenever it could be followed in the same section.

In summary the ultrastructural modifications induced by modification of the ionic environment seem to be restricted to axon terminals and synaptic boutons. They are remarquable in two points : the rapid reversibility and the interconnection between smooth membranes and a filamentous cytoplasmic network.

As compared to the other (above mentioned) systems, the response of hypothalamic neurons to potassium evoked depolarization presents many similarities : vesicular depletion, swelling of axon endings, exocytosis of synaptic vesicles, appearance of smooth cisternae and of a cytoplasmic network. The ability of such neurons to restore the vesicle population of axon terminals following return to basal potassium concentration also is in agreement with findings obtained in most systems examined, excepted however the neural lobe of the pituitary where large vacuoles instead of micro-vesicles were found to be the major route for membrane retrieval (Morris and Nordmann, 1980). A particular feature of hypothalamic neurons in culture resides in the rapidity (3 min) of the vesicle restoration, a time interval shorter than that reported for other in vitro systems of neuro-secretory neurons, although obviously longer than in the case of the neuro-muscular junction.

Perspectives

Taken together these morphological findings argue in favor of the validity of cultured hypothalamic neurons as model systems to analyze membrane traffic in relation with the release of neurohormones. The present concept that the release of neurotransmitters is accompanied by a cycle of exocytotic and endocytic events can be applied to hypothalamic secretory neurons. However once accepted this concept raises many unanswered questions : what is the role of synaptic vesicles in neuropeptide release ? what are the mechanisms of vesicle restoration within the frame of the endocytotic process ? what are the origin and nature of synaptic vesicle membranes with respect to the other membrane compartments of the endo-plasmic reticulum ? Using our model of cultured hypothalamic neurons we have started several approaches with the hope to obtain partial answers to some of these crucial questions.

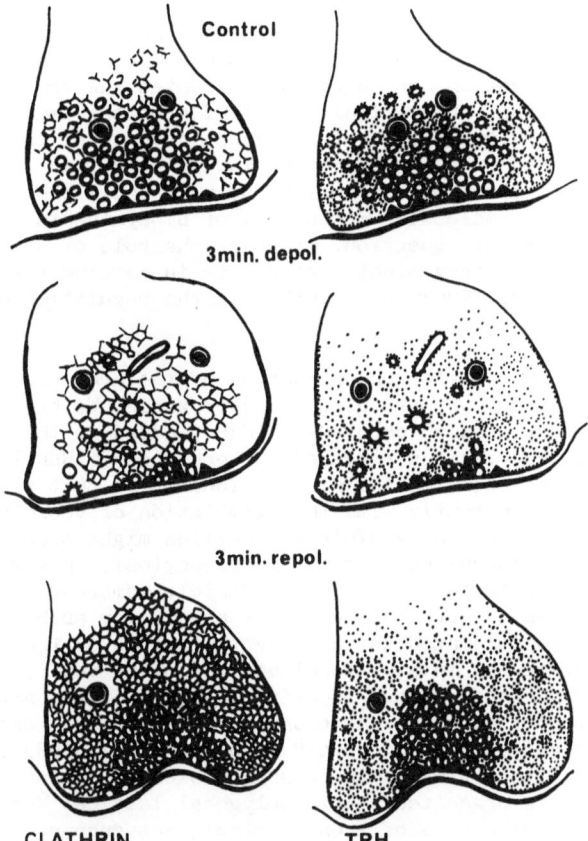

Fig. 2 – Schematic representation of vesicular traffic in axon terminals of cultured hypothalamic neurons exposed for 3 min to 60 mM K^+ and then returned for 3 min to 3 mM K^+. The right hand row schematizes the subcellular distribution of TRH as revealed by immunostaining. The left hand row schematizes the subcellular distribution of clathrin as revealed by immunostaining (see text).

TRH was immunocytochemically localized (unpublished observations) using a preembedding method previously applied to similar cultures grown in serum supplemented medium (Faivre-Bauman et al, 1980). In our serum free medium cultures about 30% of neurons are specifically stained (Faivre-Bauman et al,1981). At the electron microscope level the immunostaining was localized in the cytoplasm at the levels of pericarya, axons and axon terminals where the intensity of the cytoplasmic reaction was maximum. No reaction product was found inside membrane bound cisternae of the endoplasmic reticulum in the cell bodies as well as inside synaptic vesicles. The dense core vesicles only displayed a positive reaction in their matrix. The subcellular distribution of the TRH reaction product in synaptic boutons was not modified following exposure to high potassium concentration or return to normal potassium concentration, excepted the vesicular depletion and restoration. Thus, within the limits of the immunocytochemical techniques, the probability that synaptic vesicles serve as carrier for the release of TRH seems rather low. One is then left with the difficult problem of the subcellular origin of releasable TRH. Progress in the understanding of the molecular mechanisms of biosynthesis of TRH are needed before to answer this question. Moreover the role of synaptic vesicle remains obscure ; they might participate in calcium transport (Nordmann and Chevallier, 1980) and possibly in the regulation of cytoplasmic pH.

In order to reach a better understanding of the mechanisms of vesicle restoration, or recycling, we have used two approaches. We have started to study the uptake of horseradish peroxydase (HRP), added in the depolarizing medium. Preliminary observations revealed the presence of HRP loaded vesicles in depolarized axon terminals, thus indicating that fluid phase pinocytosis occurs concomitantly with the stimulation of vesicle exocytosis and suggesting that the vesicle restoration might occur, at least in part, as a direct consequence of membrane pinocytosis. However this does not unequivocally demonstrate that synaptic vesicle membranes are directly re-used for pinocytosis. Other tools would be needed for such a demonstration. We also attempted to identify the nature of the cytoplasmic network which was disclosed in axon terminals following vesicular depletion. For that purpose we have localized clathrin using affinity purified polyclonal antibodies which recognize both clathrin associated to membrane and the so called "soluble pool" (Louvard et al, 1983). The data strongly suggest that the cytoplasmic network possesses a clathrin-like immunoreactivity and is organized into a polygonal lattice. Moreover the intensity of the immunostaining of axon terminals was enhanced in the condition of vesicle restoration, both in the cytoplasm and around newly formed vesicles. This suggests that clathrin would play an important role in vesicular traffic through successive cycles of association to membrane and dissociation. This would provide in axon terminals a local mechanism for membrane transport (Tixier-Vidal et al, in preparation).

CONCLUSION

The comparison of two types of secretory cells - anterior pituitary cells and hypothalamic neurons - emphasizes the fact recently pointed out by M. Farquhar that "although multiple pathways of membrane traffic exist in all cell types, a given cell type can modify its pattern of membrane traffic to enable it to carry out its own specialized functions" (Farquhar et al, 1985). In contrast to prolactin cells, secretory neurons appear as highly polarized cells which display specialized domains of their territory allowing to concentrate the release of neurosecretory product at a very precise site of the cell surface : the presynaptic membrane. Are there similarities between the cellular and molecular mechanisms involved in the release of anterior pituitary hormones and of neurohormones ? At present such an assumption would be premature. However there are several favorable premises. As concerns the intracellular mechanisms recent studies with PC 12 cells (Pozzan et al, 1984, Vicentini et al, 1985) suggest that the crucial events which are known to be triggered by the activation of receptors in many systems - such as intracellular calcium rise, hydrolysis of phosphoinositides, activation of protein kinase C - might also be involved in the release of neurotransmitters. As concerns the biochemical identification of intracellular membrane compartments involved in neuro-hormone release, much remains to be done. However an important step has been recently made with the development of immunological probes specific for synaptic vesicle membrane components : a 38000 dalton membrane protein (Jahn et al, 1985) and a transmembrane glycoprotein (Mr 100,000)(Buckley and Kelly, 1985). Moreover the latter antigen could be also detected by immunofluorescence in several endocrine cells, namely the AtT-20 and GH3 pituitary cell lines. This suggests that neurons and anterior pituitary cells would share same antigenic determinants in secretory vesicles which however remain to be identified. In any case, such immunological tools offer fascinating perspectives.

REFERENCES

Buckley, K., and Kelly, R.B., 1985, Identification of a Transmembrane Glycoprotein Specific for Secretory Vesicles of Neural and Endocrine Cells. J. Cell Biol., 100:1284.

Buckley, K.M., and Landis, S.C., 1983, Morphological Studies of Neuro-transmitter Release and Membrane Recycling in Sympathetic Nerve Terminals in Culture. J. Neurocytol. 12:93.

Burke, B., and Warren, G., 1984, Microinjection of mRNA Coding for an Anti-Golgi Antibody Inhibits Intracellular Transport of a Viral Membrane Protein. Cell, 36:847.

Faivre-Bauman, A., Nemeskeri, A., Tougard, C., and Tixier-Vidal, A., 1980, Immunological Evidence for Thyroliberin (TRH) Neurons in Primary Cultures of Fetal Mouse Brain Cells. Ontogenic Aspects. Brain Res., 185:289.

Faivre-Bauman, A., Rosenbaum, E., Puymirat, J., Grouselle, D., and Tixier-Vidal, A., 1981, Differentiation of Fetal Mouse Hypothalamic Cells in Serum-Free Medium. Developm. Neurosci., 4:118.

Farquhar, M.G., Skutelsky, E.H., and Hopkins, C.R., 1975, Structure and Function of the Anterior Pituitary and Dispersed Pituitary Cells. In Vitro Studies. In: "The Anterior Pituitary" Tixier-Vidal, A. and Farquhar, M.G. eds, Academic Press, Vol. 7, p. 83.

Farquhar, M.G., Reid, J.J., and Daniell, L.W., 1978, Intracellular Transport and Packaging of Prolactin : A Quantitative Electron Microscope Autoradiographic Study of Mammotrophs Dissociated from Rat Pituitaries. Endocrinology, 102:296.

Farquhar, M.G., and Palade, G., 1981, The Golgi Apparatus (Complex) - (1954-1981) - from Artifact to Center Stage. J. Cell biol., 91:77s.

Farquhar, M.G., 1983, Intracellular Membrane Traffic : Pathways, Carriers, and Sorting Devices. Methods in Enzymol., 98:1.

Farquhar, M.G., Woods, J.W., and Brown, W.J., , 1985, Receptor Traffic to Golgi Subcompartments. In "Current Communications in Molecular Biology" "Protein Transport and Secretion" Gething, M.J. ed., CSH, p. 115.

Gourdji, D., Tougard, C. and Tixier-Vidal, A., 1982, Clonal Prolactin Strains as a Tool in Neuroendocrinology. In :"Frontiers in Neuro-Endocrinology", Ganong, W.F., and Martini, L. eds, Raven Press, New York, vol. 7, p. 317.

Green, R., and Shields, D., 1984, Somatostatin Discriminates Between the Intracellular Pathways of Secretory and Membrane Proteins. J. Cell Biol., 99:97.

Gumbiner, B., and Kelly, R.B., 1982, Two Distinct Intracellular Pathways
    Transport Secretory and Membrane Glycoproteins to the Surface of
    Pituitary Tumor Cells. Cell, 28:51.
Holtzman, E., 1977, The Origin and Fate of Secretory Packages, especially
    Synaptic Vesicles. Neurosci. 2:327.
Jahn, R., Schiebler, W., Ouimet, C., and Greengard, P., 1985, A 38,000-
    Dalton Membrane Protein (p38) Present in Synpatic Vesicles.
    Proc. Natl. Acad. Sci. USA, 82:4137.
Lescure, H., and Nordmann, J.J., 1980, Neurosecretory Granule Release and
    Endocytosis during Prolonged Stimulation of The Rat Neurohypophysis
    In Vitro. Neurosci., 5:651.
Loudes, C., Faivre-Bauman, A., Barret, A., Grouselle, D., Puymirat, J. and
    Tixier-Vidal, A. (1983a) Release of Immunoreactive TRH in Serum-Free
    Cultures of Mouse Hypothalamic Cells. Developm. Brain Res., 9:231.
Loudes, C., Faivre-Bauman, A., and Tixier-Vidal, A. (1983b) Techniques for
    culture of hypothalamic neurons. In:"Methods in Enzymology, Hormone
    Action" Part H. Neuroendocrine Peptides, Conn,M. ed., 103:313.
Louvard, D., Reggio, H., and Warren, D., 1982, Antibodies to the Golgi
    Complex and the Rough Endoplasmic Reticulum. J. Cell Biol., 92:92.
Louvard, D., Morris, C., Warren G., Stanley, K., Winkler, F., and
    Reggio, H., 1983, A Monoclonal Antibody to the Heavy Chain of
    Clathrin. The EMBO J., 2:1655.
Morin, A., Rosenbaum, E., and Tixier-Vidal, A., 1984a, Effects of Thyro-
    tropin-Releasing Hormone on Prolactin Compartments in Clonal Rat
    Pituitary Tumor Cells. Endocrinology, 115:2271.
Morin, A., Rosenbaum, E., and Tixier-Vidal, A., 1984b, Effects of Thyro-
    tropin-Releasing Hormone on Prolactin Compartments in Normal Rat
    Pituitary Cells in Primary Culture. Endocrinology, 115:2278.
Morris, J.F. and Nordmann, J.J., 1980, Membrane Recapture after Hormone
    Release from Nerve Endings in the Neural Lobe of the Rat Pituitary
    Gland. Neurosci., 5:639.
Nordmann, J.J. and Morris, J.F., 1976, Membrane Retrieval at Neuro-
    secretory Axon Endings. Nature, London, 261:723.
Nordman, J.J., and Chevallier, J., 1980, On the role of Microvesicles in
    Buffering Ca in the Neurohypophysis. Nature, London, 287:54.
Pozzan, T., Gatti, G., Dozio, N., Vicentini, L.M., and Meldolesi, J.,
    1984, Ca^{2+}-Dependent and Independent Release of Neurotransmitters
    from PC12 Cells : A Role for Protein Kinase C Activation ? J. Cell
    Biol. 99:628.
Reggio, H., Bainton, D., Harms, E., Coudrier, E. and Louvard, D., 1984,
    Antibodies against Lysosomal Membranes reveal a 100,000-mol-wt
    Protein that Cross-reacts with Purified H^+, K^+-ATPase from Gastric
    Mucosa. J. Cell Biol., 99:1511.
Saito, I., Dozio, N., and Meldolesi, J., 1985, The Effect of α-Latrotoxin
    on the Neurosecretory PC12 Cells Differentiated by Treatment with
    Nerve Growth Factor. Neurosci., 14:1163.
Schaeffer, S., and Raviola, E., 1978, Membrane Recycling in the Cone Cell
    Endings of the Turtle Retina. J. Cell Biol., 79:802.
Shoemaker, W.J., Peterfreund, R.A., and Vale, W., 1983, Methodological
    Considerations in Culturing Peptidergic Neurons. In "Hormone
    Action - Neuroendocrine Peptides" Conn, M. ed., Methods in
    Enzymology, 103:347.
Steinman, R.M., Mellman, I.S., Muller, W.A., and Cohn, Z.A., 1983,
    Endocytosis and the Recycling of Plasma Membrane. J. Cell Biol.
    96:1.
Tixier-Vidal, A., Picart, R., et Moreau, M.F., 1976, Endocytose et
    Sécrétion dans les Cellules Antéhypophysaires en Culture. Action
    des Hormones Hypothalamiques. J. Microscop. Biol. Cell., 25:159.
Tixier-Vidal, A., and Tougard, C., 1984, Recent Progress in Cell Biology
    of Secretory Process in Anterior Pituitary Cells. In : "Hormonal

Control of the Hypothalamo-Pituitary Gonadal Axis". McKerns K.W. ed., Plenum Publish. Corp., p. 169.

Tixier-Vidal, A., Picart, R., Loudes, C., and Faivre-Bauman, A., 1985, Effects of polyunsaturated Fatty Acids and Hormones on Synapto-genesis in Serum Free Medium Cultures of Mouse Fetal Hypothalamic Cells. Neurosci., in press.

Tougard, C., Picart, R., and Tixier-Vidal, A., 1980, Electron-Microscopic Cytochemical Studies on the Secretory Process in Rat Prolactin Cells in Primary Cultures., Am. J. Anat. 158:471.

Tougard, C., Picart, R., and Tixier-Vidal, A., 1982, Immunocytochemical Localization of Prolactin in the Endoplasmic Reticulum of GH3 Cells. Variations in response to Thyroliberin. Biol. Cell., 43:89.

Tougard, C., Louvard, D., Picart, R., and Tixier-Vidal, A., 1983a, The Rough Endoplasmic Reticulum and the Golgi Apparatus Visualized Using Specific Antibodies in Normal and Tumoral Prolactin Cells in Culture. J. Cell Biol., 96:1197.

Tougard, C., Picart, R., Morin, A., and Tixier-Vidal, A., 1983b, Effect of Monensin on Secretory Pathway in GH3 Prolactin Cells. A Cytochemical Study. J. Histochem. Cytochem., 31:745.

Tougard, C., Louvard, D., Picart, R. and Tixier-Vidal, A., 1985, Antibodies Against a Lysosomal Membrane Antigen Recognize a Prelysosomal Compartment Involved in the Endocytic Pathway in Cultured Prolactin Cells. J. Cell Biol., 100:786.

Tycko, B., and Maxfield, F.R., 1982, Rapid Acidification of Endocytic Vesicles Containing Alpha-2-Macroglobulin. Cell, 28:643.

Vicentini, L.M., Ambrosini, A., Di Virgilio, F., Pozzan, T. and Meldolesi, J., 1985, Muscarinic Receptor-induced Phosphoinositide Hydrolysis at Resting Cytosolic Ca^{2+} Concentration in PC12 Cells. J. Cell Biol. 100:1330.

Vila-Porcile, E., and Olivier, L., 1980, Exocytosis and Related Membrane Events. In : "Synthesis and Release of Adenohypophyseal hormones". Jutisz, M., and McKerns K.W., eds, Plenum Press, p. 67.

Walker, A.M., and Farquhar, M.G., 1980, Preferential Release of Newly Synthesized Prolactin Granules is the Result of Functional Hetero-geneity among Mammothrops, Endocrinology, 107:1095.

Zamora, A.J., and Ramirez, J.D., 1983, Structural Changes in Nerve Endings of Rat Median Eminence Superfused with Media rich in Potassium Ions. Neurosci., 10:463.

Zamora, A.J., Garosi, M., and Ramirez, V.D., 1984, Post-Stimulatory Endo-cytosis, Microvesicle Repopulation and Changes in Smooth Endo-plasmic Reticulum in Nerve Endings of the Median Eminence Super-fused in Vitro. Neurosci., 13:105.

ACKNOWLEDGEMENTS

We gratefully acknowledge the asssistance of Miss A. Bayon for the preparation of the manuscript. Works from our laboratory presented in this review have been supported by grants from the C.N.R.S. (UA 04 1115) and from the INSERM (CRE 834 018).

NEURO-STEROIDS : 3β-HYDROXY-Δ5-DERIVATIVES IN THE RAT BRAIN

P. Robel, C. Corpéchot, C. Clarke, A. Groyer,
M. Synguelakis, C. Vourc'h, and E.E. Baulieu

CNRS ER 125 and INSERM U33, Lab Hormones, Faculté de
Médecine, 94270 Bicêtre, France

INTRODUCTION

The relationship between steroid hormones and brain function has mostly been considered according to the responses elicited by the secretory products of steroidogenic endocrine glands, generally in the form of feedback control.

In the last few years, major progress has been made towards the elucidation of the molecular mechanism of action of steroid hormones in the brain and anterior pituitary. It is now recognized that steroid hormones exert their feedback control upon the pituitary secretion of their respective trophic hormones by binding to intracellular receptors and altering gene transcription in target cells within specific neuroendocrine structures (Stumpf and Grant, 1975; Warembourg, 1978; McEwen et al., 1982). These receptor systems may also account for several behavioral effects, for the biosynthesis of various enzymes and receptors, including those involved in neuromediator action, and possibly for the effects of sex steroids upon nerve cell differentiation (review in Fuxe et al., 1981).

Several metabolic conversions of steroid hormones in brain i.e. aromatization (Fishman, 1982; Naftolin et al., 1975), 5α-reduction (Celotti et al., 1979; Jaffe, 1969), catechol estrogen formation (Paul and Axelrod, 1977), etc... may supply active steroids to target cells, in adequate concentrations, possibly in a kind of paracrine arrangement.

Intravenously or locally (iontophoretically) administered steroids can elicit almost immediate cell responses, as exemplified by changes in action potentials of locally recorded neurones. Such rapid responses may result from direct steroid-plasma membrane interactions (Baulieu, 1981; Carette et al., 1979). No interaction with a specific component of nerve cell membranes has yet been documented, as is the case for the recently demonstrated progesterone receptor of Xenopus oocyte membranes (Blondeau and Baulieu, 1984), which is responsible for the inhibition of membrane bound adenylate cyclase (Finidori-Lepicard et al., 1981).

We have recently characterized pregnenolone (P), dehydroepiandrosterone (D) their sulfate esters (S), and their fatty acid esters (lipoidal derivatives, L) in the rat brain, and we also have obtained preliminary evidence for their occurence in other mammalian species (mouse, pig, monkey, and human). We proposed that their formation or accumulation in the rat brain depended on in situ mechanisms unrelated to the peripheral endocrine gland system (Corpéchot et al., 1981; Corpéchot et al., 1983). The 3β-hydroxy-Δ5-steroids (sulfates) do not interact with already described intracellular receptors, with the exception of 5-androstene-3β,17β-diol, which binds to estrogen receptor with significant affinity, and may behave as a weak estrogen (Seymour-Munn and Adams, 1983). Mammalian brain contains all the enzymes that are necessary for the conversion of P(S,L) and D(S,L) into sex steroid hormones (review in Corpéchot et al., 1981). The functional significance of such pathways remains to be established.

In the present review of our work, we comment upon the still limited amount of evidence collected upon the origin and function of brain 3β-hydroxy-Δ5-steroids.

MATERIAL AND METHODS

Characterization and Measurement of Brain Steroids

Brain samples from adult male rats of the Sprague-Dawley strain (11-12 week old) were generally used. They were collected in ice-cold isotonic saline, weighed, and homogenized in 5 ml of isotonic saline. Tracer D, DS and P were added for recovery, and free steroids were extracted with ethyl acetate (3 times 10 ml). The water phase was brought to pH 1, and 20 % sodium chloride (w:v) was added. Extraction with ethyl acetate was again performed and the extract, which contained steroid sulfates, was solvolyzed at 37°C for 12 to 16 h. Both ethyl acetate extracts were washed with NaOH and water and taken to dryness. They were taken up in 2 ml of 70 % methanol, left overnight at -20°C, and lipids were pelleted by centrifugation. This residue contained about 95 % of the lipoidal derivatives and was used for their measurements.

Partition chromatography on celite microcolumns allowed the stepwise separation of a non polar fraction, and fractions containing P and D. Saponification of the non polar fractions with methanolic KOH released steroids that were again submitted to celite chromatography (they come from "lipoïdal" conjugates – that is to say fatty acid esters) (Hochberg et al., 1977).

Table 1. D, DS, P and PS in Plasma and Organs of Male Rats

	D	DS	P	PS
Plasma	0.1 $\pm$ 0.0 (4)	0.3 $\pm$ 0.1 (9)	1.3 $\pm$ 0.2 (5)	1.4 $\pm$ 0.4 (5)
Brain :				
Anterior	0.4 $\pm$ 0.1 (4)	1.6 $\pm$ 0.2 (10)	38.4 $\pm$ 6.9 (5)	15.8 $\pm$ 3.1 (5)
Posterior	0.1 $\pm$ 0.0 (4)	4.9 $\pm$ 1.1 (11)	22.1 $\pm$ 2.9 (5)	5.7 $\pm$ 2.2 (5)
Adrenal	ND	2.9 $\pm$ 0.3 (4)	5320 $\pm$ 2717 (3)	51.5 $\pm$ 7.8 (3)
Liver	0.0 (1)	0.4 $\pm$ 0.1 (3)	4.6 $\pm$ 0.8 (4)	7.8 $\pm$ 1.7 (4)
Spleen	0.1 (1)	0.7 $\pm$ 0.1 (3)	9.4 $\pm$ 0.5 (4)	17.6 $\pm$ 1.7 (4)
Testis	ND	0.4 $\pm$ 0.0 (3)	13.1 $\pm$ 2.0 (3)	4.5 $\pm$ 0.2 (3)
Kidney	0.1 (1)	0.5 $\pm$ 0.0 (3)	11.5 $\pm$ 3.2 (3)	7.3 $\pm$ 1.4 (3)

The brain was quickly removed, divided in two parts and processed at 0-4°C. The posterior part contained the cerebellum, the pons, and the medulla oblongata (mean weight 0.37 $\pm$ 0.05 g). The anterior part contained the cortex cerebri and the mesencephalon, and excluded the bulbus olfactorius, the hypothalamus and underlying structures (mean weight 1.12 + 0.09 g). Each extract was obtained from a single animal, sacrificed 15 d after housing in controlled conditions. Results were expressed in ng steroid per g of fresh tissue or ml of plasma. Mean $\pm$ sd (n).

The preliminary identification of P and D (either unconjugated or released from their sulfate or fatty acid esters) was based on partition chromatography on celite columns and isolation of eluates containing P and D, followed by radioimmunoassay using specific antisera having minimal cross-reaction with other steroids. Definitive identification of the steroid moiety was made by mass spectrometry. After chromatography on celite and Lipidex columns, trimethylsilyl ethers were prepared and analyzed by gas chromatography/mass spectrography. Highly purified fractions were obtained, as judged by the absence of significant peaks in the total ion current chromatograms. Peaks with the retention time and all diagnostically important m/z values of authentic trimethylsilyl ethers of P and D, respectively, were observed. The amounts of steroids, calculated from the heights of the peaks were in agreement with RIA values, sometimes 10–20 % lower.

We did not measure 5-androstene-3β,17β-diol, a reduced metabolite of D. We did not find measurable amount of 17-hydroxy-pregnenolone (< 0.2 ng/g of tissue). Testosterone, corticosterone and their sulfate esters were measured by combined chromatography-RIA in a few cases.

Behavioral and electrophysiological studies were performed according to previously published procedures (Haug and Mandel, 1978; Poulain and Carette, 1981).

RESULTS

Measurement of Pregnenolone and Dehydroepiandrosterone in the Rat Brain

Pregnenolone and dehydroepiandrosterone have been found as unconjugated free steroids, sulfate esters and fatty acid esters. In the following sections, the combined amounts of the sulfate esters and free forms of either P or D or corticosterone (B) will be designated by the symbols [P], [D], and [B] respectively. Values for DL, corrected for recovery, are not yet available.

The quantitation of brain D and P appeared to depend on several environmental and methodological factors. Some have been partly documented, as lighting schedule, hour of sacrifice, housing conditions (number of animals per cage, rats of the other sex in the same room), and stress (after orchiectomy (orx) plus adrenalectomy (adx), or the corresponding sham operation). We also realized that the work up conditions for the removal and initial processing of brain tissue were critical. Therefore all physiological and pharmacological experiments were conducted under strictly defined conditions, with appropriate internal controls, and their statistical signi-

Table 2. P and PS in Rat Brain Homogenate (ng/g)

	P	PS
Control	3.3–4.5	7.6–9.7
48 h incubation at		
final pH : 5.3	19.4–21.6	0.4–0.7
6.9	9.5–11.9	6.4–8.0
8.4	9.9–11.5	9.5–ND

2.5 ml samples (660 mg wet weight) of an homogenate of male rat brain in phosphate buffer pH 7.0 were diluted 6 fold with acetate pH 5.0, or phosphate pH 7.0, or Tris pH 9.0 buffers. P and PS were measured as described, in duplicate. The total amount of P was increased by incubation at all pH; the largest increase was observed at pH 5.3 together with a large decrease of the sulfate ester form.

ficance evaluated. With these restrictions in mind, [D] concentration in the brain of adult male rats (Sprague-Dawley, 11-12 week old), was about 10 pmol per g of tissue, whereas [P] was about 45 pmol per g of tissue (Table 1), and PL concentration was about 60 pmol per g of tissue. PS concentration was always lower than that of P, whereas DS concentration was greater than that of D. Samples of a brain homogenate were incubated at 37°C for 48 h at different pH (Table 2). At pH 5.3, more P was generated than could be accounted for by the hydrolysis of PS. However, rat brain microsomes contain an enzyme which splits steroid-fatty acid esters (Kishimoto, 1973). Results are compatible with the existence of only 3 forms of P in brain : FP, PS, and PL. Although we found differences in [P] and [D] concentrations between selected brain areas, the variability between different experiments precluded any definitive statement. However, [P] and [D] were constantly higher in hypothalamus and olfactory bulbs than in any other brain structure.

The Accumulation of Brain Pregnenolone and Dehydroepiandrosterone is not Related to Peripheral Endocrine Glands

The rat adrenal secretes P, but not D, as reflected by the measurement of [P] and [D] in adrenals and in plasma. This led us to assume independent mechanisms for the formation and/or accumulation of brain 3β-hydroxy-Δ5-steroids. The relevant findings have been reported previously (Corpéchot et al., 1981; Corpéchot et al., 1983). Neither [D] nor [P] disappeared or even decreased (compared to sham operated controls) in brain 15 to 45 d after combined adx + orx (Figure 1). Brain [D] was also unchanged after administration of corticotropin or dexamethasone for 3 d, or after an acute nociceptive stimulation. In contrast, stress conditions prevailing 2 d after adx + orx or the corresponding sham operation resulted in a significant increase of brain [D] concentration (Table 3). The results were less clear cut with [P], because of additional adrenal contribution to acute changes of brain [P] concentrations.

Since neither D nor P disappeared from the blood of adx + orx animals, brain results may be explained by large accumulation of D and P arising from extra-adrenal source(s). We have tried to check this possibility by injecting radioactive P and D to intact rats. Although the concentrations of steroids were severalfold larger in brain than in plasma, both radioactive steroids were not retained for any longer time than in plasma (Corpéchot et al., 1983). We concluded that brain P and D were either produced locally or released from still undefined storage forms.

Table 3. DS in Plasma and Brain of Male Rats

	Intact	SHAM	ORX + ADX	ACTH	DXM
Plasma	0.3 + 0.1 (9)	0.6 + 0.1[+] (6)	0.4 + 0.2[+] (6)	0.5 + 0.1[+] (2)	0.2 + 0.1 (2)
Brain : anterior	1.6 + 0.2 (10)	5.8 + 1.7[+] (6)	4.4 + 1.3[+] (7)	1.4 + 0.4 (2)	2.1 + 1.3 (2)
posterior	4.9 + 1.1 (11)	13.2 + 2.2[+] (5)	9.0 + 1.5[+] (7)	4.0 + 1.1 (2)	4.5 + 1.1 (2)

Results are expressed in ng/ml or ng/g. Mean + sd (n). Adrenalectomized (ADX) and orchiectomized (ORX) rats and sham operated controls were sacrificed 2 d after operation. Corticotropin (ACTH 1-24, 0.5 mg/d x 2 d) was injected sc, and rats were sacrificed 1 d later. Dexamethasone (DXM, 1 mg/d x 3 d) was injected sc and rats were sacrificed 2 h after the injection. [+]Significantly different from value for intact rats at 5 % level.

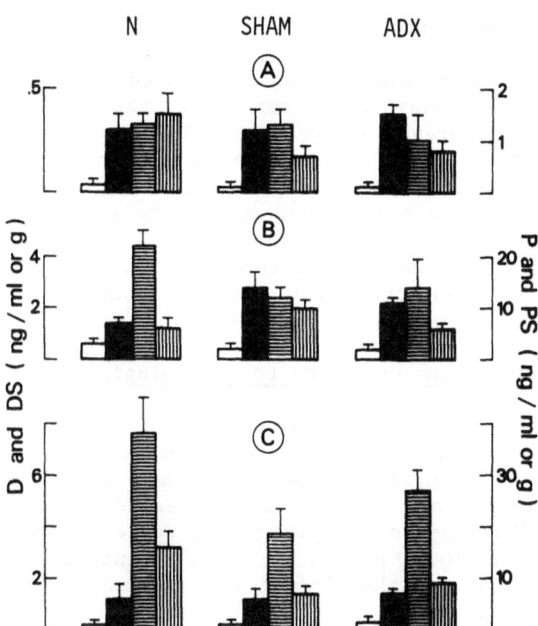

Fig. 1. D, DS, P and PS in plasma (A), posterior brain (B), and anterior
         brain (C) of intact (N), sham operated (SHAM), or castrated and
         adrenalectomized rats (ADX).
         Animals were sacrificed 15 d after housing in controlled
         conditions.
         ☐, D ; ■, DS ; ☰, P ; ▥, PS.

Pregnenolone Formation from Sterol Precursors

Attempts to demonstrate the side-chain cleavage of cholesterol by brain
slices, total homogenates or subcellular fractions were inconclusive (Clarke
et al., 1984). However, pregnenolone formation occured when an established
cell line of neural origin, the rat C6 glioma, was incubated with ^{3}H-meva-
lonolactone. Confluent cultures containing ~ 10^7 cells were incubate with
~ 15 μCi of ^{3}H-mevalonolactone (sa 24 Ci/mmol) in 5 ml of MEM containing
0.25 % gelatin, 2 g/L D-glucose, and 110 mg/L Na pyruvate, at 37°C for 24 h.
The cells were recovered, submitted to the extraction-solvolysis procedure,
and the extract was separated by partition chromatography on celite. The P
containing fraction was identified by TLC in two systems, acetylation and
TLC, and cristallisation to constant specific activity after isotopic
dilution. ^{3}H-P represented 1.52 ± 0.01 % (mean ± se, n = 6, uncorrected for
recovery) of the solvent extractible radioactivity. Its formation was
inhibited by aminogluthetimide. Similar results were obtained with C6 glioma
cells incubated with ^{3}H-cholesterol.

Fatty Acid Esters of Pregnenolone

Rat brain microsomes were prepared and incubated with ^{3}H-pregnenolone. P was converted to non polar compounds with the characteristics of endogenous brain PL or reference P stearate. The enzyme system had a pH optimum of 4.2 with acetate buffer. The apparent Km was 3.3×10^{-5} M for P and Vmax was 4.9 nmol/h/mg of microsomal protein. A high level of synthetic activity was found in the brain of young male rats (1-3 week old), which rapidly decreased with further increase in age. Saponification of the product yielded pregnenolone and a mixture of myristate (C14:0, 9 %), palmitate (C16:0, 26 %), stearate (C18:0, 11 %), oleate (C18:1, 21 %) and linoleate (C18:2, 5 %) as major fatty acid esters.

Variation of Brain Pregnenolone According to Physiological Conditions

Ontogenesis. [P] and (F + S) corticosterone [B] were measured in rat brain between birth and d 22. In newborns of both sexes, [B] and [P] were high at delivery (99 ± 71 ng/g, mean $\pm$ sd, n = 6 and 27 ± 13 ng/g, n = 6), confirming that adrenocortical activity is increased in the newborn rat (Corbier and Roffi, 1978a). Brain and plasma [B] decreased to insignificant levels between d 1 and d 10, a period of adrenal inactivity in the rat, which depends on multiple peripheral and central mechanisms (Allen and Kendall, 1967; Corbier and Roffi, 1978b; Sakly and Koch, 1983). Brain [P] also decreased steadily during the first day of life (7 ± 2 ng/g at 6 h, n = 6 and 8 ± 5 ng/g at 24 h, n = 15), but on the following days, it was in the 8-22 ng/g range. Brain [P] was always much larger than plasma [P], contrary to [B]. When 4 d old rats were treated with ACTH for 2 d, brain and plasma [B] was markedly increased, whereas [P] was unchanged (Table 4). [B] was not measurable after a 2 d dexamethasone treatment, whereas again [P] was unchanged, suggesting that during this period brain [P] is accumulated through a selective mechanism unrelated at least in part, to adrenal activity.

Circadian Variations. Male adult rats were housed in triads for 2 weeks. Lights were on between 7 am and 8 pm. The levels of [B], [P] and [D] were measured at 3 h intervals. They underwent circadian variations in the rat plasma and brain (Figure 2). When the data were interpreted by the cosinor method, the acrophases of P in brain and of D in plasma significantly preceded the acrophase of B (Table 5). The asynchrony of 3β-hydroxy-Δ5-steroid and glucocorticosteroid rhythms brings an additional argument in favor of separate regulatory mechanisms.

Table 4. P, D, and B in Brain and Plasma of 6 d Old Rats

	Control	ACTH	DEX
Brain			
P	9.5 ± 2.0	12.0 ± 1.2	10.2 ± 2.4
D	4.4 ± 0.3	3.2 ± 0.8	4.4 ± 1.0
B	4.4 ± 2.3	181 ± 102	N.M.
Plasma			
P	2.0 ± 0.3	4.8 ± 0.3	1.7-2.6
D	0.5 ± 0.3	1.2 ± 0.5	0.5 ± 0.2
B	12.0 ± 5.0	192 ± 19	N.M.

4 d old rats were treated with ACTH (50 µg on d 4 and 5) or with dexamethasone (50 µg on d 4 and 5) and sacrificed on d 6. Results are expressed in ng/g or in ng/ml, mean $\pm$ sd or range. N.M. = not measurable.

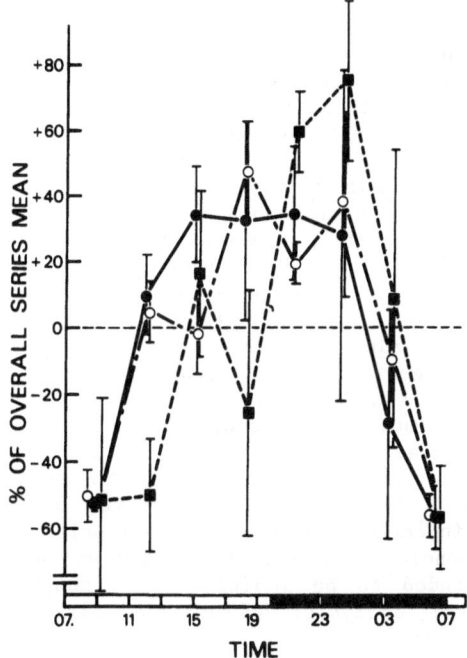

Fig. 2.   Circadian variation of pregnenolone, dehydroepiandrosterone and
           corticosterone in the rat brain.
           Measurements were performed every 3 h on groups of 3 rats. Results
           were expressed in percent of overall series mean (mean ± sem). The
           span of darkness is indicated by a black horizontal bar.

Attempts to Demonstrate the Biological Roles of Pregnenolone and Dehydroepiandrosterone in Brain

Pregnenolone and Dehydroepiandrosterone as Precursors of Sex Hormones.
Incubation of rat brain minces with ^{3}H-P yielded ^{3}H-progesterone (in addition to ^{3}H-PL). The yield of progesterone formed per 100 mg tissue was in the ∼ 1 % range in hypothalamus, amygdala, olfactory bulb and other regions of the brain, with the exception of cerebellum and the frontal cortex, where

Table 5. External Circadian Timing of P, D, and B in Rat Brain and Plasma

Site	Variable	n	% Rhythm	P	Mesor mean ± se	Amplitude mean ± se	Acrophase 95 % CL
		(light on from 7.00 to 20.00 h)					
Brain	D	24	59	< 0.001	3.5 ± 0.2	1.5 ± 0.3	−297° (−275, −319)
(ng/g)	P	24	45	0.002	19.3 ± 1.7	9.6 ± 2.4	−280° (−250, −311)
	B	21	55	0.001	14.4 ± 1.4	8.7 ± 1.9	−326° (−299, −354)
Plasma	D	23	52	0.001	0.36 ± 0.02	0.15 ± 0.03	−278° (−250, −306)
(ng/ml)	P	21	29	0.047	1.01 ± 0.08	0.28 ± 0.11	−311° (−259, −3)
	B	20	69	< 0.001	83.8 ± 6.6	57.15 ± 9.3	−325° (−305, −345)

n = number of rats; individual cosinor. For acrophase, 360° ≡ 24 h,
0° ≡ 0.00 h.

373

very little metabolism was observed (Clarke et al., 1984). Similarly, ^{3}H-D gave small amounts of ^{3}H-androstenedione and ^{3}H-DL in hypothalamic slices, but no radioactive 5-androstene-3β,17β-diol or testosterone were found. PL was also formed by the rat glioma C6 line (5-8 % per 10[7] cells per 24 h), whereas no such metabolite could be demonstrated in L 929 mouse fibroblasts.

Dehydroepiandrosterone Inhibits Aggressive Behavior of Castrated Male Mice. Group-housed triads of castrated male mice attack lactating female intruders (Haug and Mandel, 1978). Previous reports indicated that testosterone or estradiol inhibit this aggressive behavior. Male mice of the Swiss strain were castrated or sham operated (intact males) at the age of 7-8 weeks. After a 5-6 week interval, they received daily subcutaneous injections of D or vehicle alone for 15 d. Lactating female intruders were then introduced in the cages, and the latency and number of attacks were recorded during 15 min. D inhibited this aggressive behavior of castrated males in a dose dependent manner (Table 6) (Haug et al., 1983). D was also fully effective when infused through minipumps at the daily dose of 83 μg. The transformation of injected D into brain testosterone was very small; the concentration of brain T in D injected castrated male mice was almost identical to that of uninjected controls, and well below the level in intact males (Table 7). Pregnenolone infused in equimolar amount to the fully effective dose of D tended to be inhibitory, although not significantly according to statistical analysis; DS or the estrogenic 5-androstene-3β,17β-diol were not active.

Pregnenolone and Dehydroepiandrosterone Vary in Limbic Structures of Male Rats Exposed to Female Siblings. Young adult male rats (M) were exposed for several days to the scent of cycling females in absence of visual or tactile contact (M/F), and compared to males similarly exposed to other males (M/M). [P] was highest in the olfactory bulbs of M/M (Corpéchot et al., 1986) compared to other regions of the limbic system (Figure 3). It decreased more than 3 fold in M/F olfactory tubercles, amygdala, or hypothalamus (as well as in plasma, adrenals, and spleen, which was taken as a representative non-endocrine organ). In comparison with M/M levels, [D] selectively increased in the hypothalamus of M/F. These results demonstrate the selective and opposite changes of [P] and [D] upon heterosexual exposure. Therefore, they suggest regulatory mechanisms peculiar to specific parts of the brain which are not correlated with the hormone levels in blood, and could be part of the response to still undefined, presumably olfactory signals emitted by animals of the other sex.

Table 6. Effect of D upon the Aggressive Behavior of Castrated Males towards Lactating Female Intruders

| | Attacks | |
Daily dose (μg)	Number	Latency (sec)
0	50.4 + 6.9	27.1 + 12.5
20	36.1 + 7.7[+]	23.4 + 3.9[+]
40	26.1 + 5.6[+]	29.1 + 8.8[+]
80	4.0 + 1.7°	508.1 + 153[+]

Male mice of the Swiss strain were castrated at the age of 7-8 weeks and housed in triads. After a 5-6 week interval, groups of 7 triads were treated with each dose of D or vehicle alone for 15 d. A lactating female was introduced in the cage on the last day of treatment. Results are expressed as mean + sem. [+]Significantly different from control at p < 0.05. °id at p < 0.01.

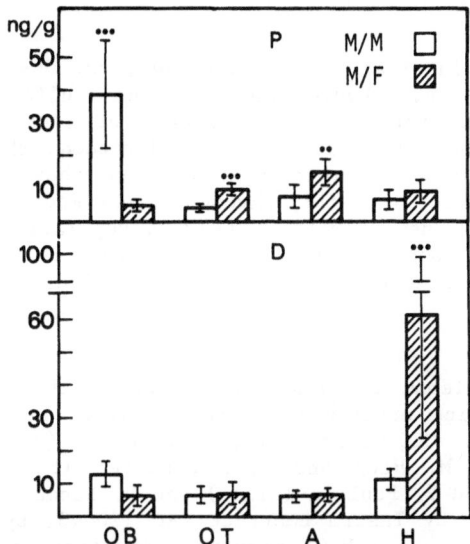

Fig. 3.   Concentrations of P and D in the brain limbic system of male rats
exposed to males (M/M) or to females (M/F).
There were 2 experimental series with 6 and 8 rats of each
category, and the 14 values obtained for [P] and [D] in olfactory
bulbs (OB), olfactory tubercles (OT), amygdalas (A) and
hypothalamus (H) were pooled. Bars indicate the standard error.
The nonparametric test of Wilcoxon (W) was applied for comparing
M/M to M/F. °°W test; $p < 0.05$; °°°W test; $p < 0.01$.

Changes in the Firing Rate of Single Neurons Induced by Direct Application of D, DS and PS. The experiments were performed in adult female
guinea pigs (400 g), anesthetized with urethane. When D, DS and PS were
applied iontophoretically or by pressure to neurons in the septo-preoptic
area, an excitatory effect was observed (Carette and Poulain, 1984). DS and
PS, applied on the same neuron, always produced a similar effect. When DS
and D were tested, some neurons were excited by both steroids whereas others
responded to DS only. DS, PS and D-induced responses displayed a short
latency in onset and offset, suggesting an action at the membrane level.

Table 7. D, DS, and Testosterone (T) in Male Swiss Mice Brain

	D	DS	T
Intact	1.8 ± 0.9 (4)	-	3.1 ± 1.1 (4)[+]
Castrated :			
untreated	2.3 ± 1.9 (7)	5.4 ± 4.5 (8)	0.04 ± 0.02 (6)
treated with D			
(80 µg/d x 15 d)	45.5 ± 14.3 (8)[+]	8.1 ± 2.6 (2)	0.13 ± 0.06 (8)[+]

Male mice of the Swiss strain, either intact or castrated at the age of
7 weeks, were thereafter housed in triads. After a 5 week interval, 27
castrated triads were treated with 80 µg D daily for 15 d, or with vehicle
alone. They were killed by decapitation 2 h after the last injection. Groups
of 9 brains were quickly removed, homogenized and processed for RIA of D(S)
and T. Results are expressed in ng/g of tissue, mean ± sd (n).
[+] Significantly different from untreated control at $p < 0.01$.

CONCLUSION

Contrary to the classical steroid hormones, the 3β-hydroxy-Δ5-steroids pregnenolone (sulfate) and dehydroepiandrosterone (sulfate) are accumulated in rat brain in concentrations far in excess of those found in plasma. Although their functions are still poorly understood, they may affect the brain by metabolism to sex steroid hormones, and they may be functionally related to sexual behavior (possibly through direct modulation of the firing rates of neurons). This new chapter of steroid regulation of the brain may not be particular to rodent, since we have preliminary evidence for the presence of P and D in other mammalian species, including primates.

ACKNOWLEDGEMENTS

We thank R. Guillemin for its hospitality at the Salk Institute (La Jolla, CA). Contributing experiments were performed by J. Sjövall and M. Axelson (Stockholm) for GLC-MS analysis, J. Pedersen (Oslo), P. Brazeau (La Jolla and Montréal), M. Haug and M.L. Schlegel (Strasbourg), J. Roffi (Orsay), B. Carette and P. Poulain (Lille) and D. Lapous (Paris). This work was supported in part by Inserm contract PRC 120048, by the MIR contract n°83.C.0924, by the R.J. and H.C. Kleberg Foundation and by the Philippe Foundation.

REFERENCES

Allen, C., and Kendall, J.W., 1967, Maturation of the circadian rhythm of plasma corticosterone in the rat, Endocrinology, 80:926.

Baulieu, E.E., 1981, Steroid hormones in the brain : several mechanisms ? in: "Steroid Hormone Regulation of the Brain," K. Fuxe, J.A. Gustafsson, and L. Wetterberg, eds., pp. 3-14, Pergamon Press, Oxford.

Blondeau, J.P., and Baulieu, E.E., 1984, Progesterone receptor characterized by photoaffinity labelling in the plasma membrane of Xenopus laevis oocytes, Biochem. J., 219:785.

Butte, J.C., Kakihana, R., and Noble, E.P., 1976, Circadian rhythm of corticosterone levels in rat brain, J. Endocrinol., 68:235.

Carette, B., Barry, J., Linkie, D., Ferin, M., Mester, J., and Baulieu, E.E., 1979, Effets de "l'oestradiol-7α-acide butyrique" au niveau des cellules hypothalamiques, C.R. Acad. Sci. Paris, 288:631.

Carette, B., and Poulain, P., (1984) Excitatory effect of dehydroepiandrosterone, its sulfate ester and pregnenolone sulfate, applied by iontophoresis and pressure, on single neurones in the septo-preoptic area of the guinea pig, Neuroscience Letters, 45:205.

Celotti, F., Massa, R., and Martini, L., 1979, Metabolism of sex steroids in the central nervous system, in: "Endocrinology," L.J. de Groot, G.F. Cahill Jr., E. Steinberger, and A.I. Winegrad, eds., Vol. 1, pp. 41-53, Grune and Stratton, New York.

Clarke, C., Groyer, A., Baulieu, E.E., and Robel, P., 1984, Pregnenolone metabolism in the rat brain, 7th International Congress of Endocrinology, Québec, Canada, abstract n°395, Excerpta Medica Inter. Congr. Series 652, Amsterdam.

Corbier, P., and Roffi, J., 1978a, Increased adrenocortical activity in the newborn rat, Biol. Neonate., 33:72.

Corbier, P., and Roffi, J., 1978b, Pituitary adrenocortical response to stress during the first day of post-natal life in the rat, Biol. Neonate, 34:105.

Corpéchot, C., Robel, P., Axelson, M., Sjövall, J., and Baulieu, E.E., 1981, Characterization and measurement of dehydroepiandrosterone sulfate in rat brain, Proc. Natl. Acad. Sci. USA, 78:4704.

Corpéchot, C., Synguelakis, M., Talha, S., Axelson, M., Sjövall, J., Vihko, R., Baulieu, E.E., and Robel, P., 1983, Pregnenolone and its sulfate ester in the rat brain, Brain Res., 270:119.

Corpéchot, C., Leclerc, P., Baulieu, E.E., and Brazeau, P., 1986, Neurosteroids : regulatory mechanisms in male rat brain during hetero-sexual exposure, Steroids, in press.

Finidori-Lepicard, J., Schorderet-Slatkine, S., Hanoune, J., and Baulieu, E.E., 1981, Steroid hormone as regulatory agent of adenylate cyclase. Inhibition by progesterone of the membrane bound enzyme in Xenopus laevis oocytes, Nature, 292:255.

Fishman, J., 1982, Biochemical mechanism of aromatization, Cancer Res., 42:3277.

Fuxe, K., Gustafsson, J.A., and Wetterberg, L., 1981, "Steroid Hormone Regulation of the Brain", Pergamon Press, Oxford.

Haug, M., and Mandel, P., 1978, Attack directed by groups of castrated male mice towards lactating or non-lactating intruders : a urine-dependent phenomenon, Physiol. Behav. 21:549.

Haug, M., Spetz, J.F., Schlegel, M.L., and Robel, P., 1983, La déhydro-épiandrostérone inhibe le comportement agressif de souris mâles cas-trées, C.R. Acad. Sci. Paris, 296:975.

Hochberg, R., Bandy, L., Ponticorvo, L., and Lieberman, S., 1977, Detection in bovine adrenal cortex of a lipoidal substance that yields pregne-nolone upon treatment with alkali, Proc. Natl. Acad. Sci. USA, 74:941.

Jaffe, R.B., 1969, Testosterone metabolism in target tissues : hypothalamic and pituitary tissues of the adult rat and human fetus, and the imma-ture rat epiphysis, Steroids, 14:483.

Kishimoto, Y., 1973, Fatty acid esters of testosterone in rat brain : identification, distribution and some properties of enzymes which synthesize and hydrolyze the esters, Arch. Biochem. Biophys., 159:528.

McEwen, B.S., Biegon, A., Davis, P.G., Krey, L.C., Luine, V.N., McGinnis, M.Y., Paden, C.M., Parsons, B., and Rainbow, J.C., 1982, Steroid hormones : humoral signals which alter brain cell properties and functions, Rec. Progr. Hormone Res., 38:41.

Naftolin, F., Ryan, K.J., Davies, I.J., Reddy, V.V., Flores, F., Petro, Z., Kuhn, M., White, R.Y., Taoka, Y., and Wolin, L., 1975, The formation of estrogens by central neuroendocrine tissues, Rec. Progr. Hormone Res., 31:295.

Paul, S.M., and Axelrod, J., 1977, Catechol estrogen-forming enzyme of brain : demonstration of a cytochrome P450 monooxygenase, Endocrinology, 101:1604.

Poulain, P., and Carette, B., 1981, Changes in the firing rate of single preoptic-septal neurons induced by direct application of natural and conjugated steroids, in: "Steroid Hormone Regulation of the Brain," K. Fuxe, J.A. Gustafsson, and L. Wetterberg, eds., pp. 191-201, Pergamon Press, Oxford.

Sakly, M., and Koch, B., 1983, Ontogenetical variations of transcortin modulate glucocorticoid receptor function and corticotropic activity in the pituitary gland, Horm. Metab. Res., 15:92.

Seymour-Munn, K., and Adams, J., 1983, Estrogenic effects of 5-androstene-3β,17β-diol at physiological concentrations and its possible impli-cation in the etiology of breast cancer, Endocrinology, 112:486.

Stumpf, W.D., and Grant, L.D., 1975, "Anatomical Neuroendocrinology," Springer Verlag, Basel.

Warembourg, M., 1978, Distribution of steroid-concentrating cells in the central nervous system, in: "Biologie Cellulaire des Processus Neuro-sécrétoires Hypothalamiques", J.D. Vincent, and C. Kordon, eds., pp. 221-237, Editions du CNRS, Paris.

ION CHANNELS AND THE CONTROL OF SECRETION

IN NORMAL ANTERIOR PITUITARY CELLS

W.T. Mason, R.J. Bicknell, P. Cobbett, D.W. Waring and
C.D. Ingram
Department of Neuroendocrinology
A.F.R.C. Institute of Animal Physiology
Babraham, Cambridge   CB2   4AT   United Kingdom

INTRODUCTION

Secretion of the anterior pituitary hormones is controlled by peptides and amines synthesized and released by discrete neurones in the hypothalamus. The action of these factors in regulating hormone release is believed to involve changes in the electrical activity of the specific target cells (Douglas and Taraskevich, 1985). Because of the exceptional variety of different hormone-containing cell types in the anterior pituitary gland, most knowledge about the membrane properties which might be important in the regulation of hormone secretion from these cells has been derived from work using pituitary tumour cell lines as models.

Our work has concentrated on the development of methods which allow the study of single, identified, non-neoplastic pituitary cells using electrophysiological techniques, and to attempt to relate these properties to the control of secretion in these cells. This approach has enabled us to ask several questions. First, what are the electrical properties of pituitary cell membranes and what ionic channels underlie these properties? Second, how do hypophysiotrophic factors secreted from the hypothalamus interact with pituitary cells to modulate release of hormone, i.e. what membrane events are triggered by binding of these factors to their receptors? Third, how do the membrane properties of these cells compare with their neoplastic counterparts?

In this paper, we shall present recent data on bovine lactotrophs and ovine gonadotrophs which respectively secrete prolactin (PRL) and the gonadotrophins luteinizing and follicle stimulating hormone (LH and FSH). We shall also discuss new approaches we are using to understand and directly observe the mechanisms by which peptide-receptor interactions in particular may activate ion channels in the pituitary cell membrane and evoke hormone secretion. These observations are providing us with new information about the regulatory processes which occur at the single cell level to modulate the secretory process.

Preparation of gonadotrophs and lactotrophs

Gonadotrophs normally constitute only about 10% of the pars distalis, but it was recently discovered that, in sheep, this cell type is the only hormone-containing cell in the pars tuberalis (PT; Gross, 1984). The isolated ovine PT secretes LH in response to repeated challenges with GnRH (Gross, Turgeon and Waring, 1984), and, as such, the gonadotrophs in this tissue appear to respond identically to those in pars distalis. The ovine PT surrounds the pituitary stalk and can be dissected with only a small amount of adhering neural tissue, thus avoiding contamination by other hormone-containing cells. We have developed techniques for dispersion of PT using trypsin and have successfully maintained these cells in culture for up to 4 weeks (Mason and Waring, 1985). Fig. 1A shows a typical cultured gonadotroph, immunocytochemically stained for LH. The refractile, granular appearance of these cells allows them to be distinguished from non-endocrine cells in the culture.

Lactotrophs have been obtained from bovine pars distalis. We have prepared enriched preparations of lactotrophs using collagenase dispersion, followed by separation of the dispersed cells on sterile isotonic Percoll density gradients. At the interface between density layers of 1.062 g/ml and 1.074 g/ml is found a fraction of relatively large cells (12-15 μm diameter). About 70% of these cells stain immunocytochemically for PRL (Fig. 1D) and the number of fibroblastic cells is greatly reduced from the

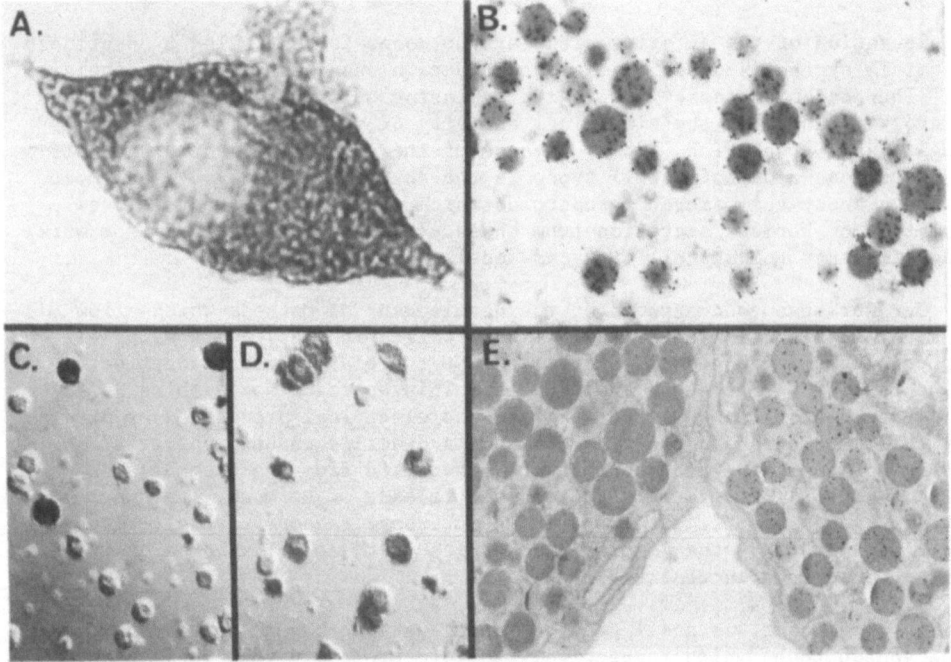

Fig. 1. (A) Light micrograph of ovine gonadotroph stained immunocytochemically for LH. (B) Electron micrograph of secretory granules
in ovine PT gonadotroph stained with immunogold technique for
LH. (C,D) Light micrographs of bovine pituitary cells in lacto-
troph-poor (C) and lactotroph-rich (D) fractions obtained by
separation of pars distalis cells, both stained for PRL. Note
only about 15% of cells in (C) are PRL-positive whereas a very
high proportion of those in (D) react for PRL, indicating a
strong enrichment of PRL cells in the fraction used for electrophysiological studies. (E) Electron micrograph of detail of two
cells stained for PRL with the immunogold technique. Note that
granules on the cell on the right are all positive for PRL,
whereas granules in the cell on the left do not stain for PRL.

initial population, enabling maintenance of these cells in culture for several weeks without overgrowth by dividing cells.

As well as using light microscopic immunocytochemistry, these gonado-troph- and lactotroph-enriched cultures have been further characterised using techniques for detection of hormone content at the ultrastructural level. With Dr. Peter Wooding of this Institute, we have used specific antibodies which recognise LH or PRL and visualised these with colloidal gold particles linked to a second antibody or protein A. Fig. 1B shows an ovine PT gonadotroph stained in this way for LH, where it can be seen that every granule is positive for LH. Likewise, bovine lactotroph fractions stained for PRL show a high proportion of PRL-containing cells. In addi-tion, studies using different diameter gold particles to differentiate staining for PRL and growth hormone (GH) have suggested that somatotrophs are the other main cell type in this fraction. However, unlike the neo-plastic GH3 and GH4 lines which synthesize and secrete both PRL and GH, none of the normal pituitary cells stain postitively for both these hormones.

Electrophysiological techniques for single cell recording

Rapid technological progress has given the experimentalist several new and important techniques for studying ionic movements across biological membranes, and we are applying many of these to study the electrical pro-perties of pituitary cells. In this section, we shall explain briefly the basis of these approaches.

The most conventional approach is intracellular recording (Fig. 2A) where, under visual control using a phase contast microscope, a very high resistance (100-300 MΩ) glass pipette is manipulated into the cell of interest. This technique allows the d.c. voltage of the cell to be mea-sured and approximate measurements of cell resistance to be made. However, in situations where isolated cells are available, this technique is rapidly becoming replaced by the 'patch clamp' technique (Hamill, Marty, Neher, Sakmann and Sigworth, 1981) which allows both voltage-clamp and current-clamp measurements to be made. To achieve this, a blunt (low resistance, i.e. 3-20 MΩ) fire-polished pipette is pressed gently against the cell mem-brane (Fig. 2B) and light suction applied (Fig. 2C). This generally results in formation of a very high resistance seal between the glass and the membrane. This seal can be 1000-5000 times the resistance of the elec-trode itself, and because the electrode resistance is so low, the recording quality is very good, and has a high frequency response.

In this mode, the background noise is exceptionally low and the mem-brane patch is essentially voltage clamped, since any potential applied to the inside of the pipette is almost totally seen to drop across the mem-brane patch. In addition, current flow due to the opening of single ion channels can be recorded. These currents are of the order of 0.5-20 pA, depending on the channel type. Variation of the driving force across the membrane patch can be effected without changing the voltage of the whole cell, and thus the conductance of the single channels can be measured. Their frequencies of opening/closing can also be observed, and rate con-stants derived for their kinetic behaviour. One can thus observe a great deal about the behaviour of isolated channels in the membrane patch which remains attached to the cell. Finally, because the pipette to membrane seal is so tight, peptides and other substances of interest can be applied to the whole cell and one can be certain these do not reach the isolated membrane patch, at least directly.

This approach is of great value in its own right, but the technique allows other types of measurements to be made. For example, if additional,

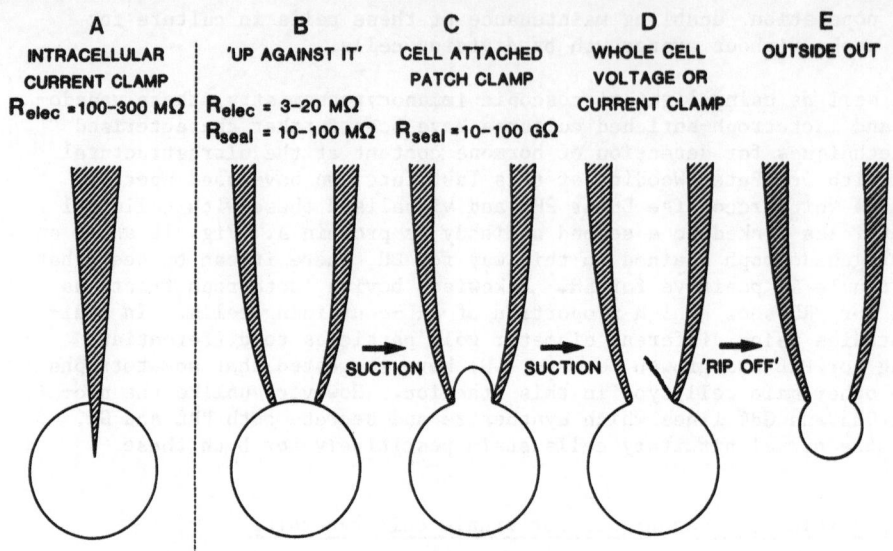

A	B	C	D	E
INTRACELLULAR CURRENT CLAMP	'UP AGAINST IT'	CELL ATTACHED PATCH CLAMP	WHOLE CELL VOLTAGE OR CURRENT CLAMP	OUTSIDE OUT
R_{elec} =100-300 MΩ	R_{elec} = 3-20 MΩ R_{seal} = 10-100 MΩ	R_{seal} =10-100 GΩ		

SUCTION   SUCTION   'RIP OFF'

Fig. 2. Diagrammatic representation of recording modes used for study
of pituitary cell electrophysiology. (A) Conventional intracellular recording where the high resistance microelectrode is
inside the cell. In patch recording techniques, however, the low
resistance electrode is pushed gently against the cell (B).
Upon suction, a high resistance seal is formed, making possible
voltage clamp recordings of single channel currents flowing
while the cell is left intact (C). In (D), further suction is
applied to break the patch of membrane, allowing the whole cell
current and voltage to be recorded under conditions where the
cell interior is perfused. If the microelectrode is withdrawn
(E), an outside/out patch is formed, permitting single channel
currents to be recorded while internal voltage is manipulated.

stronger suction is applied to the pipette, the membrane patch is disrupted
while the membrane/electrode seal remains intact, resulting in formation of
a low resistance pathway to the cell interior (Fig. 2D). This allows not
only measurement of 'whole cell' ionic currents (the macroscopic sum of all
single channel currents for that cell) to be made, but also a simple change
of recording mode allows a direct intracellular voltage measurement under
conditions where all ionic currents are clamped. Because of the ease with
which it is possible to change between these recording modes, ionic currents can be measured under voltage clamp and the consequences of this
current on cell voltage can be measured under current clamp. A final consequence of this 'whole cell' recording is that the cytoplasm rapidly
equilibrates with the pipette solution, making it possible to regulate the
ionic content of the cell interior. For the future, effects of putative
internal messengers on membrane properties can be examined in this way.

After recording, the position of the cell can be marked on the culture dish and its identity determined by immunocytochemistry.

Lactotroph electrophysiology

Basic properties. Conventional intracellular recording (as Fig. 2A)
from lactotrophs has shown them to have high resting potentials and input

resistance. Exact values for these characteristics appear to depend in part on the technique of measurement employed. Intracellular recording gives resting membrane potentials in the range of -70 to -90 mV and input resistances of 100-300 MΩ. Using whole cell recording in the current clamp mode, membrane potentials more negative than -85 mV have been measured and, because of the tight membrane-to-electrode seal, input resistances of 5-20 GΩ have been observed (Figure 3). K is the major permeant ion at normal resting potential, and elevation of K outside the cell causes a strong depolarisation, although spiking is not observed to accompany this depolarisation. Injection of depolarising current causes the cell to fire regenerative action potentials (APs), but only about 50% of these fire spontaneous (not current evoked) APs (Ingram and Mason, 1985).

Subthreshold and suprathreshold electrical activity. The dynamic electrical activity of lactotrophs can be divided into two categories, and it is useful to make this distinction at an early stage. The first is supra-threshold activity which includes APs, and the second is sub-threshold activity which appears in lactotrophs to be an important feature in AP generation, much as an excitatory postsynaptic potential evokes a spike in neurones. In the lactotroph, therefore, in addition to APs, a high degree of spontaneous voltage fluctuation is observed, and in many instances these fluctuations are large enough to exceed threshold and cause an AP to be fired. In normal recording medium containing 125 mM Na, 5 mM K and 5 mM CaCl both types of activity are reduced by divalent cations such as Co and Mn which block Ca channel currents. Using the whole cell current clamp mode it is possible to observe both types of activity (Figure 3C).

Ionic currents in lactotrophs. Since the voltage events observed in lactotrophs appear to be similar to that of neurones, it might be expected that Na, Ca and K ions would play some role in lactotroph membrane physiology. We can make direct measurements of the individual contribution of these ions by employing the whole cell voltage clamp technique, which allows us to examine the time- and voltage-dependent currents which flow across the lactotroph cell membrane. Figure 4 shows a typical family of records demonstrating that, under defined ionic conditions, lactotrophs

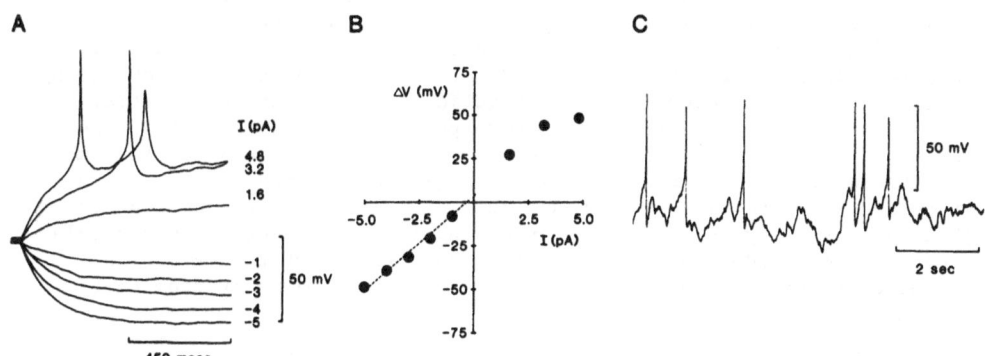

Figure 3. Current clamp recording from a bovine lactotroph, where the internal (pipette) solution contained 140 mM KCl and the external solution 140 mM NaCl and 2 mM KCl. Injection of depolarising current led to the appearance of action potentials (A). The linear portion of the steady state current to voltage relationship (B) gives an input resistance of 11.6 GΩ for this cell. Spontaneous activity consists of large subthreshold voltage fluctuations and, when these exceed the threshold, spontaneous Na spikes (C).

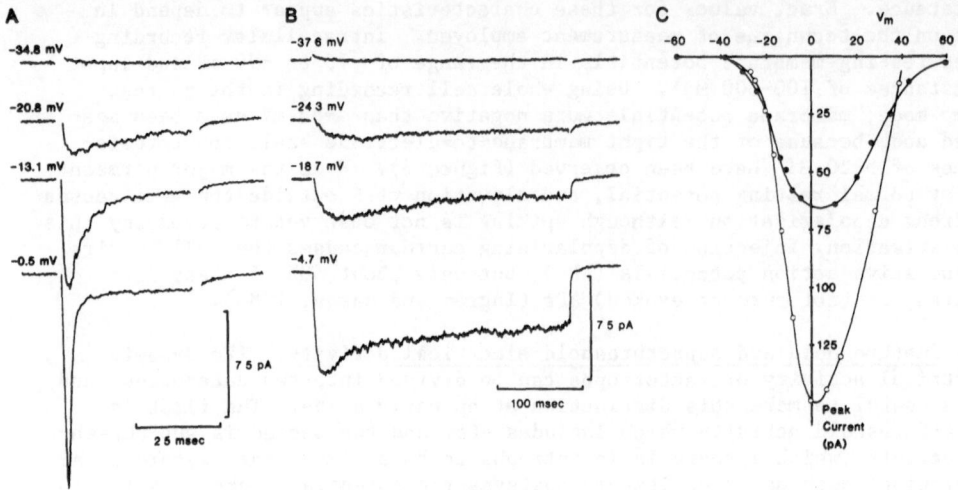

Fig. 4. Isolated ionic currents recorded from a voltage clamped lactotroph, under conditions where the internal medium contained Cs
ions and the external tetraethylammonium ions to block outward
current. (A) Na currents activated by depolarising voltages
noted on each trace, from a holding potential of -60 mV in the
absence of external Ca. (B) Following block of Na current with
TTX, Ba currents were recorded when external medium contained 25
mM Ba. Voltage clamp potential noted on each record, also from
-60 mV. Note the non-inactivating current evoked under these
conditions. (C) Current-voltage relation for Na (open circles)
and Ba (filled circles) currents.

exhibit different types of inward currents during a voltage step. When the
medium contains 125 mM Na but no Ca (Fig. 4A), a rapidly activating Na current is observed which is blocked by TTX, but this current differs from
that found in neurones, in several respects. First, it has a relatively
high threshold for activation (above -25 mV) and second, after inactivating
at these depolarised potentials, reactivation at the membrane potential at
which spiking is observed (-40 mV or more positive) has a very long time
constant. That is to say that if one attempts to evoke the Na current more
often than about every 1-2 sec, it is strongly inactivated on successive
test pulses. In practice, this means that the lactotroph cannot support
high frequency Na spikes at a rate exceeding 1-2 Hz without an appreciable
decrease in amplitude due to progressive inactivation of the Na current.
The characteristics of this inactivation thus differ markedly from that
found in most neurones.

    Another major inward current carrier detected in the lactotroph mem-
brane is that flowing through Ca channels. To date, this current has
certainly proved the most difficult to study because of its lability. An
example of this is shown in Fig. 4B, where 25 mM Ba was used as the charge
carrier for the Ca channel to maximise current flow. When a whole cell
recording is made from a lactotroph in the presence of Ba, a large inward
current (200-500 pA) is observed to be activated by voltage. The initial
rapid portion of this current inactivates very quickly with a decay time of
about 7-10 msec, but a second portion of the current shows little inactivation over a time course of 1 sec, but rather decays very slowly over tens
of seconds. Initially this non-inactivating, slowly decaying current is

large and responsible for gating of the majority of Ca current entering the cell during depolarisation, but as the inside of the cell equilibrates with the solution in the recording pipette over the course of a few minutes, this current runs down. The run-down appears not to be affected by changes in the internal free Ca (or Ba) concentration or the frequency and amplitude of applied potentials. Such a result suggests that this lactotroph Ca current may be mediated by some soluble internal factor(s) which is dialysed upon breaking through into the whole cell recording mode.

Lactotrophs also have slow, voltage-activated outward K currents (Fig. 5A) that probably promote repolarisation of the cell potential after an AP, since they activate at more depolarised potentials than the Na current. In studies where single ion channel activity has been measured, at least two main types of K channel are found. One of these is of small conductance (20-40 pS), whereas the other is of very large conductance, about 100-130 pS (Fig. 5B) and appears to be activated by internal Ca. This is probably the Ca-activated K conductance found in most other cell types examined. Interestingly, when a lactotroph is exposed to sub-micromolar concentrations of TRH and recorded in the non-invasive cell-attached mode, the activity of this large conductance K channel increases dramatically, consistent with a rise in internal Ca. In the whole cell mode, if the electrode is filled with CsCl or the bath contains a high concentration of tetraethylammonium ions, all K current is effectively blocked, making it possible to observe isolated inward currents such as those carried by Na and Ca.

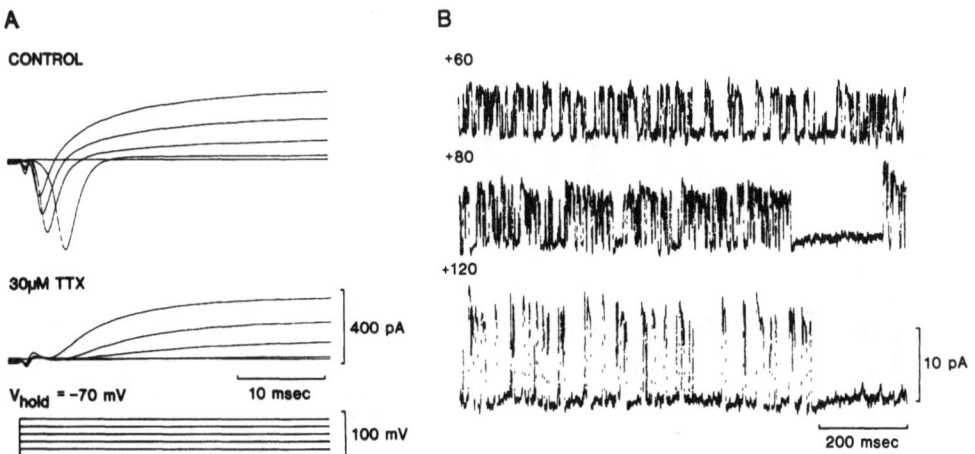

Fig. 5. Lactotroph K ion currents recorded in whole cell (A) and single channel, cell attached patch (B). In (A), the internal perfusion medium contained 140 K, permitting observation of outward K currents in addition to inward Na currents. In the control, external medium contained 125 Na as the major charge carrier and both an early inward Na current and a slowly developing, non-inactivating outward K current were observed. Addition of TTX blocked the voltage-dependent Na current. Values of depolarisation are given in the bottom trace. (B) Single K currents can also be observed with a patch recording. The single channel currents shown in (B) increased in amplitude with depolarisation of the patch by the potential shown and had a conductance of 110 pS. In cell attached mode, they showed a reversal potential about 30 mV more negative than the resting potential. This large conductance K channel is probably the Ca-dependent K channel which is activated after TRH action.

TRH action and lactotroph ion currents. The lability of Ca currents described above has caused some problems in examining the action of hypo- thalamic peptides such as TRH. When TRH is applied to a lactotroph during a conventional intracellular recording, a rapid hyperpolarisation is in- duced which is followed by a small depolarisation of 2-5 mV and occasional- ly by increased AP generation (Ingram and Mason, 1985). However, if the same experiment is carried out using the whole cell voltage clamp mode, no effects of TRH are observed. This further suggests that TRH actions might be mediated through an internal messenger which is lost through dialysis of the cell during a whole cell recording. One alternative approach, there- fore, has been to employ the non-invasive method of single channel recor- ding from the cell attached membrane patch to examine Ca channel activity.

We have made cell attached patch recordings with pipettes containing 95 mM Ba to increase the amplitude of single channel activity flowing through Ca channels. After initial patch formation only very occasional channel openings attributable to current movement through Ca channels are observed (Fig. 6). However, if TRH is applied to the medium bathing the cell, within a matter of 10-60 sec, greatly increased multiple Ca channel activity is observed in the membrane patch. Under these conditions, the activated ion channels have a single channel conductance of about 10 pS and a single channel amplitude of about 0.5-0.8 pA at the resting membrane potential. In terms of ion flow, 1 pA of current is equivalent to movement of about 3 million Ca ions/sec. Interestingly, these Ca channels show a high level of activity near the normal resting potential, and do not re- quire a large depolarisation for activation. Furthermore, if the patch

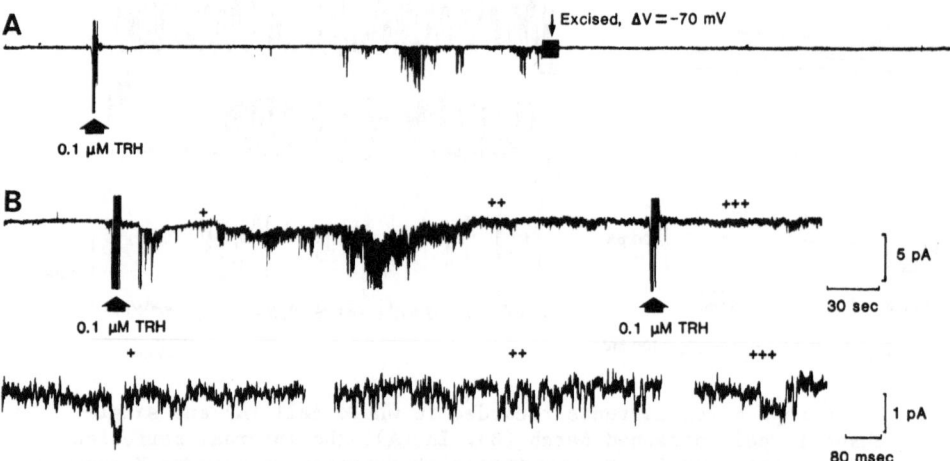

Fig. 6.   Single channel recordings of TRH-activated ion channels recorded from lactrotrophs using the cell attached patch tech- nique. In (A) and (B), the pipette medium contained 95 mM Ba as the major charge carrier for Ca channel detection. (A) TRH was added as indicated by the arrow and evoked openings of single Ca ion channels seen as downward current deflections. When the patch was excised and a -70 mV holding potential applied, no similar currents were observed, indicating loss of a soluble agent required to activate the Ca channels. (B) TRH addition elicited Ca channels when added at the arrows. The symbols mark segments of the record which can be seen below at higher gain and time base, revealing the single channel currents of 1 pA or less.

of membrane is excised from the cell such that the cytoplasmic face is exposed to the bath medium, most single channel activity due to Ca channels is abolished (Fig. 6A). This indicates that the particular type of Ca channel activated by TRH may be regulated by an internal messenger which is produced after TRH interaction with membrane receptors, and that TRH only indirectly activates lactotroph Ca channels. We have termed this an agonist-dependent Ca channel.

What are the consequences of activation of this agonist-dependent Ca channel in terms of other electrical properties of the cell? The opening of single Ca channels produces voltage 'bumps' observed during either conventional or whole cell intracellular recordings under current clamp conditions. These small depolarising potentials may sum and trigger TTX-sensitive APs. The voltage change during these APs would, in turn, activate a voltage-dependent Ca conductance, similar to that observed to persist during whole cell recording of Ca currents. However, in terms of the total Ca influx, TRH activation of agonist-dependent Ca channels would be expected to produce an influx 10 to 100 times greater than that which flows transiently through these voltage-activated channels. An additional consequence of the Ca entry induced by TRH may be the activation of Ca-dependent K channels. This would have a repolarising effect on membrane potential during TRH action.

Gonadotroph electrophysiology

__Basic electrical properties.__ In many respects, intracellular recordings from gonadotrophs have shown them to have electrical characteristics broadly similar to that found for the lactotrophs, particularly with respect to high membrane potential (-75 mV) and resistance (300 MΩ). Passage of depolarising current evokes only a single spike which is insensitive to TTX but which is completely and reversibly abolished by divalent cations which block Ca channels. The failure to evoke a TTX-dependent AP in gonadotrophs could suggest that cells in this preparation may lack conventional voltage dependent Na currents. Consistent with this, gonadotrophs were never found to fire regenerative APs spontaneously (Mason and Waring, 1985).

__GnRH effects on electrical activity of gonadotrophs.__ Ovine PT gonadotrophs have been recorded from using conventional intracellular techniques (Mason and Waring, 1985). During such recordings GnRH was applied at concentrations from 1-80 nM, the upper value being maximal for eliciting LH secretion. The major finding in these studies has been that GnRH does not evoke APs in gonadotrophs, but does bring about a large increase in the voltage 'noise' of the gonadotrophs, i.e., a large rise in the number of spontaneous voltage fluctuations. This rise can be quantitated in terms of statistical variance of membrane potential, calculated on a point to point basis using a computer. In most instances, GnRH increased the level of voltage variance by 2 to 10 fold. This membrane voltage noise is similar to that observed for transmitter action at synapses, assumed to arise from independent openings and closings of single ion channels, and the sum of which produces a large number of spontaneously occurring summations of channels giving rise to 'noise'. In gonadotrophs, the GnRH-induced noise increase is abolished by divalent cations (Co and Mn) but increased when the cell is bathed in Ba, suggesting that ion flow through the GnRH-activated channel is strongly Ca-dependent.

Detailed statistical analysis of the frequency dependence of this noise can yield a great deal of information about the temporal properties of the channels themselves. To achieve this, records are digitised by a computer and power density spectra constructed, which give an estimate of the frequency dependence of the voltage noise. In the case of

gonadotrophs, the GnRH activated channels behave as a single population
with a mean lifetime of about 40 msec. Since this value is much greater
than the membrane time constant of 5 msec or so, it provides a good estimate of at least one temporal characteristic of the GnRH-evoked channel.
The major shortfall of intracellular recording is, however, that it fails
to provide any information about electrical events with a time constant
faster than the membrane time constant. Thus, it is difficult to say
whether additional events with a much faster time constant also contribute
to the effect of GnRH. To do this, it is necessary to make high gain recordings of membrane current, where the time constant has no influence. One
way to achieve this is with the cell attached patch recording.

GnRH activation of single ion channels. Cell attached patch recordings of gonadotrophs have revealed that GnRH activates Ca channels (Figure
7). If recordings are made with GnRH in the electrode, these channels are
not activated, but application of GnRH to the bathing medium results in
rapid opening and closing of many channels which can be most conveniently
observed when the current carrier is Ba. This suggests that, like the
action of TRH on lactotrophs, GnRH probably acts through an internal messenger to open Ca channels. These channels are not observed when the
recording pipette contains only Na, or when the pipette contains 5-10 mM
Co, a Ca channel blocker, in the presence of 10-20 mM Ca. It would appear
therefore that GnRH needs to have access to the majority of the plasma membrane in order to bring about Ca channel activation. The membrane patch
recorded from contains only 1-3% of the total membrane surface and, as
such, probably contains too few receptors to maximally activate the internal messenger system.

The gonadotroph Ca channel activated by GnRH bears many similarities
to that found in lactotrophs, in that it has a small conductance of 5-9 pS,
with an amplitude of 1 pA or less near resting membrane potential. We have
estimated that a gonadotroph membrane contains about 500-1000 such channels. In addition, the gonadotroph membrane also contains large numbers of
K channels of at least two distinct conductance categories. Openings of a
large conductance channel (90-150 pS) increase dramatically following

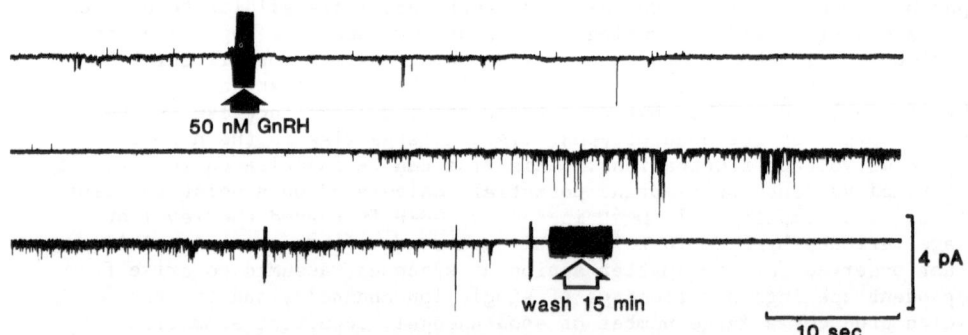

Fig. 7. Activation of inward current ion channels in the ovine PT gona-
        dotroph by GnRH. GnRH applied outside of the recording pipette
        as indicated by the arrow elicits the opening of inward current
        channels which carry predominantly Ca. If GnRH is washed away
        by use of a peristaltic pump, these ion channel openings are no
        longer observed. The fact that GnRH does not need to contact
        the ion channels in the patch pipette suggest that it functions
        to open channels through some internal messenger system.

application of GnRH to the gonadotroph, and we now know that the probability of this channel opening increases as the ionised Ca concentration on the cytoplasmic face is increased (Mason and Waring, 1986).

In contrast to the lactotroph, during cell-attached single channel recording from gonadotrophs we have never observed spontaneous electrical events which might be APs. This is consistent with a similar failure to record spontaneous or GnRH-evoked APs during conventional intracellular recording from gonadotrophs. We have also failed to find any single channel currents that carry Na and it is tempting to speculate that gonadotrophs possess mainly Ca and K channels. We are presently working to confirm this using whole cell voltage clamp recordings.

Hormone secretion from gonadotrophs and lactotrophs

Electrophysiological experiments are providing considerable detail about the membrane characteristics of these pituitary cells. It is now important to ask, what is the significance of these properties to the control of hormone secretion from the cell? We have measured hormone secretion from the cell cultures used for electrophysiological experiments, and investigated the effect of conditions used in the electrophysiological studies. It has thereby been possible to correlate the electrical responses with changes in hormone secretion.

As shown in Figure 8, PRL secretion from lactotrophs in primary culture is dependent on external Ca and is reduced by Mn which blocks Ca

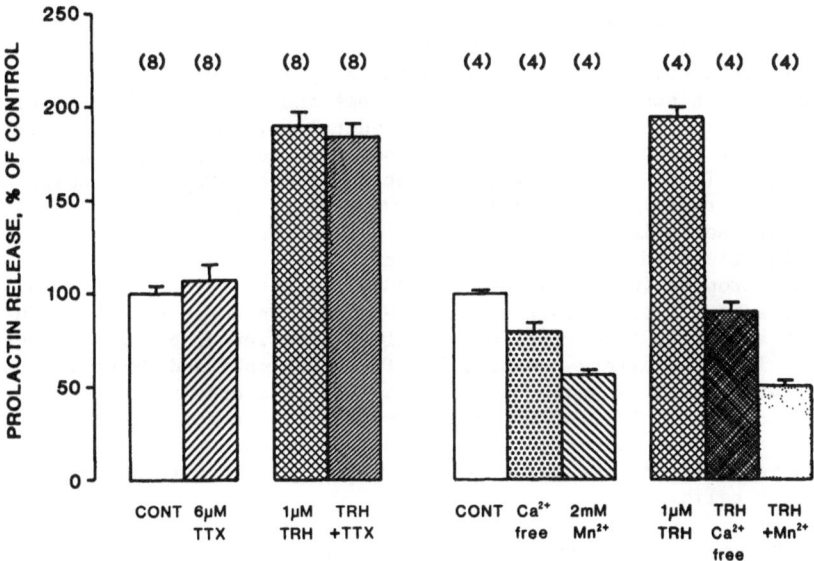

Fig. 8. Hormone secretion from bovine lactotrophs in culture, measured by radioimmunoassay of medium bathing the cells under similar conditions to those used in electrophysiological experiments. In the left two panels, it can be seen that neither basal or stimulated PRL secretion is affected by TTX which blocks voltage-dependent Na channels. However, in the right two panels it is shown that either Ca-free or Mn-containing medium inhibits PRL secretion. Thus, PRL secretion depends on Ca entry but not on Na entry via voltage-dependent Na channels responsible for the AP. Experimental numbers for each condition are given in brackets.

currents in these cells. TRH elicits PRL release from these cells, but both basal (absence of TRH) and stimulated (presence of TRH) PRL secretion are not affected by the Na channel blocker TTX. Previous studies have also indicated that TRH-induced PRL release persists in the absence of extracellular Na (Saith, Bicknell and Schofield, 1984). Likewise, secretion of gonadotrophins is strongly Ca dependent but completely unaffected by TTX (Conn and Rogers, 1980).

Thus, hormone secretion from both pituitary cell types discussed here appears to have a far stronger dependence on Ca entry than upon Na entry through conventional voltage dependent Na channels.

OVERVIEW

In this paper, we have attempted to summarise our data on the membrane properties of normal pituitary cells. We have concentrated on the lactotrophs and gonadotrophs because of the relative ease with which highly enriched fractions of these cells can be obtained. Previously, one technical problem which has impeded study of these cells has been their small size (12-20 μm), but the development of patch recording and the evolution of whole cell techniques as a derivative of the patch approach has made possible detailed study of both single and whole cell ionic currents. These same techniques should be of great use in the near future to delineate intracellular events involved in control of secretion.

One question we have addressed is the possible importance of APs in stimulus-secretion coupling in pituitary cells. Previous work with conventional intracellular recording from melanotrophs of pars intermedia and lactotrophs from the alewife fish (Douglas and Taraskevich, 1985) has suggested APs to be an important feature in the coupling process. However, a study of normal rat somatotrophs (Israel, Denef and Vincent, 1983) reported similar findings to the present work on gonadotrophs, suggesting that these cell types do not fire APs either spontaneously or in response to hypothalamic peptides. In bovine lactotrophs that do produce APs, our studies suggest that the rapidly inactivating, slowly reactivating Na currents in these cells are not able to support APs at a frequency over about 2 Hz without a significant decrease in amplitude. Furthermore, since the secretion of gonadotropins (Conn and Rogers, 1980) and PRL are unaffected by concentrations of TTX that block voltage-dependent Na currents, these currents do not appear to important for stimulus-secretion coupling. Recent work on two clonal pituitary tumour cell lines (Dubinsky and Oxford, 1984) has demonstrated that, although both are capable of secreting PRL, one has abundant Na currents (GH3), while the other (GH4) is incapable of generating voltage dependent Na currents. This corroborates our conclusion that Na currents resulting in APs do not play a major role in hormone secretion from pituitary cells.

Although some recent work on GH3 cells suggests that an initial rise in cytosolic free Ca in response to challenge by a hypothalamic peptide may be independent of external Ca (Schlegel and Wollheim, 1984), it is widely accepted that long term hormone secretion depends on entry of extracellular Ca. It is necessary therefore to examine the mechanisms by which Ca may enter the pituitary cell to support this secretion.

Previous reports have emphasized the importance of action potentials for Ca entry (see review by Douglas and Taraskevich, 1985) but several of the present observations suggest that subthreshold activity (below the level at which either Na or Ca spikes would be generated) may be a major pathway for Ca influx:

(1)   Whole cell intracellular recordings reveal a high frequency of
Ca-dependent voltage fluctuations which can be abolished with Ca channel
blockers. By comparison of the size of these fluctuations with the cell
input resistance, it can be calculated that these could result from the
opening and closing of single ion channels of about 1 pA amplitude or
less.
(2)   Cell attached patch recordings show that TRH and GnRH act, at
concentrations where secretion of respective hormones is stimulated, to
increase the probability of opening of Ca channels. Before exposure to
these peptides, only voltage-dependent channels appear to be present at
strongly depolarised potentials (around 0 mV), whereas following challenge
with the peptide agonist, many channels are activated, even at the resting
membrane potential for the cell.
(3)   Intracellular recordings show that these same peptides act on the
pituitary cells to dramatically increase the level of voltage fluctuations
without any large change in the cell potential. These channels appear,
then, to be a class of Ca channel which is agonist-activated near resting
membrane potential but which may also be voltage activated at depolarised
potentials.

One can calculate that such sustained, subthreshold Ca channel fluctuations at normal resting potential would contribute more than 97% of the
total Ca influx in a pituitary cell which fired APs at a rate of 2 Hz,
assuming that the voltage dependent Ca channels opened maximally at the
peak of the AP. This calculation is probably an overestimate, in fact,
since we know from statistical analysis of Ca channel currents in the whole
cell mode that only about 40% of the Ca channels open under these conditions. Nevertheless, subthreshold activity in the form of small depolarising potentials activated by peptide agonists must contribute significantly
to Ca influx, and may thus be the primary mechanism of membrane action of
hypophysiotrophic factors.

Finally, are there any advantages to be gained in the study of non-
neoplastic pituitary tumour cells? One obvious gain is that stable, differentiated neoplastic cell lines are not readily available for the
different pituitary hormones found in the anterior pituitary. Although
such cell lines have proved to be a valuable model in the case of lactotrophs, one major drawback to any conclusions based on a study of these
cells is that they individually synthesize both PRL and GH. This was
concluded because serial cloning produces individual clones which secrete
both hormones. On the other hand, use of the sequential hemolytic plaque
assay has shown that only about 5% of normal rat pituitary cells exist as a
mammo-somatotroph population (Leong, Lau, Sinha, Kaiser and Thorner, 1985).
On the basis of examination of granule hormone content using electron
microscopic immunocytochemistry, it appears from our studies that PRL and
GH are not usually synthesized simultaneously by the same cell. A further
complication to the use of the clonal cell lines available is that differences appear to exist between the electrical properties of different clonal lines apparently secreting the same hormone (Hagiwara and Ohmori, 1982;
Dubinsky and Oxford, 1984; Matteson and Armstrong, 1984). Techniques for
cell enrichment and post-recording immunostaining such as those used here
substantially overcome the problems of the cell type heterogeneity of the
anterior pituitary gland.

Clearly, there are many exciting features of pituitary cell physiology
remaining to be explored, but these new approaches to study of non-neoplastic cells are giving us considerable insight into the processes of stimulus-secretion coupling which play a major role in hormone release from
anterior pituitary cells.

Acknowledgements

The technical assistance of Anna Tibbs and Richard Bunting is grate-fully acknowledged. Colin Ingram was supported by a Meat and Livestock Commission studentship. Peter Cobbett was supported by a Beit Memorial Medical Research Fellowship. Dennis Waring was supported by generous grants from the Burroughs Wellcome Foundation and the Nuffield Foundation. The permanent address of Dr. Waring is Department of Human Physiology, University of California, Davis, California 95616.

References

Conn, P.M. and Rogers, D.C., 1980, Gonadotropin release from pituitary cultures following activation of endogenous ion channels, Endocrinology, 107, 2133-2137.

Douglas, W.W. and Taraskevich, P.S., 1985, The electrophysiology of adenohypophysial cells, In: The electrophysiology of the secretory cell, Poisner and Trifaro (eds) Elsevier, Amsterdam, pp. 63-92.

Dubinsky, J.M. and Oxford, G.S., 1984, Ionic currents in two strains of rat anterior pituitary tumour cells, J. Gen. Physiol., 83, 309-339.

Gross, D.S., 1984, The mammalian hypophysial pars tuberalis: a comparative immunocytochemical study, Gen. Comp. Endocrinol., 56, 283-291.

Gross, D.S., Turgeon, J.L. and Waring, D.W., 1984, The ovine pars tuberalis: a naturally occurring source of pure gonadotropes, Endocrinology, 114, 2084-2091.

Hagiwara, S. and Ohmori, H., 1982, Studies of calcium channels in rat clonal pituitary cells with patch electrode voltage clamp, J. Physiol., 331, 231-252.

Hamill, O.P., Marty, A., Neher, E., Sakmann, B. and Sigworth, F.J., 1981, Improved patch clamp techniques for high-resolution current recording from cells and cell-free membrane patches, Pflugers Arch., 391, 85-100.

Ingram, C.D. and Mason, W.T., 1985, Intracellular recordings from putative lactotrophs of bovine pituitary: modulation of spontaneous action potentials by regulators of prolactin secretion, J. Physiol., 364, 52P.

Israel, J-M., Denef, C. and Vincent, J-D., 1983, Electrophysiological properties of normal somatotrophs in culture: an intracellular study, Neuroendocrinology 37, 193-199.

Leong, D.A., Lau, S.K., Sinha, Y.N., Kaiser, D.L. and Thorner, M.O., 1985, Enumeration of lactotropes and somatotropes among male and female pituitary cells in culture: evidence in favor of a mammosomatotrope subpopulation in the rat. Endocrinology 116, 1371-1378.

Mason, W.T. and Waring, D.W., 1985, Electrophysiological recordings from gonadotrophs: evidence for Ca channels mediated by gonadotrophin-releasing hormone, Neuroendocrinology, 41, 258-268.

Mason, W.T. and Waring, D.W., 1986, Patch clamp recordings of single ion channel activation by gonadotrophin releasing hormone in ovine pituitary gonadotrophs, Neuroendocrinology, in press.

Matteson, D.R. and Armstrong, C.M., 1984, Na and Ca channels in a transformed line of anterior pituitary cells. J. Gen. Physiol. 83, 371-394.

Saith, S., Bicknell, R.J. and Schofield, J.G., 1984, Different sodium requirements for Rb efflux and for growth hormone and prolactin secretion from bovine anterior pituitary cells, Mol. Cell. Endocrinol. 35, 47-54.

Schlegel, W. and Wollheim, C.B., 1984, Thyrotropin-releasing hormone increases cytosolic free Ca in clonal pituitary cells (GH3 cells): direct evidence for the mobilization of cellular calcium. J. Cell Biol. 99, 83-87.

BIOCHEMICAL AND FUNCTIONAL CHARACTERIZATION OF GnRH RECEPTORS

Eli Hazum, Iris Schvartz and Dana Keinan

Department of Hormone Research
The Weizmann Institute of Science
Rehovot, 76100, Israel

INTRODUCTION

The primary regulator of the reproductive cycle is the hypothalamic de-capeptide gonadotropin-releasing hormone (GnRH) which stimulates gonado-tropin release from the anterior pituitary. The first step in GnRH action (Conn et al., 1981b) is its recognition by specific binding sites (recep-tors) at the surface of gonadotrope cells. The interaction of GnRH with pituitary membrane preparations or cultured pituitary cells has indicated the presence of a single class of high-affinity binding sites for both agonists and antagonists of GnRH (Clayton et al., 1978; Conne et al., 1979; Clayton and Catt, 1980, 1981; Marian and Conn, 1980; Naor et al., 1980; Hazum, 1981a; Marian et al., 1981; Meidan and Koch, 1981). Since the binding properties of GnRH receptors have been studied in detail, this review will be focussed on recent studies related to biochemical and func-tional characterization of pituitary GnRH receptors.

CHARACTERIZATION OF THE GnRH RECEPTOR

Cytochemical Characterization of Gonadotropes as Target Cells for Biotiny-lated GnRH

Dual stains that combine affinity cytochemistry to localize GnRH recep-tors and immunocytochemistry to localize storage sites of luteinizing hor-mone (LH) and follicle stimulating hormone (FSH), have been used to demon-strate that GnRH binds only to gonadotrope cells (Childs et al., 1983a,b; Hazum, 1985). In these studies, the avidin-biotin peroxidase complex technique (ABC) has been applied to localize [Biotinyl-D-Lys[6]]GnRH on the cells with the use of a dense black peroxidase substrate. The GnRH stain has been followed by immunocytochemical stains for LH β and FSH β. Most (90-100%) of the gonadotropes show stain for the biotinylated GnRH, indi-cating that GnRH binds exclusively to gonadotropes. Fig. 1 shows the co-localization of GnRH and FSH by dual staining of cultured pituitary cells.

Biochemical Characterization of Pituitary GnRH Receptors

Characterization of the GnRH-receptor in the pituitary have indicated that the receptor is a glycoprotein which contains sialic acid residues (Hazum, 1982a) and that membrane phospholipids are involved in the inter-

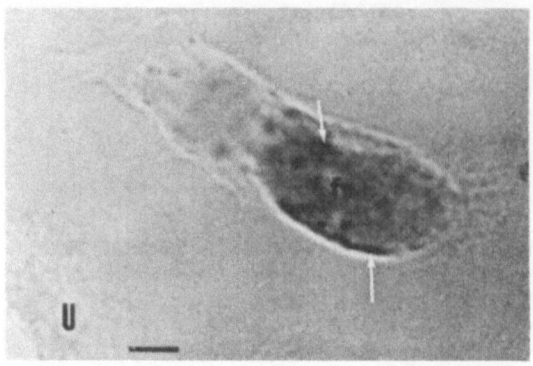

Fig. 1. Dual staining of pituitary cells. A monolayer of cells after two
days in culture has been stimulated for 1 min with
[Biotinyl-D-Lys[6]]GnRH, and stained by the ABC technique followed
by the immunocytochemical stain for FSH with 1:10,000 anti hFSH β.
The stain for GnRH is seen on the cell surface (marked by white
arrows; this stain appears as a dark gray) while the reaction
for FSH is in the center (marked by f; this stain which would
appear as an organe-red carbazole reaction, is seen here as a
light gray). Nearby is an unstained cell (U). Calibration represents 5 μm.

action between the hormone and the receptor (Hazum et al., 1982a). Recently, we have studied the effects of tunicamycin (TM) and neuraminidase,
on GnRH agonist binding and LH release in cultured rat pituitary cells
(Schvartz and Hazum, 1985). Treatment with TM, an antibiotic which inhibits protein glycosylation, abolishes the development of elongated cell
processes without any effect on cell viability. Concomitantly, TM causes
a time- and dose-dependent inhibition of LH release and GnRH agonist specific binding. Treatment with TM for 5 h does not significantly affect LH
release, whereas after 24 h there is a 50% decrease of the release. Further decrease (80%) is obtained after 48 h of incubation with TM. The effect of TM on the binding of ^{125}I-Buserelin to pituitary cells as a function of time is shown in Fig. 2. While 5-7 h of incubation with TM does
not affect the specific binding of [^{125}I]Buserelin, incubation for 20 h
causes about a 50% decrease of the specific binding. Maximal inhibition
of [^{125}I]Buserelin binding is obtained after 48 h of incubation. Longer
incubation (54 h) does not cause further alteration in the specific binding. Based on these data one can estimate that the turnover of the receptor is about 16 to 18 hr.

The inhibition of binding is due to a decrease in the number of GnRH
receptors from 3.3×10^5 receptors/gonadotrope in control cells to 2.0×10^5
receptors/gonadotrope in TM-treated cells, without any significant effect
on binding affinity (0.5 ± 0.1 nM). Protein synthesis is not affected under these experimental conditions, suggesting that the aglycosylated GnRH
receptors are probably intracellularly accumulated and are not expressed
on the cell surface. Treatment with neuraminidase also causes a dose dependent inhibition of GnRH binding. Maximal inhibition (50%) is obtained
with 100 μg/ml neuraminidase. These findings indicate that the oligosaccharide portion is essential for the functional properties of the GnRH receptor.

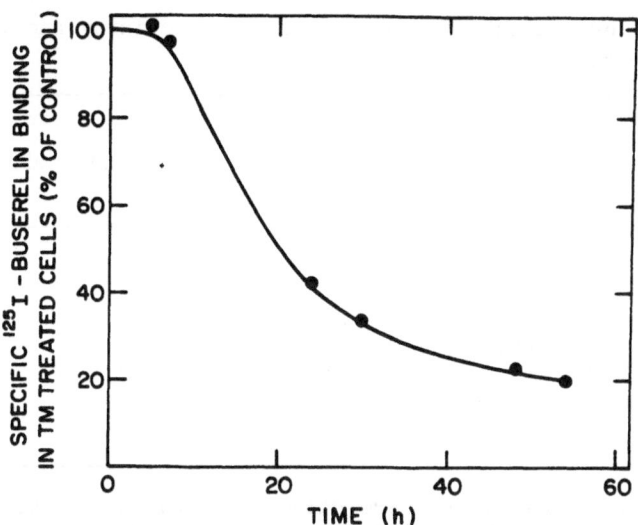

Fig. 2. Time dependent effect of TM on Buserelin binding. Cells are
treated with or without 500 ng/ml TM for various time periods.
At the time indicated, the cells are removed from the culture
dishes, incubated (0.5 - 1.5x10⁶ cells/tube) with ¹²⁵I-Buserelin
(60,000 cpm) in 0.5 ml medium 199 containing 0.1% BSA (90 min at
4°C) and the specific binding is measured.

Mapping of GnRH Receptor Binding Site

Various specific reagents have been examined for their ability to interfere with the binding of GnRH to its receptor (Keinan and Hazum, 1985).
Pretreatment of pituitary membrane preparations with sodium periodate decreases the specific binding in a dose dependent manner (IC_{50}=0.5 mM) due
to a decrease in receptor affinity (Table 1). This indicates the presence
of a sugar moiety in the binding site, as sodium periodate is known to
break the bond between vicinal hydroxyl groups of sugars. Tryptophan is
another constituent that participates in the GnRH binding site, since pretreatment of pituitary membranes with 2-methoxy-5-nitrobenzyl bromide inhibits the binding (IC_{50}=0.22 mM) by decreasing receptor affinity (Table
1). In addition, the native hormone confers on the binding site a protective effect against inactivation by 2-methoxy-5-nitrobenzyl bromide. Pretreatment of membranes with p-diazo-sulfanilic acid also inhibits the
binding of ¹²⁵I-Buserelin (IC_{50}=0.1 mM) indicating the presence of tyrosine within or near the binding site. Pretreatment of pituitary membrane
preparations with dithiothreitol also inhibits the binding (IC_{50}=14 mM)
due to a decrease in the binding affinity, which is accompanied by an increase in receptor number (Table 1). The increased number of receptors is
probably due to exposure or unmasking of receptor sites. These data suggest that there are disulfide bonds within or near the binding region.
However, it seems that the presence of sulfhydryl groups in the binding
site can be excluded, because treatment of pituitary membrane preparations
with N-ethylmaleimide and iodoacetamide does not affect the binding.
Treatment with 1-ethyl-3-(3-dimethylamino- propyl)carbodiimide and glycine
ethyl ester also prevents the binding in a dose-dependent manner (IC_{50}=25
mM) and implies that free carboxylic groups are involved in the binding
site.   Since divalent cations are more potent in inhibiting the specific

395

Table 1. Effect of reagent-treatment on binding affinity and number of GnRH receptors.

Reagent	Kd (nM)	Bmax (fmoles/pituitary)
Control	0.12±0.01	165±4
Sodium periodate (0.5 mM)	0.18±0.01	180±3
2-Methoxy-5-nitrobenzyl bromide (0.5 mM)	0.29±0.04	135±6
Dithiothreitol (10 mM)	0.22±0.02	210±7

B_{max} and K_d values (mean ± SEM) are calculated from competition binding experiments. Control and treated membranes are incubated with ^{125}I-labeled Buserelin (50.000 cpm) and various concentrations of unlabeled Buserelin (10^{-12} to 10^{-7}M) in 0.5 ml 10 mM Tris/0.1% BSA. After 90 min at 4°C, the binding is measured and plotted according to Scatchard.

binding in comparison with monovalent cations (Table 2), and since the order of potency within the divalent cations is identical to their association constants to dicarboxylic compounds, it is suggested that there are at least two carboxylic groups that participate in the binding of the hormone.

Table 2. IC_{50} values for the inhibitory effect of monovalent and divalent cations on ^{125}I-Buserelin binding.

Cation	IC_{50} (mM)
Cs^+	25
Na^+	24
K^+	20
Ca^{+2}	1.0
Mg^{+2}	0.9
Mn^{+2}	0.4
Zn^{+2}	0.15
Cu^{+2}	0.05

IC_{50} is the concentration of monovalent or divalent cations that inhibits the specific binding of ^{125}I-Buserelin to pituitary membrane preparations by 50%. The radioactive Buserelin (50,000 cpm) is incubated with various concentrations of the tested cation in a final volume of 0.5 ml 10 mM Tris/0.1% BSA, containing pituitary membranes (100 µg protein/ml). After 90 min at 4°C, the binding is measured by filtration.

Three dimensional analysis of GnRH (pGlu-His-Trp-Ser-Tyr-Gly-Leu-Arg-Pro-Gly-NH$_2$) in solution has indicated that there is a β-turn in position 6 which brings the carboxy and the amino termini of the hormone into close proximity. It has been suggested that the side chains of histidine, tyrosine and arginine form a packed unit which may play an active role in the hormone action. Tryptophan, however, is at a maximal distance from this unit and thus may act as an independent active entity (Shinitzky and Fridkin, 1976; Shinitzky et al., 1976). According to the spatial conformation of the native hormone, we can postulate a model for its interaction with the receptor. It is well known that arginyl residues on proteins may serve as positively charged loci for recognition of negatively charged anions. Thus, the driving force for the formation of the hormone-receptor complex is probably an ionic interaction between the amino acid arginine in position 8, which is positively charged, and the carboxyl groups in the binding site. In addition to the ionic interaction, the hormone-receptor complex is stabilized by aromatic $\pi - \pi$ interactions between the histidine, tryptophan and tyrosine residues in the hormone and tyrosine and tryptophan in the receptor site.

IDENTIFICATION OF THE GnRH RECEPTOR

Characterization of Photoaffinity Derivative of GnRH

In order to identify directly the postulated GnRH receptors by biochemical techniques, a photoaffinity derivative of GnRH, [azidobenzoyl-D-Lys^6]GnRH, was prepared. The photoreactivity of this analog was established by its spectral changes when irradiated with ultraviolet light and by its ability to bind covalently to pituitary membrane preparations after photoactivation (Hazum, 1981b, c). The photoaffinity derivative binds to the GnRH receptor with higher apparent affinity than GnRH and [D-Lys^6]GnRH and it is 3 to 4 times more active than GnRH in LH release in pituitary cell cultures (Hazum and Keinan, 1983a).

Photoaffinity Labeling of Pituitary and Gonadal GnRH Receptors

Photoaffinity labeling of pituitary GnRH receptors after preincubation with the iodinated photoaffinity derivative (90 min at 4°C, in the dark) results in the identification of a single specific band with an apparent molecular weight of 60,000 daltons (Hazum, 1981c; 1983a,b; Hazum and Keinan 1982, 1983b, 1984a). Our findings suggest that some differences exist between the GnRH binding sites of the pituitary and that of the gonads, as the latter have an additional specific component of 54,000 daltons (Hazum and Nimrod, 1982; Hazum and Keinan, 1984b).

The specificity of the 60K dalton band in the pituitary has been established by the following criteria: 1) GnRH analogs exhibit similar ability to inhibit photolabeling of the 60K dalton band and receptor binding (Hazum, 1981c). 2) Physiological alterations in receptor content are accompanied by similar changes in the radioactivity incorporated into the 60K dalton band (Hazum and Keinan, 1982) and 3) Covalent linking of photoreactive GnRH to gonadotropes produces a prolonged signal (Hazum and Keinan, 1983a). Similarly, solubilization and partial purification, of the GnRH receptors from bovine anterior pituitary results in the identification of a binding protein with an apparent molecular weight of 60K to 80K daltons (De Almeida Catanho et al., 1983; Winiger et al., 1983). In addition, a 60K molecular weight GnRH receptor has recently been demonstrated in the pituitary by a ligand-immunoblotting technique (Eidne et al., 1985). However, radiation inactivation of the GnRH receptor indicates an apparent molecular weight of 134K daltons (Conn and Venter, 1985).

Identification of GnRH Receptors in Control and Desensitized Pituitary Cells

Prolonged exposure of a wide variety of tissues to hormones or their agonistic analogs results in refractoriness of the tissues to further stimulation by the same hormones. Several studies have indicated that GnRH induces desensitization of pituitary cells (Smith and Vale, 1981; Loumaye and Catt, 1982; Keri et al., 1983; Smith and Conn, 1983; Smith et al., 1983; Zilberstein et al., 1983). Like others, we have shown (Hazum and Schvartz, 1984), that preincubation of GnRH with cultured rat pituitary cells, induces down regulation of GnRH receptors in a time- and dose-dependent manner. The specific binding is inhibited by 50% after 30 min and maximal inhibition (70%) is obtained after 75 min preincubation with 1 µM GnRH. Preincubation of the cells for 2 h with 10 nM GnRH inhibits the specific binding by 20%, reaching a plateau of 70% inhibition with 0.1 µM GnRH. Concomitantly, exposure of the cells to GnRH causes a time- and dose-dependent desensitization of LH release. The responsiveness of the desensitized cells is not parallel to the binding capacity and is inhibited to a greater extent (93%).

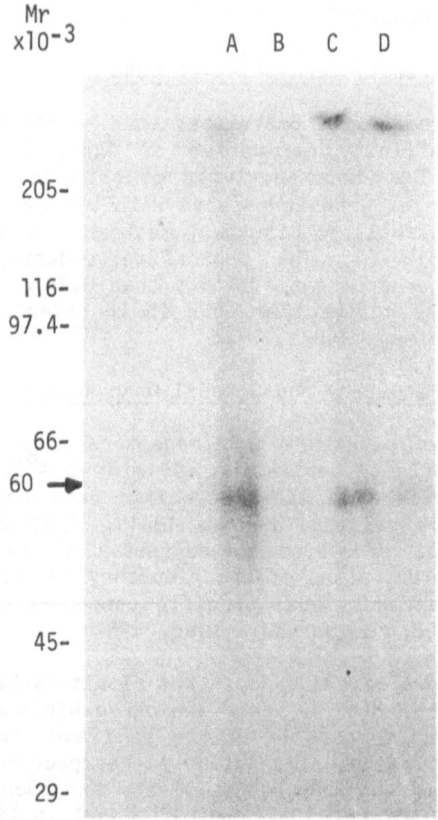

Fig. 3. Autoradiogram of sodium dodecyl sulfate polyacrylamide gel electrophoresis of GnRH receptors labeled with ^{125}I-[azidobenzoyl-D-Lys^6]GnRH in control and desensitized pituitary cells. Pituitary cultures are preincubated with medium alone (A,B) or with medium containing 1 µM GnRH (C,D). After 2 h of preincubation, the cells are washed extensively and incubated at 4°C in the dark with ^{125}I-labeled [azidobenzoyl-D-Lys^6]GnRH in the presence (B,D), or absence (A,C) of 10^{-7}M Buserelin. After 90 min the cells are photolysed (5 min at 4°C) and prepared for SDS polyacrylamide gel electrophoresis.

To evaluate whether desensitization is associated with a change in the molecular weight of the GnRH receptor, photoaffinity labeling of the receptor in control and desensitized cells has been conducted. Cells are preincubated with or without 1 µM GnRH for 2 h. At the end of the preincubation time the cells are washed and subsequently incubated with ^{125}I-[azidobenzoyl-D-Lys^6]GnRH (90 min. at 4°C). Photoactivation of the bound ^{125}I-labeled [azidobenzoyl -D-Lys^6]GnRH to control and desensitized cells results in the identification of a single specific band with the same apparent molecular weight of 60K daltons (Fig. 3). Although there is no change in the apparent molecular weight of GnRH receptors from desensitized cells, the amount of radioactivity incoporated into the 60K dalton band is 30% of that incorporated into the GnRH receptor of control cells. These results indicate that structural alterations of GnRH receptors are not associated with GnRH-induced desensitization. Therefore, desensitization may involve conformational changes in the receptor or more likely a post-receptor mechanism.

FATE OF GnRH RECEPTORS IN THE PITUITARY

The distribution of GnRH receptors in pituitary cells was first reported at the electron microscopic level by Hopkins and Gregory (1977) using a bioactive GnRH analog coupled to ferritin. Short-term incubations of this analog with pituitary cells result in an even distribution of ferritin particles over the cell surface of gonadotropes. Longer incubations result in aggregates of the bound conjugate, followed by internalization into lysosome-like structures in the Golgi area. Immunocytochemical staining of pituitary sections (Sternberger and Petrali, 1975; Dacheux, 1981) with antisera to GnRH has demonstrated that GnRH receptors are located on the plasma membranes and are also concentrated on the membranes of large secretory granules of rat and porcine gonadotropes. In vivo uptake studies (Duello and Nett, 1980; Pelletier et al., 1982; Duello et al., 1983) with ^{125}I-labeled GnRH and its agonists have also indicated an intracellular locus of radioactivity after the initial binding at the cell-membrane. Studies utilizing bioactive rhodamine derivative of GnRH (Hazum et al., 1980; Naor et al., 1981), and [biotin-D-Lys^6]GnRH as receptor probes (Childs et al., 1983a,b; Hazum, 1985) have indicated that the receptors are initially distributed evenly on the cell surface and then form clusters, which subsequently become internalized.

Recently, the photoaffinity derivative [azidobenzoyl-D-Lys^6]GnRH, has been used for localization studies (Hazum et al., 1982b, 1985), which circumvent some of the problems described in previous studies, such as rapid dissociation of the hormone from its receptor. Dispersed rat pituitary cells are incubated with the photoaffinity analog for 90 min at 4°C or for various time periods at 37°C and subsequently photolized. The distribution of the labeled hormone by light and electron microscopic autoradiography indicates that after exposure of pituitary cells to the ^{125}I-labeled hormone at 4°C (90 min), most of the labeled hormone is associated with the cell surface membrane. At 37°C (15 min) about 30% of the cell-bound labeled hormone is internalized. After 45 min at 37°C, only half of the label is associated with the plasma membrane, while the rest is found in lysosome like structures (21%), in secretory granules (20%)and in the Golgi complex. The incidence of clustered grains over the cell membrane is high both after 15 min and 45 min of incubation at 37°C, and a significant fraction of the grain is associated with coated pits of the cell membrane. A similar distribution has been observed after electron microscopy with [D-Lys^6]GnRH coupled to colloidal gold or ferritin (Jennes et al., 1983).

The process of receptor-mediated endocytosis involves binding of a ligand to a specific cell membrane receptor, concentration of ligand-receptor complexes in coated pits and transfer to intracellular components (reviewed in Hazum, 1982b). However, there are two possible mechanisms that could mediate hormone action. In the first, the binding of ligand to receptors leads to cross-linking of receptors (or receptor-effector) at the cell surface, which is by itself sufficient to trigger the subsequent biochemical events of hormone action. Although ligand internalization and degradation may occur, it is not a prerequisite for ligand function. In the second possibility, however, internalization and degradation of hormone-receptor complexes are important for biological activity.

Three different approaches have indicated that GnRH-receptor internalization, as well as cluster formation, are not required for GnRH-stimulated LH release (Conn and Hazum, 1981; Conn et al., 1981a). (i) Covalent immobilization of a GnRH agonist on agarose beads results in a derivative which stimulates LH release with full efficacy. (ii) Removal of external GnRH from the cells at different times throughout GnRH stimulation results in a prompt return of LH release to basal levels. Thus, under conditions in which internalization takes place, continuous release of LH can only occur when external GnRH is present. (iii) Comparative studies on receptor distribution and LH release have shown that under conditions when cluster formation and internalization are prevented, gonadotropin secretion by GnRH is not affected. This result suggests that internalization, as well as large-scale clusters of GnRH receptors, are not important in eliciting the biological effects of GnRH.

Recently, the conversion of a GnRH antagonist to an agonist and the potency enhancement of a GnRH agonist by bridging two molecules within a critical distance d (15 Å < d < 150 Å) suggest that receptor cross-linking as such is sufficient to activate the effector system in pituitary cells to evoke release (Conn et al., 1982a,b; Gregory et al., 1982). On the basis of these experimental data a mathematical model of hormone action that quantitatively accounts for the release of LH has been proposed (Blum and Conn, 1982).

We have used a DNP (dinitrophenyl) labeled antagonist and antibodies against DNP to indicate that receptor cross-linking is involved in the signal transduction of LH release. We have demonstrated that introduction of a dinitrophenyl group into the epsilon amino side chain of the D-Lys residue of [D-pGlu^1,D-Phe^2,D-Trp^3,D-Lys^6]GnRH alters neither the binding affinity nor the antagonist activity. Both antibodies against DNP and their Fab fragments have the ability to bind the DNP-antagonist. However, only the addition of bivalent antibodies (and not the Fab fragments) converts the DNP-antagonist to an agonist. This suggests that divalency is a critical factor in GnRH action.

These results suggest that the primary function of a GnRH agonist is to induce and maintain a state of receptor cross-linking that is by itself sufficient to trigger the subsequent biochemical events of the hormone action. A GnRH antagonist, on the other hand, is capable of binding to the receptor but is unable to induce dimerization. Recently, it has been shown that GnRH antagonists can also induce receptor clustering and internalization (Hazum et al., 1983; Jennes et al., 1984) and suggests that the GnRH antagonists can by-pass receptor cross-linking and directly induce clustering and internalization. Thus, it is possible that the antagonist is internalized through receptors that already exist in coated areas on the cell surface. The internalization of GnRH (and presumably its receptor) may have some other intracellular action or may simply be degraded or recycled.

SUMMARY

Cytochemical studies have indicated that GnRH binds exclusively to cell surface receptors on gonadotropes. Characterization of the GnRH receptor in the pituitary has indicated that the receptor is a glycoprotein and that the oligosaccharide portion is essential for the functional properties of the receptor. The amino acid residues in the receptor, essential for hormone recognition have been identified using specific chemical reagents. Based on these findings, we have proposed a model describing the interactions involved in the formation of the hormone-receptor complex. Identification of the GnRH receptors by biochemical techniques has been achieved by using a bioactive photoaffinity derivative of GnRH, [azidobenzoyl-D-Lys^6]GnRH. Photoaffinity labeling of pituitary GnRH receptors has led to the identification of a single specific band of 60K daltons in control and desensitized cells. The gonadal GnRH receptors, however, have an additional specific component of 54K daltons.

The distribution of GnRH receptors, after binding to gonadotropes, has been followed by light and electron microscopy. These studies indicate that, after exposure of gonadotropes to the hormone at 4°C, the receptors are evenly distributed. At 37°C, aggregation and subsequent internalization (via coated pits) of the hormone-receptor complex into subcellular organelles can be observed. Similar pattern of receptor redistribution is observed with a GnRH antagonist. Studies on the relationship between receptor redistribution induced by GnRH and the biological activity have indicated that receptor cross-linking per se is sufficient to activate the effector system in pituitary cells to evoke LH release.

Acknowledgments

This work was supported by the United States-Israel Binational Science Foundation, the Rockefeller Foundation, New York, the Minerva Foundation and by the fund for basic research administered by the Israel Academy of Sciences and Humanities. We are grateful to Mrs. M. Kopelowitz for typing the manuscript.

REFERENCES

Blum, J.J., and Conn, P.M., 1982, Gonadotropin-releasing hormone stimulation of luteinizing hormone release: A ligand-receptor-effector model, Proc. Natl. Acad. Sci. U.S.A., 79:7307.

Childs, G.V., Naor, Z., Hazum, E., Tibolt, R., Westlund, K.N., and Hancock, M.B., 1983a, Localization of biotinylated gonadotropin releasing hormone on pituitary monolayer cells with avidin-biotin peroxidase complexes, J. Histochem. Cytochem., 31:1422.

Childs, G.V., Naor, Z., Hazum, E., Tibolt, R., Westlund, K.N., and Hancock, M.B., 1983b, Cytochemical characterization of pituitary target cells for biotinylated gonadotropin releasing hormone, Peptides, 4:549.

Clayton, R.N., and Catt, K.J., 1980, Receptor-binding affinity of gonadotropin-releasing hormone analogs: Analysis by radioligand receptor assay, Endocrinology, 106:1154.

Clayton, R.N., and Catt, K.J., 1981, Gonadotropin-releasing hormone receptors: Characterization, physiological regulation, and relationship to reproductive function, Endocrine Rev., 2:186.

Clayton, R.N., Shakespear, R.A., and Marshall, J.C., 1978, LHRH binding to purified plasma membranes: Absence of adenylate cyclase activation., Mol. Cell. Endocrinol., 11:63.

Conn, P.M., and Hazum, E., 1981, LH release and GnRH-receptor internalization: Independent actions of GnRH, Endocrinology, 109:2040.

Conn, P.M., and Venter, J.C., 1985, Radiation inactivation (target size
analysis) of the gonadotropin-releasing hormone receptor: evidence
for a high molecular weight complex, Endocrinology, 116:1324.

Conn, P.M., Smith, R., and Rogers, D.C., 1981a, Stimulation of pituitary
gonadotropin release does not require internalization of gonadotropin-releasing hormone, J. Biol. Chem., 256:1098.

Conn, P.M., Marian, J., McMillian, M., Stern, J., Rogers, D., Hamby, M.,
Penna, A., and Grant, E., 1981b, Gonadotropin-releasing hormone action in the pituitary: A three step mechanism, Endocrine Rev.,
2:174.

Conn, P.M., Rogers, D.C., Stewart, J.M., Niedel, J., and Sheffield, T.,
1982a, Conversion of gonadotropin-releasing hormone antagonist to an
agonist, Nature, 296:653.

Conn, P.M., Rogers, D.C., and McNeil, R., 1982b, Potency enhancement of a
GnRH agonist: GnRH-receptor microaggregation stimulates gonadotropin
release, Endocrinology, 111:335.

Conne, B.S., Aubert, M.L., and Sizonenko, P.C., 1979, Quantification of
pituitary receptor sites for LHRH: Use of superactive analog as a
tracer, Biochem. Biophys. Res. Commun., 90:1249.

Dacheux, F., 1981, Ultrastructural localization of gonadotropin-releasing
hormone in porcine gonadotrophic cells, Cell Tissue Res., 216:143.

De Almeida Catanho, M.T.J., Berault, A., Theoleyre, M., and Jutisz, M.,
1983, Solubilization and partial purification of the high-affinity
gonadoliberin receptor from the bovine pituitary gland, Arch. Biochem. Biophys., 225:535.

Duello, T.M., and Nett, T.M., 1980, Uptake, localization, and retention of
gonadotropin-releasing hormone and gonadotropin-releasing hormone
analogs in rat gonadotrophs, Mol. Cell. Endocrinol., 19:101.

Duello, T.M., Nett, T.M., and Farquhar, M.G., 1983, Fate of a gonadotropin-releasing hormone agonist internalized by rat pituitary gonadotrophs, Endocrinology, 112:1.

Eidne, K.A., Hendricks, D.T., and Millar, R.P., 1985, Demonstration of a
60K molecular weight luteinizing hormone-releasing hormone receptor
in solubilized adrenal membranes by a ligand-immunoblotting technique, Endocrinology, 116:1792.

Gregory, H., Taylor, C.L., and Hopkins, C.R., 1982, Luteinizing hormone
release from dissociated pituitary cells by dimerization of occupied
LHRH receptors, Nature, 300:269.

Hazum, E., 1981a, Some characteristics of GnRH-receptors in rat pituitary
membranes: Differences between an agonist and an antagonist, Mol.
Cell. Endocrinol., 23:275.

Hazum, E., 1981b, Photo-affinity inactivation of gonadotropin-releasing
hormone receptors, FEBS Lett., 128:111.

Hazum, E., 1981c, Photoaffinity labeling of luteinizing hormone releasing
hormone receptors of rat pituitary membrane preparations, Endocrinology, 109:1281.

Hazum, E., 1982a, GnRH-receptor of rat pituitary is a glycoprotein: Differential effect of neuraminidase and lectins on agonists and antagonists binding, Mol. Cell. Endocrinol., 26:217.

Hazum, E., 1982b, Receptor regulation by hormones: Relevance to secretion
and other biological functions, in: "Cellular Regulation of Secretion and Release", P.M. Conn, ed., Academic Press, New-York.

Hazum, E., 1983a, Photoaffinity labeling in neuroendocrine tissues, Methods Enzymol., 103:58.

Hazum, E., 1983b, Photoaffinity labeling of peptide hormone receptors, Endocrine Rev., 4:352.

Hazum, E., 1985, Preparation and use of biotinylated neuroendocrine peptides, Methods Enzymol. (in press).

Hazum, E., and Keinan, D., 1982, Photoaffinity labeling of pituitary gonadotropin releasing hormone receptors during the rat estrous cycle,
Biochem. Biophys. Res. Commun., 107:695.

Hazum, E., and Keinan, D., 1983a, Covalent linking of photoreactive gonadotropin-releasing hormone to gonadotropes produces a prolonged signal, Proc. Natl. Acad. Sci. U.S.A., 80:1902.

Hazum, E., and Keinan, D., 1983b, Gonadotropin releasing hormone receptors: photoaffinity labeling with an antagonist, Biochem. Biophys.
Res. Commun., 110:116.

Hazum, E., and Keinan, D., 1984a, Characterization of GnRH receptors in
bovine pituitary membranes, Mol. Cell. Endocrinol., 35:107.

Hazum, E., and Keinan, D., 1984b, Testicular GnRH receptors: photoaffinity labeling and fluorescence distribution studies, Peptides, 5:119.

Hazum, E., and Nimrod, A., 1982, Photoaffinity labeling and fluorescence
distribution studies of gonadotropin-releasing hormone receptors in
ovarian granulosa cells, Proc. Natl. Acad. Sci. U.S.A., 79:1747.

Hazum, E., and Schvartz, I., 1984, Photoaffinity labeling of gonadotropin
releasing hormone receptors in control and desensitized pituitary
cells, Biochem. Biophys. Res. Commun., 125:532.

Hazum, E., Cuatrecasas, P., Marian, J., and Conn, P.M., 1980, Receptor-mediated internalization of gonadotropin-releasing hormone by pituitary
gonadotropes, Proc. Natl. Acad. Sci. U.S.A., 77:6692.

Hazum, E., Garritsen, A., and Keinan, D., 1982a, Role of lipids in gonadotropin releasing hormone agonist and antagonist binding to rat pituitary, Biochem. Biophys. Res. Commun., 105:8.

Hazum, E., Meidan, R., Keinan, D., Okon, E., Koch, Y., Lindner, H.R., and
Amsterdam, A., 1982b, A novel method for localization of gonadotropin
releasing hormone receptors, Endocrinology, 111:2135.

Hazum, E., Meidan, R., Liscovitch, M., Keinan, D., Lindner, H.R., and
Koch, Y., 1983, Receptor-mediated internalization of LHRH antagonists
by pituitary cells, Mol. Cell. Endocrinol., 30:291.

Hazum, E., Koch, Y., Liscovitch, M. and Amsterdam, A., 1985, Intracellular
pathways of receptor-bound GnRH agonist in pituitary gonadotropes,
Cell Tissue Res., 293:3.

Hopkins, C.R., and Gregory, H., 1977, Topographical localization of the
receptors for luteinizing hormone-releasing hormone on the surface of
dissociated pituitary cells, J. Cell Biol., 75:528.

Jennes, L., Stumpf, W.E., and Conn, P.M., 1983, Intracellular pathways of
electron opaque GnRH-derivatives bound by cultured gonadotropes, Endocrinology, 113:1683.

Jennes, L., Stumpf, W.E., and Conn, P.M., 1984, Receptor-mediated binding
and uptake of GnRH agonist and antagonist by pituitary cells, Peptides, 5:215.

Keinan, D., and Hazum, E., 1985, Mapping of gonadotropin releasing hormone receptor binding site, Biochemistry (in press).

Kéri, G., Nicolics, K., Teplán, J., and Molnár, J., 1983, Desensitization
of luteinizing hormone release in cultured pituitary cells by gonadotropin releasing hormone. Mol. Cell. Endocrinol., 30:109.

Loumaye, E., and Catt, K.J., 1982, Homologous regulation of gonadotropin-
releasing hormone receptors in cultured pituitary cells, Science,
215:983.

Marian, J., and Conn, P.M., 1980, The calcium requirement in GnRH-stimulated LH release is not mediated through a specific action on receptor binding, Life Sci., 27:87.

Marian, J., Cooper, R.L., and Conn, P.M., 1981, Regulation of the rat pituitary gonadotropin-releasing hormone receptor, Mol. Pharmacol.,
19:399.

Meidan, R., and Koch, Y., 1981, Binding of luteinizing-hormone releasing
hormone analogues to dispersed pituitary cells, Life Sci., 28:1961.

Naor, Z., Clayton, R.N., and Catt, K.J., 1980, Characterization of GnRH
receptors in cultured rat pituitary cells, Endocrinology, 107:1144.

Naor, Z., Atlas, A., Clayton, R.N., Forman, D.S., Amsterdam, A., and Catt, K.J., 1981, Interaction of fluorescent gonadotropin-releasing hormone with receptors in cultured pituitary cells, J. Biol. Chem., 256:3049.

Pelletier, G., Dube, D., Guy, J., Seguin, C., and Lefebvre, F.A., 1982, Binding and internalization of a luteinizing hormone releasing hormone agonist by rat gonadotrophic cells. A radioautographic study, Endocrinology, 111:1068.

Schvartz, I., and Hazum, E., 1985, Tunicamycin and neuraminidase effects on luteinizing hormone (LH)-releasing hormone binding and LH release from rat pituitary cells in culture, Endocrinology, 116:2341.

Shinitzky, M., and Fridkin, M., 1976, Structural features of luliberin (luteinizing hormone-releasing factor) inferred from fluorescence measurements, Biochem. Biophys. Acta, 434:137.

Shinitzky, M., Hazum, E., and Fridkin, M., 1976, Structure-activity relationships of luliberin substituted at position 8, Biochem. Biophys. Acta, 453:533.

Smith, M.A., and Vale, W.W., 1981, Desensitization to gonadotropin releasing hormone observed in superfused pituitary cells on cytodex beads, Endocrinology, 108:752.

Smith, M.A., and Conn, P.M., 1983, GnRH-mediated desensitization of the pituitary gonadotrope is not calcium dependent, Endocrinology, 112:408.

Smith, M.A., Perrin, M.H., and Vale, W.W., 1983, Desensitization of cultured pituitary cells to gonadotropin releasing hormone: evidence for a post receptor mechanism, Mol. Cell. Endocrinol. 30:85.

Sternberger, L.A., and)letrali, J.P., 1975, Quantitative immunocytochemistry of pituitary receptors for luteinizing hormone-releasing hormone, Cell Tissue Res., 162:141.

Winiger, B.P., Birabeau, M.A., Lang, U., Capponi, A.M., Sizonenko, P.C., and Aubert, M.L., 1983, Solubilization of pituitary GnRH binding sites by means of a zwitterionic detergent, Mol. Cell. Endocrinol., 31:77.

Zilberstein, M., Zakut, H., and Naor, Z., 1983, Coincidence of down-regulation and desensitization in pituitary gonadotrophs by gonadotropin releasing hormone, Life Sci., 32:663.

HORMONAL RECEPTOR PLASTICITY IN THE BRAIN AS SHOWN BY IN VITRO

QUANTITATIVE AUTORADIOGRAPHY

William H. Rostène, Denis Hervé[*], Patrick Kitabgi[**], Jocelyne Magre, and Alain Sarrieau

INSERM U.55, Hôpital Saint-Antoine, 75571 Paris Cédex 12[*]; [**]INSERM U.114, Collège de France, 75231 Paris Cédex 05; Centre de Biochimie CNRS, Université de Nice, Parc Valrose, 06034 Nice Cédex, France

INTRODUCTION

Binding of neuroactive substances to specific receptor sites initiates sophisticated events which result in specific pharmacological, behavioral and physiological responses. The complex organization and the tissular heterogeneity of most of the endocrine organs have led to the development of new methodologies for the precise localization of these binding sites. Together with constant progress obtained for the biochemical characterization of those binding sites by conventional binding assays (mainly carried out on membrane preparations), recent autoradiographic techniques provide us with new interpretations about the numerous interactions underlying brain functions.

METHODOLOGY

Autoradiographic techniques for the localization of specific receptor sites for chemical and biological substances are derived, on the one hand, from autoradiographic approaches used to visualize steroid hormone receptors (Pfaff, 1968; Stumpf, 1968), and on the other hand, on the 2-deoxyglucose technique which allows the measurement of the rate of glucose utilization in very discrete brain areas (Sokoloff, 1977). Refined adaptations of those methods for the in vitro localization of receptor sites were first described by Young and Kuhar (1979) for opiate receptors, and revealed a quite similar distribution to that found in earlier in vivo studies (Pert et al., 1975). We will first briefly describe this in vitro autoradiographic method we slightly modified for both qualitative and quantitative distribution of hormonal receptor sites, and second, we will illustrate by means of few examples its use for the study of hormonal receptor plasticity in both rat and human brains.

Both in vivo and in vitro autoradiography are based on the use of radiolabeled substances. However, in vivo studies require large quantities of radiolabeled ligand for receptor visualization and are limited to substances crossing the blood brain barrier, which is not the case, for instance, for the majority of neuropeptides. The selection of new ligands and the recent development of isotopes emetting positrons allow the in vivo visualization of receptors for neuroactive substances by means of positron emission tomography (Mazière et al., 1984; Rostène et al., 1986).

The in vitro quantative autoradiography methods though slightly different for each substance tested, is based on the same principle as that summarized in figure 1.
1) Brain or pituitary are rapidly frozen on dry ice. 2) 20 to 32 µm thin sections are cut on a cryostat at -15°C according to stereotaxic atlases. 3) They are thaw-mounted onto subbed glass slides and 4) stored at -20°C overnight and at -80°C until assay. 5) The day of the experiment,

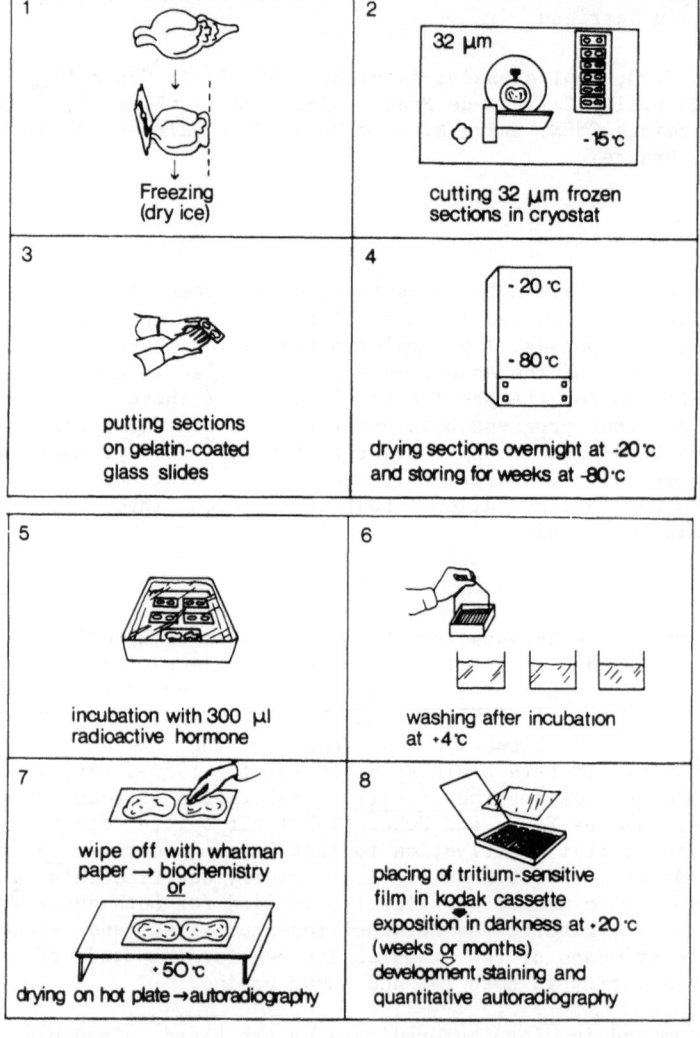

Fig. 1 Schematic representation of the in vitro autoradiographic technique on frozen-mounted brain sections.

sections were warmed at room temperature and incubated in the appropriate
buffer with the radiolabeled ligand with or without an excess of the
unlabeled substance in order to test the non specific binding. 6) At the
end of incubation time, when the equilibrium is reached, sections are
washed in consecutive changes in 50 mM Tris-HCl buffer at 4°C in order
to determine the highest ratio between specific to non specific binding.
7) An advantage of this method is that it allows to combine on adjacent
sections both biochemical characterization and topographical distribution
of the receptor sites. Biochemical data are obtained on sections wiped
off of the slides with a paper filter and counted for the determination
of the radioactivity specifically retained. Sections dried under hot or
cold air are used for autoradiography. 8) The autoradiographic procedure
we are using was first described by Palacios et al. (1981) and Rainbow et
al. (1982). We made several modifications in order to adapt this method
for quantitative autoradiography by optical density measurements
(Rostène et al., 1985a). Autoradiograms, produced by the apposition of
^{3}H-Ultrofilm (LKB, France) to the labeled sections, are obtained after an
exposure of several days, weeks or months depending on the isotope used
for the radiolabeled substance and the amount of molecules specifically
bound to the receptor sites determined by biochemical data. Using
tritium (Unnerstall et al., 1982) or iodinated standards (Rostène and
Mourre, 1985), optical densities for each structure examined is referred
to a computerized standard curve so that the results are expressed as
fmoles/mg protein.

CHANGES IN NEUROTRANSMITTER AND NEUROPEPTIDE RECEPTORS FOLLOWING BRAIN LESIONS

Dramatic modifications in binding sites can be observed after brain
lesions and may in part account for behavioral effects seen in parallel
to brain plasticity.

Neurotoxic destruction of serotoninergic (5-HT) neurons

In vitro autoradiography was recently used to study the topographical
distribution of 5-HT binding sites in the central nervous system (Young
and Kuhar, 1980; Meibach et al., 1980; Biegon et al., 1982). High densities of 5-HT binding sites are found, but not exclusively, in brain areas
where 5-HT terminals have been localized such as the hippocampus, septum,
globus pallidus, suprachiasmatic nucleus, substantia nigra and interpedun-
cular nucleus. Interestingly, both medial and dorsal raphe nuclei where
5-HT neurons originate, also contain high levels of 5-HT binding sites
(Biegon et al., 1982).

In order to test whether these 5-HT binding sites (mainly 5-HT$_1$ sites
since they are labeled with nM concentrations of ^{3}H-5-HT (Peroutka and
Snyder, 1981) are pre- or post-synaptically located, intracerebroventricular injection of a specific neurotoxin for 5-HT neurons, 5-7-
dihydroxytryptamine (5,7-DHT, 150 µg/10 µl) was carried out in nomifensin
(10 mg/kg i.p.)-treated male rats in order to block the non-specific
uptake of the drug by catecholaminergic neurons (Baumgarten et al., 1979).
As shown in figure 2, a 60% decrease in 5-HT$_1$ binding sites can be observed in the raphe dorsalis 2 weeks after 5,7 DHT. A significant decrease
(37%) is also obtained in the median raphe nucleus. Those effects are
strictly concentrated in both nuclei, since densities in adjacent brain
regions do not show any change as illustrated in figure 3a and b. Interestingly, the effect is no more observed 60 days after the lesion (figure

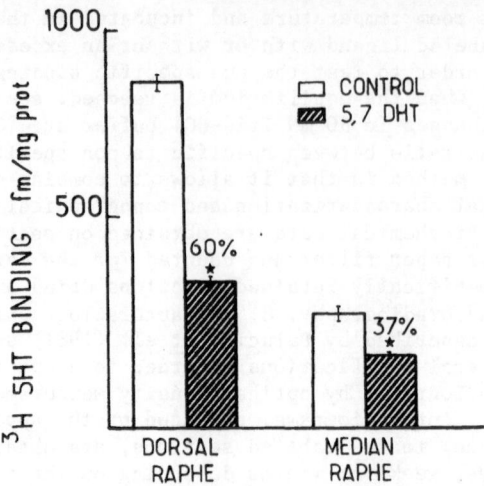

Fig. 2. Effect of i.c.v. injection of 5,7–DHT (150µg/10µl) on the
density of 5-HT$_1$ binding sites in the raphe nuclei as
measured by densitometry on autoradiograms.
* p<0.01 vs respective control.

3c). We can conclude from those data that both raphe nuclei contain 5-HT$_1$
"autoreceptors" located presynaptically, which are able to adjust themselves depending on the endogenous situation. The presence of those "autoreceptors" in raphe nuclei are confirmed by recent data using ^{3}H-5-HT or
^{3}H-PAT, a 5-HT$_{1A}$ agonist (Gozlan et al., 1983; Weissmann-Nanopoulos et
al., 1985). No modification in 5-HT$_1$ binding sites is observed after
5,7–DHT in various other brain structures with the exception of a slight
(20%) but significant increase in 5-HT$_1$ binding sites in the substantia
nigra and in the dorsal dentate gyrus of the hippocampal formation, an
effect which may be explained by a hypersensitivity of postsynaptically-
located 5-HT receptor sites in both structures (Nelson et al., 1978).

Effect of brain lesions on peptide-amine interactions at the receptor
site level

Lesions of some neurotransmitter pathways may affect receptor sites
of other neuroactive substances. A good example of such heterologous
interaction is illustrated by the close relationship which exists between
central dopaminergic (DA) pathways and neurotensin (NT) neurons.

High densities of NT binding sites have been found in the ventral
tegmental area (VMT/VTA) and in the substantia nigra (SN), brain areas
which contain the A$_{10}$ and A$_9$ DA cell groups (Quirion, 1983). Destruction
of these DA neurons by a neurotoxin, 6-hydroxydopamine (6 OHDA, 4 µg)
into the VTA, results in a large decrease in the number of NT binding
sites in the DA cell groups but also in the striatum (Hervé et al., 1985).
In contrast, similar lesions induce an increase in the number of NT binding sites in the lateral part of the prefrontal cortex with no modification in the nucleus accumbens, both structures receiving an important DA
innervation from the VTA (figure 4). However, blockade of DA neurotransmission by a long-acting neuroleptic such as piportil, induces an hyper-

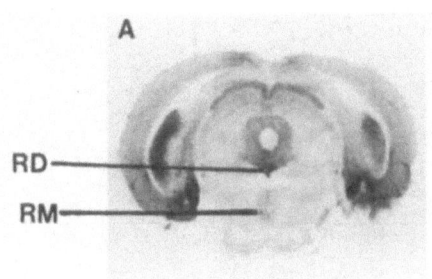

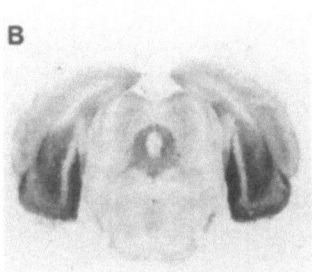
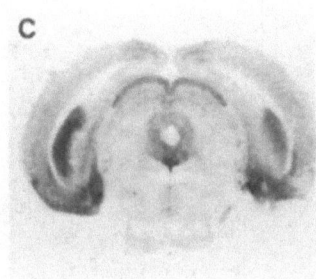

Fig. 3. Autoradiographic distribution at the level of the raphe
nuclei of 5-HT binding sites in control (A) or 14 days (B)
and 60 days (C) after i.c.v. injection of 5,7-DHT. RD: dorsal
raphe; RM: median raphe.

sensitivity of NT binding sites in the prefrontal cortex, the nucleus
accumbens and the central striatum, suggesting a postsynaptic localization of NT binding sites in those cortico-limbic structures (Hervé et
al., 1985). Such a regulation by heterologous afferent fibres has already
been proposed by several authors (Brunello et al., 1982; Tassin et al.,
1982; Fuxe et al., 1983; Rostène et al., 1983).

This interaction between DA and NT may have some clinical importance.
We recently reported a loss of ^{125}I-NT binding in the SN of Parkinsonian
subjects (figure 5) suggesting that NT receptors occur on DA cell bodies
and/or dendrites in human SN, and that NT may have a crucial role in the
regulation of DA pathways involved in Parkinsonism (Sadoul et al., 1984;
Uhl et al., 1984). Besides, treatment with neuroleptics has been shown
to enhance NT levels in some cerebral areas innervated by DA neurons
(Govoni et al., 1980) and to increase neurotensin binding sites in the
substantia nigra (Uhl and Kuhar, 1984). In addition, increases in NT
levels in cortical DA-innervated areas of schizophrenic patients
(Nemeroff et al., 1983), together with the regulation of NT receptor
density by DA as shown in the present data, may suggest that some of the
clinical effects of neuroleptics seen in schizophrenic patients are
partly mediated by changes in NT neurotransmission.

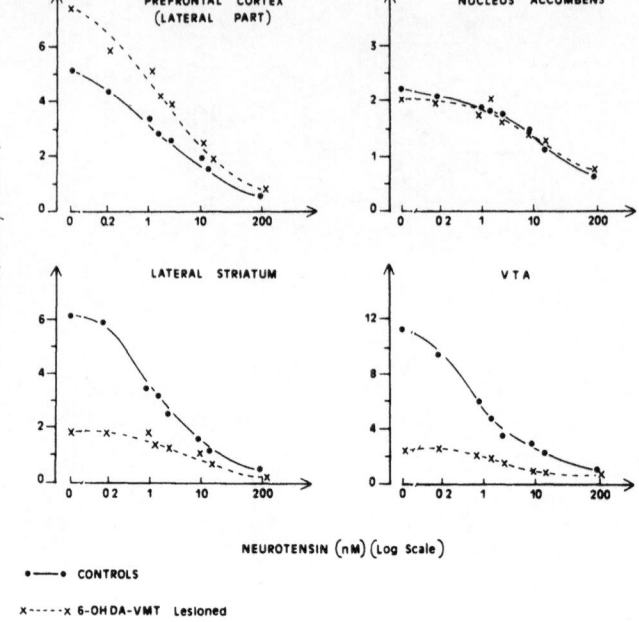

Fig. 4. Displacement curves of ^{125}I-NT binding by unlabeled NT in
control and 6 OHDA-VTA lesioned rats in various brain areas
as measured by quantification of autoradiograms. From
Hervé et al., 1985

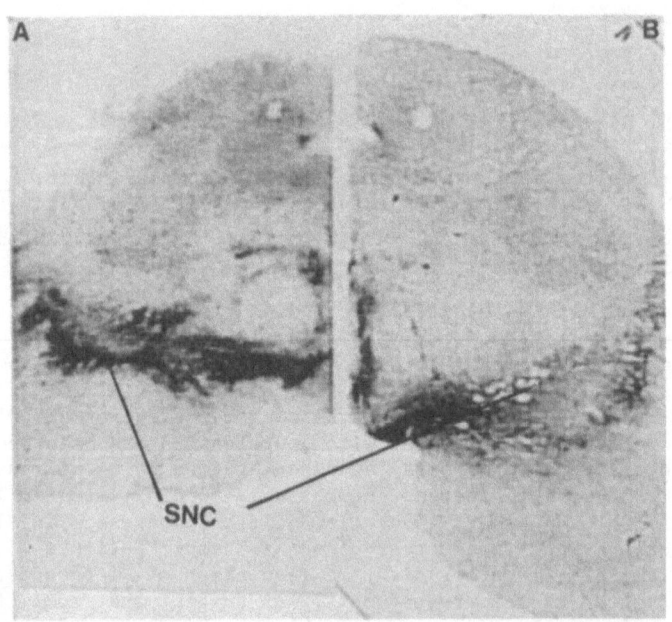

Fig. 5. Autoradiograms of in vitro ^{125}I-NT binding to human
substantia nigra from control (A) and Parkinsonian (B)
subjects. SNC: substantia nigra, pars compacta. From Sadoul
et al., 1984

HORMONAL REGULATION OF PEPTIDE-AMINE INTERACTIONS AT THE POSTSYNAPTIC
RECEPTOR SITE

An important feature of receptor modulation easily observed by means
of in vitro quantitative autoradiography is that a peptide may exert its
neuromodulatory role on brain functions by regulating receptor sensitivity of "classical" neurotransmitters.

For instance, in the suprachiasmatic nucleus (SCN), we have recently
shown that the vasoactive intestinal peptide (VIP) was able to decrease
in vitro the number of 5-HT$_1$ binding sites without affecting the affinity
of ^{3}H-5-HT (figure 6; Rostène et al., 1985b). This effect is strictly
localized in the SCN where VIP and 5-HT neurons physiologically interact
as shown by both morphological and biochemical data (Héry et al., 1984;
Rostène, 1984; Kiss et al., 1984; Bosler and Beaudet, 1985). These results
may serve as neurobiological support for the possible involvement of
VIP in the regulation of circadian rhythms of various pituitary hormones
such as prolactin (Rostène, 1984).

In contrast to what was found in the SCN, VIP is able to increase
the number of 5-HT$_1$ binding sites in the dorsal subiculum of the hippocampus, a structure known to be major site through which the hippocampal
neural activity influences the rest of the brain (Rostène et al., 1985b).
Since the hippocampus is the target brain structure for adrenal steroids
(Mc Ewen, 1982), we investigated the possibility that such steroids may
regulate the effect of VIP on 5-HT$_1$ binding sites, since adrenalectomy
(ADX) was shown to modify hippocampal VIP and 5-HT concentrations ^
(Rotsztejn et al., 1980; De Kloet et al., 1982). Indeed, ADX counteracts

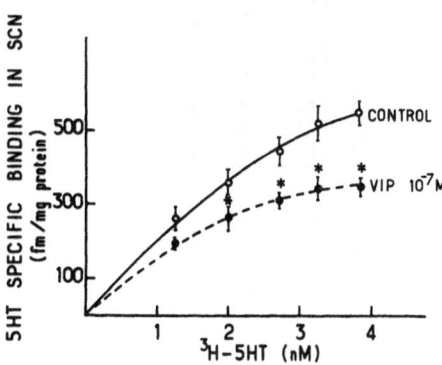

Fig. 6. Effect of VIP on specific binding of ^{3}H-5-HT in the rat
suprachiasmatic nucleus (SCN).   p<0.01 vs respective
control point. From Rostène et al., 1985b

411

the stimulatory effect of VIP on 5-HT$_1$ receptors in the dorsal subiculum
and this effect is restored in ADX-corticosterone treated rats (Rostène
et al., 1985b). Similar influences of the hormonal "milieu" on receptor
sensitivity have been reported for various neurotransmitters (Wagner et
al., 1979; Hruska and Silbergeld, 1980; Rainbow et al., 1980; Biegon et
al., 1980) suggesting that changes in neurotransmitter receptors mediate
some effects of steroids on neurotransmission.

HORMONAL RECEPTORS CHANGES FOLLOWING UP- AND DOWN-REGULATION

We have recently succeeded in using in vitro autoradiography for the
visualization on brain sections of intracellular receptors like those of
adrenal steroids (Sarrieau et al., 1984a, b). Biochemical characterization and topographical distribution of ^{3}H-corticosterone (CS) binding
sites revealed the same features as those obtained by conventional steroid
binding assays using cytosol preparations. ADX induces an increase of CS
receptors whereas ADX followed by in vivo CS treatment produces a 50%
decrease of CS binding site number (figure 7). The present data show that
plasma CS level is able to modulate the number of its own receptors in
the brain and that in vitro quantitative autoradiography allows to study
the plasticity of hippocampal adrenal steroid receptors.

Steroid receptor modulation is also observed during brain development
and aging. In the latter case, it has been recently reported substantial
decreases in the number of CS binding sites in hippocampus and amygdala

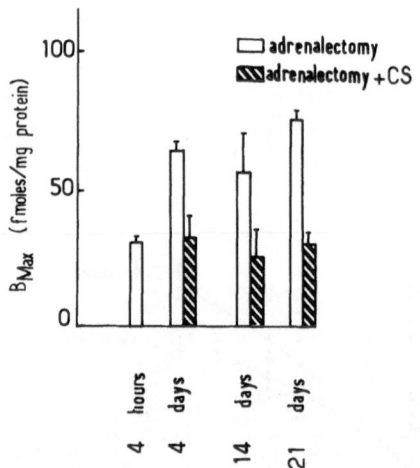

Fig.7. Modification of corticosterone (CS) binding sites in rat
hippocampus by means of in vitro incubation of brain sections
with ^{3}H-CS. The number of binding sites (B$_{max}$) as determined
by Scatchard analysis, was studied in adrenalectomized or
adrenalectomized CS-implanted rats for indicated times.

of aged rat brains (Sapolsky et al., 1983a), an effect which may explain the impairment of the adrenocortical stress response in the aged rat (Sapolsky et al., 1983b).

CONCLUSIONS

In vitro autoradiography may be very helpful for the study of the neuroendocrine system. The distribution of binding sites for various neuroactive substances involved in the regulation of pituitary hormone release both in the brain or directly in the pituitary gland, have been recently investigated. It is the case for neuropeptides such as vasopressin (Yamamura et al., 1983; Baskin et al., 1983; Biegon et al., 1984), LHRH (Reubi and Maurer, 1985), TRH (Rostène et al, 1984), opiates (Lightman et al, 1983), angiotensin II (Israel et al., 1984), somatostatin (Morel et al., 1985), CRF (Wynn et al., 1984) as well as for muscarinic receptors in rat pituitary gland (Hoover and Hancock, 1983). Future developments of this in vitro technique will lead us to an increased resolution of the autoradiograms for the ultrastructural localization of the binding sites. Furthermore, the method will provide new informations on the role of endocrine environment on central and peripheral effects of neuroactive substances, as well as in our understanding of human diseases that have correlated in receptor distributions and densities.

ACKNOWLEDGEMENTS

This work was supported by INSERM and in part by a grant from the MRT, Paris. We thank Y. Agid, M. Dussaillant, C. Fischette, F. Javoy-Agid, B.S. Mc Ewen, M. Moguilewsky, D. Philibert, J..P. Tassin and J.P. Vincent for their important contribution to various parts of this work. We are indebted to Mrs D. Lhenry for typing the manuscript.

REFERENCES

Baskin, D.G., Petracca, F., and Dorsa, D.M. 1983. Autoradiographic localization of specific binding sites for ^{3}H-Arg^8 vasopressin in the septum of the rat brain with tritium-sensitive film. Eur. J. Pharmac. 90: 155–157.

Baumgarten, H.G., Jenner, S., and Schlossberger, H.G. 1979. Serotonin neurotoxins: Effects of drugs on the destruction of brain serotoninergic, noradrenergic and dopaminergic axons in adult rat by intraventricularly, intracisternally or intracerebrally administered 5,7-dihydroxytryptamine and related compounds. In Neurotoxin, Fundamental and Clinical Advances, I.W. Chubb, L.B. Geffen eds, Adelaide University Union Press, Adelaide,pp. 221–226.

Biegon, A., Bercowitz, H., and Samuel D. 1980. Serotonin receptor concentration during the estrous cycle of the rat. Brain Res. 187: 221–225.

Biegon, A., Rainbow, T.C., and Mc Ewen, B.S. 1982. Quantitative autoradiography of serotonin receptors in the rat brain. Brain Res. 242: 197–204.

Biegon, A., Terlou, M., Voorhuis, T.D., and De Kloet, E.R. 1984. Arginine-vasopressin binding sites in rat brain: A quantitative autoradiographic study. Neuroscience Lett. 44: 229–234.

Bosler, O., and Beaudet, A. 1985. VIP neurons as prime synaptic targets for serotoninergic afferents in rat suprachiasmatic nucleus: a combined radioimmunological and immunocytochemical study. J. Cytol., in press.

Brunello, N., Barbacria, M.L., Chuang, D.M., and Costa, E. 1982. Down-
    regulation of beta adrenergic receptors following repeated desmethyl-
    imipramine injections: permissive role of serotoninergic axons.
    Neuropharmacol. 21: 1145-1149.
De Kloet, E.R., Kovacs, G.L., Szabo, G., Telegdy, G., Bohus, B., and
    Versteeg, D.H.G. 1982. Decreased serotonin turnover in the dorsal
    hippocampus of rat brain shortly after adrenalectomy: selective nor-
    malization after corticosterone substitution. Brain Res. 239: 659-663.
Fuxe, K., Agnati, L.F., Benfenati, F., Celani, M., Zini, I., Zoli, M.,
    and Mutt, V. 1983. Evidence for the existence of receptor-receptor
    interactions of monoamine receptors by neuropeptides. J. Neural
    Transm. suppl. 18: 165-179.
Govoni, S., Hong, J.S., Yang, H.Y.T., and Costa, E. 1980. Increase of
    neurotensin content elicited by neuroleptics in nucleus accumbens.
    J. Pharmacol. Exp. Ther. 215: 413-417.
Gozlan, H., El Mestikawy, S., Pichat, L., Glowinski, J. and Hamon, M.
    1983. Identification of presynaptic serotonin autoreceptors using
    a new ligand: ^{3}H-PAT. Nature 305: 140-142.
Hervé, D., Tassin, J.P., Studler, J.M., Dana, C., Kitabgi, P., Vincent,
    J.P., Glowinski, J., and Rostène, W. 1985. Dopaminergic control of
    ^{125}I-neurotensin binding site density in corticolimbic structures of
    the rat brain. Proc. Natl. Acad. Sci. USA, in press
Héry, M., Faudon, M. and Héry, F. 1984. Effect of VIP on serotonin relea-
    se in the suprachiasmatic area of the rat: Modulation by estradiol.
    Peptides 5: 313-317.
Hoover, D.B., Hancock, J.C.. 1983. Autoradiographic localization of
    quinuclidinyl benzilate binding to rat pituitary gland. Neuroendo-
    crinology 37: 297-301.
Hruska, R.E., and Silbergeld, E.K. 1980. Estrogen treatment enhances
    dopamine receptor sensitivity in the rat striatum. Eur. J. Pharmacol.
    61: 397-400.
Israel, A., Correa, F.M.A., Niwa, M., Saavedra, J.M. 1984. Quantitative
    determination of angiotensin II binding sites in rat brain and pitui-
    tary gland by autoradiography. Brain Res. 322: 341-345.
Kiss, J., Léranth, Cs, and Halasz, B. 1984. Serotoninergic ending on
    VIP-neurons in the suprachiasmatic nucleus and on ACTH-neurons in the
    arcuate nucleus of the rat hypothalamus. A combination of high resolu-
    tion autoradiography and electron microscopic immunocytochemistry.
    Neuroscience Lett. 44: 119-124.
Lightman, S.L., Ninkovic, M., Hunt, S.P., and Iversen L.L. 1983. Evidence
    for opiate receptors on pituicytes. Nature 305: 235-237.
Mazière, B., Loc'h, C., Hantraye, P., Guillon, R., Duquesnoy, N.,
    Soussaline, F., Naquet, R., Comar, D., and Mazière, M. 1984. ^{76}Br-
    bromospiroperidol: A new tool for quantitative in vivo imaging of
    neuroleptic receptors. Life Sci. 35: 1349-1356.
Mc Ewen, B.S., 1982. Glucocorticoids and hippocampus: Receptors in search
    of a function. In: Current Topics in Neuroendocrinology, D. Ganten,
    D.W. Pfaff eds, Springer-Verlag, Berlin, pp. 1-22.
Meibach, R.C., Maayani, S., and Green J.P. 1980. Characteriezation and
    radioautography of ^{3}H-LSD by rat brain slices in vitro: the effect of
    5-hydroxytryptamine. Eur. J. Pharmacol. 67: 371-382.
Morel, G., Leroux, P., and Pelletier, G. 1985. Localization and characte-
    rization of somatostatin-14 and somatostatin-28 receptors in the rat
    pituitary as studied by slide-mounted frozen sections. Neuropeptides
    6: 41-52.
Nelson, D.L., Herbet, A., Bourgoin, S., Glowinski, J., and Hamon, M.
    1978. Characteristics of central 5-HT receptors and their adoptive

changes following intracerebral 5,7-dihydroxytryptamine administration in the rat. Mol. Pharmacol. 14: 983-995.

Nemeroff, C.B., Youngblood, W.W., Manberg, P.J., Prange, A.J. Jr, and
Kizer, J.S. 1983. Regional brain concentrations of neuropeptides in
Huntington's chorea and schizophrenia. Science 221: 972-975.

Palacios, J.M., Niehoff, D.L., and Kuhar, M.J. 1981. Receptor autoradiography with tritium-sensitive film: potential for computerized densitometry. Neuroscience Lett. 25: 101-105.

Peroutka, S.J., and Snyder, S.H. 1981. Two distinct serotonin receptors:
regional variations in receptor binding in mammalian brain. Brain
Res. 208: 339-347.

Pert, C.B., Kuhar, M.J., and Snyder, S.H. 1975. Autoradiographic localization of binding of the opiate receptor in rat brain. Life Sci. 16:
1849-1854.

Pfaff, D.W. 1968. Autoradiographic localization of radioactivity in rat
brain after injection of tritiated sex hormones. Science 161: 1355-
1356.

Quirion, R. 1983. Interactions between neurotensin and dopamine in the
brain: an overview. Peptides 4: 609-615.

Rainbow, T.C., Bleisch, W.V., Biegon, A., and Mc Ewen, B.S. 1982. Quantitative densitometry of neurotransmitter receptors. J. Neurosci.
Meth. 5: 127-138.

Rainbow, T.C., De Groff, V., Luine, V.N., and Mc Ewen, B.S. 1980. Estradiol 17 beta increases the number of muscarinic receptors in hypothalamic nuclei. Brain Res. 198: 239-243.

Reubi, J.C., and Maurer, R. 1985. Visualization of LHRH receptors in the
rat brain. Eur. J. Pharmacol. 106: 453-454.

Rostène, W.H. 1984. Neurobiological and neuroendocrine functions of the
vasoactive intestinal peptide (VIP). Prog. Neurobiology 22: 103-129.

Rostène, W., Besson, J., Broer, Y., Dussaillant, M., Grouselle, D.,
Kitabgi, P., Lhiaubet, A.M., Morgat, J.L., Sarrieau, A., and Vial, M.
1985a. Localisation par radioautographie des récepteurs des neuropeptides dans le système nerveux central. Ann. Endocrinol. 46: 27-33.

Rostène, W.H., Fischette, C.T., Dussaillant, M., Mc Ewen, B.S. 1985b.
Adrenal steroid modulation of vasoactive intestinal peptide effect on
serotonin binding sites in the rat brain shown by in vitro quantitative autoradiography. Neuroendocrinology 40: 129-134.

Rostène, W.H., Fischette, C.T.,, Rainbow, T.C., and Mc Ewen, B.S. 1983.
Modulation by vasoactive intestinal peptide of serotonin receptors
in the dorsal hippocampus of the rat brain: An autoradiographic
study. Neuroscience Lett. 37: 143-148.

Rostène, W.H., Morgat, J.L., Dussaillant, M., Rainbow, T.C., Sarrieau, A.,
Vial, M., and Rosselin, G. 1984. In vitro biochemical characterization and autoradiographic distribution of ^{3}H-thyrotropin-releasing
hormone binding sites in rat brain sections. Neuroendocrinology
39: 81-86.

Rostène, W., and Mourre, C. 1985c. Préparation de standards iodés pour
radioautographie quantitative in vitro à l'aide d'un film sensible
au tritium. C.R. Acad. Sci. Paris 301: 245-250.

Rostène, W., Quirion, R., Beaudet, A., and Mazière, B. 1986. New brain
imaging approaches for the visualization of receptors for chemical
messengers and drugs. Médecine Sciences, in press

Rotsztejn, W.H., Besson, J., Briaud, B., Gagnant, L., Rosselin, G., and
Kordon, C. 1980. Effect of steroids on vasoactive intestinal peptide
in discrete brain regions and peripheral tissues. Neuroendocrinology
31: 287-291.

Sadoul, J.L., Checler, F., Kitabgi, P., Rostène, W., Javoy-Agid, F. and
Vincent, J.P. 1984. Loss of high affinity neurotensin receptors in

substantia nigra from Parkinsonian subjects. Biochem. Biophys. Res. Commun. 125: 395-404.

Sapolsky, R.M., Krey, L.C., and Mc Ewen, B.S. 1983a. Corticosterone receptors decline in a site-specific manner in the aged rat brain. Brain Res. 289: 235-240.

Sapolsky, R.M., Krey, L.C., and Mc Ewen, B.S. 1983b. The adrenocortical stress-response in the aged male rat: Impairment of recovery from stress. Exp. Gerontol. 18: 55-64.

Sarrieau, A., Vial, M., Philibert, D., and Rostène, W. 1984a. In vitro autoradiographic localization of ^{3}H-corticosterone binding sites in rat hippocampus. Eur. J. Pharmacol. 98: 151-152.

Sarrieau, A., Vial, M., Philibert, D., Moguilewsky, M., Dussaillant, M., Mc Ewen, B.S., and Rostène, W. 1984b. In vitro binding of tritiated glucocorticoids directly on unfixed rat brain sections. J. Ster. Biochem. 20: 1233-1238.

Sokoloff, L. 1977. Relation between physiological function and energy metabolism in the central nervous system. J. Neurochem. 29: 13-26.

Stumpf, W.E. 1968. Estradiol-concentrating neurons: Topography in the hypothalamus by dry-mount autoradiography. Science 162: 1001-1003.

Tassin, J.P., Simon,H., Hervé, D., Blanc, G., Le Moal, M., Glowinski, J., and Bockaert, J. 1982. Non-dopaminergic fibres may regulate dopamine-sensitive adenylate-cyclase in the prefrontal cortex and nucleus accumbens. Nature 295: 696-698.

Uhl, G.R., and Kuhar, M.J. 1984. Chronic neuroleptic treatment enhances neurotensin receptor binding in human and rat substantia nigra. Nature 309: 350-352.

Uhl, G.R., Whitehouse, P.J., Price, D.L., Tourtelotte, W.W., and Kuhar, M.J. 1984. Parkinson's disease: depletion of substantia nigra neurotensin receptors. Brain Res. 308: 186-190.

Unnerstall, J.R., Niehoff, D.L., Kuhar, M.J., and Palacios, J.M. 1982. Quantitative receptor autoradiography using ^{3}H-Ultrofilm: application to multiple benzodiazepine receptors. J Neurosci. Meth. 6: 59-73.

Wagner, H.R., Crutcher, K.A., and Davies, J.N. 1979. Chronic estrogen treatment decreases beta adrenergic responses in rat cerebral cortex. Brain Res. 171: 147-151.

Weissmann-Nanopoulos, D., Mach, E., Magre, J., Blaquiere, B., and Pujol, J.F. 1985. Evidence for 5-HT$_1$ binding sites on 5-HT containing neurons in the raphe dorsalis and centralis of the rat brain. Neurochem. Intern., in press

Wynn, P.C., Hauger, R.L., Holmes, M.C., Millan, M.A. , Catt, K.J., and Aguilera, G. 1984. Brain and pituitary receptors for corticotropin releasing factor: Localization and differential regulation after adrenalectomy. Peptides 5: 1077-1084.

Yamamura, H.I., Gee, K.W., Brinton, R.E., Davis, T.P., Mac Hadley, and Wamsley, J.K. 1983. Light microscopic autoradiographic visualization of ^{3}H-arginine vasopressin binding sites in rat brain. Life Sci. 32: 1919-1924.

Young, W.S., III, and Kuhar, M.J. 1980. Serotonin receptor localization in rat brain by light microscopic autoradiography. Eur. J. Pharmacol. 62: 237-239.

CYTOSOLIC FACTORS MODULATING PROLACTIN BINDING

T. A. Bramley and A. S. McNeilly

Department of Obstetrics & Gynaecology
Centre for Reproductive Biology
and MRC Unit for Reproductive Biology
37 Chalmers Street
Edinburgh EH3 9EW

INTRODUCTION

Prolactin exerts a variety of effects on a number of tissues, most notably the liver, mammary gland, adrenal, kidney and gonads. Our studies have been concerned with the lactogenic receptors of the pig and sheep corpus luteum. Following the rupture of the preovulatory follicle and shedding of the oocyte at ovulation, the remaining follicle cells rapidly differentiate and hypertrophy. The corpus luteum gland so formed secretes large amounts of progesterone, which transforms the oestrogen-primed endometrium to the secretory state necessary for implantation of an embryo.

Experiments in vivo have demonstrated that prolactin is required for full luteal function in both the sheep (Denamur, Martinet & Short, 1973; Kann & Denamur, 1974; Martal, 1981) and pig (du Mesnil du Buisson & Denamur, 1969; du Mesnil du Buisson, 1973), though the amounts required are minimal (Niswender, 1974; Louw, Lishman, Botha & Baumgartner, 1974). Furthermore, sheep and pig corpora lutea (CL) have specific, high affinity receptors for lactogenic hormones (Rolland, Gunsalus, & Hammond, 1976; Chan, Robertson & Friesen, 1978; Jammes, Schirar & Djiane, 1985). Moreover, prolactin can stimulate progesterone secretion by the pregnant pig CL in vitro (Grinwich, McKibbin & Murphy, 1983).

Prolactins and growth hormones form a closely-related family of hormones with similar structures, which probably evolved from an early gene duplication (Wallis, 1981; Miller & Eberhardt, 1983). Despite their structural similarities (Niall, Hogan, Tregean, Segre, Hwang & Friesen, 1973; Shome & Parlow, 1977), their actions are quite distinct. Thus, growth hormones act via specific, high affinity somatotropic receptors which bind growth hormones, but not prolactins. In contrast, prolactins act via high affinity lactogenic receptors, which bind prolactins but not growth hormones (Shiu & Friesen, 1976). However, human growth hormone (hGH) has both somatotropic and lactogenic activity, and hGH stimulates lactogenesis in mammary gland explants of several species (Forsyth, Folley & Chadwick, 1965; Kleinberg, & Todd, 1980). Furthermore, radiolabelled human growth hormone and sheep prolactin tracers bind equivalently to lactogenic receptors of rat and mouse liver and rabbit mammary gland (Shiu & Friesen, 1974; Posner, Kelly, Shiu & Friesen, 1974; Posner, 1976; Herington, Vieth & Burger, 1976), and both hormones mutually cross - react with equivalent potencies for binding sites in isolated membrane fractions

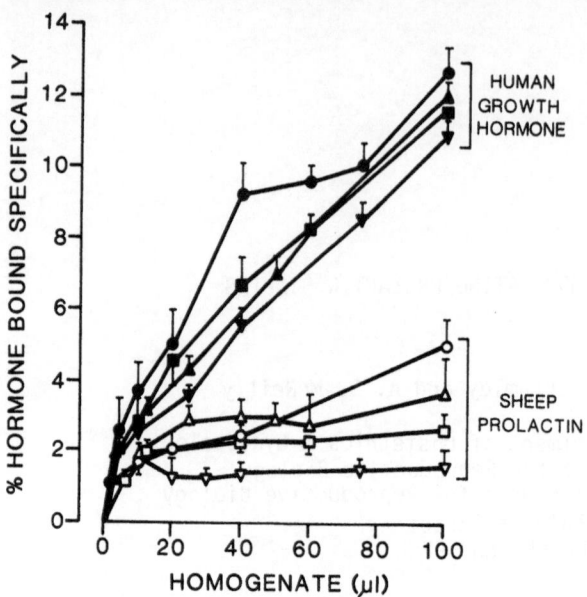

Fig. 1. Binding of [^{125}I]-hGH and [^{125}I]-sheep PRL to homogenates of sheep
corpus luteum (Redrawn from McNeilly, Bramley & Friesen, 1986).

from these tissues. Binding sites of hGH are therefore, usually
differentiated on the basis of their cross-reactivity to either growth
hormones (somatotropic) or prolactins (lactogenic).

Binding of [^{125}I]-hGH and [^{125}I]-sheep PRL to sheep corpus luteum

Because of its ready availability in highly-purified form, and the
stability of the radioiodinated hormone, many studies have used [^{125}
I]-labelled hGH as binding ligand for studies of lactogenic hormone
receptors. However, sheep corpus luteum receptors should interact in vivo
with the homologous hormone, sheep prolactin (PRL). We therefore compared
the binding of sheep PRL and human GH to homogenates of sheep CL (Fig. 1).

Specific binding of hGH tracer was high, and increased linearly with
increasing homogenate concentration. In contrast, binding of sheep PRL to
sheep CL was low, and often showed little or no increase in specific
binding with increasing homogenate concentration. Similar results were
obtained for the binding of [^{125} I]-hGH and [^{125} I]-sheep PRL to sheep CL
and sheep liver microsomes, even though both tracers had similar binding
activities to pregnant rat liver membrane receptors.

Poor binding of homologous prolactin tracers compared to heterologous
tracers has been described for other lactogenic receptors (Posner, 1976;
Posner, Patel, Vezinhet & Charrier, 1980; Servely, Emarre, Houdbine,
Djiane, DeLouis & Kelly, 1983; Gertler, Ashkenazi & Madar, 1984; Jammes,
Schirar & Djiane, 1985), and has often been ascribed to damage incurred
during radioiodination of the prolactin molecule.

However, when the specific binding of fresh hGH and sheep PRL tracers
(with similar specific activities of binding to rabbit mammary gland
receptors) was measured with a number of sheep and pig CL homogenates, a
clear difference was seen. Whereas sheep PRL once more bound poorly to
sheep CL binding sites compared to hGH (ca 2.5 molecules of hGH were bound
for each sheep PRL, Table 1), binding of sheep PRL to (heterologous) pig CL

receptors demonstrated a binding activity equivalent to that of hGH. Furthermore, the number of binding sites and binding affinities of the two tracers were identical for the heterologous receptors of the rat or pig ovary, (Table 2), whereas the same sheep PRL tracer consistently bound less than hGH to sheep CL or sheep liver microsomes (Table 2).

Table 1.Binding of $[^{125}I]$-hGH and $[^{125}I]$-sheep PRL to sheep and pig CL homogenates.

	n	Specific binding ratio (^{hGH}/oPRL)	r
Sheep CL	40	2.42	0.79
Pig CL	40	0.86	0.90

The specific binding of fresh radioiodinated hGH and sheep PRL tracers to homogenates of sheep and pig CL was measured and analysed by least square regression analysis.

Table 2. Binding of $[^{125}I]$-hGH and $[^{125}I]$-sheep PRL to homologous and heterologous receptors.

	$[^{125}I]$-hGH		$[^{125}I]$-sheep PRL	
	Ka ($\times 10^{10}$ M^{-1})	Number of sites (fmoles)	Ka ($\times 10^{10}$ M^{-1})	Number of sites (fmoles)
Rat ovarian homogenates	1.0	67	1.0	67
Pig CL homogenates	6.0	12	6.0	12
Sheep CL microsomes	1.4	79	2.2	18
Sheep liver microsomes	2.1	78	1.9	26

Radioiodination may induce a subtle change in the sheep PRL molecule which alters a domain which is necessary for binding to homologous, but not heterologous receptors. However, iodination of sheep PRL with non-radioactive iodide, or exposure of hormone to lactoperoxidase and hydrogen peroxide under the same conditions used for $[^{125}I]$-labelling failed to affect either the displacement of $[^{125}I]$-hGH from sheep CL receptors, (Fig. 2b) or prolactin bioactivity (Fig. 2a) in the Nb$_2$ tumour cell bioassay, (Klindt, Robertson & Friesen, 1982). Hence, poor binding of sheep PRL to homologous receptors cannot be attributed to iodination damage.

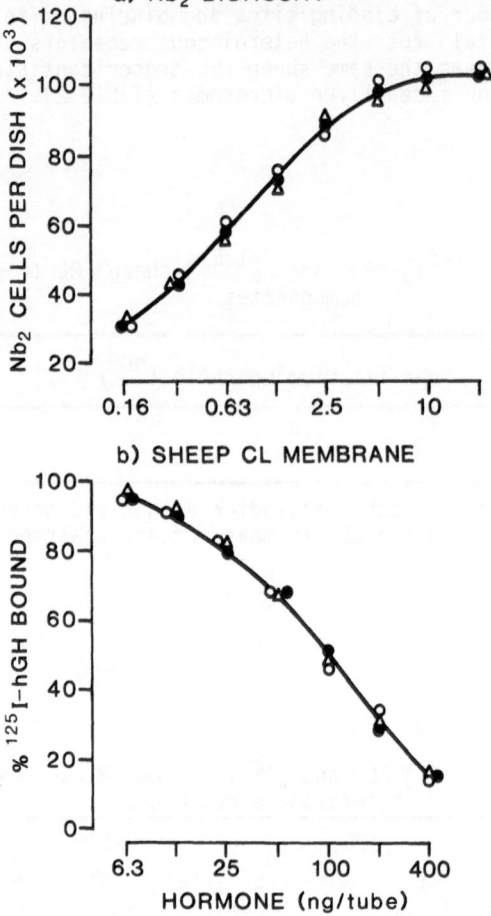

Fig. 2.   Lack of effect of PRL iodination on (a) Nb_2 tumour cell
bioactivity or (b) displacement of [^{125}I]-hGH from sheep CL
receptors.   (Redrawn from McNeilly, Bramley & Friesen, 1976).

Specificity of binding sites

     As mentioned above, hGH can bind to both somatotropic and lactogenic
receptors.   Sheep PRL should bind only to lactogenic binding sites in the
sheep: however, sheep PRL may be somatotropic in other species because GH
receptors see the foreign prolactin as a growth hormone (Nicoll, 1982).
Thus, the discrepancy between binding of hGH and sheep PRL could be
explained if the majority of pig and sheep luteal receptors were
somatotropic rather than lactogenic.   However, [^{125}I]-hGH binding to both
pig and sheep CL was completely displaceable by low (ng) concentrations of
hGH, sheep PRL (and hPRL), whereas 1000-fold greater concentrations of
sheep growth hormone were required (Table 3).   Indeed, even this low
cross-reactivity could be accounted for by contamination of the sheep GH
preparation by sheep PRL (0.1% by RIA). Furthermore, both unlabelled hGH
and sheep PRL could completely displace the binding of either [^{125}I]-hGH
or [^{125}I]-sheep PRL to sheep CL receptors (Fig. 3), indicating that
luteal hGH binding sites were very largely lactogenic.

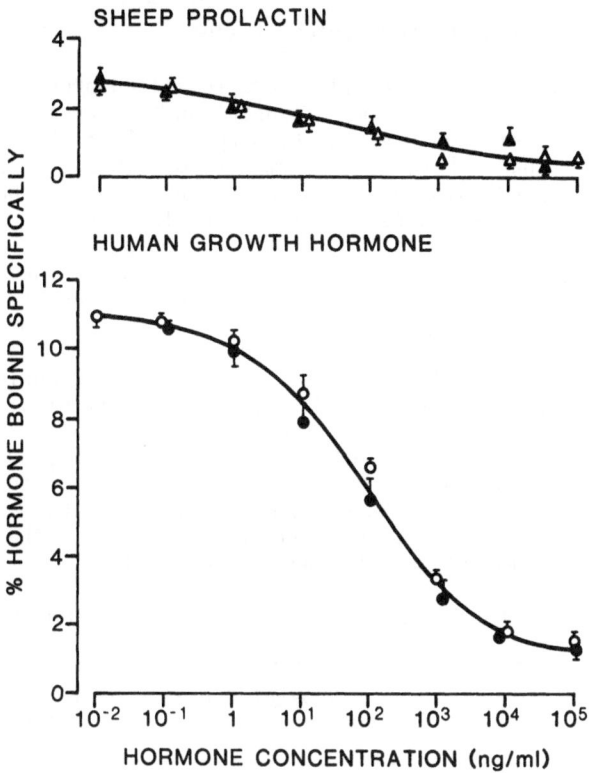

Fig. 3. Displacement of hGH or sheep PRL binding to sheep CL homogenates by unlabelled hGH or sheep PRL (Redrawn from Bramley & Menzies, 1986).

Table 3. Hormone specificity of sheep and pig CL [^{125}I]-hGH binding sites

CL Receptor	Tracer	Apparent Kd (ng/ml)		
		hGH	Sheep PRL	Sheep GH
Sheep	[^{125}I]-hGH	12±3(8)	51±15(6)	1250±50(4)
	[^{125}I]-sheep PRL	13±4(8)	12± 5(7)	
Pig	[^{125}I]-hGH	9±4(5)	11± 5(4)	750±30(4)
	[^{125}I]-sheep PRL	25±4(5)	3± 1(8)	

Apparent Kd values were calculated from displacement curves of hGH or sheep PRL binding to pig or sheep CL homogenates.

Subcellular fractionation of sheep luteal tissue

hGH binding sites in the rat liver are localised predominantly in the Golgi membranes, and only a small proportion are expressed on the cell surface membrane (Herington, Vieth & Burger, 1976; Posner, Bergeron, Josefsberg, Khan, Khan, Patel, Sikstrom & Verma, 1981). The discrepancy between hGH and sheep PRL binding could be accounted for if sheep PRL bound only to cell-surface receptors, whilst hGH also bound to intracellular forms of the receptor. Fractionation of sheep luteal homogenates on surcrose gradients indicated that the peaks of hGH and sheep PRL binding both equilibrated in the same regions of the gradient (d= 1.10 - 1.15 g/cm^3) which were enriched in other cell-surface membrane markers (LH-receptor, [Na$^+$ + K$^+$]-ATPase, 5'- nucleotidase, 5'- phosphodiesterase). Markers for the other major intracellular organelles were enriched in other regions of the gradients (Table 4). Moreover, pretreatment with digitonin, which distinguishes cholesterol-rich (Golgi & cell-surface) membranes from cholesterol-poor (intracellular) membranes (Bramley & Ryan, 1979) increased the buoyant densities of both hGH and sheep PRL binding to the same extent as the other cell-surface markers tested (Table 4). Thus, binding sites for both hormones appeared to be similarly localised in the same regions (Golgi/cell-surface membrane) of the sheep luteal cell.

Table 4   Buoyant densities of sheep CL lactogenic binding sites

Organelle	Buoyant density (g/cm^3)	
	Control	Digitonin-treated
[^{125}I]-hGH binding	1.10-1.15	1.14-1.17
[^{125}I]-sheep PRL binding	1.10-1.15	1.14-1.17
Cell surface membranes	1.10-1.14	1.14-1.17
Mitochondria	1.15-1.17	1.15-1.17
RER	1.10-1.12	1.09-1.12
SER	1.10-1.14	1.07-1.13
GERL	1.10-1.20	1.13-1.19
Lysosomes	1.16-1.20	Solubilised
Nuclei	>1.23	>1.23

Buoyant densities of sheep Cl lactogenic receptors and subcellular organelles. Values are ranges of density (mode $\pm$ 68%) for 3-8 fractionations. (Data from Bramley & Menzies, 1986).

Prolactin degrading enzymes

If sheep luteal cells possessed a protease activity which could degrade prolactin specifically (Posner, Kelly, Shiu & Friesen, 1974), prolactin tracer would be inactivated, and PRL binding would decrease markedly relative to that of hGH.

However, preincubation of sheep CL or liver microsomes with hGH or sheep PRL tracers, followed by incubation of the unbound hormone fraction with fresh sheep or pig CL receptors indicated minimal degradation of either tracer. Furthermore, attempts to block proteinase activities by prolonged preincubation of membranes with a range of protease inhibitors with different specificities for the major classes of protease failed to restore sheep PRL binding to the levels observed with hGH (Table 5). These agents had no effect on hGH/PRL binding ratios in the pig CL. Hence, the discrepancy between sheep PRL and hGH binding to homologous receptors was not due to prolactin degradation.

Furthermore, this discrepancy was not due to the assay conditions employed (pH, activating metal ion, etc), nor to the presence of a soluble PRL-specific receptor which could complex sheep PRL tracer, preventing its recovery in the binding assay. (Sheep CL and liver cytosol preparations had negligible PRL- or hGH-binding activity, and the polyethyleneglycol precipitation technique used to separate bound and free hormone precipitates both particulate and soluble receptors; Shiu & Friesen, 1974b).

Table 5. Lack of effect of protease inhibitors on specific binding of hG and sheep PRL

| Protease specificity | Binding ratio (hGH/oPRL) | | |
	Sheep CL	Sheep Liver	Pig CL
Control	4.42	8.5	1.3
Serine protease			
PMSF	4.45	7.5	1.06
TPCK	5.7	8.6	1.23
BAME	5.1	11.2	1.40
Aprotinin	5.3	8.5	1.5
Soya bean trypsin inhibitor	5.9	11.1	1.5
Chymostatin	5.6	8.5	1.2
Thiol protease			
N-Ethylmaleimide	4.6	10.6	0.90
Iodoacetate	4.7	8.1	0.82
Metalloprotease			
EDTA	4.35	8.6	1.2
o Phenanthroline	4.5	9.0	1.4
Carboxypeptidase			
Pepstatin	4.8	7.3	1.0

Inhibitors were preincubated for 20h at 20°C with membranes before measurement of specific binding of [^{125} I]-hGH and [^{125} I]-sheep PRL.

Effects of cytosol on luteal lactogenic receptors

Sheep. We observed that the discrepancy between hGH and sheep PRL binding increased with purification of sheep luteal membranes: moreover, the abilities of unlabelled hGH and sheep PRL to compete for [^{125} I]-hGH binding diverged with increasing purification of luteal membranes (Table 6). Addition of cytosol could both narrow the difference between hGH and sheep PRL potencies to that seen with homogenates and inhibit [^{125}I]-hGH binding to luteal microsomes (Table 6). The latter effect of cytosol appeared to be tissue-specific: [^{125}I]-hGH binding to sheep CL microsomes was inhibited by cytosol fractions from CL, but not cytosol from liver, kidney, lung or muscle (Table 7). In contrast, hGH binding to liver microsomes was inhibited by liver cytosol but not by cytosol from CL and other tissues. However, luteal cytosol fractions also inhibited the PRL-dependent growth of Nb_2 tumour cells, demonstrating that luteal cytosol could act on the lymphoma cell receptor, and indicating that inhibition of hGH binding was not due to the release of endogenously bound PRL released into the cytosol during homogenisation and tissue fractionation.

   Pig.   Since hGH and sheep PRL were bound similarly by pig luteal receptors, we tested the effects of cytosol on prolactin binding to pig CL microsomes as a negative control. To our surprise, we found a marked stimulation of prolactin binding which was dependent on the amount of cytosol added (Fig. 4). Moreover, the effects of cytosol were strongly dependent on divalent metal ions. In the absence of divalent ions, cytosol inhibited PRL-binding in a dose-related manner, whereas the same cytosol fraction stimulated PRL-binding in the presence of Mn^{2+} (or Mg^{2+}).

Table 6. Effect of cytosol on binding of hGH and sheep PRL to sheep membrane fractions.

	Specific Binding Ratio (hGH/oPRL)	Ratio of Apparent Kd (oPRL/hGH)
Homogenate	4.6 (14)	8.9 (4)
100,000g pellet	19.0 (7)	41.2 (4)
100,000g pellet + cytosol	6.2 (7)	8.6 (3)

Table 7. Tissue specificity of effects of cytosol on sheep [^{125}I]-hGH binding.

Microsomes	Cytosol	Specific binding of [^{125}I]-hGH (% control)
CL	CL	71+4 (8)***
	Liver	92+5 (5)
	Other+	93+6 (5)
Liver	CL	86+5 (6)
	Liver	54+6 (8)***
	Other+	93+8 (6)

*** p <0.001   + Lung, muscle, kidney

Chromatography of a highly-active, stimulatory cytosol fraction on Sephacryl S 200, followed by assay of each fraction for its ability to enhance the binding of PRL to washed, mid-luteal pig microsomes demonstrated two fractions with metal ion-dependent, stimulatory activity with molecular weights of 100,000 and 12,000 to 15,000 respectively. It is not yet clear whether the two peaks are related or not. In the absence of divalent metal ions, there was no stimulation of binding, but a broad region with some inhibitory activity was apparent (Fig. 5).

If the effects of cytosol are important physiologically, one would predict that the levels of these cytosol factors might vary with the stage of luteinization of the corpus luteum. The effects of cytosol fractions prepared from CL at different stages of the luteal phase and pregnancy on prolactin binding to washed pig mid-luteal microsomes are shown in Table 8.

Early luteal cytosol preparations were most frequently inhibitory, whereas cytosol from mid- and late-luteal CL were usually stimulatory. Cytosol from CL of pregnancy had little or no effect.

Conclusions

We have shown that the poor binding of prolactin to homologous receptors in the sheep may be due to removal of factors (present in the cytosol after homogenization and fractionation) which can modulate the binding of PRL to its receptors. Our evidence suggests that these factors may be tissue-specific, and vary with the stage of the luteal phase and pregnancy. Tissue-specific modulation of prolactin receptors (and action?) may be one way in which one particular target organ for PRL can be turned on or off, without affecting other prolactin-sensitive tissues.

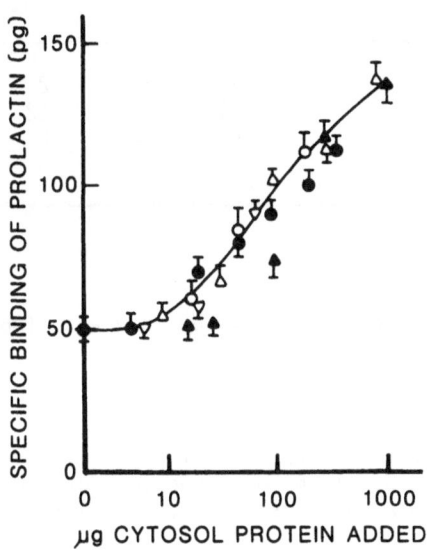

Fig. 4. Effects of pig cytosol on binding of $[^{125}I]$-sheep PRL to washed pig mid-luteal microsomes. (Redrawn from Bramley & Menzies, 1986). Symbols represent different cytosol preparations.

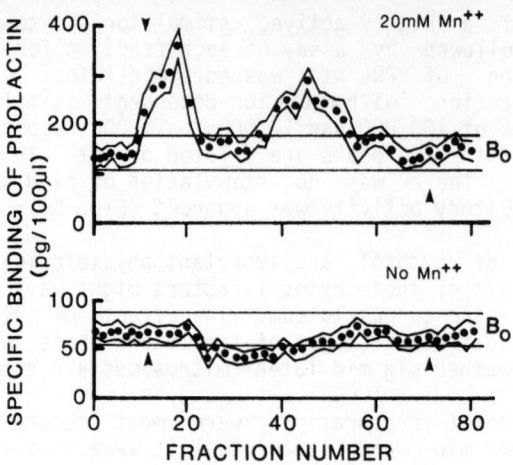

Fig. 5.   Chromatography of a stimulatory pig mid-luteal cytosol fraction on binding of [^{125} I] -sheep PRL to washed mid-luteal microsomes in the absence or presence of divalent cations.   (Redrawn from Bramley & Menzies, 1986).   Arrows indicate V_0 and V_t.

Table 8   Effects of cytosol fractions from CL at different stages of the luteal phase or pregnancy on binding of [^{125} I]-sheep PRL to washed pig mid-luteal microsomes.

Stage of Luteal Phase	Effect on Prolactin Binding (% control)		
	Stimulation (>30%)	No effect (70-130%)	Inhibition (>30%)
Early luteal	2	2	6
Mid luteal	17	12	6
Late luteal	6	2	0
Pregnancy (30 days)	1	9	1

Acknowledgements

This work was supported in part by a grant (TAB) from the MRC (G 8209157 SB).

REFERENCES

Bramley, T.A. & Ryan, R.J. (1979) Endocrinology, 104: 979-988.
Bramley, T.A. & Menzies, G.S. (1986) J. Endocrinology (in preparation).
Chan, J.S.D., Robertson, H.A. & Friesen, H.G. (1978) Endocrinology 102: 632-680.
Denamur, R., Martinet, J. & Short, R.V. (1973) J. Reprod. Fert. 32: 207-218.

du Mesnil du Buisson, F. (1973) in "Le Corps Jaune", pp 225-237,
(R. Denamur & A. Netter eds) Mason et Cie, Paris.

du Mesnil du Buisson, F. & Denamur, R. (1969) Exerpta Med. Int. Cong.
Series 184: 928-934.

Forsyth, I.A., Folley, S.J. & Chadwick, A. (1965) J. Endocrinology 31:
115-126.

Gertler, A., Ashkenazi, A. & Madar, Z. (1984) Molecular & Cellular
Endocrinology 34: 51-57.

Grinwich, D.L., McKibbin, P..E. & Murphy, B.O. (1983) in "Factors
regulating ovarian function" pp 123-128 (G. S. Greenwald & P. F.
Terranova eds) Raven Press, New York.

Herington, A.C., Vieth, N. & Burger, H.G. (1976) Biochemical J. 158:
61-69.

Jammes, H., Schirar, A. & Djiane, J. (1985) J. Reprod. Fert. 73: 27-35.

Kann, G. & Denamur, R. (1974) J. Reprod. Fert. 39: 473-483.

Kleinberg, D.L. & Todd, J. (1980) J. Clin. Endocrinol. Metab. 51:
1009-1013.

Klindt, J., Robertson, M.S. & Friesen, H.G. (1982) Endocrinology 111:
350-352.

Louw, B.P., Lishman, A.W., Botha, W.A. & Baumgarther, J.P. (1974) J.
Reprod. Fert. 40: 455-458.

Martal, J. (1981) J. Reprod. Fert. 30: 210-210.

Miller, W.L. & Eberhardt, N.L. (1983) Endocrine Revs. 4: 97-130.

McNeilly, A.S., Bramley, T.A. & Friesen, H.G. (1986) J. Endocrinology.
(Submitted).

Niall, H.D., Hogan, M.L., Tregean, G.W., Segre, G.V., Hwang, P. & Friesen
H.G. (1973) Rec. Prog. Horm. Res. 29: 387-416.

Nicoll, C.S. (1982) Perspect. Biol. Med. 25: 369-381.

Niswender, G.D. (1974) Endocrinology 94: 612-615.

Posner, B.I. (1976) Endocrinology 98: 645-654.

Posner, B.I., Kelly, P.A., Shiu, R.P.C. & Friesen, H.G. (1974)
Endocrinology 95: 521-531.

Posner, B.I., Patel, B., Vezinhet, A. & Charrier, J. (1980)
Endocrinology 107: 1954-1958.

Posner, B.I., Bergeron, J.J.M., Josefsberg, Z., Khan, M.N., Khan, R.J.,
Patel, B.A., Sikstrom, R.A. & Verma, A.K. (1981) Rec. Bog. Horm. Res.
37: 539-579.

Rolland, R., Gunsalus, G.L. & Hammond, J.M. (1976) Endocrinology
98: 1083-1091.

Servely, J-L., Emane, M.N'G., Houdbine, L-M., Djiane, J., Delouis, C. &
Kelly, P.A. (1983) General & comp. Endocrinol. 51: 255-266.

Shiu, R.P.C. & Friesen, H.G. (1974a) Biochem. J. 140: 301-311.

Shiu, R.P.C. & Friesen, H.G. (1974b) J. Biol. Chem. 249: 7902-7911.

Shiu, R.P.C. & Friesen, H.G. (1976) in "Basic Applications & Clinical
Uses of Hypothalamic Hormones" (A.L. Charro Selgado, R.
Fernandez-Durango, & J.G. Lopez del Campo eds) pp 71-75 Exerpta Medica.
Amsterdam.

Shome, B. & Parlow, A.F. (1977) J. Clin. Endocrinol. Metab 45: 1112-1115.

Wallis, M. (1981) J. Mol. Evol. 17: 10-18.

PITUITARY GnRH RECEPTORS - RECENT STUDIES AND THEIR FUNCTIONAL SIGNIFICANCE

R. N. Clayton*, L. S. Young, S. I. Naik, A. Detta and
S. D. Abbot

Department of Medicine, University of Birmingham
Edgbaston, B15 2TH, U.K.

INTRODUCTION

The demonstration of GnRH receptor changes in a variety of physiological and abnormal states in animals has led to the general belief that these changes are functionally related to gonadotroph function. This idea was based largely on the apparently 'tight' qualitative positive correlation between GnRH receptor changes and changes in serum gonadotrophin levels in animals (reviewed in 1-3). That this may be an oversimplification was suggested by several lines of evidence derived either from *in vivo* studies in mice, or from *in vitro* studies designed to explore the cellular biochemical events involved in GnRH receptor regulation and action. This does not deny that GnRH receptors are a prerequisite for the hormone's action in respective target tissues.

This review will summarise some of our recent studies on GnRH receptor regulation in mice and in primary cultures of rat pituitary cells, attempting to define their functional significance in the context of subsequent responsiveness of the gonadotroph to GnRH challenge.

RECEPTOR REGULATION

a) In vivo studies in mice

Our studies of negative steroid feedback regulation of GnRH receptors in mice, in contrast to rats, showed a consistent and persistent (up to 3 months) 50% decrease in GnRH receptors at a time when serum gonadotrophin levels were elevated (4,5). The GnRH receptor fall could be prevented by replacement with either testosterone or oestrogen plus progesterone (4,5) suggesting that these steroids were required for maintenance of a 'normal" receptor complement in mice. The converse is true of the rat since sex steroid replacement, in doses that maintain accessory gland weights, reduce high post-castration GnRH receptor levels (1-3). On the premise that GnRH receptor levels are regulated primarily by the degree of exposure to endogenous GnRH we reasoned that the post-gonadectomy fall in receptors was due to either (a) receptor down-regulation consequent upon excessive GnRH secretion or (b) reduced endogenous GnRH secretion, an unlikely event when serum gonadotrophins are elevated. Exogenous GnRH treatment of ovariectomised mice neither reversed nor further reduced GnRH receptors, although the same regimen increased GnRH receptors in pituitaries of intact female mice. Additionally, GnRH immunoneutralisation further reduced low receptor levels in ovariectomised as well as intact female mice (6).

These data implied that the receptor fall after ovariectomy was not primarily dependent upon changes in endogenous GnRH secretion, although the homologous ligand was required to some extent for receptor maintenance in mice, as in rats (1-3). Indeed studies in GnRH deficient hypogonadal (hpg) mice indicate a clear role for the up-regulatory role of GnRH on its own receptors in this species (7,8) At least acutely in GnRH deficient animals homologous ligand GnRH receptor up-regulation is not dependent on the presence of the gonads (9), although in animals with 'normal' GnRH secretion the ovaries are important in this regard (10). Thus, as far as hormonal regulation of GnRH receptors in mice is concerned the post-gonadectomy state is unusual. To avoid difficulties in interpretation relating to possible changes in endogenous GnRH secretion we proceeded to investigate the direct effects of oestradiol (E_2) and progesterone (P) using hpg female mice. The combination of these steroids increased GnRH receptors, though levels did not reach those of normal female mice. However, when E_2 and P were combined with exogenous GnRH normal female values were obtained indicating the dual and synergistic hormonal regulation of GnRH receptors acting at the level of the gonadotroph (11). These studies with gonadal steroid treatment of hpg and normal female mice also highlighted the divergent effects of the steroids on inhibition of gonadotrophin secretion and biosynthesis while stimulating GnRH receptor levels. However it should be pointed out that the 50% GnRH receptor fall post-ovanectomy does not limit the pituitary response to maximal stimulation by GnRH since a prompt and large discharge of both LH and FSH is observed (10).

We thus observed, for the first time from *in vivo* studies, that a reduction in GnRH receptors of this magnitude did not adversely affect gonadotroph function. Subsequent *in vitro* studies confirmed this conclusion (vide infra). On the other hand, GnRH receptor increases in hpg mice treated with GnRH seemed to precede the major rise in FSH secretion and synthesis (7) suggesting that these changes were required for activation of gonadotroph function. However, an alternative interpretation is that these receptor changes are simply another consequence of GnRH action rather than a prerequisite for it.

From these *in vivo* studies in mice it is evident that hormonal regulation of GnRH receptors is complex and varies between species, this variation being related principally to the extent by which regulation depends upon gonadal steroid effects at the pituitary level. Thus, no one species can be used as a universal animal model and caution must be exercised in extrapolating from one species to another. Further, the relationship between GnRH receptor changes and gonadotroph function is complicated and it bears mention that no study has yet shown that when GnRH receptors increase after castration of rats the responsiveness of the gonadotroph to GnRH is altered.

b)   In vitro studies: rat pituitary cells in culture

The advantage(s) of the *in vitro* cell culture model for receptor regulation studies are primarily that the hormonal environment can be precisely regulated with respect to a) combination of different hormones b) concentration c) duration of exposure. Also simultaneous assessment of receptor changes and functional responses is possible. The cellular biochemical mediators of receptor up-and down-regulation can also be analysed. Considering gonadal hormone regulation of receptors *in vitro* Giguere et al. (12) showed that culture of cells in the presence of either testosterone or dihydrotestrosterone decreased GnRH receptors and that this was accompanied by reduced sensitivity (ie. shift to the right in the dose-response curve) to GnRH. In contrast, oestradiol has long been known to sensitise the gonadotroph response to GnRH *in vitro* and this is accompanied by a modest (25-50%) increase in GnRH receptors (13,14). Thus, in these circumstances there did seem to be a good correlation between altered receptor content and sensitivity to the ligand, this being similar to the *in vivo* studies in rats (as opposed to mice). Our own studies (Abbot & Clayton, unpublished) with oestradiol have been inconclusive showing either no change or up to 100% increase in GnRH receptors after 2-3 days in the presence of the steroid. We have no clear explanation for the inconsistency of our results between experiments.

The first studies to directly show homologous ligand induction of

GnRH receptors in vitro were those of Loumaye & Catt (15). The increase (100-200% above control) is dose dependent and blocked by a GnRH antagonist so is mediated through the GnRH receptor itself, and occurs after a lag period of some 6-12 hours, well after LH release has ended (15-18). In some studies (15-17) there was a transient (first hour) small down-regulation of receptors prior to the later increase. This was not observed in short-term (up to 12 hours) GnRH antagonist-treated cells, indicating that antagonist-receptor complexes do not become internalised or processed. Antagonists therefore simply occupy receptor sites preventing access of the agonist ligand. Eventually, GnRH antagonist-receptor complexes are internalised but this is probably the result of generalised turnover of the cell membrane. Homologous ligand receptor up-regulation is dependent upon protein synthesis, and intact microtubule function (16,19), though these agents do not alter unstimulated GnRH receptor levels over 12 hrs. Non - GnRH receptor mediated activation of gonadotrophs by membrane depolarisation with KCl induces GnRH receptors by a similar magnitude and with a similar time-course as GnRH itself but, as predicted, is not blocked by the GnRH antagonist (18). Receptor induction by GnRH, membrane depolarisation, or exogenous cyclic adenosine nucleotides is blocked by the Ca^{2+} channel antagonist verapamil and calcium chelators (EGTA) (16,19). Further evidence for the Ca^{2+} dependence of GnRH receptor up-regulation is provided by studies using the Ca^{2+} ionophore A23187. At low concentrations (0.01 and 0.1μM), which do not release LH, receptors are increased though with higher concentrations (10μM) a 50% decrease is observed. (16,19). Thus, as with the releasing action of GnRH receptor up-regulation is a calcium dependent process.

In the early years of searching for the intracellular 'second messenger' for GnRH-stimulated LH release much attention was focused on activation of adenylate cyclase and cAMP generation. The present consensus view is that while cAMP may increase after GnRH stimulation of gonadotrophs this is a late event and is not implicated in the immediate LH releasing action of GnRH. However, cAMP might be involved in other aspects of GnRH action, such as LH biosynthesis, GnRH receptor up-regulation and morphological changes. Certainly in granulosa cells cAMP analogues can induce LH receptor formation (20). We have shown that dibutyryl cAMP,8 Br cAMP, but not inactive analogues or cGMP (all 1mM), increased receptors by about 100% between 4-18 hrs after exposure, an effect surprisingly blocked by a GnRH antagonist. It appears that in some way cAMP analogues interact with the GnRH receptor in cells and at 1mM concentrations active analogues can inhibit ^{125}I-GnRH analogue binding to pituitary cells (Young LS & Clayton RN-unpublished observations), pituitary membranes (21), and solubilised ovarian and pituitary GnRH receptors (22). The precise nature of this interaction has not yet been determined but it appears specific and consistent. Interestingly, the phosphodiesterase inhibitor IBMX (0.2mM) enhanced receptor induction by GnRH, but not LH release, though IBMX inhibited the receptor induction by dbcAMP. It is difficult to pinpoint the precise cellular locus of cAMP action but it clearly allows the conclusion that this nucleotide might be involved in the mechanism of homologous ligand GnRH receptor up-regulation. This would be compatible with cAMP being involved in the longer-term trophic actions of GnRH and might reconcile some of the conflicting interpretations regarding the role of adenylate cyclase in gonadotrophs. Perhaps the cyclase is activated subsequent to other steps in GnRH action such as Ca^{2+}- calmodulin or protein kinase C activation.

FUNCTIONAL CONSEQUENCES OF GnRH RECEPTOR CHANGES

From the foregoing it is evident that many hormonal and biochemical manipulations can alter the GnRH receptor concentration on gonadotrophs. However it is important to determine of what significance these are for subsequent responses to the hormone. We have addressed this question in two ways: firstly by producing GnRH receptor up-regulation *in vitro* and then, after thorough washing, rechallenging the cells with increasing concentrations of GnRH; secondly by preparing cell cultures from pituitaries with either increased or decreased GnRH receptors *in vivo* (ovariectomised rats and mice, respectively) and determining GnRH dose-response characteristics.

a)   In vitro receptor up-regulation and gonadotroph responses

After 3 days in culture 2×10^5 pituitary cells were exposed to 1nM GnRH, 1mM dbcAMP, 58 mM KCl, or $0.1\mu M$ A23187 which resulted in a 70-100% increase in GnRH receptors after 10 hrs. Cells were thoroughly washed and incubated for a further 4 hrs with GnRH (0.01-100nM) and LH release determined. Prior to the challenge with GnRH some cells from each treatment group were lysed for determination of LH content to measure the degree of LH depletion during the 10 hr pretreatment period. Despite the 70-100% increase in receptors those cells treated with dbcAMP and $0.1\mu M$ A23107 produced GnRH dose-response curves superimposable on those of untreated cells. Neither maximal LH release nor the GnRH ED_{50} concentration was affected (23). There was no depletion of cellular

LH in those cells. In contrast cells pretreated with GnRH, KCl, or 10μM A23187 showed marked reduction in LH release at all GnRH concentrations without change in ED_{50}. There was marked depletion of LH content and when corrected for this (24) the dose-response curves were the same as for control cells. That the resultant decrease in absolute LH release is due to reduction in cellular content of the hormone was indicated by normal dose-responses to other LH secretagogues like KCl and calcium ionophore (23). However, pretreatment with a higher GnRH concentration ($10^{-8}M$) caused complete desensitisation to subsequent GnRH challenge, despite receptor-up regulation, this result being similar to that described previously (16,24,25).

Our studies reach the same conclusion as those of others (16,24,25) that pituitary desensitisation is not only dependent upon the concentration and duration of exposure to GnRH but is the result primarily of disruption of as yet undetermined post-receptor steps in the signal transduction cascade.

We were unable, therefore, to show any sensitisation of gonadotroph function as a consequence of GnRH receptor up-regulation *in vitro*.

b) In vivo receptor changes and subsequent gonadotroph resonses

A possible reason for our failure to demonstrate any change in gonadotroph function after *in vitro* up-regulation of GnRH receptors may relate to the relatively short time of treatment (10hrs) being inadequate for induction of the functional coupling of these 'new' receptors to the signal transduction events. In an attempt to exclude this possibility pituitary cell cultures were prepared from rats that had been ovarietomised 10-14 days previously in which, from previous studies, we knew that GnRH receptors would have been increased for at least 7 days. The responses to GnRH of ovariectomised rat pituitary cells were compared with those derived from pituitaries of 10 day ovariectomised mice in which the GnRH receptor content is reduced by 50%. After three days in culture, the cells were challenged with GnRH (0.01-100nM) and LH release measured after 4 hrs and compared with the responses of cells derived from intact rats or mice. In all experiments the relative GnRH receptor changes found in pituitaries homogenised immediately after sacrifice (70-100% increase relative to intact for ovx rats and 50% decrease for ovx mice) were retained both immediately after enzymic dispersion and after 3 days in culture immediately prior to GnRH challenge (Abbot & Clayton, submitted for publication).

The GnRH dose-response curves from the ovx rat and mouse pituitary cells showed increased LH release of between 2-3 fold at all concentrations of hormone, including the basal. There was no change in the GnRH ED_{50} concentration. The increased amount of LH release was probably the result of more hormone being available for release since in both whole pituitaries prior to dispersion and in cells after 3 days of culture the LH content was 1.5 – 3 times that in glands and cells from intact pituitaries.

Thus, as with *in vitro* GnRH receptor up-regulation increasing GnRH receptors by 70-100% *in vivo* did not sensitise the gonadotroph to the ligand. Likewise a 50% decrease in receptors did not impair gonadotroph responsiveness.

INTERPRETATION OF GnRH RECEPTOR CHANGES AND THEIR FUNCTIONAL SIGNIFICANCE

It is necessary to try and explain the failure to demonstrate any functional alteration in gonadotrophs with altered receptor concentration.

First it is possible that the experimental system was not sufficiently sensitive to demonstrate small changes in GnRH ED_{50} concentration. It probably cannot reliably detect shifts of less than 2-3 fold. With an increased GnRH receptor concentration we would predict (by the law of mass action [H] + [R]   [HR]) a greater formation of ligand-receptor complexes at lower ligand concentration. This predicts a shift to the left in the GnRH dose response curve without change in maximum LH release, provided all these receptors are coupled to the release machinery of the cell. That this was not observed may suggest that the extra receptors are not 'functional' (with the proviso above that our methods are sufficiently sensitive). There is no means of distinguishing between functional and non-functional receptors until a single receptor-coupled biochemical marker can be identified, which has not so far been possible in this ligand-receptor system.

Secondly, the *in vitro* assays measured LH accumulation over 4 hrs. It is possible that increased GnRH receptor levels alter the kinetics of LH release such that the first phase (15-30 mins) is enhanced. This remains to be tested experimentally.

Thirdly, it has been shown that only 20% of GnRH receptors need be occupied to elicit maximal LH release (26) (ie. the 'spare' receptor concept (27). This conclusion is obtained because the GnRH dose-response curve for LH release lies to the left of the GnRH receptor occupancy curve. On this basis it is not surprising that an increase from 100% of receptors to 200% does not alter maximal LH output.

Conversely, a reduction from 100% to 50% is unlikely to impair secretion. It would be necessary to reduce GnRH receptors to well below 20% of 'control' values before responsiveness was impaired, and this was not achieved in ovariectomised mice. Such a severe reduction in GnRH receptors is probably not achieved in any *in vivo* situation, for even in hpg mice GnRH receptors values are 30% of those of controls (7) and these pituitaries can release LH *in vivo* though the absolute amount is limited by very low pituitary LH content. The only practical way to produce marked loss of GnRH binding, without disrupting other intracellular events, would be with a GnRH antagonist *in vitro* or *in vivo*.

Of the possibilities discussed the spare receptor explanation seems the most reasonable, though there may be others that are currently unknown or untested.

The available evidence indicates that modulation of gonadotroph GnRH receptor concentration is not a major determinant of the cells responsivity, which is determined at other, post-receptor, sites. These might include (i) amount of LH in the 'readily-releasable' pool (ii) movement of LH secretory granules to the marginal zone of the cell (28) (iii) alteration in the signal-transduction events such as membrane phospholipid turnover and/or protein kinase C redistribution. The only possible exception(s) to this view are the *in vitro* studies with either androgen or oestrogen treatment (v.s), but here the receptor changes were relatively small (25-50%) while the sensitivity changes to GnRH were more marked, again implying a component of post-receptor modification of function.

Since we have been unable to define any major consequences of GnRH receptor changes for gonadotroph function - what is the meaning of the receptor changes that have been demonstrated both *in vivo* and *in vitro*? The majority of changes reported have been of receptor increases, either by homologous ligand up-regulation, or by steroid hormones. These occur slowly over hours, are dependent upon protein synthesis and microtubules, and occur in response to exogenous cyclic adenosine nucleotides which produce morphological changes in other cell types (20). Further, the delayed receptor increase makes it most unlikely that the homologous ligand exposes hidden binding sites already present in the cell membrane. These facts all point to the conclusion that GnRH receptor up-regulation represents a 'trophic' effect of GnRH independent of its releasing ability. As such the receptor changes are another expression of GnRH action in gonadotrophs. Thus, the earlier concept (1-3) that *in vivo* GnRH receptor levels reflect the extent of preceeding (hours not mins) pituitary exposure to endogenous GnRH remains valid. While this may well be correct for the rat, it must be remembered that in other species (e.g. mouse) changes in pituitary exposure to gonadal hormones must also be considered.

Acknowledgments

The authors are grateful of the M.R.C., Govt of Pakistan, and Central Birmingham Health District for financial support, and Dr H.M. Charlton for supplies of normal and hpg mice.

References

1) R.N. Clayton & K.J. Catt (1981) Gonadotrophin-releasing hormone receptors: characterisation, physiological regulation, and relationship to reproductive function Endocrine Reviews 2: 186

2) J.C. Marshall, A. Barkan, G.A. Bourne, J.A. Duncan, A. Garcia-Rodriguez, D.R. Pieper & S. Regiani. (1983) Physiology of pituitary gonadotrophin-releasing hormone receptors. Ch 22 p 205 In Recent Advances in Male Reproduction: Molecular Basis and Clinical Implication. Ed D'Agata, Lipsett, Polosa & Van der Molen Raven Press New York.

3) R.N. Clayton, S.I. Naik, L.S. Young & H.M. Charlton (1984) Physiological regulation of pituitary GnRH receptors. Biochemical Endocrinology: Hormonal Control of the Hypothamo-Pituitary Gonadal axis Ed. McKerns K. W. & Naor Z. Plenum Press New York. p141.

4) S.I. Naik, L.S. Young, H.M. Charlton & R.N. Clayton. (1984) Pituitary gonadotrophin-releasing hormone receptor regulation in mice. I males. Endocrinology 115: 106

5) S.I. Naik, L.S. Young; H.M. Charlton & R.N. Clayton. (1984) Pituitary gonadotrophin-releasing hormone receptor regulation in mice II. Females. Endocrinology 115: 106.

6) S.I. Naik, L.S. Young, G. Saade, A. Kujore, H.M. Charlton & R.N. Clayton. (1985) Role of GnRH in the regulation of pituitary GnRH receptors in female mice. J. Reprod. Fertil. 74: 605.

7) L.S. Young, A. Speight, H.M. Charlton, & R.N. Clayton. (1983) Pituitary gonadotrophin releasing hormone receptor regulation in hypogonadotrophic hypogonadal (hpg) mouse Endocrinology 113:55.

8) L.S. Young, A. Detta, R.N. Clayton, A. Jones & H.M. Charlton (1985) Pituitary and gonadal function in hypogonadotrophic hypogonadal (hpg) mice bearing hypothalamic implants. J.Reprod. Fert 74: 247.

9) A. Detta, S.I. Naik, H.M. Charlton, L.S. Young & R.N. Clayton. (1984). Homologous ligand induction of pituitary gonadotrophin releasing hormone receptors in vivo is protein synthesis dependent. Molec. Cell. Endocrinology 37: 139.

10) S.I. Naik, G. Saade, A. Detta & R.N. Clayton (1985) Homologous ligand regulation of gonadotrophin-releasing hormone receptors in vivo: relationship to gonadotrophin secretion and gonadal steroids. J. Endocrinol. 107:41.

11). S.I. Naik, L.S. Young, H.M. Charlton, & R.N. Clayton. (1985) Evidence for a pituitary site of gonadal steroid stimulation of GnRH receptors in female mice. J. Reprod. Fertil. 74: 615.

12) V. Giguere, F-A Lefebvre, & F. Labrie. (1981) Androgens decrease LHRH binding sites in rat anterior pituitary cells in culture. Endocrinology 108: 350.

13) E. Loumaye & L. Forni. (1982) Regulatory actions of 17β estradiol and progesterone upon pituitary GnRH receptors in vitro Endocrinology 110 Supl. p286 Abstr. 825.

14) L.K. Tang, A.C. Martellock & J.K. Honuchi. (1982) Estradiol stimulation of LH response to LHRH and LHRH binding in pituitary cultures. Am. J. Physiol 242: E392.

15) E. Loumaye & K.J. Catt. (1982) Homologous regulation of gonadotrophin releasing hormone receptors in cultured pituitary cells. Science 215: 983.

16) E. Loumaye & K. J. Catt. (1983) Agonist-induced regulation of pituitary receptors for gonadotrophin-releasing hormone. J. Biol. Chem. 258: 12002.

17) P.M. Conn, D.C. Rogers & S.G. Seay. (1984) Biphasic regulation of the GnRH receptor by receptor microaggregation and intracellular Ca^{2+} levels. Mol Pharmacol 25: 51.

18) L.S. Young, S.I. Naik & R.N. Clayton. (1984) Adenosine 3'5' monophosphate derivatives increase gonadotrophin releasing hormone receptors in cultured pituitary cells Endocrinology 114: 2114.

19) L.S. Young, S.I. Naik & R.N. Clayton. (1985) Pituitary gonadotrophin releasing hormone receptor up-regulation in vitro: dependence on calcium and microtubule function. J. Endocrinol 107: 49.

20) M. Knecht & K.J. Catt. (1982) Induction of luteinising hormone receptors by adenosine 3'5'- monophosphate in cultured granulosa cells. Endocrinology 111: 1192.

21) M.A. Smith, M.H. Perrin & W.W. Vale. (1982) Interaction of adenosine 3'5' monophosphate derivatives with the gonadotrophin releasing hormone receptor on pituitary and ovary. Endocrinology 111: 1981.

22) A.M. Capponi, M.L. Aubert & R.N. Clayton (1984) Solubilised active ovarian gonadotrophin-releasing hormone receptors retain binding properties for adenosine 3'5' cyclic monophosphate derivatives. Life Sciences 34: 2139

23) L.S. Young, S.I. Naik & R.N. Clayton. (1985) Increased gonadotrophin releasing hormone receptors on pituitary gonadotrophs: effect on subsequent LH secretion. Mol. Cell. Endocrinol. 41: 69.

24) M.A. Smith, M.H. Perrin & W.W. Vale. (1983) Desensitisation of cultured pituitary cells to gonadotrophin releasing hormone: evidence for a post-receptor mechanism. Mol. Cell. Endocrinol 30: 85.

25) Z. Naor, M. Katikineni, E. Loumaye, A.G. Vela, M.L. Dufau & K.J. Catt. (1982) Mol. Cell. Endocrinology. 27: 213.

26) Z. Naor, R.N. Clayton & K.J. Catt. (1980) Characterisation of gonadotrophin-releasing hormone receptors in cultured rat pituitary cells. Endocrinol 107: 1144.

27) K.J. Catt, J.P. Harwood, G. Aguilera & M.L. Dufau (1979) Hormonal regulation of peptide receptors and target cell responses. <u>Nature</u> 280: 109.

28) J.F. Morris, C.E. Lewis & G. Fink. LHRH priming in gonadotrophs: a model system for the analysis of neuroendocrine mechanisms at the cellular level.   Proceedings of IFBE meeting Edinburgh (1985).

EFFECTS OF AN LHRH ANTAGONIST ON LUTEAL FUNCTION IN THE MACAQUE

Hamish M. Fraser, Mairwen Abbott and Neil C. Laird

MRC Reproductive Biology Unit
Centre for Reproductive Biology
37 Chalmers Street
Edinburgh EH3 9EW
Scotland

INTRODUCTION

The role of luteinizing hormone releasing hormone (LHRH) and the pituitary
gonadotrophins in controlling progesterone secretion from the corpus luteum
of the human and the non-human primates is one of the most important
fundamental questions concerning the hypothalamo-pituitary-ovarian axis
still to be resolved. An understanding of this inter-relationship would
help in the development of a way of reliably preventing progesterone
production from the corpus luteum necessary for the establishment and
maintenance of early pregnancy.

The variety of approaches which have been used to investigate the role of
the pituitary in control of luteal function have produced some conflicting
results. Studies using antibodies to human chorionic gonadotrophin (hCG)
or ovine LH to neutralize LH during the early luteal phase in monkeys
demonstrated a shortening of cycle length (Moudgal et al 1972; Groff et al
1984). Investigations of hypophysectomized women treated with LH or hCG
for different periods after ovulation suggested that while the quota of LH
required for ovulation could maintain progesterone secretion for a few
days, the further daily administration of LH was necessary for normal
luteal function (Van de Wiele et al 1970). In contrast, evidence that
luteal function is independent of continued pituitary luteotrophic support
has been provided by observations in rhesus monkeys hypophysectomized 1 day
after ovulation (Asch et al 1982). The result was supported by further
studies in which LH release was suppressed by blocking the LHRH receptor
by administration of an LHRH antagonist throughout the luteal phase
(Balmaceda et al 1983). In both experiments progesterone secretion
continued as normal.

The situation is complicated by the possibility that the dependence of the
corpus luteum on the pituitary may change with age, there being an initial
phase during which progesterone production is maintained by the LH produced
by the mid-cycle surge, after which it becomes dependent on further LH
secretion from the pituitary gland.

In recent years there has been considerable progress made in the
understanding of the neuroendocrine control of the menstrual cycle.
Although the transport of LHRH in the confines of the hypophysial portal

system has all but precluded detailed analysis of the LHRH release pattern
at different stages of the cycle, much has been learnt from meticulous
studies on the profile of the frequency and amplitude of LH pulses
occurring in the peripheral blood and believed to closely reflect the
number of LHRH pulses from the hypothalamus. In addition to those
physiological studies, the LHRH signal can be manipulated in several ways.
Surgical destruction of the arcuate region of the medial basal hypothalamus
of rhesus monkeys prevented endogenous LHRH secretion allowing the
examination of the profile of exogenous LHRH necessary to re-initiate
ovulatory cycles (Knobil, 1980). In the normal animal the action of LHRH
can be immediately neutralized at a specific time by i.v. administration of
an LHRH antiserum (Fraser et al 1984; Lincoln et al 1985). More recently
antagonists capable of blocking the pituitary LHRH receptor at reasonable
doses in primates have been developed (Nestor et al 1983).

Use of these approaches has been largely centred around the control of
follicular development and induction of the preovulatory LH surge.
Latterly, however, they have been employed to study the role of LHRH in
luteal function and are resolving the issue of the hypothalamic-pituitary
control of the corpus luteum.

CHANGES IN LH PULSE FREQUENCY DURING THE LUTEAL PHASE

While it has been known for many years that as the luteal phase progresses
LH pulse frequency decreases and amplitude increases, it is only recently
that the approach of detailed evaluation of these changes and their
relationship to progesterone secretion has been exploited in studies in the
rhesus monkey (Ellinwood et al 1984; Healy et al 1984) and in women
(Filicori et al 1984). These studies have revealed that during the early
luteal phase LH pulse frequency is highest, with low amplitude and high
basal value. There is no apparent relationship between progesterone
secretion and LH pulses, the progesterone concentrations being low and
without episodic pattern. During the mid luteal phase, the majority of LH
pulses are associated with progesterone pulses, although some progesterone
rises do not appear to be assoicated with an LH pulse. As the late luteal
phase is entered the LH pulses are infrequent, but of high amplitude and
are followed by a large episode of progesterone release from the corpus
luteum.

The apparent reduction in LHRH pulse frequency as the luteal phase
progresses is presumably due to a suppressive action of the increasing
concentrations of progesterone on the hypothalamic LHRH pulse generator
(Ellinwood et al 1984; Lincoln et al 1985). The relationship between LH
pulses and progesterone secretion during the mid-to-late luteal phase
strongly indicate that the corpus luteum is being governed largely by acute
changes in pituitary LH release at this time. The absence of a correlation
between LH pulses and progesterone during the early luteal phase still
leaves open the possibility that the corpus luteum is autonomous of LH at
this time or is maximally stimulated by the high basal LH. Studies
described below in which the LHRH signal is blocked during different stages
of the luteal phase are beginning to unravel this relationship.

EFFECTS OF LHRH WITHDRAWAL

In an elegant study, Hutchison and Zeleznik (1984) induced ovulatory cycles
with uniform pulses of LHRH in rhesus monkeys in which endogenous hormone
release was prevented by hypothalamic lesions, and examined the effects of
stopping the pulses on day 3 or day 8 of the luteal phase. By stopping
LHRH pulses on day 8 progesterone in the blood fell to non-detectable
levels by the following afternoon and even more crucial, stopping pulses on

day 3 resulted in non-detectable progesterone by the afternoon of day 5. Both treatments resulted in early menstruation. The results provide compelling evidence that the primate corpus luteum requires the presence of pituitary gonadotrophin both during the early as well as the mid-luteal phase.

Our own studies have involved the use of the stumptailed macaque (Macaca arctoides) which has a luteal phase lasting 14–17 days (figure 1). In studies in which an ovine gamma globulin to LHRH of high titre and potency was administered i.v. at various stages of the cycle to neutralize LHRH we made the preliminary observation that in 2 of 3 animals in which the antibody was given on the day after the preovulatory LH surge that the rise

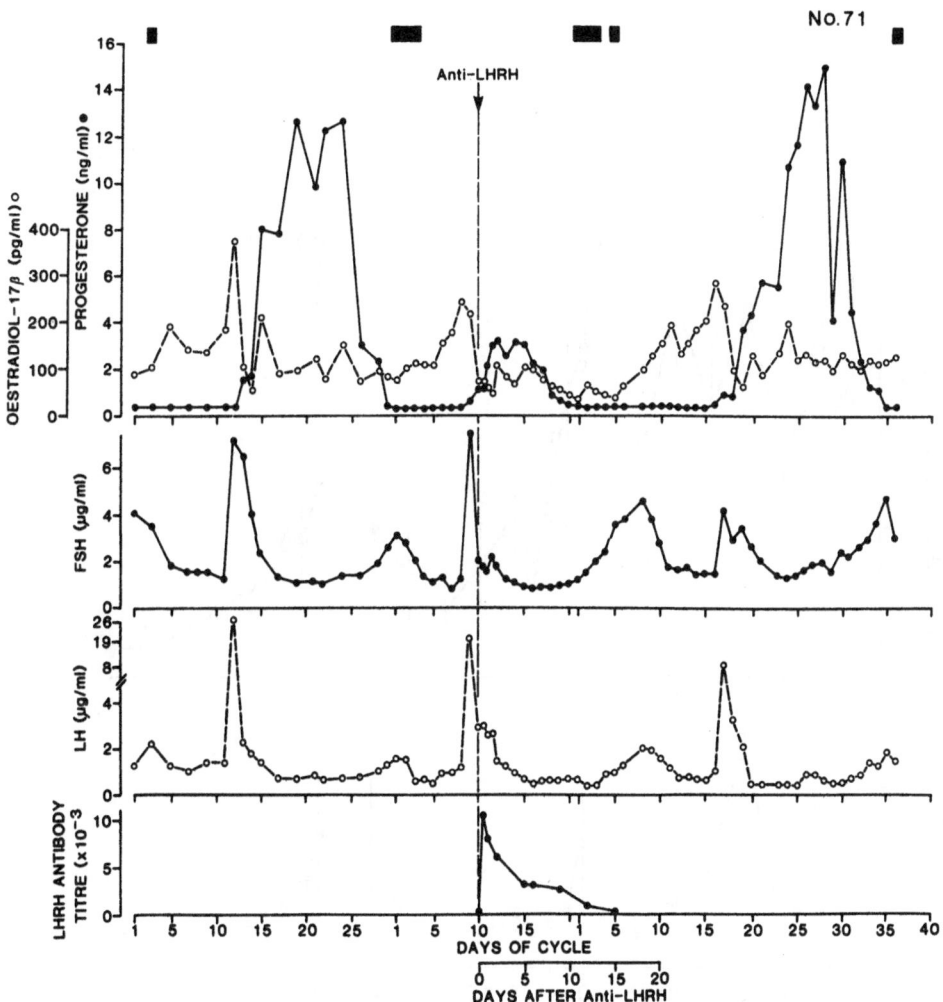

Fig. 1 Changes in serum concentrations of oestradiol, progesterone, LH and FSH as measured by radioimmunoassay during the menstrual cycle of the stumptailed macaque and the effect of an i.v. injection of an ovine gamma globulin (no. 94) to LHRH on the day after the LH surge during the second cycle. Note low progesterone levels and shortening of luteal phase. Ten days post-injection LHRH antibodies titres have declined and LH, FSH and oestradiol secretion resume. Bars show menstrual bleeding.

in progesteone concentrations during the early luteal phase was smaller than normal and that after 5 or 6 days progesterone concentrations fell to non-detectable values and premature menstruation occurred (figure 1).

These preliminary results provided evidence that the corpus luteum of this species is dependent on LHRH and LH release from a few days after the LH surge. The question of the attenuated progesterone rise during the early luteal phase could be the result of stimulation from the LH surge.

The main drawback in injecting these monkeys with LHRH antibodies was that the animals could not be used for repeat injections in further investigations because of the risk of hypersensitive reaction to the sheep gamma globulins. The availability of a highly potent selective antagonist analogue of LHRH allowed us to test the sensitivity of acute withdrawal of pituitary LH by a more acceptable approach.

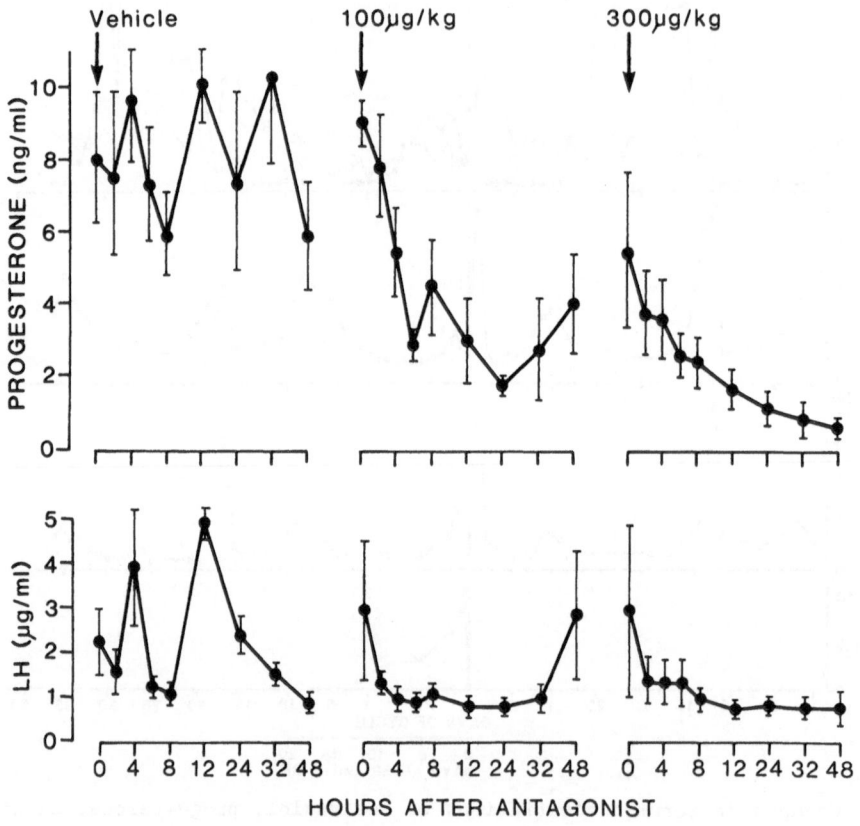

Fig. 2 Acute effect of a single s.c. injection of vehicle alone, 100 ug LHRH antagonist/Kg or 300 μg LHRH antagonist/Kg on days 9/10 after the preovulatory LH surge, on serum concentrations of bioactive LH and progesterone in the stumptailed macaque. The same three monkeys were treated on each occasion. Values are mean ± S.E.M.

For our initial testing of the effects of the LHRH antagonist
[N-Ac-D-Nal(2)^1,D-pCl-Phe^2,D-Trp^3,D-hArg(Et$_2$)^6,D-Ala^{10}] LHRH
we postulated that the corpus luteum of the stumptailed macaque would be
most susceptible to LH withdrawal during the late luteal phase (Fraser et
al 1985). On day 9/10 of the luteal phase 3 monkeys were injected s.c.
with 100 μg LHRH antagonist/kg body weight. In order to observe the acute
effects of the antagonist, blood samples were taken at 0,2,4,6,8,12,24 and
32h after administration and the effects on serum concentrations of LH and
progesterone compared to values obtained at the same period during a
control cycle in the same animals. In the control cycles in which the
monkeys were treated with vehicle on days 9/10 of the luteal phase, the
pattern of bioactive LH concentrations during the following 32h period was
characterised by one to three marked peaks of activity which were
associated with high serum progesterone concentrations of between 10 and 14
ng/ml (figure 2). When the three monkeys were treated with 100μg
antagonist/kg on days 9/10 of the luteal phase their serum bioactive LH
concentrations fell during the following 32h. Serum progesterone
concentrations also declined, values falling to less than 0.6 ng/ml on at
least one occasion during this period (figure 2). However, two animals
showed a small increment of LH bioactivity associated with a rise in serum
progesterone concentrations at 8h suggesting a blunting rather than
complete suppression of the pituitary gonadotroph. By 2 days after
antagonist administration, serum progesterone concentrations had risen to
within normal range in all animals. Since this experiment had revealed a
temporary suppression of progesterone concentrations, the animals were
treated with 300μg antagonist/kg on the same schedule. After the higher
dose serum LH concentrations remained at basal values during the following
32h. In all three monkeys serum progesterone concentrations fell steadily
during this period (figure 2) and by 48h non-detectable (less than 0.4
ng/ml) values were recorded.

To test the susceptibility of the corpus luteum to acute withdrawal of LH
during the early or mid luteal phase, 300 μg/Kg antagonist was administered
as a single injection on days 1,3,4 and 7 after the preovulatory LH peak.

Serum concentrations of bioactive LH fell to low levels for 24-48h after
treatment in all monkeys. Even during the early luteal phase the treatment
reduced progesterone secretion showing that the corpus luteum is dependent,
at least in part, on continuous gonadotrophin support during this period.
This suppression of progesterone after antagonist was temporary apart from
some monkeys treated during the late luteal phase. Monkeys treated in the
early-mid luteal phase demonstrated a recovery of luteal function after the
action of antagonist had subsided (figure 3).

To investigate if more prolonged suppression of LH release by LHRH
antagonist could induce a permanent suppression of luteal progesterone
secretion, 6 monkeys were treated with 3 consecutive daily injections of
300 μg LHRH antagonist/kg body weight. Treatment was commenced between days
1-5 of the luteal phase (figure4). All monkeys showed a fall in serum
progesterone concentrations which was followed by a premature menstruation.
In 5 of the 6 monkeys this suppression in progesterone was maintained
throughout the duration of the normal luteal phase (e.g. figure 4). These
results show conclusively that suppression of pituitary LH secretion via
receptor block by LHRH antagonist can suppress progesterone secretion from
the corpus luteum. The failure of a previous study (Balmaceda et al 1983)
to alter serum progesterone concentrations in rhesus monkeys by daily
administration of another LHRH antagonist may have been the result of
inadequate suppression of LH by this less potent compound.

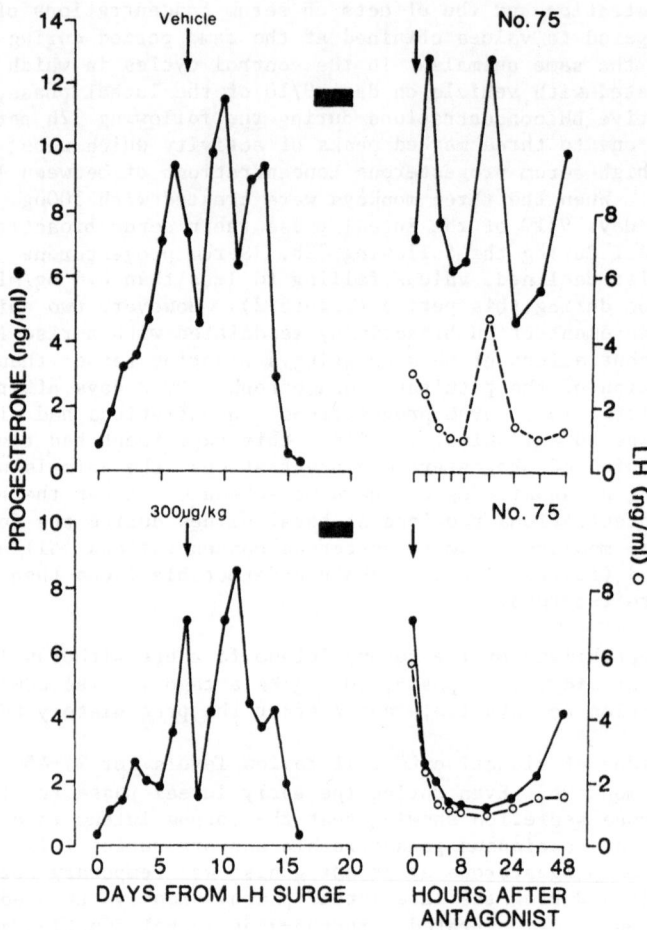

Fig. 3 Changes in serum concentrations of progesterone and bioactive LH after injection of either vehicle or 300 µg LHRH antagonist/Kg administered on day 7 after the LH surge in consecutive cycles in a stumptailed macaque.

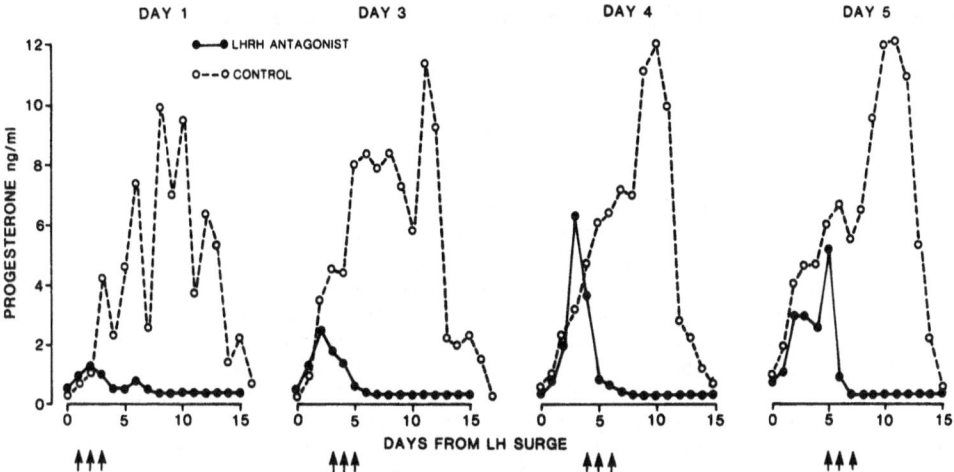

Fig. 4 Serum concentrations of progesterone during a control cycle (open circles) and in a subsequent cycle during which the monkeys were treated with 3 consecutive daily injections of 300 µg LHRH antagonist/Kg (arrows) beginning at different days during the early luteal phase.

DO LHRH ANALOGUES EXERT DIRECT ACTIONS ON THE PRIMATE CORPUS LUTEUM?

The presence of high affinity receptors for LHRH agonists on the rat ovary, including luteal tissue, is now well documented (Clayton et al 1979). Through these receptors LHRH and its agonists can exert direct effects on steroidogenesis (Hsueh & Jones 1981). The characteristics of binding are identical to those found with anterior pituitary tissue. In contrast, there are reports that luteal tissue from the human (Clayton & Huhtaniemi 1982) and rhesus monkey (Asch et al 1981) fails to bind labelled LHRH agonist. However, our group have clearly demonstrated low affinity binding sites for LHRH agonist both on human corpus luteum (Popkin et al 1983) and placenta (Currie et al 1981).

The significance of these binding sites is unknown and studies have failed to demonstrate any effect on steroidogenesis of incubating LHRH analogues with human luteal tissue in vitro (Casper et al 1984). On the other hand, in vitro studies with human placenal cultures incubated with LHRH have demonstrated a stimulatory effect on hCG release (Khodr & Siler-Khodr, 1978; Belisle et al 1984) which can be overcome by LHRH antagonist (Siler-Khodr et al 1983).

Data available indicates that the antagonists do not bind to the same sites as LHRH agonists on the human corpus luteum (T.A. Bramley, personal communication) or placenta (Belisle et al 1984) since antagonists do not displace labelled agonist.

It also appears that the action of LHRH antagonists on the rat gonad has been confined to blocking the actions of LHRH agonists rather than exerting an independent effect. Thus the evidence that LHRH antagonists will exert a direct action on the primate corpus luteum is extremely limited at present, but in view of the milligram doses still required to suppress pituitary function in vivo, the possibility of the high circulating levels

induced acting via a binding site on luteal tissue should not yet be
dismissed.

CONCLUSIONS

Recent investigations on the role of LHRH in control of the pituitary-
ovarian axis during the luteal phase have helped considerably in answering
some of the unsolved questions of the relationship between the pituitary
gland and luteal function in the primate. During the mid-late luteal
phase, the one-to-one relationship between LH pulses and progesterone
pulses suggests that the corpus luteum is largely governed by the pituitary
at this time. This has been confirmed by studies on withdrawal of LHRH
stimulation in monkeys with hypothalamic lesions, LH antibodies or by
specific block of the LHRH receptor by a highly potent LHRH antagonist in
the present study. These studies have established that progesterone
secretion can be abolished when LH release is prevented.

During the early luteal phase LH pulses do not seem to drive progesterone
secretion. This might suggest that the corpus luteum is independent of the
pituitary gland at this time. This is supported by the failure of hCG to
induce progesterone secretion. On the other hand, it may well be
significant that basal LH levels are high during this period and this could
mean the luteal cells are being maximally stimulated. Administration of
LHRH antagonist also suppressed progesterone secretion when given during
the early luteal phase. Although the corpus luteum recovered its function
after a single administration, 3 consecutive daily antagonist injections
caused a sustained suppression of serum progesterone.

A suppression of LH release followed by a marked effect on luteal function
might have potential to prevent pregnancy by withdrawing the essential
source of progesterone. However, suppression of progesterone during the
late luteal phase may not be sufficient to cause luteolysis during a
fertile cycle since the corpus luteum would be 'rescued' by hCG from the
trophoblast. Prevention of pregnancy at this stage of the cycle would
probably depend upon extra-pituitary actions which have yet to be
established. It may be necessary to administer LHRH antagonist during the
early luteal phase in order to induce luteolysis as has been achieved in
the present study. It will be important to establish whether this
treatment can induce a permanent suppression of progesterone secretion or
whether the CG produced in increasing amounts from around day 7 of the
luteal phase during a fertile cycle is able to reactivate the corpus
luteum.

ACKNOWLEDGEMENTS

We are grateful to Drs. J.J. Nestor and B.H. Vickery (Syntex, Palo Alto)
for the gift of the LHRH antagonist.

REFERENCES

Asch, R.H., Abou-Samra, M., Braunstein, G.D. and Pauerstein, C.J. Luteal
    function in hypophysectomized rhesus monkeys. J. Clin. Endocr. Metab.
    55:154 (1982).

Asch, R.H., Sickle, M.V., Rettori, V., Balmaceda, J.P., Eddy, C.A., Coy, D.H. and Schally, A.V. Absence of LH-RH binding sites in corpora lutea from rhesus monkeys (Macaca mulatta). J. Clin. Endocr. Metab. 53:215 (1981).

Balmaceda, J.P., Borghi, M.R., Coy, D.H., Schally, A.V. and Asch, R.H. Suppression of postovulatory gonadotrophin levels does not affect corpus luteum function in rhesus monkeys. J. Clin. Endocr. Metab. 57:866 (1983).

Belisle, S., Guerin, J.F., Bellabarba, D. and Lehoux, J.G. Luteinizing hormone-releasing hormone binds to enriched human placental membranes and stimulates in vitro the synthesis of bioactive human chorionic gonadotripin. J. Clin. Endocr. Metab. 59:119 (1984).

Casper, R.J., Erickson, G.F. and Yen, S.S.C. Studies on the effect of gonadotropin-releasing hormone and its agonist on human luteal steroidogenesis in vitro. Fertil. Steril. 42:39 (1984).

Clayton, R.N., Harwood, J.P. and Catt, K.J. Gonadotrophin-releasing hormone analogue binds to luteal cells and inhibits progesterone production. Nature 282:90 (1979).

Clayton, R.N. and Huhtaniemi, I.T. Absence of gonadotrophin-releasing hormone receptors in human gonadal tissue. Nature 299:56 (1982).

Currie, A.F., Fraser, H.M. and Sharpe, R.M. Human placental receptors for luteinizing hormone releasing hormone. Biochem. Biophys. Res. Commun. 99:323 (1981).

Ellinwood, W.E., Norman, R.L. and Spies, H.G. Changing frequency of pulsatile luteinizing hormone and progesterone secretion during the luteal phase of the menstrual cycle of rhesus monkeys. Biol. Reprod. 31:714 (1984).

Filicori, M., Butler, J.P. and Crowley, W.F. Neuroendocrine regulation of the corpus luteum in the human. Evidence for pulsatile progesterone secretion. J. Clin. Invest. 73:1638 (1984).

Fraser, H.M., Baird, D.T., McRae, G.I., Nestor, J.J. and Vickery, B.H. Suppression of luteal progesterone secretion in the stumptailed macaque by an antagonist analogue of luteinizing hormone releasing hormone. J. Endocr. 104:R1 (1985).

Fraser, H.M., McNeilly, A.S. and Popkin, R.M. In: "Immunological Aspects of Reproduction in Mammals" ed B. Crighton. Butterworths. pp399-418 (1984).

Groff, T.R., Madhwa Raj, H.G., Talbert, L.M. and Willis, D.L. Effects of neutralization of luteinizing hormone on corpus luteum function and cyclicity in Macaca fasicularis. J. Clin. Endocr. Metab. 59:1054 (1984).

Healy, D.L., Schenken, R.S., Lynch, A., Williams, R.F. and Hodgen, G.D. Pulsatile progesterone secretion: its relevance to clinical evaluation of corpus luteum function. Fertil. Steril. 41:114 (1984).

Hutchison, J.S. and Zeleznik, A.J. The rhesus monkey corpus luteum is dependent on pituitary gonadotrophin secretion throughout the luteal phase of the menstrual cycle. Endocrinology 115:1780 (1984).

Hsueh, A.J. and Jones, P.B.C. Extrapituitary actions of gonadotrophin-releasing hormone. Endocr. Rev. 2:437 (1981).

Knobil, E. The neuroendocrine control of the menstrual cycle. Rec. Prog. Horm. Res. 36:53 (1980).

Kohdr, G. and Siler-Khodr, T.M. The effect of luteinizing hormone-releasing factor on human chorionic gonadotrophin secretion. Fertil. Steril. 30:301 (1978).

Lincoln, D.W., Fraser, H.M., Lincoln, G.A., Martin, G.B. and McNeilly, A.S. Hypothalamic pulse generators. Rec. Prog. Horm. Res. 41:369 (1985).

Moudgal, N.R., MacDonald, G.J. and Greep, R.O. Role of endogenous primate LH in maintaining corpus luteum function in the monkey. J. Clin. Endocr. Metab. 35:113 (1972).

Nestor, J.J., Tahilramani, R., Ho, T.L., McRae, G., Vickery, B.H. and
    Bremner, W.J. In: "Peptides: Structure and Function". Proceedings of
    the 8th American Peptide Symposium, pp861-864. Eds V.J. Hruby and
    D.H. Rich. Rockford, Illinois: Pierce Chemical Co (1983).
Popkin, R., Bramley, T.A., Currie, A., Shaw, R.W., Baird, D.T. and Fraser,
    H.M. Specific binding of luteinizing hormone releasing hormone to
    human luteal tissue. Biochem. Biophys. Res. Commun. 14:750 (1983).
Siler-Khodr, T.M., Khodr, G.S., Vickery, B.H. and Nestor, J.J. Inhibition
    of hCG, hCG and progesterone release from human placental tissue in
    vitro by a GnRH antagonist. Life Sci. 32:241 (1983).
Van de Wiele, R.L., Bogumil, J., Dyrenfurth, I., Ferin, M., Jewelerwicz,
    R., Warren, M., Rizkallah, T. and Mikhail, G. Mechanisms regulating
    the menstrual cycle in women. Rec. Prog. Horm. Res. 26:63 (1970).

COMPARISON OF PROTEIN AND RIBONUCLEIC ACID COMPOSITION OF <u>POST MORTEM</u>

BRAIN TISSUE BETWEEN CONTROLS AND CASES OF ALZHEIMER TYPE DEMENTIA

J.C. Pascall, N.M. Borthwick, A. Curtis,
V. Lyons and C.M. Yates

MRC Brain Metabolism Unit, Royal Edinburgh Hospital
Morningside Park, Edinburgh, EH10 5HF

Alzheimer disease (AD) is an early onset dementia (patient under the age of 65 at the time of onset) characterized clinically by an initial loss of memory for recent events, with a later deterioration of cognition, memory and personality (Katzman, 1976; Terry & Davies, 1980; Wiskniewski & Iqbal, 1980). At a gross anatomical level, the brains of patients with AD are generally smaller than those of controls and usually show atrophy of the temporal and frontal lobes. At a histological level, the disorder is characterized by the presence of plaques, consisting of abnormal neurites associated with extracellular amyloid (Wisniewski & Terry, 1976; Tomlinson, 1982) and neurofibrillary tangles, comprised of paired helical filaments, which are concentrated in the perikarya (cell bodies) of neurones (Kidd, 1963; Tomlinson, 1982).

Neurochemically, the disease is characterized by decreases in the levels of choline acetyltransferase and acetylcholinesterase (White et al., 1977; Davies, 1979), dopamine β-hydroxylase (Cross et al., 1981), noradrenaline (Yates et al., 1983) and somatostatin (Davies et al., 1980). In contrast to the large number of neurotransmitter studies (for review see Hardy et al., 1985), there have been few reports on proteins in brains from cases of AD (Selkoe, 1980; Kosik et al., 1982). Recently the abnormal proteinaceous plaques and neurofibrillary tangles associated with AD have been investigated biochemically and immunohistochemically. Plaque amyloid has been shown to contain a 4kd protein which shows no homology with any known protein and is not found in brains from control subjects (Glenner & Wong, 1984a,b; Masters et al., 1985), suggesting that plaque amyloid may be a consequence of an abnormal gene or defective transcription. By contrast, tangles have been proposed to be aberrant post-translational products of normal neurofilaments (Sternberger et al., 1985).

In this Unit, we have adopted a more general approach to the study of proteins in AD. Initially, we compared brain proteins separated by sodium dodecylsulphate polyacrylamide gel electrophoresis (SDS-PAGE) from homogenates of temporal cortex from AD and control cases. Cytosol and pellet fractions were obtained by 100,000g centrifugation of homogenates of temporal cortex from five AD cases and three age-matched controls with no clinical signs of a central nervous system disorder. Proteins in

these fractions were analysed by SDS-PAGE and the gels scanned
densitometrically.   The protein profiles obtained from the pellet
fractions were the same in the AD, as in the control, cases.   Soluble

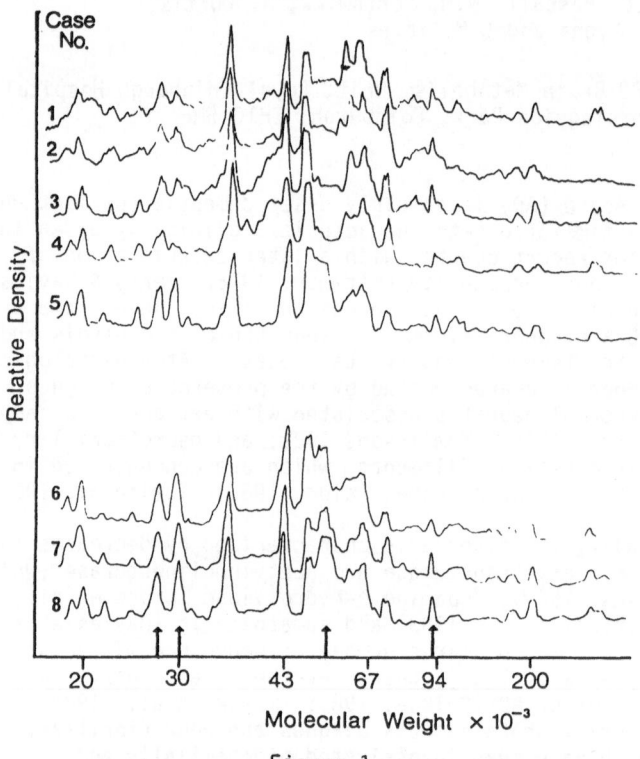

Figure 1

Analysis of soluble proteins in the cytosolic fraction of homogenates
of temporal cortex from five AD cases and three controls.

Tissue samples (80-100mg) were homogenized in ice-cold 50mM Tris-HCl
pH 7.5, containing 1mM-EGTA, 1mM-MgCl$_2$ and 0.01% sodium azide.   The
homogenate was centrifuged at 10,000g for 10 min at 4°C and the
supernatant re-centrifuged at 100,000g for 60 min.   SDS polyacrylamide
gel electrophoresis of 50µg protein of the supernatant was carried out as
described by Lamelli (1970) on 6-15% gradient polyacrylamide gels.
Proteins were stained with Coomassie Brilliant Blue R and the gel tracks
scanned using a Quick Scan R & D scanner (Helena Laboratories).

Cases 1 to 5 are from AD
Cases 6 to 8 are from controls

proteins in the AD cytosol fractions, however, differed from proteins in
control cytosol fractions (Fig. 1).   A 55,000 Mr protein was dramatically
reduced in four, of the five AD cases.   Reductions were also seen in
proteins of Mr 28,000, 30,000, 92,000 and 200,000.   One AD case had a
normal protein profile.   The 55,000 Mr protein was identified, by
Western blot analysis, as tubulin but the identities of the other
proteins is unknown (Borthwick et al., 1985).

    In order to determine if these protein losses were due to altered
mRNA or to post-translational abnormalities, proteins translated from

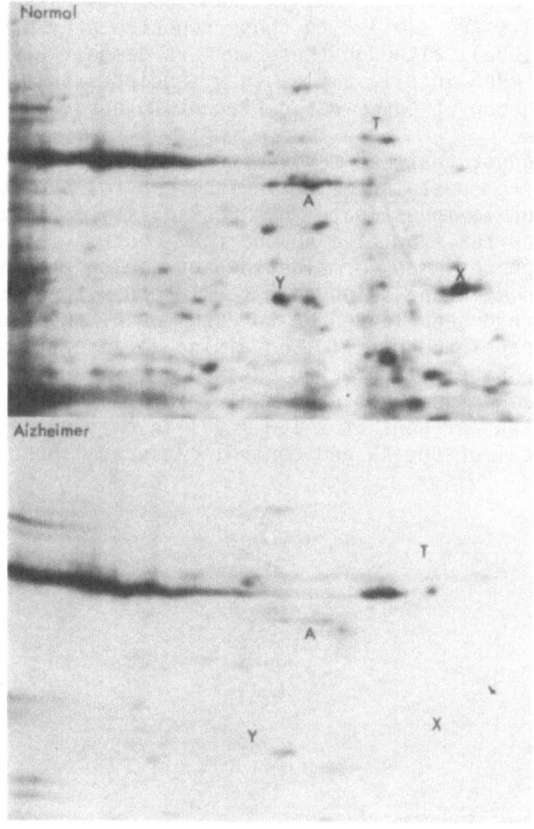

Figure 2

    Two dimensional SDS-PAGE of proteins synthesized by a rabbit
reticulocyte lysate cell-free system in response to polyadenylated RNA
isolated from temporal cortex from an AD and a control case.

    Polyadenylated RNA (1µg) isolated from total RNA by oligo
(dT)-cellulose affinity chromatography, was translated in a
nuclease-treated rabbit reticulocyte lysate cell-free system (Pelham &
Jackson, 1976) in the presence of [^{35}S] methionine.   The radiolabelled
proteins were analysed by two-dimensional polyacrylamide SDS gel
electrophoresis (O'Farrell, 1975; Curtis et al., 1985).
Proteins indicated are: T - tubulin;   A - actin
                 x and y - unknown proteins greatly reduced in AD
                 compared with controls.

mRNA isolated from AD and control cases were compared. Polyadenylated RNA was prepared from temporal cortex from 4 AD cases and from 2 controls. The polyadenylated RNA was translated in the presence of [35S] methionine by a reticulocyte lysate protein synthesising system. The labelled translation products were separated by 2-dimensional gel electrophoresis and the gels autoradiographed. Autoradiographs typical of the AD and control cases are shown in Fig. 2. Approximately equal amounts of polyadenylated mRNA gave rise to fewer proteins in the AD, than in the control, samples. We have not yet been able to definitely identify any of the proteins which are absent from the cases of AD but present in controls, although a reduction in translatable tubulin mRNA was seen in some but not all AD cases (the example shown in figure 3 illustrates a case where tubulin mRNA loss was essentially total). These results are similar to those reported previously (Marotta & Sajdel-Sulkowska, 1983), although those workers demonstrated a reduction in tubulin mRNA in all samples they studied; this may reflect the fact that their group of cases was different to ours.

These results suggest that RNA derived from post-mortem brain from cases of AD differs from post-mortem brain from control subjects. We have, therefore, begun to investigate this possibility. Total RNA from temporal or frontal cortex from five AD and four control cases was separated by electrophoresis on formaldehyde-containing agarose gels and blotted on to a nylon membrane "Biodyne A". In order to check that equal amounts of RNA had been loaded on to each track, the blots were probed with a nick-translqted plasmid containing an insert coding for 7SL-RNA. This probe hybridized to two RNA species, A and B, in the four controls and to three RNA species, A, B and C, in all five AD cases. Species C of 7SL-RNA was present at a low level in only one control. Autoradiographs typical of the AD and control cases are shown in

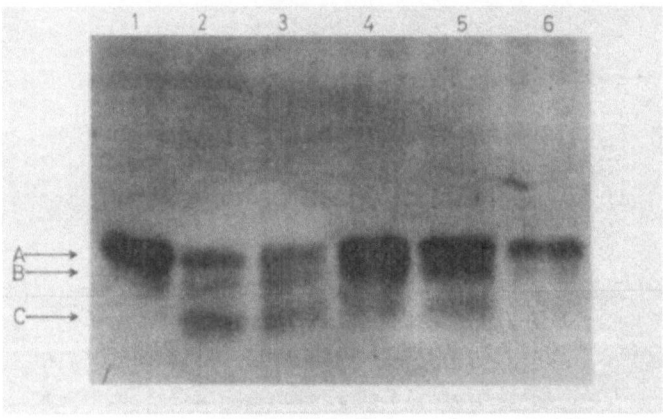

Figure 3

Northern blot analysis of the multiple forms of 7SL RNA isolated from cortical grey matter from AD and control cases.

Total RNA (10µg) isolated (Kaplan et al., 1979) from cortical grey matter from two control (tracks 1 & 6) and four AD cases (tracks 2-5) was separated by electrophoresis on formaldehyde-containing 1.1% (w/v) - agarose gels (Maniatis et al., 1982), blotted onto Biodyne A membranes and probed with nick-translated plasmid containing a cDNA encoding mouse 7SL RNA (Balmain et al., 1982) under the conditions described by Taylor et al. (1984).

Fig. 3.   The additional 7SL-RNA species found in brains from cases of AD could reflect a change in the pattern of gene transcription or an increase in RNA degradation.   The latter explanation is consistent with the reduction in translational activity of polyadenylated RNA observed by ourselves and Marotta & Sajdel-Sulkowska (1983), although the latter authors showed clear reductions in mRNA encoding tubulin.   Our data could provide further evidence for increased ribonuclease activity (Sajdel-Sulkowska & Marotta, 1984) in neuropathologically affected cerebral cortex in AD.   Whilst certain RNA species, such as 7SL-RNA, appear to be susceptible to degradation, other species such as the RNA from which actin is translated, may be resistant to degradation. Further studies are in progress to compare the translatability of mRNA, in particular tubulin mRNA, isolated from AD and control brain, with the aim of elucidating the mechanism and significance of the reduction in soluble tubulin in AD cerebral cortex.   2-dimensional analyses of homogenates of AD and control cerebral cortex are also underway in an attempt to identify the other soluble proteins which are reduced in neuropathologically affected areas of AD brain.

REFERENCES

Balmain, A., Krumlauf, R., Vass, J.K. and Birnie, G.G. (1982)   Nucl.
      Acids Res. 10 4259-4277.
Borthwick, N.M., Yates, C.M. and Gordon, A. (1985)   J. Neurochem. 44
      1436-1441.
Cross, A.J., Crow, T.J., Perry, E.K., Perry, R.H., Blessed, G. and
      Tomlinson, B.E. (1981)   Br. Med. J. 282 93-94.
Curtis, A., Lyons, V. and Fink, G. (1985)   J. Endocrinol. 105 163-168.
Davies, P. (1979)   Brain Res. 171 319-327.
Davies, P., Katzman, R. and Terry, R.D. (1980)   Nature 288, 279-280
Hardy, J., Adolfsson, R., Alafuzoff, I., Bucht, G., Marcusson, J.,
      Nyberg, P., Perdahl, E., Wester, P. and Winblad, B. (1985)
      Neurochem. Int. 7, 545-563.
Kaplan, B.B., Bernstein, S.L. and Gioios, A.E. (1979)   Biochem. J. 183
      181-184.
Katzman, R. (1976)   Arch. Neurol. 33 217-218
Kidd, M. (1963)   Nature 197, 192-193
Kosik, K.S., Gilbert, J.M., Selkoe, D.J. and Strocchi, P. (1982).   J.
      Neurochem. 39 1529-1538.
Maniatis, T, Fritsch, E.F. and Sambrook, J. (1982)   In: Molecular
      Cloning pp. 202-203, Cold Spring Harbour Laboratory, New York.
Marotta, C.A. and Sajdel-Sulkowska, E.M. (1983).   J. Neurochem. 41,
      suppl. 520A.
O'Farrell, P.H. (1975)   J. Biol. Chem. 250 4007-4021.
Pelham, H.R.B. and Jackson, R.J. (1976)   Eur. J. Biochem. 17 328-338.
Sajdel-Sulkowska, E.M. and Marotta, C.A. (1984)   Science 225 947-949.
Selkoe, D.J. (1980)   Ann. Neurol. 8 465-478.
Taylor, J.B., Craig, R.K., Beale, D. and Ketterer, B. (1984).   Biochem.
      J. 219, 223-231.
Terry, R.D. and Davies, P. (1980)   Ann. Rev. Neurosci. 3 77-95.
Tomlinson, B.E. (1982)   Psychol. Med. 12 449-459.
White, P. Hiley, C.R., Goodhardt, M.J., Carrasco, L.H., Keet, J.P.,
      Williams, I.E.I. and Bowen, D.M. (1977)   Lancet i 668-671.
Wisniewski, H.M. and Iqbal, K. (1980)   Trends Neurosc. 3 226-228.
Yates, C.M., Harmar, A.J., Rosie, R., Sheward, J., Sanchez de Levy, G,
      Simpson, J., Maloney, A.F.J., Gordon, A. and Fink, G. (1983)
      Brain Res. 258 45-52.

NEUROENDOCRINE CHANGES IN ALZHEIMER'S DISEASE: RAISED PLASMA
CONCENTRATIONS OF GROWTH HORMONE AND THYROID STIMULATING HORMONE AND
REDUCED CONCENTRATIONS OF OESTROGEN-STIMULATED NEUROPHYSIN

J.E. Christie, L. J. Whalley, J. Bennie, H. Dick,
D.H.R. Blackwood and G. Fink

MRC Brain Metabolism Unit
Royal Edinburgh Hospital
Morningside Park
Edinburgh, EH10 5HF

INTRODUCTION

The neuropathological (Ishii, 1965) and some of the neurochemical
features of Alzheimer-type dementia (ATD) affect not only the cerebral
cortex but also subcortical structures including the hypothalamus in
which there are reductions in the concentrations of acetylcholine
(Davies, 1979), noradrenaline (Yates et al, 1981) and somatostatin
(Davies et al, 1982).   Somatostatin is also reduced in CSF in patients
with ATD (Wood et al, 1982).   Acetylcholine, noradrenaline and
somatostatin are involved in the control of secretion of pituitary
hormones (Reichlin, 1985) and the aim of the present study was to
determine whether plasma hormone concentrations would 1) reflect the
changes in central neurotransmitters that occur in ATD, and 2) prove
useful in the differential diagnosis of ATD from other causes of dementia
and from depression.

METHODS

The plasma hormone concentrations in patients with presenile ATD were
compared with those in major depressive disorders (MDD), other types of
dementia and control subjects.   All subjects were in good physical
health with no evidence of renal, hepatic or endocrine disease on routine
clinical investigation and were free of antidepressant and neuroleptic
drugs for at least six months and throughout the duration of the study.

Alzheimer's type dementia

Eighteen patients with symptom onset before the age of 65, met the
following criteria: 1) steadily progressing dementia with initial
dysmnesia, 2) no history of another type of dementia or focal
neurological signs or hypertension, 3) CT scan which was either normal
or showed cerebral atrophy and no other pathology, 4) no focal EEG
abnormalities 5) normal ECG, haematological, biochemical and CSF
clinical investigations.

Other dementias

Nine demented patients did not meet the criteria for ATD but were retained in the study to form a second dementia group unlikely to have ATD. Five patients had focal CT changes suggestive of vascular disease, two patients had features of Pick's disease, one had Huntington's Chorea and one had dementia associated with lower limb spasticity thought to be disseminated sclerosis.

Depressed patients

Sixteen depressed patients were studied. The Present State Examination (PSE) (Wing et al, 1974) and the Hamilton Depression Rating Scale (HRS) (Hamilton, 1960) were completed within a few days of admission to hospital. Data from the PSE and case records were assessed independently by two psychiatrists, and classified according to the Research Diagnostic Criteria (RDC) (Spitzer et al, 1978). All 16 met criteria for major depressive disorder (MDD) sub-typed as follows: 10 MDD psychotic, 4 endogenous and 2 probable endogenous.

Control subjects

Control subjects (N = 37) were recruited from hospital staff and other volunteers.

Blood sampling

An indwelling cannula with a heparin lock was inserted at 06.40h and left in place until 24.00h. Blood samples taken at 07.00, 07.30 and 08.00h (AM), 15.00, 15.30 and 16.00h (PM), and 23.00, 23.30 and 24.00h (EV) were collected in lithium heparin-coated tubes containing 100 k. i.u. Trasylol and kept at 4°C before and during centrifugation. Plasma was stored at -40°C until assay.

Hormone assays

Plasma hormone concentrations of GH, TSH, PRL, LH, ESN and cortisol were determined by radioimmunoassay as described previously (Whalley et al, 1985). The sensitivity of the assays (90% B/Bo) were: 1.6mU/l for GH (100µl sample); 0.8mU/l for TSH (200µl sample); 70mU/l for PRL (50µl sample); 0.60U/l for LH (100µl sample); 85pM/l (1µg/l) for ESN (100µl sample); 11.6nM/l (4.2µg/l) for cortisol.

Statistical analysis

Distributions of mean AM, PM and EV plasma concentrations of cortisol, were log-transformed to approximate normality and the significance of differences between ATD, MDD and control subjects determined by one way analysis of variance and the multiple range test of Duncan. The distributions of the other hormones, however, could not be normalized and the Mann-Whitney U test was, therefore, used to determine the significance of differences. Plasma TSH concentrations and prolactin concentrations increased with age and plasma GH concentrations decreased with age in the control population. Thus, for these hormones, the ATD patients were compared with similarly aged MDD patients and control subjects. Plasma GH concentrations were higher in female compared with male control subjects in the EV and plasma prolactin concentrations were higher in female compared with male controls in the PM.

Table 1

PLASMA GH CONCENTRATIONS IN ALZHEIMER PRESENILE DEMENTIA,
OTHER DEMENTIAS AND AGE-MATCHED CONTROL SUBJECTS

	N	Age Mean(SD)	Growth Hormone mU/l, mean (SE)		
			AM	PM	EV
Alzheimer	17	61 ± 5	7.7 ± 1.0*	7.4 ± 1.3	5.2 ± 0.7
Depression	7	57 ± 7	9.3 ± 2.8	7.5 ± 1.5	4.8 ± 0.7
Other dementias	9	65 ± 9	5.1 ± 0.9	4.5 ± 0.8	6.0 ± 1.1
Controls	11	57 ± 6	4.1 ± 0.5	3.9 ± 0.3	6.9 ± 2.3
Alzheimer (female)	12	61 ± 5	8.8 ± 1.1**	8.6 ± 1.6*	5.6 ± 0.8
Controls (female)	9	58 ± 5	4.1 ± 0.5	4.0 ± 0.4	7.7 ± 2.8

ATD v Controls * $p < 0.05$
            ** $p < 0.02$

RESULTS

Growth Hormone

Table 1 shows that plasma GH concentrations in the AM were
significantly higher ($p < 0.02$) in ATD compared with age-related control
subjects but were not different from the concentration in age-related
MDD. Plasma GH concentrations in female ATD patients were higher in the
AM ($p < 0.02$) and PM ($p < 0.05$) than age-related female control subjects
but similar in other dementias and control subjects of both sexes.

Thyroid Hormones

Table 2 shows that plasma TSH concentrations were higher, at all
three time periods in ATD compared with age-related MDD (AM, $p < 0.002$;
PM, $p < 0.02$; EV, $p < 0.05$).   In the AM, 13 ATD patients (72%) but none
of the MDD patients had TSH concentrations greater than 4.5mU/l.   Female
ATD patients had higher plasma TSH concentrations throughout the day
compared with age-related female control subjects (AM, $p < 0.05$; PM, $p <$
0.05; EV, $p < 0.02$).   Plasma TSH concentrations were similar in other
dementias and control subjects.   Plasma triiodothyronine (T_3) and
thyroxine (T_4) and reverse triiodothyronine (rT_3) concentrations were
similar in ATD and age-related control subjects (ATD, mean ± SE, $T_4 =$
96.4 ± 5.7nM/l, $T_3 = 1.58 ± 0.09$nM/l, $rT_3 = 0.30 ± 0.05$nM/l, control
subjects T_4 99.5 ± 4.5nM/l, $T_3 = 1.73 ± 0.07$nM/l, $rT_3 = 0.22 ±$
0.01nM/l).   All subjects had T_3 and T_4 concentrations within the
normal laboratory range.   One ATD patient had very high plasma TSH
concentrations (mean for the day 98.5mU/l) but normal T_3, T_4 and
rT_3 concentrations and no clinical features of hypothyroidism.   This
patient's thyroid status was reassessed over a four year period during
which she developed no features of hypothyroidism, her T_3 and T_4
remained within the normal range and plasma TSH concentrations gradually
declined to a value of 10.9mU/l.

Table 2

PLASMA TSH CONCENTRATIONS IN ALZHEIMER PRESENILE DEMENTIA,
OTHER DEMENTIAS AND AGE-MATCHED CONTROL SUBJECTS

	N	Age Mean(SD)	TSH mU/l, mean (SE)		
			AM	PM	EV
Alzheimer	18	61 ± 5	12.2 ± 6.6+++	7.5 ± 3.0++	12.2 ± 6.1+
Depression	7	57 ± 7	3.3 ± 0.2	3.1 ± 0.2	3.7 ± 0.5
Other dementias	9	65 ± 9	4.7 ± 0.6	4.0 ± 0.4	5.1 ± 0.5
Controls	11	57 ± 6	4.4 ± 0.2	3.9 ± 0.2	4.9 ± 0.5
Alzheimer (female)	13	61 ± 5	15.4 ± 9.0*	8.9 ± 4.1*	15.3 ± 8.4**
Controls (female)	9	58 ± 5	4.2 ± 0.1	3.8 ± 0.2	4.5 ± 0.2

ATD v Controls   * $p < 0.05$
                 ** $p < 0.02$

ATD v MDD   + $p < 0.05$
             ++ $p < 0.02$
           +++ $p < 0.002$

Oestrogen-stimulated neurophysin (ESN)

Plasma concentrations of ESN were significantly lower in ATD compared
with MDD patients (AM, $p < 0.01$; PM, $p < 0.01$; EV, $p < 0.05$) and
control subjects (AM, $p < 0.001$; PM, $p < 0.01$; EV, $p < 0.001$) and were
also lower in the AM ($p < 0.05$) in ATD compared with other dementia
patients (Table 3). When the nine values for each subject were
averaged, mean plasma ESN concentrations were significantly lower in ATD
compared with MDD ($p < 0.001$) other dementias ($p < 0.05$) and control
subjects ($p < 0.001$). All ATD patients had a mean plasma ESN
concentration below 170pM(2µg/l). When AM plasma TSH and ESN
concentrations (TSH > 4.5mU/l; ESN < 170pM/l) were used to discriminate
between ATD, MDD, other dementias and control subjects, 75% of the ATD
patients were correctly identified together with one patient with
multi-infarct dementia and one control subject but no MDD patients, a
specificity of 86%.

Table 3

PLASMA OESTROGEN STIMULATED NEUROPHYSIN CONCENTRATIONS
IN ALZHEIMER PRESENILE DEMENTIA, OTHER DEMENTIAS AND CONTROL SUBJECTS

	N	Age Mean(SD)	oestrogen stimulated neurophysin pM/l, mean (SE)		
			AM	PM	EV
Alzheimer	16	61 ± 5	105 ± 6**	102 ± 6*	107 ± 9**
Depression	14	46 ± 13	230 ± 41++	239 ± 45++	202 ± 37+
Other dementias	9	65 ± 9	192 ± 44°	173 ± 36	163 ± 34
Controls	36	40 ± 13	191 ± 24	185 ± 25	239 ± 38

ATD v Controls   * p < 0.01
                     ** p < 0.001

ATD v MDD   + p < 0.05
                    ++ p < 0.01

ATD v other dementias ° p < 0.05

Conversion: SI to traditional units
ESN: pM/l ≈ 0.012µg/l

Cortisol

Plasma cortisol concentrations were higher throughout the day in patients with MDD than in control subjects ($p < 0.05$) but not significantly greater than in patients with ATD (Table 4). Plasma cortisol concentrations in ATD and other dementias patients were significantly higher in the EV than control subjects ($p < 0.05$).

Prolactin and luteinising hormone

Prolactin concentrations (mean ± SE) were similar in ATD (AM = 493 ± 123 mU/l, PM = 355 ± 27mU/l, EV = 360 ± 26 mU/l), age-related MDD (AM = 687 ± 304mU/l, PM = 678 ± 313mU/l, EV = 527 ± 158mU/l), other dementias (AM = 403 ± 57mU/l, PM = 318 ± 30mU/l, EV = 425 ± 36mU/l) and control subjects (AM = 344 ± 36mU/l, PM = 291 ± 20mU/l, EV = 335 ± 28mU/l).

Plasma LH concentrations (mean ± SE) did not differ significantly between ATD post-menopausal women (N = 11; AM = 16.4 ± 1.6U/l, PM = 14.3 ± 1.1U/l, EV = 15.0 ± 1.4U/l), and female patients with other dementias (N = 5; AM = 11.9 ± 2.3U/l, PM = 12.0 ± 1.6U/l, EV = 12.7 ± 1.6U/l) and post-menopausal female control subjects (N = 9; AM = 14.6 ± 2.2U/l, PM = 14.3 ± 1.9U/l, EV = 13.9 ± 1.4U/l).

Table 4

PLASMA CORTISOL CONCENTRATIONS
IN ALZHEIMER PRESENILE DEMENTIA, OTHER DEMENTIAS AND CONTROL SUBJECTS

	N	Age Mean(SD)	AM	Cortisol nM/l, mean (SE) PM	EV
Alzheimer	17	61 ± 5	469 ± 23	244 ± 22	210 ± 25*
Depression	16	46 ± 13	511 ± 31*	301 ± 39*	214 ± 24*
Other dementias	9	65 ± 9	530 ± 41*	274 ± 41	249 ± 34*
Controls	37	40 ± 13	436 ± 25	230 ± 13	154 ± 16

* v Control subjects p < 0.05

Conversion: SI to traditional units
Cortisol: 1nM/l ≈ 0.36µg/l

DISCUSSION

The present study has identified selective changes in plasma
concentrations of TSH, GH and ESN that are specific to ATD and which can
be related to the known neurochemical deficits in ATD. Although raised
plasma GH concentrations could be due to increased release of
GH-releasing hormone (GHRH) it is more likely in ATD to be due to
decreased release of somatostatin which is known to be deficient in
ATD. Since somatostatin also inhibits TSH a similar mechanism may
account for the increased plasma TSH concentrations in ATD.
Hypothalamic thyroid releasing hormone (TRH) concentrations are normal in
post-mortem tissue in ATD (Yates et al, 1983) and it is, therefore,
unlikely that raised plasma TSH are due to increased release of TRH.
The increase in plasma GH and TSH concentrations is more marked in female
ATD patients but the reasons at present are unknown.

The highly significant reduction of plasma ESN concentrations in ATD
may be related to the marked deficiency in ATD brain of acetylcholine
which is known to stimulate oxytocin release (Poulain & Wakerley, 1982)
and, therefore, presumably the oxytocin associated ESN. The decreased
ESN concentrations may, however, also be due to decreased synthesis of
ESN itself since in rats the staining intensity for the neurohypophysial
hormones decreases progressively with age, especially in the
paraventricular nucleus and in oxytocin-containing neurones (Watkins &
Choy, 1980) but recent studies of human brain have failed to show any
change in the area of oxytocin cells with increasing age (Fliers et al,
1985). Scrapie in sheep, which has some neuropathological features in
common with those of ATD, causes degeneration of magnocellular neurones
with loss of neurophysin in the posterior pituitary gland (Parry &
Livett, 1976).

The majority of studies report that concentrations of dopamine and
homovanillic acid are normal in ATD (Yates et al, 1979; Mann et al,
1980). Consistent with this finding, plasma prolactin concentrations
were similar in ATD patients and age-related controls in the present

study as well as in the study of Balldin et al, (1982).   The raised EV plasma cortisol concentrations in ATD compared with control subjects are not useful diagnostically because the AM and EV plasma cortisol concentrations are increased in other dementias and increased throughout the day in MDD as has also been shown to be the case in many previous studies of depression (Gibbons, 1964; Carpenter & Bunney, 1971).   Raised plasma cortisol concentrations are found in many psychotic patients irrespective of the presence of depressed mood (Christie et al, 1986).

Plasma TSH and ESN concentrations are useful in discriminating between ATD, other causes of dementia and depression.   Morning TSH concentrations greater than 4.5mU/l, identified 72% of ATD and no depressed patients and the additional use of AM ESN concentrations improved the discrimination between ATD, other dementias, MDD and control subjects to a sensitivity of 75% and specificity of 86%.

These results show significant increases in the concentrations of GH, TSH and cortisol and a decrease in concentration of ESN in the plasma of patients with ATD.   Increased plasma GH and TSH concentrations in ATD are consistent with reports of reduced somatostatin in post-mortem neurochemical studies, while decreased ESN may be caused by decreased cholinergic activity.   These characteristic hormonal changes in ATD provide the basis for psychopharmacological analysis of the central neurochemical abnormalities of ATD and offer important biological markers for discriminating between ATD, other dementias and major depression.

ACKNOWLEDGEMENTS

The authors would like to thank the pituitary hormone programme of the NIADDK (Baltimore, USA) and the Scottish Antibody Production (Carluke) for the generous provision of radioimmunoassay materials, the Department of Clinical Chemistry, the Royal Infirmary of Edinburgh for measurement of T_3 and T_4 and the Clinical Research Centre (Harrow) for measurement of rT_3; Mrs. M. Arkley, Mr. M. Surtees, Mrs. G. Sanchez-Watts and Mrs. H. Wilson for their technical assistance; Mr. J. Sloane-Murphy and Mrs. P. Ritchie for their expert care of the patients and research assistance; Norma Brearley for the preparation of the manuscript.

REFERENCES

Balldin, J., Gottfries, C.-G., Karlsson, I. AND   1983, Dexamethasone suppression test and serum prolactin in dementia disorders, Br. J. Psychiatry 143: 277-281.

Carpenter, W.T. Jr. and Bunney, W.E., 1971, Adrenal cortical activity in depressive illness, Am. J. Psychiat. 128: 31-40.

Christie, J.E., Whalley, L.J., Dick, H., Blackwood, D.H.R., Blackburn, I.M. and Fink, G., 1985, Raised plasma cortisol concentrations a feature of drug-free psychotics and not specific for depression, Br. J. Psychiat. (in press Jan. 1986).

Davies, P., 1979, Neurotransmitter related enzymes in senile dementia of the Alzheimer type, Brain Research 171: 319-327.

Davies, P., Katz, D.A. and Crystal, H.A., 1982, Choline acetyltransferase, somatostatin, and substance P in selected cases of Alzheimer's disease. Corkin, S., Davis, K.L., Growdon, J.H., Usdin, E., Wurtman, J. eds. in: Alzheimer's disease: A report of progress in research, New York: Raven Press.

Fliers, E., Swaab, D.F., Pool, C.W. and Verwer, R.W.H., The vasopressin and oxytocin neurons in the human supraoptic and paraventricular nucleus; changes with aging and in senile dementia. Brain Research (in press).

Hamilton, M., 1960, A rating scale for depression, J. Neurol. Neurosurg. Psychiat. 23: 56-62.

Ishii, T., 1965, Distribution of Alzheimer's neurofibrillary changes in the brain stem and hypothalamus of senile dementia, Acta. Neuropathologica 6: 181-187.

Mann, D.M.A., Lincoln, J., Yates, P.O., Stamp, J.E. and Toper, S., 1980, Changes in the monoamine containing neurones of the human CNS in senile dementia, Br. J. Psychiat. 136: 533-541.

Parry, H.B. and Livett, B.G., 1976, Neurophysin in the brain and pituitary gland of normal and scrapie-affected sheep-I, Neuroscience 1: 275-299.

Poulain, D.A., Wakerley, J.B., 1982, Electrophysiology of hypothalamic magnocellular neurones secreting oxytocin and vasopressin, Neuroscience 7: 773-808.

Reichlin, S., 1985, Neuroendocrinology, in: Williams Textbook of Endocrinology, 7th edition, London: Saunders, 492.

Spitzer, R.L., Endicott, J. and Robins, E., 1978, Research diagnostic criteria: rationale and reliability, Arch. Gen. Psychiat. 36: 773-782.

Watkins, W.B. and Choy, V.J., 1980, The impact of aging on neuronal morphology in the rat hypothalamo-neurohypophysial system: An immunohistochemical study. Peptides 1: 239-245.

Whalley, L.J., Christie, J.E., Bennie, J., Dick, H., Blackburn,I.M., Blackwood, D., Sanchez Watts, G. and Fink, G., 1985, Selective increase in plasma luteinising hormone concentrations in drug free young men with mania, Br. Med. J. 290: 99-102.

Wing, J.K., Cooper, J.E. and Sartorius, N., 1974, The measurement and classification of psychiatric symptoms. London: Cambridge University Press.

Wood, P.L., Etienne, P., Lal, S., Gauthier, S., Cajal, S. and Nair, N.P.V., (1982), Reduced lumbar CSF somatostatin levels in Alzheimer's disease, Life Sciences 31: 2073-2079.

Yates, C.M., Ritchie, I.M. and Simpson, J., 1981, Noradrenaline in Alzheimer-type dementia and Down syndrome. Lancet ii: 39-40.

Yates, C.M., Harmar, A.J., Rosie, R., Sheward, J., Sanchez De Levy, G. Simpson, J., Maloney, A.F.J., Gordon, A. and Fink, G., 1983, Thyrotropin-releasing hormone, luteinizing hormone-releasing hormone and substance P immuno-reactivity in post-mortem brain from cases of Alzheimer-type dementia and Down's syndrome, Brain Research 258: 45-52.

Yates, C.M., Allison, Y., Simpson, J., Maloney, A.F.J. and Gordon, A., 1979, Dopamine in Alzheimer's disease and senile dementia, Lancet ii: 851-852.

THE USE OF AN LHRH AGONIST IN THE TREATMENT AND INVESTIGATION OF THE PRE-

MENSTRUAL SYNDROME

John Bancroft, Harry Boyle[*] and Hamish Fraser

MRC Reproductive Biology Unit
[*]Reproductive Endocrinology Laboratories
37 Chalmers Street, Edinburgh EH3 9EW

The premenstrual syndrome (PMS) remains an enigma. It has as yet eluded satisfactory definition[+]. The symptoms are protean, involving psychological states such as irritability, depression, tiredness, feelings of abdominal bloatedness, tender swollen breasts, headaches, appetite change, especially carbohydrate craving, changes in bowel and bladder function, allergic reactions, acne, reduced resistance to infections and others. Their temporal relationship with the menstrual cycle is also variable, though symptoms are usually confined to the second half of the cycle declining soon after the onset of menstruation.

It remains uncertain whether we are considering one basic cyclical pattern or a number of patterns which have in common some form of temporal relationship with the menstrual cycle and which combine in various ways. Good scientific research into the phenomenon is still limited and the clinical literature abounds with unsubstantiated opinions about causation and treatment. It is nevertheless an issue of considerable importance. Not only because of the degree of associated suffering but also because of its theoretical relevance to the interaction between the hypothalamic-pituitary-gonadal axis and central modulators of mood.

The precise temporal relationship between the cyclicity of symptoms and the ovarian cycle remains a crucial aspect which has received insufficient research attention. We are approaching this in three ways.

(1) By measurement of both symptoms and ovarian hormones throughout the cycle (Backstrom et al. 1983). This shows a striking association of symptoms with the luteal phase. There is no simple relationship, however, between oestradiol or progesterone concentrations and the level of symptoms as the latter increase throughout the luteal phase whilst the luteal steroids both rise and fall. It nevertheless remains important to establish whether symptoms depend on corpus luteum activity.

(2) By assessing cyclical symptoms during natural variations in the ovarian cycle, e.g. during spontaneous anovular cycles, short luteal phases and lactational ovarian suppression.

[+]Dalton's (1984) definition is "the recurrence of symptoms in the pre-menstruum with absence of symptoms in the postmenstruum.

(3) By manipulating the ovarian cycle and assessing the effect on cyclical symptoms (e.g. with oral contraceptives or LHRH analogues)

It is this use of LHRH agonists that we will be discussing in this paper.

In late 1984 Dr. Yen and his colleagues (Muse et al. 1984) reported the results of a placebo-controlled treatment study of PMS using a Salk Institute LHRH agonist given by daily subcutaneous injections of 50 µg. Active and placebo injections were given each for three months in 8 women with PMS. After a brief stimulatory phase ovarian activity was suppressed immediately by the agonist. Physical premenstrual symptoms were also significantly reduced during the first and subsequent months of active agonist but not by placebo. Psychological symptoms were not effectively suppressed until the second and third months. This was a convincing demonstration of therapeutic efficacy, unusual in the PMS literature. But its clinical usefulness is still uncertain as daily injections are not a practical form of treatment and the long term consequences of the marked ovarian oestrogen suppression are likely to be a disadvantage.

For the past 4 years we have been using Buserelin, the Hoechst LHRH agonist administered by nasal spray, in the treatment, and more particularly the investigation of PMS. We have so far treated 20 women for periods up to 2 years. In this first phase of research we have included in the study any woman with clear cyclical symptoms who has not responded to conventional treatments. We have also varied the dosage and time of onset, though in most cases we have used 600 µg daily in three divided doses and have usually started during the first 3 days of the menstrual cycle. Each woman in the study completed daily ratings of mood and physical symptoms, using visual analogue scales (Sanders et al. 1983) and collected early morning urine on three mornings a week (Monday, Wednesday, Friday) for assay of total urinary oestrogens and pregnanediol (Brown et al. 1968; Chamberlain & Contractor, 1968) throughout two untreated cycles and for the duration of buserelin treatment.

Therapeutic results will be reported fully elsewhere but can be briefly summarised as follows. In no case was the treatment without effect on the PMS. In 10 women (50%), cyclical symptoms were effectively reduced and treatment was continued for periods ranging from 5 to 15 months per course (see Table 1). Reasons for stopping in these 10 women varied. In 5 cases there was concern about the long term effects of hypoestrogenisation. In one woman the dose was reduced from 600 to 400 µg because of hot flushes. Following this she ovulated and became pregnant. She has since given birth to a normal infant. Another woman stopped in order to get pregnant and did so within 4 months. In a further case, after six months of amenorrhoea and good symptom relief treatment was stopped when the woman was sterilised. Symptoms returned and treatment was restarted. For some reason it was never again possible to suppress her ovarian activity or symptoms and buserelin treatment was accompanied by prolonged menstrual bleeding and was eventually stopped.

The other 10 women discontinued treatment after 10 weeks or less (see Table 2). In 7 cases this was because 'premenstrual' symptoms were both intensified and made persistent. In 5 of these women it was mainly psychological symptoms of irritability or depression that were aggravated. In one case it was cyclical acne and in another cyclical eczema, in both cases severely affected. One woman changed from a cyclical pattern of symptoms to one of extreme lability, which was probably worse. One experienced relief of premenstrual symptoms but stopped because of loss of sexual interest. One woman with severe premenstrual depression was undoubtedly triggered by the buserelin into a hypomanic state which

Table 1. Therapeutic Outcome, Side-Effects and Reasons for Stopping in 20 Women receiving Buserelin Treatment

		"Successes"						"Failures"		
Patient	Age	Length of Treatment MONTHS	No. of Menstrual Bleeds	Side-Effects	Patient	Age	Treatment WEEKS	No. of Menstrual Bleeds	Reasons for Stopping
McD	39	16	8	Nil	Wa	33	12	+	Loss of libido
D	34	14	1	Eczema worsened	Re	29	10	0	'PMS' worse and persistent
H	36	14	4	Fatigue	Bt	35	10*	+	Severe Acne
P	32	11	5	Flushes (pregnancy)	McPh	32	9	0	'PMS' worse and persistent
McP	32	9	6	Flushes	Ro	35	7*	0	Labile mood
C	36	9	8	Flushes	McG	30	7*	1	Premenstrual depression worse
B	29	7*	1	Nil	Ly	28	7*	0	Premenstrual irritability worse and persistent
W	38	7	5	Nil	McI	40	5	2	Premenstrual depression worse and persistent
Pa	31	6	1	Flushes	Lo	39	3	0	Premenstrual eczema worse and persistent
R	35	5	3	Flushes Loss of libido	G	26	3	1	Hypomanic illness

*Dose 400 µg daily, remainder 600 µg daily   +Hysterectomy

persisted and required hospitalisation. This was the first evidence she
had shown of a bi-polar affective disorder. All of these negative reactions
are of theoretical interest.

Apart from treatment, this use of buserelin offers a number of
opportunities for investigation of the process of PMS and we will give
some examples.

The introduction of placebo

We have not routinely used placebo control. The length of time
required in many cases to achieve ovarian suppression, the extent of the
initial stimulatory phase and the carry over effects would make a cross-over
comparison of buserelin and placebo, as used by Muse et al. (1984)
difficult to interpret. In 3 cases we have introduced placebo, blind, after
several months of treatment and stable symptom control. In each case there
has been return of symptoms during placebo administration. This method of
placebo use offers an interesting method of investigation in its own right,
as shown in Fig.1. Placebo was started blindly after 9 months of treatment
and 8 months of amenorrhoea and symptom suppression. Fifteen days after
stopping buserelin urinary oestrogens started to rise. After a further
two weeks 'premenstrual' symptoms returned reaching a peak in 5 days and,
because of their clinical severity, required a return to active treatment.
On the first day of resumption, menstrual bleeding occurred and the urinary
pregnanediol started to rise. There was then a further period of menstrual
bleeding followed by a return to the previous state of symptom control.
This was a clear example of 'premenstrual' symptoms occurring in the
absence of ovulation or progesterone rise.

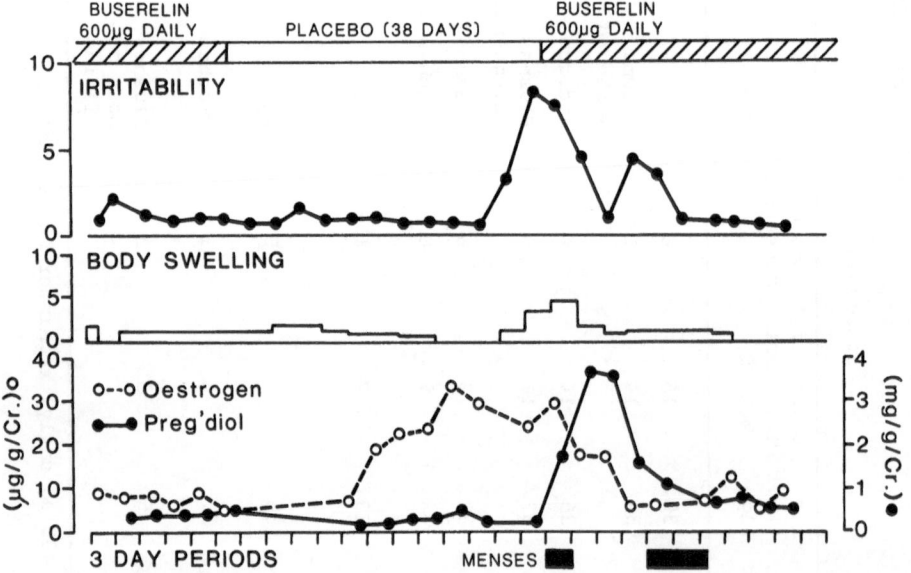

Fig. 1.   Single blind introduction of placebo for a 34 year old woman after
          9 months of buserelin 600 mg daily (patient D, see text for
          details)

Initial stimulation effects

The negative effects of treatment are also of considerable interest. Many forms of treatment have been tried for PMS usually with unpredictable results, made more difficult to interpret because of placebo effects. It is most unusual for a treatment to relieve symptoms in some cases and aggravate them in others. Oral contraceptives are the other main example.

The initial stimulation effects of LHRH agonists on ovarian function are well recognised. In this study in 13 women this initial hormonal stimulation was accompanied by intensification of symptoms. A typical example is shown in Fig. 2. A more striking example is shown in Fig. 3. In this case the stimulation effects did not appear until a few days after starting buserelin. There was a massive rise in urinary oestrogens, far greater than occurred during this woman's normal follicular development, and accompanied by severe and prolonged symptoms. After a further menstrual bleed, she settled down with suppression of both ovarian activity and symptoms. This is a further clear example of 'premenstrual' symptoms occuring in the absence of ovulation or corpus luteum activity.

In an attempt to minimise this initial stimulation effect we have started buserelin treatment towards the end of the luteal phase. Evidence from stump-tail macques, which have menstrual cycles similar to the human female, suggested that in this way undue follicular stimulation could be avoided (Fraser & Sandow, 1985). Fig. 4 shows an example of this late luteal onset with apparently good effect. In another case the results were very different. Fig. 5 shows onset of treatment on day 25 of the cycle. This was followed by marked stimulation of the corpus luteum and a rise in pregnanediol with intensification of symptoms. Menstrual bleeding did not start until 12 days later. This was followed by a temporary respite but then followed return of symptoms to a severe degree which persisted in association with ovarian suppression. Treatment was continued in the expectation that this effect would be transient but was eventually stopped after 10 weeks with no sign of improvement and no return of ovarian activity or menstruation. This state continued for a further seven weeks after stopping buserelin until a course of progesterone suppositories was given, menstrual bleeding followed and her symptoms returned to their previous cyclical pattern. This was the most severe example of symptom aggravation.

The premenstruum during buserelin treatment

Whereas suppression of ovulation occurred in all women, usually within the first month, suppression of follicular development, as shown by urinary oestrogen rise, was much more variable. This is reflected by the variable frequency of menstrual bleeding as shown in Table 1. All but one of the women who menstruated during treatment showed a return of symptoms during the premenstrual days even though the bleeding was highly irregular and unpredictable and not preceded by ovulation or luteal activity. In most cases these symptoms were mild in comparison with before treatment, but Fig. 6 shows a striking example of this in which premenstrual symptoms developed to a severe degree in the absence of any obvious follicular activity. Investigation of changes preceding this type of menstrual bleeding might be informative.

Conclusions

This study represents the first journey into a highly complex system. Whilst no firm conclusions can be drawn at this stage some suggestions are warranted.

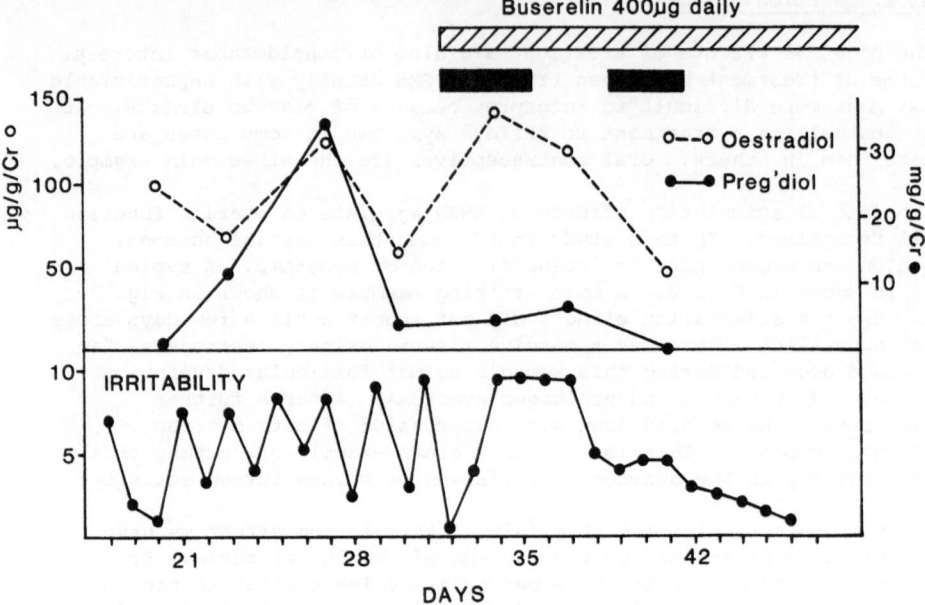

Fig. 2. Initial stimulation effects of buserelin in a 29 year old woman with PMS (patient BL - see text for details)

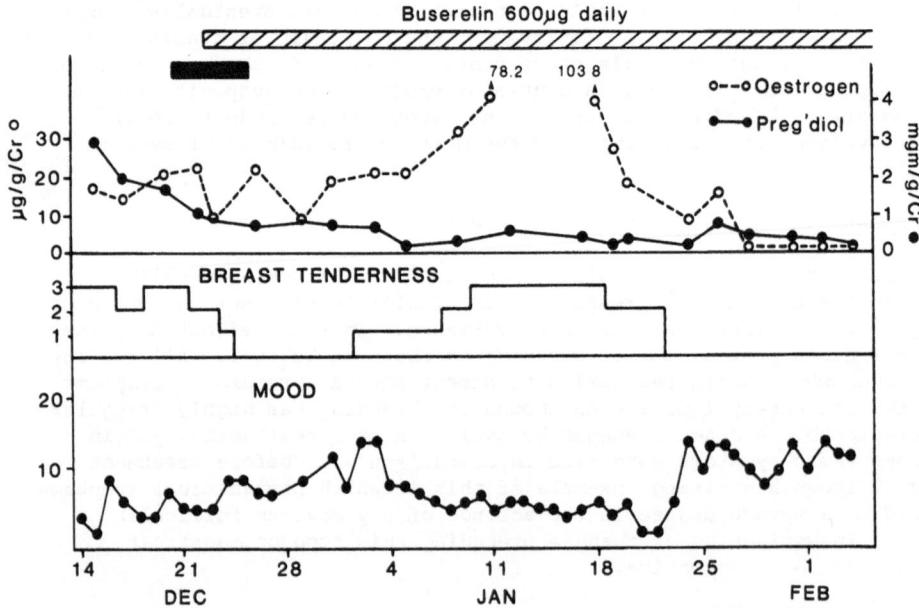

Fig. 3. Onset of buserelin treatment during menstruation showing relatively late stimulatory phase with substantial follicular stimulation, no ovulation and associated worsening of symptoms (patient H)

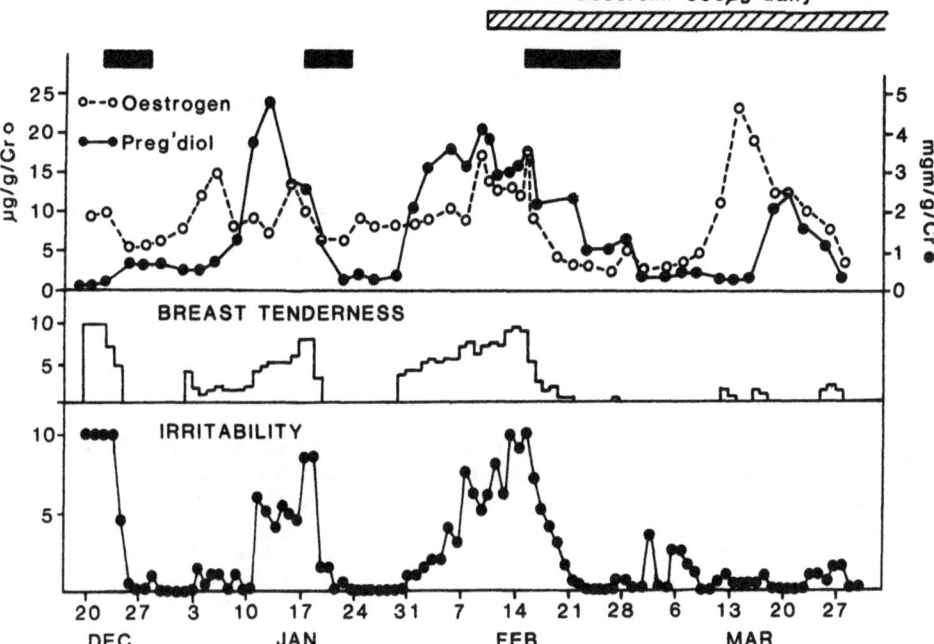

Fig. 4. Starting buserelin in the late luteal phase. There was slight
aggravation of symptoms at the onset but thereafter symptoms were
almost eliminated in spite of a further period of follicular
stimulation and incomplete corpus luteum activity. Symptoms
remained well controlled for 11 months (patient P)

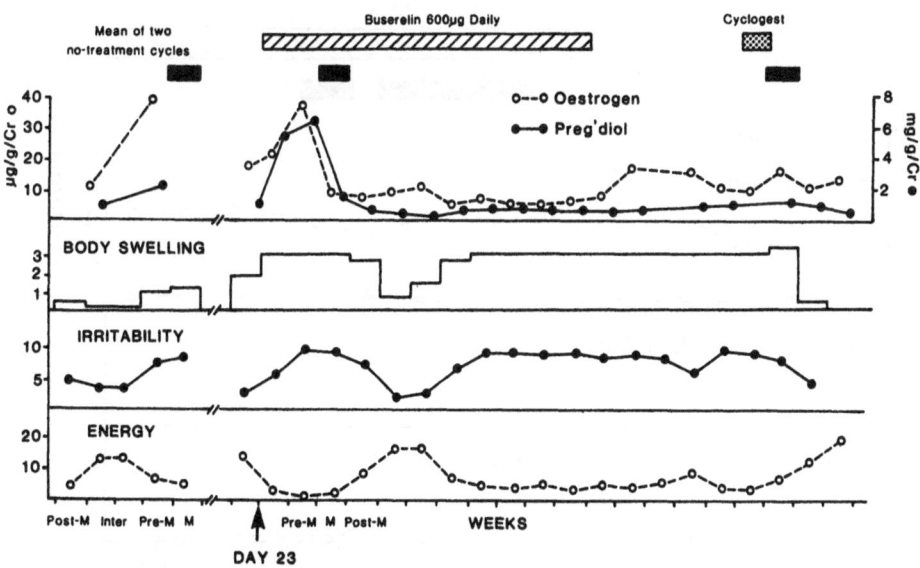

Fig. 5. Starting buserelin on day 23 of the cycle. Mean symptom levels
(per week) for two pre-treatment cycles are shown (patient Re,
see text for details)

(a) Buserelin treatment, whilst it may be effective in some women, is clearly not suitable for general use until we have better indications of good and bad response. The long term problems of hypo-estrogenisation may be avoidable by careful adjustment of dosage as it was evident that in some cases worthwhile symptom improvement occurred in the presence of some follicular activity.

(b) The mechanism underlying the aggravation of symptoms by buserelin deserves further study. A comparison of the effects of an agonist with an effective antagonist, when clinically available, or with an antigonadotrophic agent such as danazol may be informative.

(c) The results indicate that premenstrual symptoms can occur in the absence of a corpus luteum. It is unlikely that they are caused by direct effects of either oestradiol or progesterone or any particular combination of the two.

(d) The principal adverse effects of buserelin treatment have involved psychological symptoms, either depression or irritability. It is possible that these mood states in PMS are related to the ovarian cycle simply in terms of the 'ovarian clock' and are not caused by ovarian factors (Bancroft & Backstrom, 1985). If this were so then suppression of the ovarian cycle would 'free' the mood state, allowing it to be manifested in an irregular or even persistent fashion. Further treatment studies should control for the presence or absence of premenstrual mood changes, especially depression. For this purpose a rational method of categorising different sub-types of PMS is required and should be given high priority in future research.

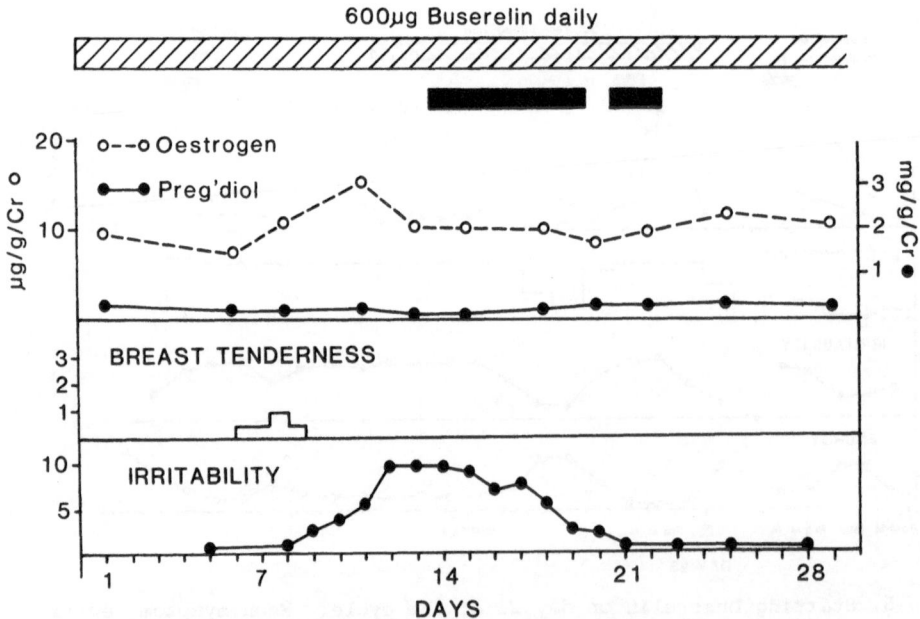

Fig. 6.   Premenstrual symptoms occurring during buserelin treatment with no ovulation and minimal follicular activity (patient McD)

References

Bäckström, T., Sanders, D., Leask, R., Davidson, D., Warner, P. and
   Bancroft, J., 1983, Mood, Sexuality, Hormones and the Menstrual Cycle.
   II. Hormone Levels and their Relationship to the Premenstrual Syndrome,
   Psychosom. Med., 45(6):503.
Bancroft, J. and Bäckström, T., 1985, Premenstrual Syndrome, Clin.
   Endocrinol., 22:313.
Brown, J. B., Macleod, S. L., MacNaughton, C., Smith, M. A. and Smyth, B.,
   1968, A rapid method for estimating oestrogens in urine using a semi-
   automated extractor, J. Endocrinol., 42:5.
Chamberlain, J. and Contractor, S. F., 1968, A gas liquid chromatographic
   method for the rapid estimation of pregnanediol and alopregnanediol in
   non-pregnant urine, Am. J. Obstet. Gynecol., 101:649.
Dalton, K., 1984, Premenstrual Syndrome and Progesterone Therapy, 2nd ed,
   Heinemann, London.
Fraser, H. M. & Sandow, J., 1985, Suppression of follicular maturation by
   infusion of a luteinizing hormone releasing hormone agonist starting
   during the late luteal phase in the stumptailed macaque monkey.
   J. Clin. Endocr. Metab. 60:579-584.
Muse, K. N., Cetel, N. S., Futterman, L. A. and Yen, S. S. C., 1974, The
   Premenstrual Syndrome. Effects of "medical ovariectomy". N. Engl. J.
   Med., 22:1345,
Sanders, D., Warner, P., Bäckström, T. and Bancroft, J., 1983, Mood,
   Sexuality, Hormones and the Menstrual Cycle. I. Changes in Mood and
   Physical State: Description of Subjects and Method, Psychosom. Med.
   45(6):487.

Acknowledgements

We are grateful to Dr. Patrick Magill and Hoechst (UK) Ltd. for
the supply of Buserelin nasal spray.

DIRECT EFFECTS OF LHRH AND AGONISTS ON HUMAN BREAST CANCER CELLS

W. R. Miller*, W. N. Scott*, R. Morris+, H. M. Fraser°,
and R. M. Sharpe°

*University Department of Clinical Surgery, Royal Infirmary
Edinburgh, +Department of Pathology, University Medical
School, Edinburgh, °MRC Unit of Reproductive Biology
Chalmers Street, Edinburgh

INTRODUCTION

The growth of the human female breast is clearly under endocrine control.
In the absence of stimulatory hormones, the breast remains rudimentary
whilst major development normally only occurs at times of increased hormone
production such as puberty and pregnancy. It is, therefore, not
surprising that certain breast cancers should also require hormones for
their continued growth. Whilst polypeptide hormones are capable of
influencing events within breast cancers, (Dilley and Kister, 1975, Leung
and Shiu, 1981) the most convincing evidence for the involvement of endocrine agents is for steroid hormones, particularly oestrogens. This
includes the observations that, a) the growth of cell lines of human breast
cancer may be stimulated by oestrogen both in culture (Butler et al, 1981,
Benz et al, 1983) and in immunosuppressed animals (Shafie 1980, Seibert et
al, 1983); b) the administration of oestrogen to breast cancer patients
with metastatic deposits in bone may cause increased urinary calcium
excretion due to accelerated tumour growth (Pearson et al, 1954); c)
oestrogen deprivation therapy causes regression of breast cancers in about
one-third of patients with advanced disease (Henderson and Canellos, 1980);
d) these effects are invariably associated with tumours which possess
receptors for oestrogen (Jensen et al, 1974, McGuire et al, 1975). Endocrine intervention therefore represents a major treatment option for advanced breast cancer especially against tumours which are rich in oestrogen
receptors. In premenopausal women, therapy usually takes the form of
castration which classically is achieved by surgical or radiological
ablation of the ovaries. However, with the observation that chronic administration of agonist analogues of LHRH suppresses the pituitary-ovarian
axis (Schally et al, 1970, Petrosa et al, 1980) and may reduce circulating
oestrogen to castrate levels (Nicholson and Maynard, 1979, Klijn and de
Jong, 1982), the concept has developed of using LHRH agonists in the treatment of breast cancer as a form of chemical ovariectomy (Corbin, 1982, Furr
and Nicholson, 1982). Whilst clinical studies with LHRH agonists have
largely been performed in premenopausal women (Klijn et al, 1984) there
are reports of beneficial anti-tumour effects in postmenopausal patients
who were without measurable ovarian function (Harvey et al, 1984). Such
observations raise the likelihood of direct effects of LHRH agonists on
breast tumours, a possibility supported by the increasing number of reports
describing effects of LHRH agonists on extra pituitary tissues (Khodr and

Siler Khodr, 1978, Lamberts et al, 1982, Tureck et al, 1982, Tan and
Rousseau, 1983).

The evidence will therefore be presented that LHRH agonists have
direct inhibitory effects on the growth of breast cancer cells in culture
and that tumour cells possess specific binding sites for LHRH and its agonist
in addition to the data that indicate that LHRH agonists produce a therapeutic castration and thereby are of benefit to premenopausal women with
advanced breast cancer.

LHRH agonists as therapy for premenopausal patients with advanced breast cancer

The use of chronic LHRH agonist treatment as a contraceptive to
suppress follicular development and ovulation is well established (Bergquist
et al, 1981, 1982a- Schmidt-Gollwitzer et al, 1981, 1982).   The exact
mechanism by which this is achieved is not fully elucidated but pituitary
desensitization (Belchetz et al, 1978; Sandow et al, 1978; Bergquist et al
1982b) which reduces gonadatrophin drive to the ovaries, is primarily
implicated.   Chronic administration of LHRH agonists to premenopausal
women therefore reduces circulating oestrogens to castrate levels, although
this may take several months to achieve (Nicholson et al, in press).   These
observations drew attention to the possibility of using LHRH agonists to
treat premenopausal women with oestrogen dependent breast cancer (Corbin,
1982, Furr and Nicholson, 1982).   So that a consistent suppression of
gonadal steroids may be maintained, higher doses of agonist are required
than for contraception.   This can be achieved via nasal spray (Hardt and
Schmidt-Gollwitzer 1983) but daily injections or infusions result in more
reliable suppression (Klijn and de Jong, 1984).

To date, only small numbers of patients with advanced breast cancer
have been treated with LHRH agonists.   Klijn and de Jong (1984) have used
the LHRH agonist, Buserelin, in a variety of regimes involving parental
administration followed by either intranasal or subcutaneous treatment alone
or in combination with other endocrine agents.   Of a total of twenty two
women receiving agonist alone, four patients had an objective response and
another four had stabilization of disease.   Treatment was associated with
anovulation in all patients although transient peaks of plasma oestradiol
were detected in most.   Beneficial effects of other LHRH agonists have
also been documented.   Nicholson et al (1985) reported responses in three
of fifteen and Harvey et al (1984) in two of four premenopausal women with
advanced breast cancer treated with Zoladex.

This overall response rate to LHRH agonists is similar to that which
would have been obtained by ovariectomy and there is reason to believe that,
in the majority of premenopausal women, the benefits of LHRH agonists were
achieved by a castration-like effect.   Thus, in general, patients with the
best responses to LHRH agonist had complete endocrine castration and
possessed oestrogen receptor-positive tumours whilst non-responders to LHRH
agonists rarely responded to subsequent removal of their ovaries.   Only
single cases have been reported of (a) a response in a patient with per-
sistant plasma oestrogen following LHRH agonist therapy (Klijn et al, 1984),
(b) remission in an oestrogen receptor negative tumour (Klijn et al, 1984)
and (c) response to surgical castration in a patient previously failing
agonist therapy (Nicholson et al, 1985).   Treatment with LHRH agonist is,
however, unlikely to replace surgical removal of the ovaries as the primary
endocrine therapy in premenopausal women with advanced breast cancer - at
least at the present time.   Although ovariectomy is irreversible, it is a
surgical procedure with very little morbidity and, because of the age-
related incidence of breast cancer (Thomas, 1983), the operation is usually

being performed in late premenopausal women. This means that many patients already have the children they desire and, in the face of potentially life-threatening disease, irreversible loss of reproductive function becomes a minor consideration. Furthermore, there are drawbacks with the present LHRH agonist therapies including the need for repeated treatment, the initial stimulatory effects on oestradiol secretion, especially when therapy is commenced during the follicular phase (Fraser and Sandow, 1985), the significant length of time before circulating oestrogens are reduced to castrate levels (Nicholson et al, 1984) and the intermittent break-through peaks of ovarian oestrogen (Klijn et al, 1984). However, the introduction of implants of agonist in a slow-release formulation which can be administered on a once per month basis should increase the usefulness and acceptance of agonist therapy.

LHRH agonists as therapy for postmenopausal patients with advanced breast cancer

Less attention has been given to the use of LHRH agonists in postmenopausal patients. This is because the initial concept of using the agent as a form of medical castration would preclude benefits in women without ovarian function. However, there are three reports which are worthy of mention. The first is that of Harvey et al (1984) who treated 72 postmenopausal patients with advanced breast cancer, with the agonist, leuprolide. Of these women, 10% obtained a partial response and 22% experienced stabilization of progressive disease with therapy. In a further investigation (Nicholson et al, in press), one of six postmenopausal patients responded to the agonist, ICI 118630 and Mathe is cited as having observed tumour regression in a postmenopausal women with metastatic disease following treatment with D-Trp-6-LH-RH (Schally et al, 1984). Whilst the response rate in postmenopausal patients appears to be significantly less than in premenopausal women, it ought to be noted that most of the postmenopausal group have previously been heavily treated with other endocrine therapies and would be expected to have a poorer prognosis than the premenopausal patients, many of whom received LHRH agonist as a first treatment option. The importance of these investigations in postmenopausal women is that they clearly indicate the possibility of obtaining anti-tumour effects with LHRH agonists in patients without overtly functioning ovaries and in whom treatment does not significantly change circulating oestrogens (Nicholson et al, in press). In order to explain the mechanism by which these beneficial effects are achieved it is necessary to look beyond the pituitary-ovarian axis and consider more direct effects on the tumour itself.

Effects of LHRH and its analogues on the growth of breast cancer cells in culture

Cell lines have now been established from many human breast cancers and some, such as the MCF-7, T-47D and MDA-MB-231 are well characterized (Lippman, 1981, Keydar et al, 1979, Cailleau et al, 1974). These lines form useful systems in which to study regulation of cell growth. We have therefore investigated the effects of LHRH and its analogues on such cultures, particularly the MCF-7 cell line, which has already been extensively studied for its sensitivity to steroid hormones (Lippman et al, 1976 a,b,c). In the present studies, the cell line was between its 183rd and 190th passage. Its growth in culture was sensitive to oestrogen, being stimulated by concentrations of oestradiol as low as 10^{-10}M and inhibited by the anti-oestrogen, tamoxifen (Figure 1). The cells had a human karyotype and were tumourogenic when innoculated into immunosuppressed mice; the resulting xenografts were oestrogen dependent (appearing only in animals supplemented with oestrogen) and had a histology consistent with breast carcinoma, including evidence of vascular invasion.

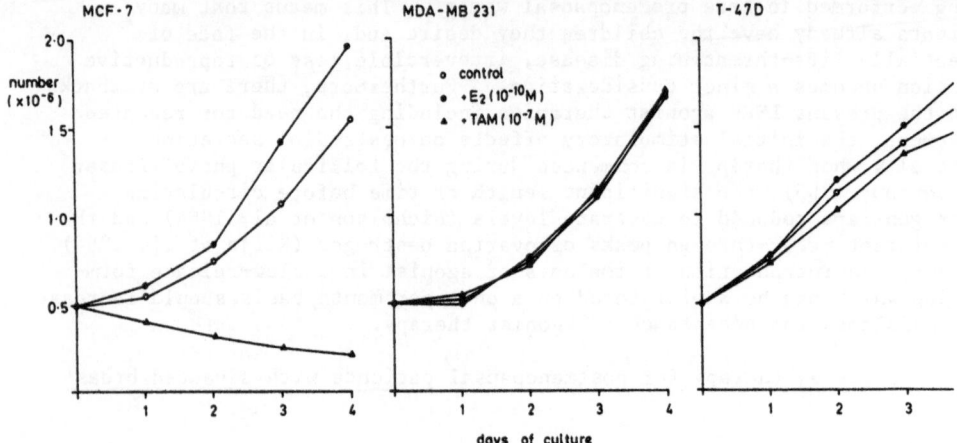

Fig. 1 The effects of oestradiol (E_2) and antioestrogen, tamoxifen (TAM)
on the growth of human breast cancer cells in culture. Each
point represents the mean value of triplicate systems.

To study effects on growth in culture, cells were grown at 37°C in
Dulbecco's minimal essential medium containing 10% heat-inactivated foeatal
calf serum, under a humidified atmosphere of 5% CO_2:95% air. Once growing
in log phase, cells were collected and plated out in 60mm Petri dishes
(0.5×10^6 cells in 4ml culture media). Sufficient dishes were set up so
that each test system could be studied in triplicate at each time point.
Hormones were added to the appropriate test systems after being dissolved
in culture media. Cells were incubated for either 1, 2, 3 or 4 days,
media being replenished each day. Growth was monitored by counting
harvested cells in a haemocytometer after collection by trypsinization.
Further experimental detail has already been published (Miller et al, 1985).

The effects of the LHRH agonist, buserelin, on the growth of MCF-7
cells is shown in Figure 2. All concentrations of the agonist other than
10^{-11}M caused significant ($p < 0.001$) inhibition of growth in comparison
with control cells. These inhibitory effects were dose-related and concentrations of buserelin in excess of 10^{-9}M produced a net decrease in
cell numbers after four days of culture. Levels in excess of 10^{-7}M
agonist resulted in a progressive decrease in cell numbers over the whole
test period. Inhibitory effects were highly reproducible and could be
detected microscopically (see Plate) as early as day 2 of culture.
Although cell numbers are decreased even after short exposure to LHRH
agonist, the remaining cells appear viable and, after an initial delay,
will show increased growth if cultured in media without LHRH agonist. The
specificity of these inhibitory effects was studied by performing similar
studies with native LHRH, the 3-10 fragment of LHRH and the LHRH antagonist,
(N-Ac-D-Na(2)[1], D-pCL-Phe[2], D-Trp[3], D-hArg(Et$_2$)[6], D-Ala[10])LHRH. Native
LHRH was also capable of inhibiting cell growth although much higher concentrations were required than with the LHRH agonist (Figure 3). The
3-10 fragment of LHRH did not significantly affect the growth of the MCF-7
cell line. The major inhibitory effects of LHRH agonist on cell growth
were completely blocked by addition of the LHRH antagonist, the combination
of the two peptides having no greater effect than the antagonist alone
(Figure 4); the latter only caused minor but nevertheless significant
suppression of cell growth. These results suggest that LHRH and its

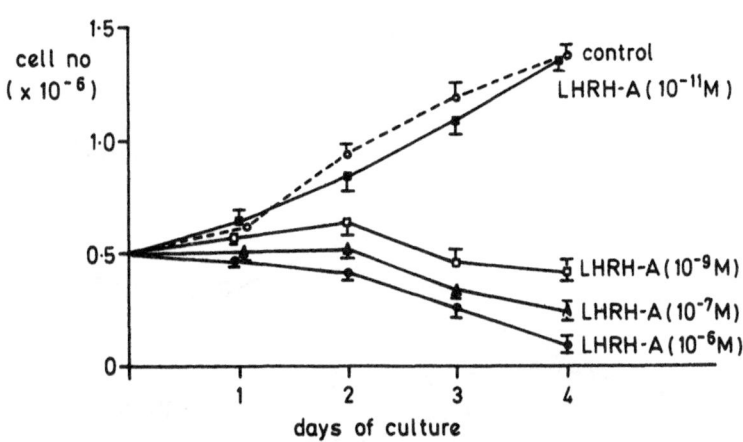

Fig. 2   Inhibition of the growth of MCF-7 breast
          cancer cells by the LHRH agonist (LHRH-A),
          buserelin.   Broken line represents growth
          pattern of control cells grown in the absence
          of agonist, solid line cells cultured with
          agonist at concentrations indicated.   Each
          point is the mean of triplicate cultures and
          vertical lines the standard deviation of the
          mean.

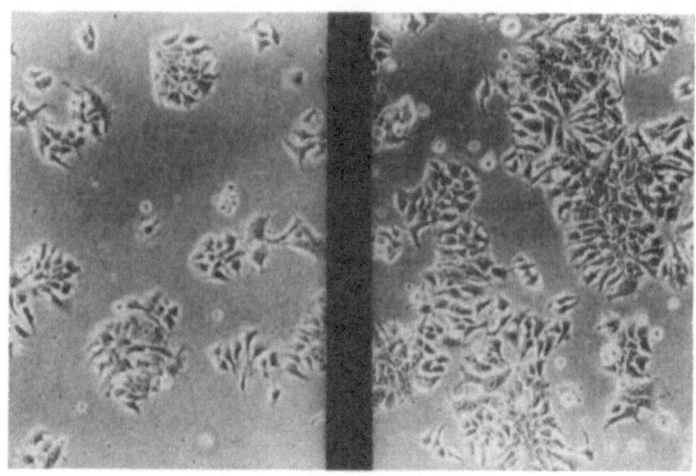

Plate   MCF-7 human breast cancer cells during the 4th
         day of culture in the absence (left panel) and
         presence of the LHRH agonist, buserelin, at a
         concentration of $10^{-9}M$ (right panel)

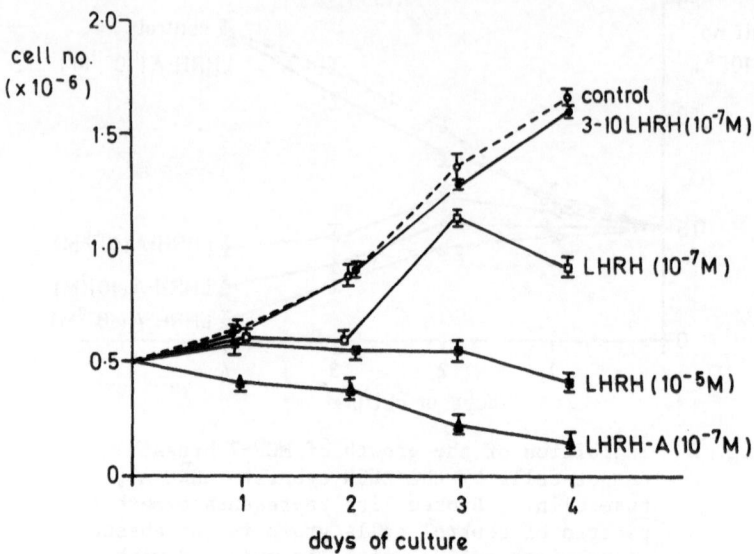

Fig. 3 The comparative effects of native LHRH, the
3-10 fragment of LHRH and LHRH agonist (LHRH-A)
on the growth of MCF-7 breast cancer cells in
culture. Broken line represents growth
pattern of control cells in the absence of
hormone addition. Each point is the mean of
triplicate cultures and vertical lines the
standard deviation of the mean.

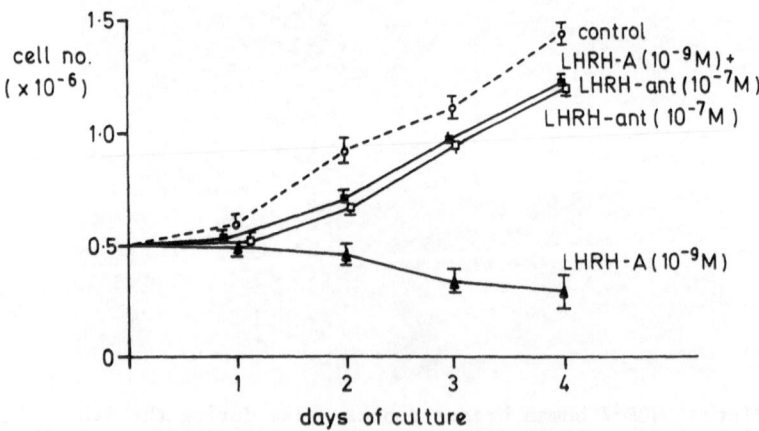

Fig. 4 The effects of LHRH agonist (LHRH-A) or LHRH
antagonist (LHRH-ant) alone and in combination
on the growth of MCF-7 breast cancer cells in
culture. Each point is the mean of
triplicate cultures and vertical lines the
standard deviation of the mean.

analogues are inhibiting the growth of the MCF-7 breast cancer cell line by means of a specific recognition mechanism.

To determine whether inhibitory effects may be observed in other breast cancer cells, similar studies have been performed with the T-47D and MDA-MB-231 lines. These cell lines have a different sensitivity to oestrogen when compared with MCF-7 cells (Figure 1), the growth of MDA-MB-231 cells being resistant to both oestradiol and tamoxifen and that of T-47D cells having only limited sensitivity to these agents. The effects of LHRH agonist on the growth of MDA-MB-231 and T-47D cells are shown in Figure 5. Concentrations of agonist which were inhibitory in MCF-7 cells were totally ineffective against MDA-MB-231 and produced minimal effects in the T-47D (although some inhibition was evident after 4 days of culture). It is clear, therefore, that the inhibitory action of LHRH agonist is not a non-specific effect on the growth of cells in general. Although it is pre-mature to draw firm conclusions on studies in only 3 cell lines, it is interesting that the sensitivity to LHRH agonist paralleled that to oestrogen. It may, therefore, be relevant that the inhibition by LHRH agonist in the MCF-7 cell line may be partially reversed by the simultaneous inclusion of oestrogen in the culture medium (Figure 6). Anti-steroidal actions have been reported for LHRH agonists in other tissues (Petroza et al, 1980, Sundaram et al, 1981).

These direct inhibitory effects on tumour cells, if reflected in vivo, would have important implications in the treatment of breast cancer. First of all, there would be no reason to restrict the use of LHRH agonists to premenopausal patients. Indeed the responses reported in postmenopausal women with advanced breast cancer may have resulted from direct anti-tumour effects of LHRH agonist. Furthermore, effects against oestrogen independ-ent tumours cannot be excluded. Although we could only clearly detect inhibition by LHRH agonists in the oestrogen receptor positive MCF-7 line but not in the oestrogen receptor negative MDA-MB-231 and T-47D lines, others have described inhibitory effects in T-47D cells using longer culture times (Wiznitzer and Benz, 1984). It may be relevant that although largely unresponsive to oestrogen, the T-47D line is sensitive to prolactin and progestogens (Shiu and Paterson, 1984, Wiznitzer and Benz, 1984, Bardon et al, 1985).

The regimes of LHRH agonist which are now being conventionally employed have been designed for use as a form of medical ovariectomy. It has to be appreciated that the doses of LHRH agonist necessary to suppress the pituitary-ovarian axis may not be the same as those to inhibit directly breast tumour growth. Methods of administration will have to be developed so that optimal amounts of agonist can reach metastatic deposits of tumour systemically and be maintained in the local environment of the cancer cells. Infusion or implantation techniques will probably be required.

LHRH-binding sites on breast tumour cells

The demonstration that the inhibitory effect of the LHRH agonist on growth of MCF-7 cells in-vitro was blocked completely by addition of an LHRH antagonist implies that the described effects are mediated via stereo-specific recognition sites i.e. receptors. To test this conclusion, MCF-7 cells grown to confluence were harvested by mechanical agitation and repeated aspirations into a Pasteur pipette and these cells then incubated with ^{125}I-LHRH agonist in-vitro. Initial binding studies were performed at 21°C using a 10 min incubation period (Miller et al, 1985) but the results described below were obtained by incubating cells on ice for 90 mins followed by rapid separation of bound and free hormone by centri-fugation at 4°C for 5 min at 1000g. Comparable results were obtained using either of these incubation procedures.

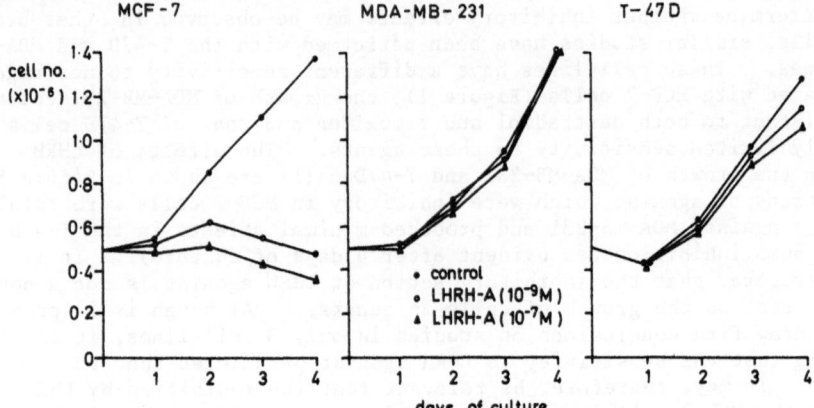

MCF-7   MDA-MB-231   T-47D

Fig. 5 The effects of LHRH agonist (LHRH-A), buserelin
on the growth of human breast cancer cells in
culture.   Each point represents the mean
value from triplicate systems.

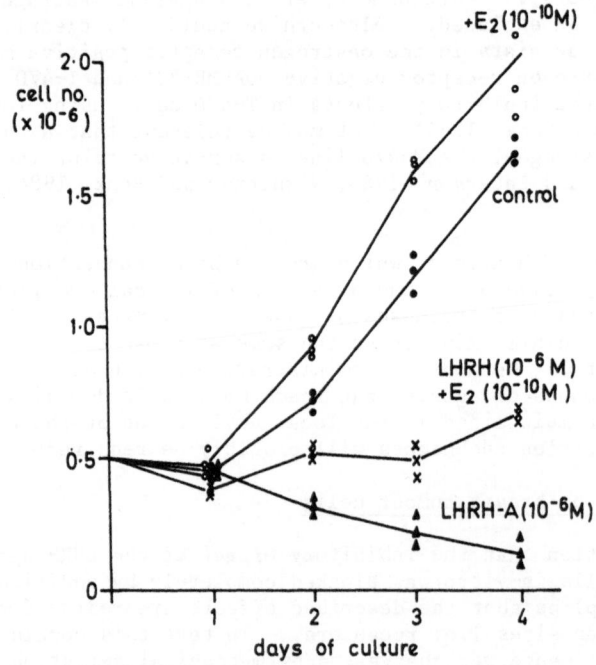

Fig. 6 The effects of LHRH agonist (LHRH/LHRH-A) or
oestradiol (E_2) alone and in combination on
the growth of MCF-7 breast cancer cells in
culture

As judged by the displacement of labelled LHRH agonist by unlabelled LHRH agonist, the MCF-7 cells contained binding sites with a low affinity (K_d ~4 x 10^{-5}M) for the LHRH agonist (Figure 7). However, these binding sites showed a high degree of specificity in that only peptides which were LHRH-like competed for binding with ^{125}I-LHRH agonists. Thus, two LHRH antagonists ({Ac3-4 dehydro-Pro^1, D-pFPhe^2, D-Trp^{3,6}}LHRH and {N-A$_c$-D-Nal(2)^1, D-pCl-Phe^2, D-Trp^3, D-hArg(Et$_2$)^6, D-Ala^{10}}LHRH) both competed for binding with higher affinity (Kd ~ 10^{-6}M) than did the LHRH agonist whilst native LHRH competed very poorly (Kd >5 x 10^{-4}). Surprisingly, $^{3-10}$LHRH, which is completely devoid of any biological activity in-vitro, displaced binding of ^{125}I-LHRH agonist nearly as effectively as did the unlabelled agonist itself. Of the other peptides tested, at concentrations up to 1mM, only bradykinin caused any displacement of binding and this was considered to be due to alteration of the pH of the incubation medium which contained a pH indicator.

Whilst these results demonstrate the presence on MCF-7 cells of specific binding sites for LHRH-like peptides, the low affinity of these sites for the LHRH agonist is puzzling in that biological effects of the same agonist on growth of the MCF-7 cells in-vitro was obtained with doses (10^{-9}M) far lower than the apparent affinity (Kd) of the receptor (4 x 10^{-5}M) for this peptide. We still have no explanation for this disparity, but it should be emphasized that it has been evident in every binding experiment performed. One possibility is that the MCF-7 cells contain two classes of LHRH-receptors, one of high affinity but low capacity and the other of low affinity and high capacity, with the former mediating the biological effects on growth. Unfortunately, because of the low numbers of LHRH-receptors present and other technical limitations it is not feasible to test directly this possibility.

It is equally intriguing that LHRH-receptors with relatively low affinity (Kd's of 10^{-6} to 10^{-7}M) for the LHRH agonist used in the present studies, have also been reported in other human extra-pituitary tissues such as the placenta (Currie et al, 1981) and corpus luteum (Popkin et al, 1983, Bramley et al, 1985). Moreover, in the placenta, these receptors have been shown to mediate biological effects of LHRH-like peptides (Belisle et al, 1984), and the gene for LHRH has been isolated from the human placenta (Seeburg and Adelman, 1984).

Therefore, it appears that several human extra-pituitary tissues contain LHRH-receptors which have a considerably lower affinity for LHRH agonists than do the respective extra-pituitary LHRH-receptors in the rat testis (Sharpe and Fraser, 1980) and ovary (Clayton et al, 1979). This may indicate that the human extra-pituitary LHRH-receptors are designed to detect a locally-produced LHRH-like peptide, the structure of which is different from native hypothalamic LHRH. In this respect it is of interest that LHRH has been reported in human breast milk (Amarant et al, 1982) and LHRH-like activity has been located immunohistochemically in certain ductal carcinomas of the breast (Seppala and Wahlstrom, 1980). Most recently, LHRH-receptors have been identified in a high percentage of ductal breast carcinomas (Eidne et al, 1985), although it is puzzling that these receptors were shown to have a relatively high affinity (Kd ~ 10^{-8}M) for an LHRH agonist, a finding which contrasts with the present findings in cultured breast tumour cells. The only major differences in technique used by Eidne et al (1985), when compared with the present studies, were the use of a membrane preparation and the use of an LHRH nonapeptide agonist with modifications at positions 6, 7 and 9 of the molecule, as opposed to the agonist used in our studies which had modifications only at positions 6 and 9. In preliminary studies using membrane preparations from breast cancer tissue, we have again found binding sites with low affinity (~ 10^{-5}M) for our LHRH agonist (data not shown), which suggests that the important

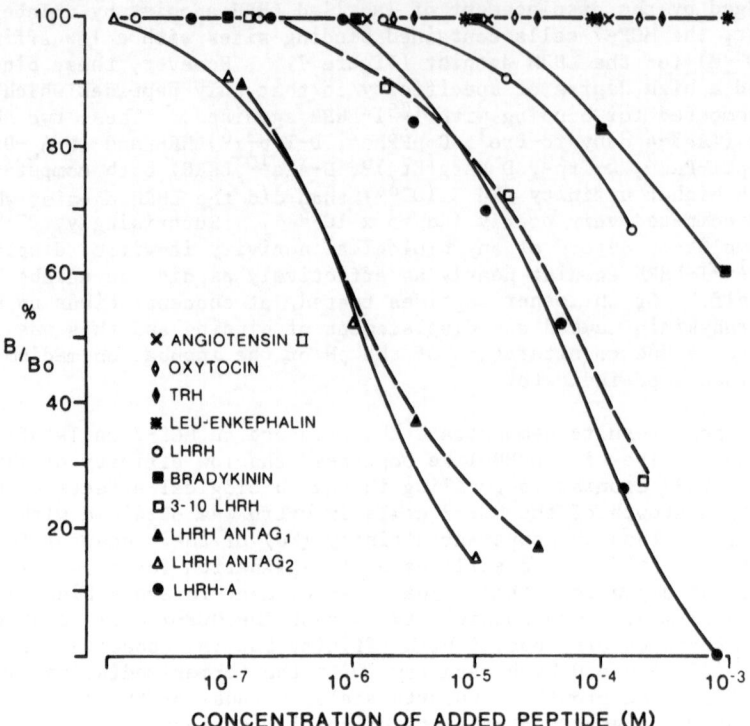

Fig. 7 Displacement of the binding of ^{125}I-labelled LHRH
agonist to isolated MCF-7 cells by increasing
concentrations of unlabelled LHRH agonist (LHRH-A),
two LHRH antagonists (ANTAG$_1$ and ANTAG$_2$), LHRH and
$^{3-10}$LHRH as well as several other peptides.   Non-
specific binding was determined by addition of
7 x 10^{-4}M unlabelled LHRH-A.   Each point is the
mean of triplicate incubations.

LHRH ANTAG$_1$ = (Ac3-4 dehydro-Pro^1, D-pFPhe^2,

D-Trp^{3,6})LHRH

LHRH ANTAG$_2$ = (N-Ac-D-Nal(2)^1, D-pCl-Phe^2, D-Trp^3,

D-hArg(Et$_2$)^6, D-Ala^{10})LHRH

LHRH-A = (D-Ser-t-bu^6, des-Gly-NH$_2$10)LHRH ethylamide

difference between our studies and those of Eidne et al (1985) is the use
of an LHRH agonist with an additional modification at position 7.

Taken together, our findings and those of Eidne et al (1985) suggest
that LHRH-like molecules may be capable of exerting direct effects on some
human breast tumours, although what proportion of tumours might respond,
and the nature of the response, remains to be determined.   In our studies,
we have tested for inhibitory effects of LHRH agonists on the growth of 3
breast tumour cell lines and only in the MCF-7 cells were major effects
obtained; in the MDA-MB-231 cells no effects on growth were observed whilst
in the T-47D line, only minor inhibition of cell growth was obtained (see
Figure 5).   However, a recent report has described major inhibitory

effects of an LHRH agonist on prolactin-stimulated growth of T-47D cells (Witzner and Benz, 1984), which perhaps suggests that LHRH exerts its inhibition only when growth of the cancer cells is <u>stimulated</u>, the nature of the stimulus being specific for different cell lines, i.e. oestrogen for MCF-7 cells and prolactin to T-47D cells. With respect to these findings, we have also identified low affinity binding sites for LHRH agonist in cells from the T-47D (K_D ~ 6 x 10^{-5}M) and MDA-MB-231 (K_D ~ 1.5 x 10^{-4}M) lines (data not shown).

SUMMARY

    In premenopausal women with advanced breast cancer, treatment with LHRH agonists may cause a medical ovariectomy and thereby induce the regression of oestrogen-dependent tumours. However, there is good evidence from laboratory studies of breast cancer cells in culture and beneficial results of LHRH agonist therapy in postmenopausal women which suggest a more direct anti-tumour effect. The inhibitory action of LHRH and its analogues on cultured tumour cells appears to be mediated by a specific recognition mechanism and the presence of specific binding sites on breast cancer cells can be demonstrated. If the anti-tumour effects of LHRH agonist are to be fully exploited in clinical practice, future research should be directed towards elucidating the mechanism by which LHRH agonists inhibit tumour growth and to determining the optimal regime for administration of the agent.

REFERENCES

Amarant, T., Fridkin, M., and Koch Y., 1982, Luteinizing hormone-releasing hormone and thyrotropin-releasing hormone in human and bovine milk, Europ J Biochem, 127:647.

Bardon, S., Vignon, F., Chalbos, D., and Rochefort, H., 1985, RU 486, a progestin and glucocorticoid antagonist, inhibits the growth of breast cancer cells via the progesterone receptor, J Clin Endocrinol Metab, 60:692.

Belchetz, P.E., Plant, T.M., Nakai, Y., Keogh, E.J., and Knobil, E., 1978, Hypophyseal responses to continuous and intermittent delivery of hypothalamic gonadotrophin-releasing hormone, Science, 202:631.

Belisle, S., Guevin, J-F., Bellabarba, D., and Leloux, J-G., 1984, Luteinizing hormone-releasing hormone binds to enriched human placental membranes and stimulates in vitro the synthesis of bioactive human chorionic gonadotropin, J Clin Endocrinol Metab, 59:119.

Benz, C., Cadman, E., Givin, J., Wu, T., Amara, J., Eisenfeld, A., and Dannies, P., 1983, Tamoxifen and 5-fluoroucil in breast cancer: Cytotoxic synergism in vitro, Cancer Res., 43:5298.

Bergquist, C., Nillius, S.J., Wide, L., and Lindgren, A., 1981, Endometrial patterns in women on chronic luteinizing hormone-releasing agonist treatment for contraception, Fert Steril, 36:339.

Bergquist, C., Nillius, S.J., and Wide, L., 1982a, Long-term intranasal luteinizing hormone-releasing hormone agonist treatment for contraception in women, Fert Steril, 38:190.

Bergquist, C., Nillius, S.J., and Wide, L., 1982b, Failure of positive feedback of oestradiol during chronic intranasal luteinizing hormone-releasing hormone agonist treatment, Clin Endocrinol, 16:147.

Bramley, T.A., Menzies, G.S., and Baird, D.T., 1985, Specific binding of LHRH and an agonist to human corpus luteum homogenates: Characterization, properties and luteal phase levels, J Clin Endocrinol Metab, 41:in press.

Butler, W.B., Kelsey, W.H., and Goran, N., 1981, Effects of serum and insulin on the sensitivity of the human breast cancer cell line MCF-7 to estrogen and antiestrogens, Cancer Res, 41:82.

Cailleau, R., Young, R., Olive, M., and Reeve, W.J., 1974, Breast tumour cell lines from pleural effusions, JNCI, 53:661.

Clayton, R.N., Harwood, J.P., and Catt, K.J., 1979, Gonadotropin-releasing hormone analogue binds to luteal cells and inhibits progesterone production, Nature, 282:90.

Corbin, A., 1982, From contraception to cancer: a review of the therapeutic applications of LHRH analogues as antitumour agents, Yale J Biol Med, 55:27.

Currie, A.J., Fraser, H.M., and Sharpe, R.M., 1981, Human placental receptors for luteinizing hormone releasing hormone, Biochem Biophys Res Commun, 99:332.

Dilley, W.G., and Kister, S.J., 1975, In vitro stimulation of human breast tissue by human prolactin, JNCI, 55:35.

Eidne, K.A., Flanagan, C.A., and Millar, R.P., 1985, Gonadotropin-releasing hormone binding sites in human breast carcinoma, Science, in press.

Fraser, H.M., and Sandow, J., 1985, Suppression of follicular maturation by infusion of a luteinizing hormone releasing hormone agonist starting during the luteal phase in the stumptailed macaque monkey, J Clin Endocrinol Metab, 60:579.

Furr, B.J.A., and Nicholson, R.I., 1982, Use of analogues of luteinizing hormone-releasing hormone for the treatment of cancer, J Reproduct Fertil, 64:529.

Hardt, W., and Schmidt-Gollwitzer, M., 1983, Sustained gonadal suppression in fertile women with the LHRH agonist, buserelin, Clin Endocrinol, 19:613.

Harvey, H.A., Lipton, A., and Max, D.T., 1984, LHRH analogs for human mammary carcinoma in: "LHRH and its Analogs, Contraceptive and Clinical Application", B.H. Vickery, J.J. Nestor and E.S.E. Hafez, eds, MTP, Lancaster, p 329.

Henderson I.C., and Canellos, G.P., 1980, Cancer of the breast: the past decase, N Engl J Med, 302:17,78.

Jensen, E.V., Mohlar, S., Gorell, T.A., and De Sombre, E.R., 1974, The role of estrophilin in oestrogen action, Vit and Horm, 32:89.

Keydar, I., Chen, L., Karby, S., Weiss, F.R., Delarea, J., Radu, M., Chaitcik, S., and Brenner, H.J., 1979, Establishment of a cell line of human breast carcinoma origin, Eur J Cancer, 15:659.

Khodr, G., and Siler-Khodr, T.M., 1978, The effect of luteinizing hormone-releasing factor on human chorionic gonadotrophin secretion, Fert Steril, 30:301.

Klijn, J.G.M. and de Jong, F.H., 1982, Treatment with a luteinizing-hormone-releasing-hormone analogue (buserelin) in premenopausal patients with metastatic breast cancer, The Lancet, i:1213.

Klijn, J.G.M., and de Jong, F.H., 1984, Long-term treatment with the LHRH-agonist buserelin (HOE 766) for metastatic breast cancer in single and combined drug regimens, in: "LHRH and its analogues", F. Labrie, A. Belanger and A. Dupont, eds, Elsevier Science, Amsterdam, New York, p 425.

Klijn, J.G.M., de Jong, F.H., Blankenstein, M.A., Docter, R., Alexieva-Figusch, J., Blonk-van-der Wijst, J., and Lamberts, W.S.J., 1984, Anti-tumour and endocrine effects of chronic LHRH agonist treatment (buserelin) with or without tamoxifen in premenopausal metastatic breast cancer, Breast Cancer Res and Treat, 4:209.

Lamberts, S.W.J., Timmers, J.M., Oosterom, R., Verleun, J., Rommerts, F.G., and de Jong, F.H., 1982, Testosterone secretion by cultured arrhenoblastoma cells:Suppression by a luteinizing hormone-releasing hormone agonist, J Clin Endocrinol Metab, 54:450.

Leung, C.K.H., and Shiu, R.P.C., 1981, Required presence of both estrogen and pituitary factors for the growth of human breast cancer cells in athymic nude mice, Cancer Res, 41:546.

Lippman, M., 1981, Hormonal regulation of human breast cancer cells in vitro, in: "Banbury Report No 8 Hormones and Breast Cancer", M.C. Pike, P.K. Siiteri and C.W. Welsch, eds, Cold Spring Harbor Laboratory, U.S.A., p 171.

Lippman, M., Bolan, G., and Huff, K., 1976a, The effects of estrogens and antioestrogens on hormone-responsive human breast cancer in long-term tissue culture, Cancer Res, 36:4595.

Lippman, M.E., Bolan, G., and Huff, K., 1976b, The effect of glucocorticoids and progesterone on hormone-responsive human breast cancer in long-term tissue culture, Cancer Res, 36:4602.

Lippman, M.E., Bolan, G., and Huff, K., 1976c, The effects of androgens and anti-androgens on hormone-responsive human breast cancer in long-term tissue culture, Cancer Res, 36, 4610.

McGuire, W.L., Pearson, O.H., and Segaloff, A., 1975, Predicting hormone responsiveness in human breast cancer, in: "Oestrogen receptors in human breast cancer", W.L. McGuire, P.P. Carbone and E.D. Vollmer, eds, Raven Press, New York, p 17.

Miller W.R., Scott, W.N., Morris, R., Fraser, H.M., and Sharpe, R.M., 1985, Growth of human breast cancer cells inhibited by a luteinizing hormone-releasing agonist, Cancer Res, 36:3610.

Nicholson, R.I., and Maynard, P.V., 1979, Anti-tumour activity of ICI 118630, a new potent luteinizing hormone-releasing hormone agonist, Brit J Cancer, 39:268.

Nicholson, R.I., Walker, K.J., Turkes, A., Turkes, A.O., Dyas, J., Blamey, R.W., Campbell, R.F., Robinson, M.R.G., and Griffiths, K., 1984, Therapeutic significance and the mechanism of action of the LHRH agonist ICI 118630 in breast and prostatic cancer, J Steroid Biochem, 20:129.

Nicholson, R.I., Walker, K.J., Turkes, A., Dyas, J., Plowman, P.N., Williams, M., and Blamey, R.W., 1985, Endocrinological and clinical aspects of LHRH action (ICI 118630) in hormone dependent breast cancer, J Steroid Biochem, in press.

Pearson, O.H., West, C.D., Hollander, V.P., and Treves, N.E., 1954 Valuation of endocrine therapy for advanced breast cancer, J.A.M.A., 154:234.

Petroza, E., Vilchez-Martinex, J.A., Coy, D.H., Arumura, A., and Schally, A.V., 1980, Reduction of LHRH pituitary and estradiol uterine binding sites by a superactive analog of luteinizing hormone-releasing hormone, Biochem Biophys Res Commun, 95:1056.

Popkin, R., Bramley, T.A., Currie, A., Shaw, R.W., Baird, D.T., and Fraser, H.M., 1983, Specific binding of luteinizing hormone-releasing hormone to human luteal tissue, Biochem Biophys Res Commun, 114:750.

Sandow, J., von Rechenberg, W., Jerzabek, G., and Stoll, W., 1978, Pituitary gonadotrophin-inhibition by a highly active analog of luteinizing hormone-releasing hormone, Fert Steril, 30:205.

Schmidt-Gollwitzer, M., Hardt, W., Schmidt-Gollwitzer, K., von der Ohe, M., and Nevinng-Stickel, J., 1981, Influence of the LHRH analogue buserelin on cyclic ovarian function and on endometrium. A new approach to fertility control?, Contraception, 23:187.

Schmidt-Gollwitzer, M., Hardt, W., Schmidt-Gollwitzer, K., and von der Ohe, M., 1982, The contraceptive use of buserelin, a potent LHRH agonist: clinical and hormonal findings, in: "LHRH peptides as male and female contraceptives", G. Zatuchni, ed, Lippincott, U.S.A., p 199.

Seeburg, P.H., and Adelman, J.P., 1984, Characterization of cDNA for precursor of human luteinizing hormone-releasing hormone, Nature, 311:666.

Seibert, K., Shafie, S.M., Triche, T.J., Whang-Peng, J.J., O'Brien, S.J., Toney, J.H., Huff, K.K., and Lippman, M.E., 1983, Clonal variations of MCF-7 breast cancer cells in vitro and in athymic nude mice, Cancer Res, 43:2223.

Seppala, M., and Wahlstrom, T., 1980, Identification of luteinizing hormone-releasing factor and alpha subunit of glycoprotein hormones in ductal carcinoma of the mammary gland, Int J Cancer, 26:267.

Shafie, S.M., 1980, Estrogen and growth of breast cancer:New evidence suggests indirect action, Science, 209:701.

Schally, A.V., Arimura, A., and Coy, A.H., 1980, Recent approaches to fertility control based on derivatives of LHRH, Vit and Horm, 38:257.

Schally, A.V., Redding, T.W., and Comaru-Schally, A.M., 1984, Potential use of analogs of LHRH in the treatment of hormone sensitive neoplasias, Cancer Treatment Reports 25th Anniversary Issue (NIH), 68:No 1.

Sharpe, R.M., and Fraser, H.M., 1980, Leydig cell receptors for luteinizing hormone releasing hormone and its agonists and their modulation by administration or deprivation of the releasing hormone, Biochem Biophys Res Commun, 45:256.

Shiu, R.P.C., and Paterson, J.A., 1984, Alteration in cell shape, adhesion and lipid accumulation in human breast cancer cell (T-47D) by human prolactin and growth hormone, Cancer Res, 44:1178.

Sundaram, K., Cao, Y-C., Wang, N-G., Bardin, C.W., Rivier, J., and Vale, W., 1981, Inhibition of the action of sex steroids by gonadotropin-releasing hormone (GnRH) agonists: a new biological effect, Life Sciences, 28:83.

Tan, L., and Rousseau, P., 1983, The chemical identity of the immuno-reactive LHRH-like peptide biosynthesized in the human placenta, Biochem Biophys Res Commun, 109:1061.

Thomas, D.B., 1983, Factors that promote the development of human breast cancer, Environmental Health Perspectives, 50:209.

Tureck, R.W., Mastroianni, L., Blasco, L., and Strauss, J.R., 1982, Inhibition of human granulosa cell progesterone secretion by a gonadotropin-releasing hormone agonist, J Clin Endocrinol Metab, 54:1078.

Wiznitzer, I., and Benz, C., 1984, Direct growth inhibiting effects of the prolactin antagonists Buserelin and Pergolide on human breast cancer, Proc Ann Am Assoc Cancer Res, 25:208.